Medical and Health Genomics

"Physician who fails to enter the body of a patient with the lamp of knowledge and understanding can never treat diseases"

- Charaka, a noted Ayurveda practitioner, wrote the famous treatise "Charak Samhita" on Ayurvedic medicine in Ancient India c.1000 BC. The ancient Indian text refers to genetic factors determining the sex of the child and the origin of congenital blindness in the sperm or ovum rather than the mother or the father.

" …εἰ γὰρ ἐκ φλεγματώδεος φλεγματώδης, καὶ ἐκ χολώδεος χολώδης γίνεται, καὶ ἐκ φθινώδεος φθινώδης, καὶ ἐκ σπληνώδεος σπληνώδης, τί κωλύει ὅτῳ πατὴρ καὶ μήτηρ εἴχετο, τούτῳ τῷ νοσήματι καὶ τῶν ἐκγόνων ἔχεσθαί τινα; ὡς ὁ γόνος ἔρχεται πάντοθεν τοῦ σώματος, ἀπό τε τῶν ὑγιηρῶν ὑγιηρὸς, ἀπό τε τῶν νοσερῶν νοσερός…"

Περὶ ἱερῆς νούσου

"…For if a phlegmatic person be born of a phlegmatic, and a bilious of a bilious, and a phthisical of a phthisical, and one having spleen disease, of another having disease of the spleen, what is to hinder it from happening that where the father and mother were subject to this disease, certain of their offspring should be so affected also? As the semen comes from all parts of the body, healthy particles will come from healthy parts, and unhealthy from unhealthy parts…"

"On the Sacred Disease" Hippocrates of Kos (Ἱπποκράτης; c.460–c.370 BC)

Dedication

To,
The Late Shri Anand Swarup Kumar, Our Father
and Shrimati Hardevi Kumar, Our Mother.
Dhavendra, Anju, Ashish, Jaime, Jaya, Nikita and Mayank

To,
Grigoria, Emmanuel, Gregory, Alexander, Christina and their spouses
Stylianos, Athena, Stylianos, Anne-Grigoria, Sophia, Raphael, Elisabeth

Medical and Health Genomics

Edited by

Dhavendra Kumar
Institute of Cancer & Genetics University Hospital of Wales
Cardiff University School of Medicine Cardiff, UK
Genomic Policy Unit Faculty of Life Sciences and Education
University of South Wales Pontypridd, UK

Stylianos Antonarakis
Department of Medical Genetics
University Hospitals of Geneva, Switzerland
Institute of Genetics and Genomics of Geneva,
Geneva, University of Switzerland

AMSTERDAM • BOSTON • HEIDELBERG • LONDON • NEW YORK • OXFORD • PARIS
SAN DIEGO • SAN FRANCISCO • SINGAPORE • SYDNEY • TOKYO
Academic Press is an imprint of Elsevier

Academic Press is an imprint of Elsevier
125 London Wall, London EC2Y 5AS, UK
525 B Street, Suite 1800, San Diego, CA 92101-4495, USA
50 Hampshire Street, 5th Floor, Cambridge, MA 02139, USA
The Boulevard, Langford Lane, Kidlington, Oxford OX5 1GB, UK

Notices
Knowledge and best practice in this field are constantly changing. As new research and experience broaden our understanding, changes in research methods, professional practices, or medical treatment may become necessary.

Practitioners and researchers must always rely on their own experience and knowledge in evaluating and using any information, methods, compounds, or experiments described herein. In using such information or methods they should be mindful of their own safety and the safety of others, including parties for whom they have a professional responsibility.

To the fullest extent of the law, neither the Publisher nor the authors, contributors, or editors, assume any liability for any injury and/or damage to persons or property as a matter of products liability, negligence or otherwise, or from any use or operation of any methods, products, instructions, or ideas contained in the material herein.

British Library Cataloguing-in-Publication Data
A catalogue record for this book is available from the British Library

Library of Congress Cataloging-in-Publication Data
A catalog record for this book is available from the Library of Congress

ISBN: 978-0-12-420196-5

For information on all Academic Press publications
visit our website at https://www.elsevier.com/

Working together
to grow libraries in
developing countries

www.elsevier.com • www.bookaid.org

Publisher: Mica Haley
Acquisition Editor: Peter Linsley
Editorial Project Manager: Lisa Eppich
Production Project Manager: Edward Taylor
Designer: Matthew Limbert

Typeset by TNQ Books and Journals
www.tnq.co.in

Contents

List of Contributors   xi
Foreword   xiii
Preface   xv

1. The Human Genome

D. Kumar

Introduction   1
Hereditary Factors, Genes, Genetics, and
  Genomics   1
Structure and Organization of Nucleic Acids   2
Human Genome Variation and Human Disease   5
  Measuring Genetic and Genomic Variation   6
  Genome Variation and Human Disease   7
The Mitochondrial Genome   8
Functional Genomics, Transcriptomics,
  and Proteomics   9
Translational Human Genomics   10
Human Genomics for Socioeconomic
  Development   11
Conclusions   12
References   12

2. Genomic Technologies in Medicine and Health: Past, Present, and Future

Y.-H. Rogers and C. Zhang

Introduction   15
Sequencing Technologies   16
Computational and Information Technologies   16
Applications of Genomic Technologies   17
The Microbiome and Human Health   18
  The Pediatric Microbiome   18
  The Microbiome and Oral Health   18
  The Microbiome and Gastrointestinal Health   18
The 1000 Genomes Project and Structural
  Variations in the Human Genomes   18
Noninvasive Prenatal Testing by Sequencing
  of Cell-Free Fetal DNA in the Maternal Blood   19
Prenatal Diagnosis by Whole Genome
  Sequencing of "Jumping Libraries"   19
Postnatal Diagnosis   20

Genome Sequencing in Newborn Healthcare   21
Genome Sequencing in Other Research Areas   22
The Cancer Genome Atlas and Cancer
  Genomics   22
Cancer Diagnosis   22
Precision Medicine   23
  Development of Targeted Therapeutics for
    Genetic Disorders   23
  Development of Targeted Therapeutics for
    Cancers   24
  Genomic Profiling and Counseling   24
  The Precision Medicine Initiative   24
Policy and Regulatory Issues   25
Genomic Education   25
References   25

3. Genomic Databases, Access Review, and Data Access Committees

M. Shabani, B.M. Knoppers and P. Borry

Introduction   29
Underlying Principles, Policies,
  and Guidelines   30
Examples of Controlled-Access Databases
  and Pertinent Data Access Committees   30
  European Genome-phenome Archive   30
  Database of Genotypes and Phenotypes   31
Ethical and Legal Challenges   32
  Consent   32
  Oversight Mechanisms on Downstream
    Uses and Enforceability   33
  Data Producers and Data Users' Interests
    and Benefits   33
Conclusion   34
Acknowledgments   34
References   34

4. Diagnostic Genomics and Clinical Bioinformatics

A. Haworth, H. Savage and N. Lench

Introduction   37
  Current Clinical Practice: Diagnosis   37
  Chromosomal Microarrays   37

Single Gene Testing   38
Gene Panel Testing   38
Exome Sequencing   39
Whole Genome Sequencing   39
Diagnostic Testing Strategy   39
Inherited Cardiac Conditions   40
Hypertrophic Cardiomyopathy   40
Aortopathies   40
Intellectual Disability   41
Deciphering Developmental Disorders   41
The UK100K Genomes Project:
Large-Scale Implementation of
Whole Genome Sequencing   42
Clinical Bioinformatics   42
Clinical Phenotyping   42
Bioinformatics Pipelines   43
Reference Genome   43
Interpretation of DNA Sequence Variants   44
Protein-Altering Variants   45
Variants Causing Aberrant Messenger
   RNA Splicing   45
Nucleotide Conservation   45
Inheritance and Mutation Type   45
Segregation Analysis   45
Mutational Mechanism and Variant Spectrum   46
De Novo Mutations   46
Functional Evidence   46
Databases and Literature   46
Literature Databases   47
Online Mendelian Inheritance in Man   47
Incidental Findings   47
Data Sharing   48
Conclusion   48
References   48

5. Epigenetics and Epigenomics in
Human Health and Disease

R. Festenstein

Introduction   51
Epigenotype and Regulation of Gene
Expression   51
Epigenotypes and Human Disease   53
Trinucleotide Repeats and Human Disease   54
Friedreich Ataxia   54
Myotonic Dystrophy   55
Fragile X Syndrome   56
Facioscapulohumeral Dystrophy   57
Immunodeficiency, Centromeric Region
   Instability, and Facial Anomalies Syndrome:
   Genetic Mutations Causing Methylation   59
Rett Syndrome   60
α-Thalassemia X-Linked Mental Retardation
   Syndrome   61

Epigenetic Mechanisms in Cancer   62
Conclusion   63
Acknowledgment   63
References   63

6. Mitochondrial Genomics:
Emerging Paradigms and
Challenges

V. Singh, P. Gupta, N.K. Rajput and A. Bhardwaj

Introduction   75
Nongenetic Diagnostics for Mitochondrial
Dysfunction   75
Genome Data Deluge   78
DNA Testing, Next Generation Sequencing,
and Mitochondrial Disease Diagnosis   79
Next Generation Sequencing   80
Next Generation Sequencing–Based
Diagnostic Assay Design for Mitochondrial
Disorders   80
Translational Advancements   81
Semantic Data Standards for Community
Collaboration   84
The Road Ahead   85
Conclusions   85
References   86

7. The Significance of Metabolomics
in Human Health

D.F. Gomez-Casati, M. Grisolía and M.V. Busi

Introduction   89
Metabolomics in Human Diseases   90
Biomarker Detection   93
Group Discrimination (Diagnosis)   93
Inborn Metabolic Disorders   93
Cancer Metabolomics   94
Biomarkers for Psychiatric Diseases   94
Metabolomics in Environmental and
Public Health   95
Human Nutrition   96
Intestinal Metabolome   96
Conclusions   97
Acknowledgment   98
References   98

8. Microbial Genomics: Diagnosis,
Prevention, and Treatment

A. Mutreja

Background   101
Control and Prevention   101
Treatment and Diagnostics   103
References   105

9. A Metagenomic Insight Into the Human Microbiome: Its Implications in Health and Disease

R. Saxena and V.K. Sharma

Outline of the Chapter	107
Introduction	107
Human-Associated Microflora	109
Alterations in the Healthy Human Microbiome: Association With Diseases	110
The Oral Microbiome	110
The Skin Microbiome	111
The Gut Microbiome	111
Diseases Associated With Variation in the Gut Microbial Community	112
Novel Therapeutic Strategies Based on the Human-Associated Microbiome	115
Conclusion	116
Glossary	117
List of Acronyms and Abbreviations	117
References	117

10. Pharmacogenetics and Pharmacogenomics

A. Alfirevic and M. Pirmohamed

Introduction	121
Pharmacogenomic Information in Drug Labeling	122
Pharmacogenetic Guidelines	124
Pharmacogenetic Study Design	124
Clinical Areas	126
Conclusions	133
References	134

11. Medical and Health Aspects of Genetics and Genomics

D. Kumar

Introduction	139
Chromosomal Disorders	141
Mendelian (Single-Gene) Disorders	142
Polygenic or Multifactorial Disorders	142
Mitochondrial Genetic Disorders	144
Genomic Disorders	145
Disorders of Genomic Imprinting: Epigenetic Diseases	145
Disorders of the Genome Architecture	150
Disorders With Trinucleotide (Triplet) Repeats	155
Complex Genomic Diseases	156
Disease Spectrum, Biological Pathways, and Genotypes	157
Summary	159
References	159

12. Content and Variation of the Human Genome

S.E. Antonarakis

Introduction: "The Genome Anatomy"	161
Protein-Coding Genes	165
Noncoding, RNA-Only Genes	166
Regions of Transcription Regulation	168
Conserved Elements Not Included in the Previous Categories	168
Repetitive Elements	168
Segmental Duplications	169
Special Genomic Structures Containing Selected Repeats	170
Human Centromeres	170
Human Telomeres	171
Short Arms of Human Acrocentric Chromosomes	171
Mitochondrial Genome	172
Genomic Variability	172
Genome Variation as a Laboratory Tool to Understand the Genome	176
References	176

13. Spectrum of Genetic Diseases and Management

B. Kerr

Introduction	179
Molecular Approach to Genetic Disease Nosology	179
From Marfan Syndrome to Fibrillinopathies	180
Diagnosis of Marfan Syndrome	182
Differential Diagnosis	182
Management of Marfan Syndrome	183
Treatment	183
The Emergence of RASopathies	184
Malignancy Risk in RASopathies	185
Making a Diagnosis of a Genetic Disorder in the Era of Molecular Pathways	186
Conclusion	187
References	187

14. Genomic Analysis in Clinical Practice: What Are the Challenges?

A. Lucassen and I. van Langen

Introduction	191
Consent for Genome Testing in Clinical Practice	191
Incidental Findings	195
Management of Incidental Findings	195
The Data Interpretation Problem	197

What Constitutes a Result in Genomic
    Medicine?   197
Diagnostic Versus Population Screening   198
Conclusions   198
References   198

15. Genomic Perspective of Genetic Counseling

C.L. Gaff and I. Macciocca

Introduction   201
Ethos and Principles of Genetic Counseling   201
Elements of Genetic Counseling   202
Counseling Issues   203
    Uncertainty   203
    Confidentiality and Privacy of Genomic
        Data   203
    Secondary Findings   204
    Family Communication   205
    The "Healthy Genome"   205
Case Study   205
Practicalities of Counseling for Genomic
    Tests   206
    Genetic Information Is Key   209
    Patient Autonomy Must Be Supported   210
    Patient Emotions Make a Difference   210
    Relationship Is Integral to Genetic
        Counseling   210
Conclusion   210
Acknowledgment   210
References   210

16. Genetics and Genomics of Reproductive Medicine and Health

D. Kumar

Introduction   213
Female Infertility   213
    Congenital Anomalies of the
        Reproductive Tract   213
    Disorders of Abnormal Ovulation   214
    Disorders of the Endometrium   217
Male Factor Infertility   218
    Congenital Anomalies of the
        Male Reproductive Tract   218
    Molecular Genetics of Male Factor
        Infertility   219
    Mitochondrial Genes in Male Factor
        Infertility   221
    New Genomic Applications in Male
        Factor Infertility   221
Genetic Factors in In Vitro Fertilization   221
    Preimplantation Genetic Testing and
        Screening   222

Recent Advances in Prenatal Diagnosis   223
    Aneuploidy (Trisomy 21) Pregnancy   223
    Cell-Free Fetal DNA   223
Summary   224
References   224

17. Stratified and Precision Medicine

D. Kumar

Introduction   227
Molecular, Genetic, and Genomic
    Revolutions in Medicine   227
Personalized Medicine   228
Stratified Medicine   230
    Challenges for Stratified Medicine   232
    The Future of Stratified Medicine   233
Summary   234
References   234

18. Teaching and Training Medicine in Genomic Era

B. Korf

Introduction   237
Integration of Genomics into Medical
    Practice   237
    The Practice of Genomic Medicine   238
    The Roles of the Geneticist and
        Nongeneticist Health Providers   240
    Competencies Required in
        Genomic Medicine   241
Genomics and the Medical Education
    Landscape   242
    Premedical Education   242
    Medical School   242
    Residency Education   242
    Postgraduate Education   243
Approaches to Medical Genomics
    Education   243
    Medical School   243
    Residency Education   244
    Continuing Medical Education   244
Conclusions and Final Comments   244
References   244

19. Genomics, New Drug Development, and Precision Medicines

D. McHale and M. Penny

Introduction   247
    The Drug Discovery and Development
        Process   248
    Preclinical Testing   248
    Clinical Development   249

Applying Genomics to Drug Discovery   250
  Choosing the Best Drug Targets   250
  Complex Trait Genetics   251
  Single Gene Disorders/Traits   251
  Drug-Specific Targets Approach   252
Applying Pharmacogenetics to Drug
  Development   253
  Pharmacodynamic Variability   253
  Pharmacokinetic Variability   254
Predicting Safety   255
  Predicting Type B Adverse Events   255
  Predicting Type A Adverse Events   255
  Individualized Therapy: An Integrated
    Response   256
  Improving Disease Classification:
    Stratified Medicines   256
  Adverse Drug Reactions   257
Summary   257
References   258

20. Cancer Genetics and Genomics

J. Whitworth and E. Maher

Introduction   261
Inherited Cancers: Germ Line and
  Somatic   261
  Oncogenes, Tumor Suppressor
    Genes, and the Identification of
    Cancer Predisposition Genes   261
  Identifying Cancer Predisposition
    Genes   262
  High, Medium, and Smaller Risks   263
Genetic Testing in Cancer Predisposition
  Syndromes   264
  Risk Information as Therapy   264
  Cancer Surveillance   265
  Prophylactic Surgery   265
  Pharmacological Management   266
Inherited Cancer Genes   266
  Phenotypic Effects of Mutations in
    Cancer Predisposition Genes   266
  Genotype–Phenotype Correlation   280
Genetic Testing and Mainstreaming   281
  Genetic Testing and Next Generation
    Sequencing   281
  Reasons for Nondetection of Causative
    Genetic Changes Through
    Clinical Testing   282
Conclusion   283
References   283

**21. The Provision of Medical and Health
Genetics and Genomics in the
Developing World**

*N. Sirisena, D. Sumathipala, K. Wettasinghe
and V.H.W. Dissanayake*

Introduction   285
Public Health Programs in Genetics and
  Genomics   287
Medical Services Incorporating
  Genetics and Genomics   290
Summary   293
References   293

**22. Genomic Applications in Forensic
Medicine**

C. Børsting and N. Morling

Introduction   295
The Basics of Next Generation Sequencing   297
Single Molecule Sequencing   299
Next Generation Sequencing Solutions in
  Forensic Genetics   300
  Short Tandem Repeat Sequencing   302
  The First Commercial Next Generation
    Sequencing Kits for Forensic Genetics   303
  New Frontiers in Forensic Genetics   305
Concluding Remarks   306
Acknowledgment   307
References   307

**23. Public and Population Health
Genomics**

A.L. Wise and T.A. Manolio

Introduction   311
Breast Cancer   311
Colorectal Cancer   312
  Bronchial Asthma   313
Crohn Disease   313
Alzheimer Dementia   313
Cystic Fibrosis   315
Cross-Cutting Issues of Population
  Genomics   315
Summary   316
References   316

Glossary   319
Index   327

List of Contributors

A. Alfirevic University of Liverpool, Liverpool, United Kingdom

S.E. Antonarakis University of Geneva Medical School, Geneva, Switzerland; University Hospitals of Geneva, Geneva, Switzerland; iGE3 Institute of Genetics and Genomics of Geneva, Geneva, Switzerland; HUGO (Human Genome Organization), Geneva, Switzerland

A. Bhardwaj Institute of Microbial Technology, Council of Scientific and Industrial Research, Chandigarh, India

P. Borry University of Leuven, Leuven, Belgium

C. Børsting Section of Forensic Genetics, Department of Forensic Medicine, Faculty of Health and Medical Sciences, University of Copenhagen, Copenhagen, Denmark

M.V. Busi Universidad Nacional de Rosario, Rosario, Argentina

V.H.W. Dissanayake University of Colombo, Colombo, Sri Lanka

R. Festenstein Imperial College, London, United Kingdom

C.L. Gaff Melbourne Genomics Health Alliance, Melbourne, Australia; The University of Melbourne, Melbourne, Australia

D.F. Gomez-Casati Universidad Nacional de Rosario, Rosario, Argentina

M. Grisolía Universidad Nacional de Rosario, Rosario, Argentina

P. Gupta Institute of Microbial Technology, Council of Scientific and Industrial Research, Chandigarh, India; Bhaskaracharya College of Applied Sciences, University of Delhi, New Delhi, India

A. Haworth Congenica Ltd., Hinxton, United Kingdom

B. Kerr Manchester Academic Health Sciences Centre (MAHSC), Manchester, United Kingdom

B.M. Knoppers McGill University, Montreal, QC, Canada

B. Korf University of Alabama at Birmingham, Birmingham, AL, United States

D. Kumar The University of South Wales, Pontypridd, Wales, United Kingdom; Cardiff University School of Medicine, University Hospital of Wales, Cardiff, United Kingdom

N. Lench Congenica Ltd., Hinxton, United Kingdom

A. Lucassen University of Southampton Medical Centre, Southampton, United Kingdom; University Medical Centre Groningen and Rijksuniversiteit Groningen, Groningen, The Netherlands

I. Macciocca Victorian Clinical Genetics Service, Melbourne, Australia

E. Maher University of Cambridge, Cambridge, United Kingdom

T.A. Manolio National Institutes of Health (NIH), Bethesda, MD, United States

D. McHale UCB, Braine L'Alleud, Belgium

N. Morling Section of Forensic Genetics, Department of Forensic Medicine, Faculty of Health and Medical Sciences, University of Copenhagen, Copenhagen, Denmark

A. Mutreja MSD-Wellcome Trust Hilleman Laboratories, New Delhi, India; Wellcome Trust Sanger Institute, Cambridge, United Kingdom

M. Penny Biogen, Cambridge, MA, United States

M. Pirmohamed University of Liverpool, Liverpool, United Kingdom

N.K. Rajput Institute of Microbial Technology, Council of Scientific and Industrial Research, Chandigarh, India

Y.-H. Rogers The Jackson Laboratory for Genomic Medicine, Farmington, CT, United States

H. Savage Congenica Ltd., Hinxton, United Kingdom

R. Saxena Indian Institute of Science Education and Research Bhopal, Madhya Pradesh, India

M. Shabani University of Leuven, Leuven, Belgium

V.K. Sharma Indian Institute of Science Education and Research Bhopal, Madhya Pradesh, India

V. Singh Amity University, Noida, India

N. Sirisena University of Colombo, Colombo, Sri Lanka

D. Sumathipala University of Colombo, Colombo, Sri Lanka

I. van Langen University Medical Centre Groningen and Rijksuniversiteit Groningen, Groningen, The Netherlands

K. Wettasinghe University of Colombo, Colombo, Sri Lanka

J. Whitworth University of Cambridge, Cambridge, United Kingdom

A.L. Wise National Institutes of Health (NIH), Bethesda, MD, United States

C. Zhang The Jackson Laboratory for Genomic Medicine, Farmington, CT, United States

Foreword

Although inherited diseases in man have been known since biblical times, and the first clear scientific descriptions of genetic disorders by Garrod and others date back to the early years of the 20th century, genetics was only introduced into regular medical practice with the advent of genetic counseling clinics in several countries, about 40 or 50 years ago. The discovery, in 1959, that Down syndrome was caused by a chromosome abnormality was an enormous step toward recognition that laboratory-based genetics had a role to play in medicine. In the early 1970s, few members of the public (or of the health professions) had heard of genetics and it was considered pretty irrelevant to healthcare practice.

The change since then has been astonishing. Genetics and genomics are now fully on the radar of most practicing physicians, researchers, and health managers, as well as journalists (who enjoy exaggerating the expected rate of progress) and politicians (who are commonly enthusiastic and sometimes surprisingly well informed). There can no longer be any doubt that the inherited components of disease are firmly embedded in medical practice and popular culture. When I say at parties that I am a medical geneticist, people more often look interested than mystified and bored.

Although it is true that much of this change has been driven by advancing technology, that is an oversimplification. The single biggest event was the sequencing of the human genome. That achievement and the decision of those involved to make their data freely available to all others, in academia and in industry, at an early stage of the research process, were transformative. Its echoes have completely changed not only medical research, but also anthropology, agriculture, and even historical research—witness the use of mitochondrial sequencing to confirm the identity of King Richard III several 100 years after his burial.

But the technology that we now enjoy was not there when the Human Genome Project started, and arguably would not be there today without that initiative. At the time, the genome project was an act of scientific imagination; a few people with extraordinary vision were able to see, long before most others and against a significant amount of scientific and public opposition, the immense value such sequence information would be. The technology was not confined to sequencing methodologies. The volume of data generated could not have been handled without commensurate improvements in computer hardware and software. Today it is arguable that the real block to understanding genetics and applying it clinically has more to do with our inability to understand and manage the vast amounts of data being generated, than to a need for better and cheaper sequencing.

Despite this meteoric progress, there is still actually rather little that modern genetics can do in the clinic, outside of defining the molecular basis of rare inherited disorders. But, particularly in cancer studies, that is changing rapidly. It is now very likely indeed that, despite all the previous overoptimistic statements that failed to materialize, genetics will pay big health dividends within a modest period of time. We can see it happening around us, eg, in new targeted cancer therapies. It is very exciting indeed.

Since the completion of the human genome sequence and rapid advances in genomic diagnostic methods, many avenues for diagnostic, therapeutic, and preventative intervention have emerged with promising medical and health applications. Those working in public and population health are considering many aspects of genomics for improving population health, particularly for the benefit of less developed or developing nations. Many national and international genetic and genomic communities and organizations are now actively engaged in furthering these objectives.

Against that background, this new book could not be better timed. It reviews many aspects of the field, and tries to help both professionals and interested observers to understand some of the core principles and complexities. Contributions from a distinguished group of authors, led by experienced and professional medical geneticists, provide an excellent resource of information. I am confident it will be of great value.

Martin Bobrow
Cambridge, England, August 2015

Preface

GENOMIC MEDICINE AND HEALTHCARE

In any sociocultural and geographic setting the provision of medical and healthcare are dependent upon several factors, including societal, financial, and political denominators. However, in the background of all these factors, the fact remains that peoples for whom any medical and healthcare system is aimed for are fundamentally different. The physical and psychological variation of the majority and minority population groups in any society or country determine the outcomes of medical and healthcare provision, however basic or sophisticated. Assessing human variation solely for the purposes of assessing the outcomes of medicine and healthcare practices would require many different approaches. Many such approaches include conventional anthropological and sociocultural variables. Since the discovery of the ABO, Rhesus, and other blood group systems in the early 20th century, the human variation has been assessed using a number of different biomarkers. However, despite many years' work and enormous data, a meaningful and scientific plausible correlation has not been possible. Following the Human Genome Project and the subsequent sequencing of thousands of individual human genomes, and the discovery of the extensive genomic variation and individuality, a major paradigm shift is taking place with considerable impact on the nature and pattern of medicine and healthcare. The people and the society at large have high expectations from the genomic-led contemporary and future medical and healthcare practices. There are huge expectations and hype surrounding the idea of genomic medicine (also known as *personalized/precision medicine*), which is based on the individual genomic variation. We are probably not fully prepared for this; there is a lot more to be learned and accomplished before the society and medical/health professions could offer genomic healthcare with robust scientific confidence while making this efficient and cost-effective.

Genetic and genomic variation among peoples and population groups are one of the many variables that influence the outcomes of any medical and healthcare practice. This needs to be linked with the specific "gene–molecule" systems that operate at the cell and tissue levels. Following the successes of clinical genetics, along with related genetic laboratory techniques, a number of specific genetic diseases (chromosomal, single gene, and rare genetic syndromes) are now causally linked to pathogenic changes in many genes and molecules that operate in conjunction with many other biological systems. Some of these rare conditions result from specific mutations or pathogenic sequence variation within one particular gene or loss of function of other genes that belong to a multigene family encoding many peptides with overlapping structural similarity and physiological functions. Examples include transcription factors, nuclear envelope genes, *RAS-MAPK* genes, *TBX* genes, genes for many inherited metabolic disorders, and a large number of gene–molecule families for sophisticated neuronal functions. Equipped with the knowledge from genetic and molecular advances in uncommon and extremely rare diseases, researchers and clinicians are now looking into solving the molecular complexities of common medical diseases with a considerable heritability. The scientific information from studying specific system and multisystem rare genetic disorders has given us insight to many fundamental molecular biological processes that are not only important for understanding the pathogenesis but also govern the outcomes of specific therapeutic interventions. Thus successful outcomes of genetic and genomic applications in medicine and healthcare practices would be dependent upon understanding the molecular biology of sequential diseases processes and their overall clinical impact governed by individual genomic variation. This is in essence the basis of *medical and health genomics*, rapidly emerging along with the practice of specific genetic and genomic medicine.

In keeping with the many dilemmas and predicaments surrounding the genetic and genomic applications in medicine and healthcare, this new book sets out to collate basic facts and information that could form the core of *genomic (personalized/precision) medicine* and *genomic healthcare*. A number of scientists, clinicians and healthcare professionals have contributed to this high profile work under the broad meaning title of *Medical and Health Genomics*.

The core concepts of human genomics are presented with emphasis on new emerging genomic technologies, burgeoning genomic databases with enormous amount classified and unclassified data, genomic applications, and translations in clinical medicine and public health, particularly the diagnostic genomics and clinical bioinformatics contributing to deciphering many complex phenotypes, citing the specific example of the model of Victor McKusick's *Online Mendelian Inheritance in Man (OMIM)*.

Some additional areas of particular importance are highlighted, including epigenetics modifications in human health and disease; metabolomics with its many applications; introduction to metagenomics and ecogenomics; wide-ranging applications of microbial genomics in diagnosis, treatment, and prevention of microbial diseases; and the personalized pharmacotherapy based on the genetic variation of each individual (pharmacogenomics).

Medical and health aspects of genetics and genomics are discussed in detail with evidence on the relevance of genome variation in human health and disease. This is further supported by critical information on multidisciplinary medical management using examples of systemic rare genetic diseases, genetic and genomic testing and screening, and genomic perspectives of genetic counseling. The section on personalized and stratified medicine includes information on novel genomics-led drug discovery and development and discussion on novel genomics-led therapeutic approaches. The organization, availability, and harmonization of genetic and genomic healthcare are reviewed with emphasis on medical and health burden of rare genetic diseases, multidisciplinary genetics and genomics-led reproductive healthcare, and common and complex genetic cancer. This section includes a separate chapter on teaching and training genetics/genomics for medical and healthcare professionals. The detailed glossary lists a number of key phrases, definitions, abbreviations, and acronyms that the reader might find useful.

Editors and contributors hope that the book will convey the core concepts of medical and health genomics, highlighting the specialist genomic fields, emphasis on delivery of genetic/genomic medicine and healthcare, drawing attention to specific issues and requirements of human genomics in developing countries and emerging perspectives of human and medical genomics in the context of public and population health.

The society and many people have high hopes and expectations from many recent new developments and progress in human (medical) genomics in the diagnosis and management of both rare and common medical and health problems. These sentiments were echoed by Mr. Anthony Charles Lynton (Tony) Blair, the British Prime Minister, on the occasion of the White House announcement of the completion of the First Survey of the Entire Human Genome Project, broadcast (jointly with the United States President Bill Clinton) on the day of the publication of the first draft of the human genome, "Ever so often in the history of human endeavor, there comes a breakthrough that takes humankind across a frontier into a new era. ... today's announcement is such a breakthrough, a breakthrough that opens the way for massive advancement in the treatment of cancer and hereditary diseases. And that is only the beginning."

We are delighted to present this book at a critical historical phase of genome science, with the new global wave to incorporate strengths and unlimited potential of genomics for much needed advances in healthcare. This is reflected in President Obama's State of the Union address on 30 January, 2015, launching the new United States Precision Medicine initiative, "*To enable a new era of medicine through research, technology, and policies that empower patients, researchers, and providers to work together toward development of individualized treatments.*"

Dhavendra Kumar, Cardiff, Wales, UK
Stylianos Antonarakis, Geneva, Switzerland
Editors
December 2015

Chapter 1

The Human Genome

D. Kumar[1,2]

[1]The University of South Wales, Pontypridd, Wales, United Kingdom; [2]Cardiff University School of Medicine, University Hospital of Wales, Cardiff, United Kingdom

Chapter Outline

Introduction	1	The Mitochondrial Genome	8
Hereditary Factors, Genes, Genetics, and Genomics	1	Functional Genomics, Transcriptomics, and Proteomics	9
Structure and Organization of Nucleic Acids	2	Translational Human Genomics	10
Human Genome Variation and Human Disease	5	Human Genomics for Socioeconomic Development	11
Measuring Genetic and Genomic Variation	6	Conclusions	12
Genome Variation and Human Disease	7	References	12

INTRODUCTION

Toward the end of the last millennium, tremendous growth in the sophistication of the biological sciences was harnessed in medicine, the food industry, and related bioindustries. New discoveries and innovations in biological sciences during the five decades leading up to the 21st century have centered on genetics and genomics. It took just over 50 years after the unraveling of the structure of the molecule of nucleic acids, the key unit of the biological life, for scientists to embark on sequencing of major organisms' entire genetic constitution or genome. The word *genome* includes *gene* and *-ome*, implying complete knowledge of all genes and related elements in any single organism. Inevitably, this led to enthusiastic expansion of the whole science and thence to the emergence of *genomics* [1]. The suffix *-omic*, derived from the ancient Greek, refers to in-depth knowledge. Not surprisingly, genomics was followed by a plethora of related *-omics*; for example, proteomics, metabolomics, transcriptomics, and so on [2]. Currently, we have over 30 such disciplines with the *-omics* suffix.

The ultimate goal of any scientific discipline is its translation for the benefit of all humans, crossing all possible barriers and boundaries. Major advances in medicine and health were only possible through understanding basic principles and mechanisms underlying disease processes. This was facilitated by rapid applications of physical and chemical sciences in medicine and health; for example, radiographical diagnosis, ultrasound diagnosis, microbiology diagnosis, immunohistochemical diagnosis, and finally, molecular diagnosis. Developments and advances in genetics have led to a better understanding of the principles governing heredity and the familial transmission of physical characteristics and diseases, better understanding of the pathophysiology of diseases, the development of new methods of clinical and laboratory diagnosis, and innovative approaches to making early diagnoses (eg, prenatal diagnoses and newborn screening) and offering reproductive choices, including preimplantation genetic diagnoses. All these developments are now accepted within the broad fields of human genetics, medical genetics, clinical genetics, genetic medicine, and the new emerging field of genomic medicine. Not surprisingly, the field remains wide open, encompassing the massive field of human genomics, broadly focusing on medical and health genomics [3].

This chapter leads the book, providing the basic factual information for grasping the concepts of heredity, genes, genetics, and genomics. It is expected that the reader will proceed to subsequent chapters better equipped with the introduction to genetic/genome sciences, genetic diseases, genetics and genomics in medicine, applications in public health, and specific issues related to society, ethics, and law [4].

HEREDITARY FACTORS, GENES, GENETICS, AND GENOMICS

The concepts of *heredity* and *hereditary factors* date back several hundred and probably even thousands of years. The popular darwinian theory of natural selection rests on the core concept of the *transmission of hereditary factors*[5]. For several thousand years, various descriptions and explanations have been put forward to define the

Medical and Health Genomics. http://dx.doi.org/10.1016/B978-0-12-420196-5.00001-0

physical shape and functional nature of hereditary factors. In the historical context, the concept of the *gene* was introduced only recently as the most acceptable answer to explain one of the hereditary factors. However, it remains unclear when and by whom this term was first introduced. It does not matter, as the term *gene* (from the Greek *genos*, race) is now universally accepted and used in the context of understanding heredity and hereditary factors, and is probably the single most important biological factor regulating biological life, ranging from single-cell organisms to multicellular mammals. Rapid and extraordinary scientific progress made during the 19th and 20th centuries has led to the development of *genetics*, the science of heredity. This has now been transformed into the broader field of *genomics* that includes all genes with all possible heritable biologically active or inactive regulatory and evolutionary genetic elements, whether recent or extending back through several thousand years of life on our planet.

In biological terms, genes, genetics, and genomics are keys to procreation, development, growth, function, and survival. The health of any living organism is judged by its physical and functional existence. Thus genes, genetics, and genomics are central to all forms of biological health, including that of humans. Human health depends not only on its own *genetic* or *genomic* constitution, but on that of other organisms whose well-being is also essential to human health—for example, food (plants, fish, and animals), shelter (homes made of wood from trees), the environment (water, trees, and plants), protection (clothes from cotton and animal skin), and transportation (animals and vehicles made of wood from trees). From a medical perspective, the science of genetics or genomics offers deep insight into and evidence for a number of human diseases, including infectious diseases resulting from either lack of protection and/or failure in controlling the spread of microbial infections or parasitic infestations. This chapter introduces the reader to some of the basic facts about genes, genetics, and genomics, and discusses how these impact human health and that of the plants, crops, and animals necessary for human health and survival. This is obviously more relevant to millions of people in the developing and less-developed countries, where limited resources and lack of infrastructure limit the optimal use of the science of genetics and genomics in applications to eradicate poverty and ensure optimal health. The reader will find cross-references to separate chapters in the book containing detailed information and further discussion of each subject.

A detailed description of the basic principles of genetics and human genetic diseases is beyond the scope of this chapter. Some of these facts are explained in subsequent chapters and various other information resources on basic genetics and medical genetics. However, some basic principles and relevant information are outlined in this section to assist the reader with limited understanding of basic genetics.

STRUCTURE AND ORGANIZATION OF NUCLEIC ACIDS

Living organisms are divided into two large classes—the *eukaryotes* and *prokaryotes*. The cells of the eukaryotes have a complex compartmentalized internal structure, the nucleus; these include algae, fungi, plants, and animals. Prokaryotes, on the other hand, are single-celled microorganisms without any specific part harboring the genetic material or genome; examples include bacteria and other related microorganisms. The other types of living organisms are viruses, which are intracellular obligate parasites living in both eukaryotes and prokaryotes, and are composed of short dispersed nucleic acid [deoxyribonucleic acid (DNA) or ribonucleic acid (RNA)] sequences.

Genetic information is transferred from one generation to the next by small sections of the nucleic acid, DNA, which is tightly packaged into subcellular structures called *chromosomes*. Prokaryotes usually have a single circular chromosome, while most eukaryotes have more than two, and in some cases up to several hundred. In humans, there are 46 chromosomes arranged in 23 pairs, with one of each pair inherited from each parent (Fig. 1.1A and B). Twenty-two pairs are called *autosomes*, and one pair is called *sex chromosomes*, designated as X and Y; females have two X chromosomes (46, XX) and males have an X and a Y (46, XY).

A chromosome consists of a tightly coiled length of DNA and the proteins (eg, chromatins) that help define its structure and level of activity. DNA consists of two long strands of nucleotide bases wrapped round each other along a central spine made up of phosphate and sugar (Fig. 1.2). There are four bases: adenine (A), guanine (G), cytosine (C), and thymine (T). Pairing of these bases follows strict rules: A always pairs with T, and C with G. Two strands are therefore complementary to each other.

Genes are made up of specific lengths of DNA that encode the information to make a protein, or RNA product. RNA differs from DNA in that the base thymine (T) is replaced by uracil (U), and the sugar is ribose. It acts as a template to take the coded information across to ribosomes for final assembly of amino acids into the protein peptide chain (Fig. 1.3). The bases are arranged in sets of three, referred to as *codons*. Each codon "codes" for a specific amino acid; hence the term *genetic code*. Codons are located in *exons*, which contain the coding sequences. A gene may consist of several such coding DNA segments. Exons are separated from each other by noncoding sequences of DNA, called *introns*. Although they are not yet known to be associated with any specific function, it is likely that some of these introns might be of evolutionary significance or associated with other fundamental biological functions. During

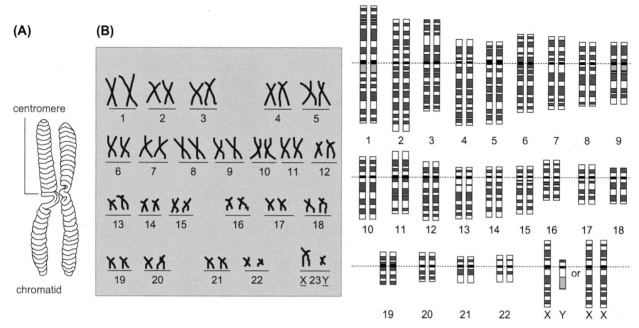

FIGURE 1.1 Human chromosomes. (A) Diploid set in a male (46, XY). (B) Complete set of human chromosomes map.

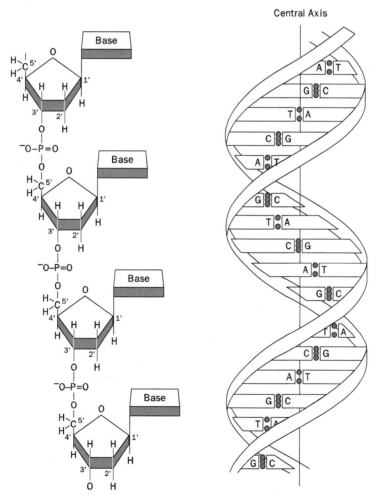

FIGURE 1.2 The Watson-Crick model of the double helix structure of the nucleic acid molecule. *Adopted with permission from Turnpenny P, Ellar S, editors. Emery's elements of medical genetics. 14th ed. Edinburgh: Elsevier Churchill Livingstone; 2012.*

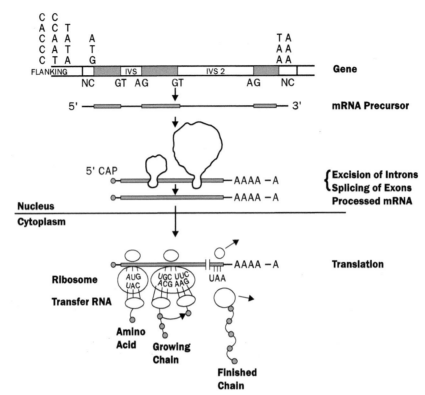

FIGURE 1.3 The synthesis of the peptide chain from the coding sequences in the exon. *mRNA*, messenger RNA; *CAP*, adenyl cyclase associated protein; *IVS*, intervening sequence (Turnpenny and Ellard, 2011).

the transcription of DNA, the introns are spliced out, and the exons then attach to messenger RNA (mRNA) to start the process of protein synthesis.

Proteins are one of the major constituents of the body's chemistry. These are remarkably variable in their structure, ranging from tough collagen that forms connective tissue and bone, through the fluid hemoglobin that transports oxygen, to thousands of enzymes, hormones, and other biological effectors and their receptors that are essential for the structures and functions of the body. Each protein is made up of one or more peptide chains consisting of series of amino acids, of which only 20 occur in living organisms. The different structures and functions of proteins depend on the order of amino acids as determined by the genetic code.

DNA has the remarkable property of self-replication. The two strands of a DNA molecule separate as chromosomes divide during cell division. There are two types of cell division; *mitosis* in all body cells, and *meiosis*, which is specifically confined to the gonads in making sperm and eggs (Fig. 1.4). During mitosis, no reduction of the number of chromosomes takes place (*diploid*, or 2n), while meiosis results in half the number of chromosomes (*haploid*, or 1n). The new pairs of DNA are identical to those from which they were synthesized. However, sometimes mistakes or mutations occur. These usually result from substitution of a different base, or are caused by extensive structural changes to genes. In other words, any "spelling mistake" in the letters A–T or C–G could result in either absence of coded

information (*nonsense mutation*) or a different message (*missense mutation*). However, not all mutations or spelling mistakes have an adverse effect (*neutral mutations*). Conversely, some changes in the genes might result in a favorable property; for example, resistance to disease or other environmental hazard. This is the basis for the gradual changes in species over millions of years of evolution. On the other hand, mutations may result in defective gene functions, leading to a disease or susceptibility to a disease as a result of qualitative or quantitative changes in the gene product, the peptide chain. However, these changes may also result from epigenetic mechanisms, abnormal RNA molecules, and posttranslational modifications (see Glossary). A brief introduction to these molecular processes is provided elsewhere in this chapter; interested readers are advised to consult dedicated texts on cell and molecular biology.

Studies on human genomic variations in different population groups and the resemblance of several genome sequences to other genomes (*comparative genomics*) have offered wide-ranging evidence to support the followers of Charles Darwin. Apart from reproduction, genes, gene-sequence variation, genomic variation, and epigenetic factors are important in growth, development, aging, and senescence. Some of these may be evolutionarily conserved across species, but relevant to human health. Mutations and alterations in several of these genomic elements are linked to a broad range of medical conditions.

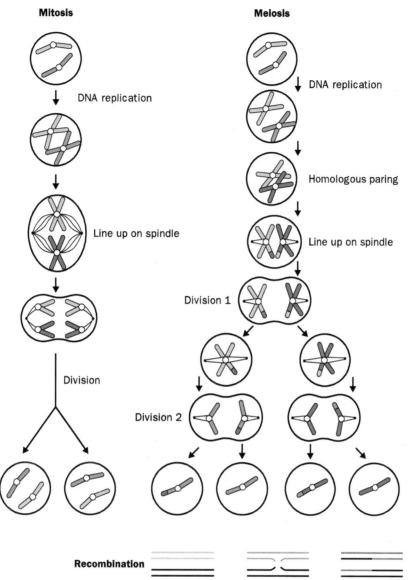

FIGURE 1.4   Steps in mitosis and meiosis during a eukaryotic cell division; note *(bottom)* exchange of the genetic material (recombination) through homologous pairing (Turnpenny and Ellard, 2011).

HUMAN GENOME VARIATION AND HUMAN DISEASE

The advent of recombinant DNA technology in the 1970s revolutionized our ability to characterize and capitalize on the molecular basis of human genetic disease. This laid the foundation of eventually mapping and deciphering the DNA sequence of all the structural and functional genes of the human genome. The Human Genome Project (HGP) was therefore a natural progression from all previous developments in the field of human genetics. Such a mammoth task could not have been accomplished without the international collective efforts supported by generous funding from governmental and nongovernmental sources [6].

The project (HGP) has helped map and provide nucleotide sequences of around 23,000 nuclear genes, which, along with a number of other sequence variations, compose the whole human genome. Although a large number of the nuclear genes have been assigned with a structural or functional link, the precise roles of other parts of the genome are not yet fully understood. However, HGP provides the basis for "functional genomics" to explore further the genome's functional role and understand the complex mechanisms through which genes and their products interact to affect biological function and influence disease processes. The development of new therapeutic agents is now possible on the basis of genomic arrangement and its designated functional role. This approach also helps characterize

TABLE 1.1 DNA and Gene Content of the Human Reference Genome (GRCh37, February 2009)

Chromosome	Length (bp)	Protein-Coding Genes (Known + Novel)	Pseudogenes	RNA Genes	Genes per Mb (mega base)
1	249,250,621	2037	1131	672	8.17
2	243,199,373	1259	947	526	5.18
3	198,022,430	1066	719	430	5.38
4	191,154,276	758	698	363	3.97
5	180,915,260	874	675	343	4.83
6	171,115,067	1042	726	358	6.09
7	159,138,663	907	800	350	5.70
8	146,364,022	731	568	288	4.99
9	141,213,431	803	714	260	5.69
10	135,534,747	762	498	295	5.62
11	135,006,516	1320	774	290	9.78
12	133,851,895	1051	582	336	7.85
13	115,169,878	326	323	173	2.83
14	107,349,540	652	472	310	6.07
15	102,531,392	605	471	329	5.90
16	90,354,753	867	384	229	9.60
17	81,195,210	1197	255	273	14.74
18	78,077,248	277	56	157	3.55
19	59,128,983	1418	180	198	23.98
20	63,025,520	546	213	189	8.66
21	48,129,895	233	150	69	4.84
22	51,304,566	455	308	105	8.87
X	155,270,560	836	780	351	5.38
Y	59,373,566	53	327	44	0.89
Total	**3,286,906,385**	**21,099**	**15,520**	**11,960**	**6.42**

The overall totals are derived from a slightly different analysis from the individual chromosome totals, so the figures do not exactly add up.
Data from Ensembl Release 66, March 20, 2012.

the genomes of various pathogens and other organisms, an invaluable tool in realizing the full potential of this field to improve human health [7].

Measuring Genetic and Genomic Variation

Humans have two genomes: nuclear and mitochondrial. Normal diploid cells contain two copies of the nuclear genome and a much larger but variable number of copies of the mitochondrial genome. The nuclear genome is approximately 2×10^5 times larger than the mitochondrial genome (3×10^9 vs 16,569 bp), and contains more than 1500

times the number of protein-coding genes (approximately 21,000 vs 13), including many required for mitochondrial functions. Genetic and genomic variation is abundant in both genomes. However, in general this implies to nuclear genome.

The "finished" human genome sequence was published in 2004 [7]. Table 1.1 shows the current best estimates of the size and gene content of each chromosome. These figures are for the human reference genome. There are striking differences between the nuclear and mitochondrial DNA (Table 1.2). They do not correspond precisely to the genome of any actual individual, because the genomes of healthy

TABLE 1.2 Comparison of the Human Nuclear and Mitochondrial Genomes

	Nuclear Genome	Mitochondrial Genome
Size	3×10^9 bp	16,659 bp
Topology	23 linear molecules	1 circular molecule
Number of genes	Approximately 21,000	37
% Coding sequence (including genes for functional RNAs)	Approximately 1.4%	93%
Average gene density	Approximately 1 per 125 kb (variable)	1 per 0.45 kb
Introns	Average 8 per gene (variable)	None
Repetitive DNA	Approximately 50%	None

normal individuals vary somewhat in chromosome sizes and numbers of genes, as described below. Nor is the reference genome in any sense an "ideal" human genome. It is simply an arbitrary and reasonably typical reference point for comparing human genome sequences. Uncertainties in the figures relate primarily to the highly repetitive DNA of centromeres and telomeres and to the number of RNA genes, which are difficult to identify from sequence data.

The most direct way to measure genetic differences, or genetic variation, is to estimate how often two individuals differ at a specific site in their DNA sequences (that is, whether they have a different nucleotide base pair at a specific location in their DNA). First, DNA sequences are obtained from a sample of individuals. The sequences of all possible pairs of individuals are then compared to see how often each nucleotide differs. When this is done for a sample of humans, the result is that individuals differ, on average, at only about one in 1300 DNA base pairs. In other words, any two humans are about 99.9% identical in terms of their DNA sequences.

During the past several years, a new type of genetic variation has been studied extensively in humans: *copy-number variants* (*CNVs*), comprising of DNA sequences of 1000 base pairs or larger, fairly distributed across the genome [8]. In some instances, CNVs could be deleted, duplicated, or inverted in some individuals with mild phenotypic effects. Several thousand CNVs have been discovered in humans, indicating that at least 4 million nucleotides of the human genome (and perhaps several times more) vary in copy number among individuals. CNVs thus are another important class of genetic variation and contribute to at least an additional 0.1% difference, on average, between individuals. Despite significant progress, the medical and health implications of CNVs are not entirely clear [9].

Comparisons of DNA sequences can be done for pairs of individuals from the same population or for pairs of individuals from different populations. Populations can be defined

in various ways; one common way is to group individuals into populations according to the continent of origin. Using this definition, individuals from different populations have roughly 10–15% more sequence differences than do individuals from the same population [this estimate is approximately the same for both *single nucleotide polymorphisms* (SNPs) and CNVs]. In other words, people from different populations are slightly more different at the DNA level than are people from the same population. The slightness of this difference supports the conclusion that all humans are genetically quite similar to one another, irrespective of their geographic ancestry [10].

Because it is still fairly expensive to assess DNA sequences on a large scale, investigators often study genetic variations at specific sites that are known to vary among individuals. Suppose that a specific site in the DNA sequence harbors an A in some individuals' DNA sequences and a G in others'. This is an SNP, where *polymorphism* refers to a genetic site that exists in multiple forms. The proportion of individuals who have an A and the proportion with a G give the frequency of each form, or *allele*, and this frequency can be estimated for a sample of individuals from a population. If the frequencies of A in three different populations are 0.10, 0.20, and 0.50, the *genetic distance* between the first two populations is smaller than that between the third population and the first two. On the basis of this assessment, the first two populations are genetically more similar than either is to the third. To get a more accurate picture of genetic differences, hundreds or thousands of SNP frequencies would be assessed to yield the average genetic difference among pairs of populations [11].

Genome Variation and Human Disease

A number of genes have direct or indirect influence on most human diseases. Because individuals have different variants of genes, it follows that the risk of developing various diseases will also differ among individuals. Consider a simple example: Jim Fixx, a well-known runner and fitness enthusiast, died of a heart attack at the age of 52. Sir Winston Churchill, who was renowned for his abhorrence of exercise and his love of food, drink, and tobacco, lived to the age of 90. It is plausible that genetic differences between Fixx and Churchill were responsible, at least in part, for the paradoxical difference in their life spans. (Indeed, Jim Fixx's father had a heart attack at the age of 35 and died of a second heart attack at the age of 43.)

Because genes are passed down from parents to offspring, diseases tend to "cluster" in families. For example, if an individual has had a heart attack, the risk that his or her close relatives, offspring, or siblings will have a heart attack is two to three times higher than that of the general population. Similar levels of increased risk among family members are seen for colon cancer, breast cancer, prostate cancer, type 2 diabetes mellitus, and many other diseases. This clustering in families is partly the result of shared nongenetic factors (eg, families tend to be similar in terms of their dietary and exercise habits) and partly the result of shared genes. As we have seen, populations differ somewhat in their genetic backgrounds. It is thus possible that genetic differences could be partly responsible for differences in disease prevalence. For many disorders caused by genetic changes in single genes, these differences are readily apparent. Cystic fibrosis, for example, is seen in about one in 2500 Europeans, but only in one in 90,000 Asians. Sickle cell disease is much more common in individuals of African and Mediterranean descent than in others, although it is found at lower rates in many other populations because of migration and intermarriage.

These differences in prevalence can be attributed to the evolutionary factors that influence genetic variation in general. *Mutation* is the ultimate source of all genetic variation. In some cases, such as hemochromatosis in Europeans and sickle cell disease in Africans, the responsible mutations have arisen within the last few thousand years, helping to account for a fairly restricted distribution of the disease. *Natural selection* also plays a role in population differences in some genetic diseases. For sickle cell disease and related diseases known as the *thalassemias*, *heterozygotes* (those who carry a single copy of a disease-causing mutation) are relatively resistant to the malaria parasite. Cystic fibrosis heterozygotes are resistant to typhoid fever, and hemochromatosis heterozygotes absorb iron more readily, perhaps protecting them against anemia. Also, the process of *genetic drift*, which is accentuated in small populations, can raise the incidence of disease-causing mutations quickly just by chance (eg, Ellis-van Creveld syndrome, a reduced-stature disorder, is unusually common among the Old Order Amish of Pennsylvania) [12]. In contrast to the effects of natural selection and genetic drift, which tend to promote population differences in disease prevalence, *gene flow* (the exchange of DNA among

populations) tends to decrease differences among populations. With the enhanced mobility of populations worldwide, gene flow is thought to be increasing steadily.

These same factors can affect common diseases such as cancer, diabetes mellitus, hypertension, and heart disease, but the picture is more complex, because these diseases are influenced by multiple genetic and nongenetic factors. Common diseases do vary in incidence among populations: hypertension occurs more often in African Americans than European Americans, and type 2 diabetes mellitus is especially common among Hispanic and Native American populations [13]. Although genes clearly play a role in causing common diseases, it is less clear that genetic differences between populations play a significant role in causing differences in prevalence rates among populations. Consider another example: the Pima Native American population in the southwestern United States now has one of the highest known rates of type 2 diabetes mellitus in the world. About half of adult Pimas are affected. Yet this disease was virtually unknown in this population before World War II. Obviously, the Pimas' genes have not changed much since the middle of the 20th century. Their environment, however, has changed dramatically with the adoption of a "Western" high-calorie, high-fat diet, and a decrease in physical exercise. In this case, it is almost certain that the rapid increase in type 2 diabetes mellitus prevalence has much more to do with nongenetic than genetic causes [14].

But why does a Western diet seem to have a greater effect on some populations than others? Perhaps differences in genetic background, interacting with dietary and other lifestyle changes, help account for this variation. As additional genes that influence susceptibility to common diseases are discovered, and as the roles of nongenetic factors are also taken into account, it is likely that this picture will become clearer.

THE MITOCHONDRIAL GENOME

The mitochondrial genome is very different from the nuclear genome (Fig. 1.5; Table 1.2). In many respects, it has more in common with bacterial genomes than the eukaryotic nuclear genome. This is consistent with the idea that mitochondria originated as endosymbiotic bacteria within some ancestral eukaryotic cell. If this theory is correct, then over the years, the mitochondria have gradually transferred more and more of their functions to the nucleus. This is evident from the fact that a number of nuclear genes encode the great majority of mitochondrial proteins. Cells contain many mitochondria (typically 100–1000; maybe 100,000 in an oocyte), so mitochondrial DNA (mtDNA) might be formally classified among the repetitive DNA in a cell. Although the mitochondrial genome is very small compared with its nuclear counterpart, because there are many copies, mtDNA often makes up 1% or so of total cellular DNA.

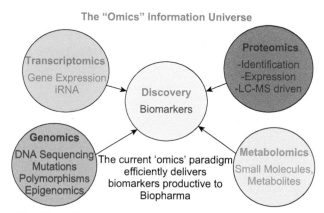

The "Omics" Information Universe

FIGURE 1.5 The *-omics* paradigm, showing four major branches. *iRNA*, Informational RNA; *LC-MS*, liquid chromatography–mass spectrometry.

As in bacteria, the mitochondrial genome is circular and closely packed with genes. There are no introns and little intergenic noncoding DNA. Some genes even overlap. In the nuclear genome, it is not uncommon for genes on opposite strands to overlap. However, in this case, genes on the same strand overlap, using the same template but read in different reading frames. Twenty-four of the 37 genes specify functional RNAs [two ribosomal RNAs and 22 transfer (tRNAs)]; the other 13 genes encode components of the electron transport pathway.

A short segment of the mitochondrial genome is triple-stranded, on which the displacement loop (D-loop) is non-coding produced by replication forks overlapping as they travel in opposite directions around the circular DNA. The D-loop contains the only significant amount of noncoding DNA in the mitochondrial genome. Perhaps because of this, it is the location of many of the DNA polymorphisms that are such useful tools for anthropologists researching the origins of human populations. Because there is no recombination among mtDNA, complete haplotypes of polymorphisms are transmitted through the generations, modified only by recurrent mutation. This makes mtDNA a highly informative marker of ancestry, at least along the maternal line.

Mitochondrial DNA replication and transcription use nuclear-encoded polymerases. Transcription proceeds in both directions round the circle. The initial products are two large multicistronic RNAs, which are subsequently cleaved to make the individual mtRNAs. All the protein components of the translation machinery are nuclear-encoded, but mitochondria exclusively encode the tRNAs, and these use a coding scheme slightly different from the otherwise universal code. There are four stop codons: UAG, UAA, AGG, and AGA. UGA encodes tryptophan, and AUA specifies isoleucine, rather than arginine as normally. Presumably, with only 13 protein-coding genes, the mitochondrial system could tolerate mutations that modified the coding scheme in a way the main genome could not.

Mutations in mtDNA are important causes of disease, and perhaps also of aging [15]. Phenotypes caused by variation in mtDNA are transmitted exclusively down the maternal line (matrilineal inheritance), but most genetic diseases where there is mitochondrial dysfunction are caused by mutations in nuclear-encoded genes, and so follow normal mendelian patterns. As cells contain many copies of the mitochondrial genome, they can be heteroplasmic, containing a mix of different sequences. Unlike mosaicism for nuclear variants, heteroplasmy can be transmitted by a mother to her children.

FUNCTIONAL GENOMICS, TRANSCRIPTOMICS, AND PROTEOMICS

Functional genomics, specifically transcriptomics, is a systematic effort to understand the function of genes and gene products by high-throughput analysis of gene transcripts in a biological system (cell, tissue, or organism) with the use of automated procedures that allow scale-up of experiments classically performed with single genes [15]. Functional genomics can be conceptually divided into *gene-driven* and *phenotype-driven* approaches. Gene-driven approaches rely on genomic information to identify, clone, and express genes, as well as to characterize them at the molecular level. Phenotype-driven approaches rely on phenotypes, either identified from random mutation screens or associated with naturally occurring gene variants, such as those responsible for mouse mutants or human diseases, to identify and clone the responsible genes without prior knowledge of the underlying molecular mechanisms [16]. The tools of functional genomics have enabled the development of systematic approaches to obtaining basic information for most genes in a genome, including when and where a gene is expressed and what phenotype results if it is mutated, as well as the identification of the gene product and the identity of other proteins with which it interacts. Functional genomics aspires to answer such questions systematically for all genes in a genome, in contrast to conventional approaches that address one gene at a time.

Within the context of functional genomics, an important part of functional arrangement of all genomes consists of areas that are external to any coding gene sequences. These play a crucial role in gene expression. Collectively this is known as *epigenome*. Studies in epigenetics and epigenomics have established core principles that are responsible for modulating development and differentiation and respond to external changes. Patterns of cell- and tissue-specific gene expression are established and maintained by the patterns of epigenetic marks on the genome. These consist of DNA methylation and a variety of specific covalent modifications of histones. The epigenetic marks or signatures are established by a large series of "writers": DNA methyltransferases, histone methyltransferases and demethylases, histone acetyltransferases and deacetylases, histone kinases and phosphatases, and so on. In some cases, small RNA molecules help ensure sequence specificity. The effects on gene expression are mediated by epigenetic "readers," including methylated DNA-binding proteins, chromodomain and bromodomain proteins that bind methylated and acetylated histones respectively, and a large number of other proteins [17].

As a result of epigenetic modifications, chromatin exists in a variety of epigenetic "flavors." The basic distinction is between heterochromatin (inactive, repressed) and euchromatin (potentially active), but subtypes define transcriptional activity and regulatory elements such as promoters, enhancers, and insulators. The flavor depends on a combination of types and relative quantities of marks rather than a simple histone code.

Central to functional genomics is the complex organization of RNA molecules that occupies bulk of the intergenic parts of the genome. Apart from well characterized coding RNA (cRNAs), messenger RNA (mRNA), ribosomal RNA (rRNA), and transfer RNA (tRNA), there are many other noncoding RNAs (ncRNAs) [18]. These can be divided into "classical" ncRNAs and long intergenic ncRNAs (lincRNAs). The classical ncRNAs are small molecules, typically 16–30 nt, derived by processing much longer precursors. There has been an explosion in our knowledge of the numbers and classes of these molecules, and this is still a very active research area. The main well-established classes are:

- Small nuclear RNAs (snRNAs) form part of the spliceosomal machinery.
- Small nucleolar RNAs (snoRNAs) act as sequence-specific guides for enzymes that chemically modify specific bases in ribosomal and other RNAs.
- MicroRNAs (miRNAs) control translation of many mRNAs by binding to sequences in the 3′ untranslated region.
- Piwi-associated RNAs (piRNAs) act in gametes to ensure stability of the genome. There appear to be many thousands of piRNA genes, grouped in around 100 clusters.

Table 1.3 lists the numbers of genes encoding these molecules, but these are subject to major revision because it is very difficult to identify functional ncRNAs and distinguish them from the large number of nonfunctional variants present in the genome [19].

Analysis and application of the rapid accumulation of highly sophisticated genome and proteome data necessitated development of powerful computational programs and relevant hardware tools. Storage, retrieval, and assimilation of enormous amounts of data require fast and accurate computational skills. *Bioinformatics* deals with these requirements within the broad biomedical and biotechnology sectors. There are several literature and online resources with detailed descriptions of the role and scope of bioinformatics [20].

TABLE 1.3 RNA Genes in the Human Genome

RNA Species	Number of Functional Genes
Ribosomal RNA	150–200
Transfer RNA	496
Small nuclear RNA	91
Small nucleolar RNA	375
MicroRNA	1733
Piwi-associated RNA	114 clusters

miRNAblog.com, piRNAbank.ilab.ac.in.

A number of biomedical and biotechnology disciplines have emerged during the last two decades, all ending with the suffix -*omics*. -*Omics* is derived from *ome* (Greek, *omoyous*), which refers to *complete knowledge*. The ancient language Sanskrit has a similar word, *ohm,* with similar meaning and expression. A number of these -*omics* have direct or indirect links to the fundamentals of genome science and technology. A number of biological models have been developed and tested using genomic, transcriptomic, proteomic, and metabolomic approaches (Fig. 1.5). *Systems biology* refers to developing and testing biological models based on -*omic* sciences [21]. The central dogma of the *systems biology* is the computational analysis of complex and enormous data at all biological levels: gene, molecule, cell, tissue, organ, and whole body.

TRANSLATIONAL HUMAN GENOMICS

The potential of application of genome science and technology in medicine and health has led to the emergence of *genomic medicine*, a natural outcome of the tremendous progress made in medical genetics and genomics [22]. However, final endpoints in genomic medicine will largely depend upon judicious and efficacious application and utilization of the diagnostic and therapeutic potential of genome-based technologies; for example, clinical applications of microarray technology. This process requires multifaceted systematic and analytical research efforts to translate the basic scientific information into practical and pragmatic applications following the principles of good medical practice (Fig. 1.6). There is no disagreement that this *translational genome research* is vital for the successful and efficient delivery of promises made by researchers and physicians behind the genomic medicine movement.

The process for translational genome research includes the participation of several researchers drawn from different disciplines. The multidisciplinary model for translational genome research is widely accepted, and includes several key elements. Informatics and computational networks

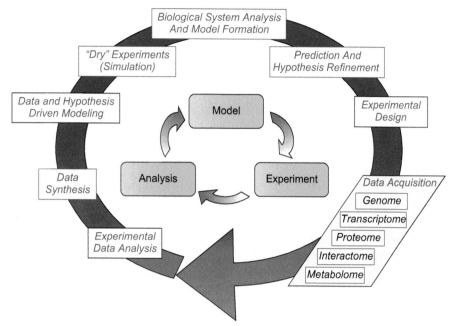

FIGURE 1.6 Informatics as the central dogma for systems biology and genome sciences.

remain the core element for translational genomics research and systems biology [23]. A framework for the continuum of multidisciplinary translation research is recommended to utilize previous research outcomes in genomics and related areas of health and prevention [24].

The whole process of *translational genome research* includes four phases and revolves around the development of evidence-based guidelines. Phase 1 translation research seeks to move a basic genome-based discovery into a candidate health application, such as a genetic test or intervention. Phase 2 translation research assesses the value of genomic applications for health practice, leading to the development of evidence-based guidelines. Phase 3 translation research attempts to move evidence-based guidelines into health practice through delivery, dissemination, and diffusion research. Phase 4 translation research seeks to evaluate the "real world" health outcomes of a genomic application in practice. It is important to appreciate that the whole process of translation research leading to evidence-based guidelines is a dynamic one, with considerable overlap between the different stages. The process should be able to accommodate new knowledge that will inevitably arrive during translation research.

The role of translational genome research, including that of clinical trials, is crucial in developing evidence-based, good-practice guidelines [25]. The aim should be to obtain vital genetic and genomic information, including laboratory material for research, from the patient, family, and community, and then use this scientific data and information for clarification and ratification. The outcomes of translational gene research should be valid and

deliverable in the clinic for diagnostic and therapeutic applications [26].

HUMAN GENOMICS FOR SOCIOECONOMIC DEVELOPMENT

Since 2001, rapid progress has been made in harnessing the huge potential of genome science and technology for its economic and health benefits globally, in particular in less- and least-developed nations [27]. Apart from the World Health Organization, other international and national institutions engaged in this endeavor include the Human Genome Organization, the Organization for Economic Cooperation and Development, the McLaughlin-Rotman Center for Global Health (The University of Toronto, Canada), the Mexican Health Foundation, the Beijing Genomics Institute, the Department of Science and Technology (Government of India), and many more. All these institutions are focused on supporting and exploiting the huge potential of genomic technologies and related bioinformatics developments on the global economy and on health [28]. The impact of genome sciences and technologies will manifest in the following wide-ranging areas:

- personalized medicine and health approaches that will help people and societies shift the focus from "sick care" to "well care and prevention."
- biotechnology methods to produce environmentally clean and efficient fuel and chemicals to accelerate transition from petroleum-based economies

- genome-driven plant- and crop-growing methods for producing affordable food for less- and least-developed economies
- promoting genomic science and technology in animal breeding and livestock improvement
- supporting genome research for new drug discovery and drug development for enhancing pharmaceutical efficacy
- applications of genomic biotechnologies in the study and monitoring of environmental health.

CONCLUSIONS

Developments in genetics and the subsequent sequencing of human and other genomes have provided us with an opportunity to review the role of genes and genomes in all aspects of health and disease. Human health, including causation of disease, is not exclusively dependent on the human genes and genome. Evolutionary links with other genomes and ecologically relevant and beneficial parts of other genomes play crucial roles in the maintenance of human health and, to some extent, in morbidity and mortality. Understanding genomes of microbes, parasites, animals, plants, and crops is an acknowledged priority of current biomedical and biotechnology research.

Conventionally, the causation of human disease includes malformations, trauma, infection, immune dysfunction, metabolic abnormality, malignancy, and degenerative conditions associated with aging. Genetic factors have long been recognized in all of these disease groups. The traditional genetic categories of diseases include chromosomal disorders, single-gene or mendelian diseases, and several forms of multifactorial/polygenic conditions. In addition, somatic genetic changes and mutations of the mitochondrial genome probably account for a small, albeit important, number of diseases. These groups of disorders are well recognized and have an established place in the classification of human disease. Recent developments in genome research have provided vast data indicating different genomic mechanisms to explain complex pathogenesis in some disorders. The spectrum of these disorders is wide and includes both acute and chronic medical and surgical diseases. Perhaps it is reasonable to identify these disorders on the basis of underlying molecular pathology, including genomic imprinting, genomic rearrangements, and gene–environment interactions involving multiple genes and genomic polymorphisms.

This chapter has reviewed the fundamental aspects of genetics and genomics that are closely related to human health and morbidity. The genomic approaches to understanding and managing human disease are rapidly being incorporated in the practice of clinical medicine, rapidly emerging as *genomic medicine*. In addition, applications of genome science and technology are also reforming biotechnologies in a number of industries, including pharmaceutical, agricultural, and ecological bioengineering. The enormous impact of genome sciences and technologies on the economy of the developing world will be judged on applications in a number of areas, including biofuels, accelerated breeding of crops and livestock, personalized health products, pharmaceutical efficacy, and genomic monitoring of environmental health.

REFERENCES

[1] Temple LK, McLeod RS, Gallinger S, Wright JG. Defining disease in the genomics era. Science 2001;293(5531):807–8.

[2] Lay Jr JO, Liyanage R. Problems with the "omics". TrAC Trends Anal Chem 2006;25(11):1046–56.

[3] Feero WG, Guttmacher AE. Genomic medicine: an updated primer. N Engl J Med 2010;362(21):2001–11.

[4] Kumar D, Weatherall DJ. Genomics and clinical medicine. Oxford University Press; 2008.

[5] Fisher RA. The genetical theory of natural selection: a complete variorum edition. Oxford University Press; 1999.

[6] Collins FS, Morgan M, Patrinos A. The Human Genome Project: lessons from large-scale biology. Science 2003;300(5617):286–90.

[7] Consortium I.H.G.S. Finishing the euchromatic sequence of the human genome. Nature 2004;431(7011):931–45.

[8] Conrad DF, Pinto D, Redon R, Feuk L, Gokcumen O, Zhang Y, et al. Origins and functional impact of copy number variation in the human genome. Nature 2009;464(7289):704–12.

[9] Pinto D, Pagnamenta AT, Klei L, Anney R, Merico D, Regan R, et al. Functional impact of global rare copy number variation in autism spectrum disorders. Nature 2010;466(7304):368–72.

[10] Freeman JL, Perry GH, Feuk L, Redon R, McCarroll SA, Altshuler DM, et al. Copy number variation: new insights in genome diversity. Genome Res 2006;16(8):949–61.

[11] Akey JM, Zhang G, Zhang K, Jin L, Shriver MD. Interrogating a high-density SNP map for signatures of natural selection. Genome Res 2002;12(12):1805–14.

[12] Puffenberger E. Genetic heritage of the Old Order Mennonites of southeastern Pennsylvania. Am J Med Genet Part C Semin Med Genet (Wiley Online Library) 2003;121C(1):18–31.

[13] Harris MI, Eastman RC, Cowie CC, Flegal KM, Eberhardt MS. Racial and ethnic differences in glycemic control of adults with type 2 diabetes. Diabetes Care 1999;22(3):403–8.

[14] Harris MI, Klein R, Cowie CC, Rowland M, Byrd-Holt DD. Is the risk of diabetic retinopathy greater in non-Hispanic blacks and Mexican Americans than in non-Hispanic whites with type 2 diabetes? A US population study. Diabetes Care 1998;21(8):1230–5.

[15] Yaspo M-L. Taking a functional genomics approach in molecular medicine. Trends Mol Med 2001;7(11):494–501.

[16] Steinmetz LM, Davis RW. Maximizing the potential of functional genomics. Nat Rev Genet 2004;5(3):190–201.

[17] Ram O, Goren A, Amit I, Shoresh N, Yosef N, Ernst J, et al. Combinatorial patterning of chromatin regulators uncovered by genome-wide location analysis in human cells. Cell 2011;147(7):1628–39.

[18] Mercer TR, Mattick JS. Structure and function of long noncoding RNAs in epigenetic regulation. Nat Struct Mol Biol 2013;20(3):300–7.

[19] Taft RJ, Pang KC, Mercer TR, Dinger M, Mattick JS. Non-coding RNAs: regulators of disease. J Pathol 2010;220(2):126–39.

[20] Rehm B. Bioinformatic tools for DNA/protein sequence analysis, functional assignment of genes and protein classification. Appl Microbiol Biotechnol 2001;57(5–6):579–92.

[21] Gehlenborg N, O'Donoghue SI, Baliga NS, Goesmann A, Hibbs MA, et al. Visualization of omics data for systems biology. Nat Methods 2010;7:S56–68.

[22] Burke W, Kuszler P, Starks H, Holland S, Press N. Translational genomics: seeking a shared vision of benefit. Am J Bioeth 2008;8(3):54–6.

[23] Karp PD, Paley SM, Krummenacker M, Latendresse M, Dale JM, Lee TJ, et al. Pathway Tools version 13.0: integrated software for pathway/genome informatics and systems biology. Brief Bioinform 2010;11(1):40–79.

[24] Khoury MJ, Clauser SB, Freedman AN, Gillanders EM, Glasgow RE, Klein WM, et al. Population sciences, translational research, and the opportunities and challenges for genomics to reduce the burden of cancer in the 21st century. Cancer Epidemiol Biomarkers Prev 2011;20(10):2105–14.

[25] Kumar D. Clinical medicine in the genome era: an introduction. Genomics Clin Med 2008;(53):145.

[26] Cooper DN, Chen JM, Ball EV, Howells K, Mort M, Phillips AD, et al. Genes, mutations, and human inherited disease at the dawn of the age of personalized genomics. Hum Mutat 2010;31(6):631–55.

[27] Singer PA, Daar AS. Harnessing genomics and biotechnology to improve global health equity. Science 2001;294(5540):87–9.

[28] Kumar D. Genomics and health in the developing world, vol. 62. Oxford University Press; 2012.

Chapter 2

Genomic Technologies in Medicine and Health: Past, Present, and Future

Y.-H. Rogers, C. Zhang

The Jackson Laboratory for Genomic Medicine, Farmington, CT, United States

Chapter Outline

Introduction   15
Sequencing Technologies   16
Computational and Information Technologies   16
Applications of Genomic Technologies   17
The Microbiome and Human Health   18
   The Pediatric Microbiome   18
   The Microbiome and Oral Health   18
   The Microbiome and Gastrointestinal Health   18
The 1000 Genomes Project and Structural Variations
in the Human Genomes   18
Noninvasive Prenatal Testing by Sequencing of Cell-Free
Fetal DNA in the Maternal Blood   19
Prenatal Diagnosis by Whole Genome Sequencing of
"Jumping Libraries"   19

Postnatal Diagnosis   20
Genome Sequencing in Newborn Healthcare   21
Genome Sequencing in Other Research Areas   22
The Cancer Genome Atlas and Cancer Genomics   22
Cancer Diagnosis   22
Precision Medicine   23
   Development of Targeted Therapeutics for Genetic
Disorders   23
   Development of Targeted Therapeutics for Cancers   24
   Genomic Profiling and Counseling   24
   The Precision Medicine Initiative   24
Policy and Regulatory Issues   25
Genomic Education   25
References   25

INTRODUCTION

The decoding of the first human genome in 2001 [1,2] marked the dawn of the genomic era. It remains the world's largest collaborative research project in biology [3]. This landmark accomplishment has changed how scientists approach biological problems related to humans by providing a means of rendering the human genetic blueprint in its entirety. This ability to decipher the code of life has paved the way for genomic exploration across many fields such as medicine, agriculture, bioenergy production, synthetic biology, environmental science, computational science, information technology, and health and wellness management. Sequencing the human genome was hailed as the "holy grail" of biology—a remarkable accomplishment, heralding comparisons to discovering penicillin and landing on the moon—and was predicted to spurn a subsequent "genomic revolution" that would transform the practice of medicine.

Despite such promise and enthusiasm, the impact of this scientific achievement on medicine and the healthcare system has been minuscule. Since 2001, only a very small fraction of the US patient population has benefited from this advancement. The first two versions of the human genome took over 14 months to complete, at the high cost of $3 billion (International Human Genome Sequencing Consortium version) [1] and $300 million (Celera Genomics version) [2]. At the turn of the 21st century, advancements in genomic science and medicine were hampered, awaiting innovations in DNA sequencing methodologies, computational analyses, and information technologies in order to increase speed and efficiency, while simultaneously decreasing costs. Over the next decade, the advent of next generation sequencing (NGS) capabilities brought unprecedented advances in DNA sequencing throughput and speed as well as significant cost reduction, making whole genome sequencing (WGS) a possibility for the average researcher or clinical investigator.

Medical and Health Genomics. http://dx.doi.org/10.1016/B978-0-12-420196-5.00002-2

SEQUENCING TECHNOLOGIES

Genomic science depends heavily on innovations in DNA sequencing, computation analysis, and information technologies. To overcome the speed and cost challenges, momentous advancements in these areas must occur. Driven by the demand for much more affordable and higher throughput genome sequencing needs, remarkable advancements have been made in DNA sequencing technology development. As a result, DNA sequencing technologies have undergone a complete transformation since the completion of the first human genome. Although it is still considered the gold standard and being used for small scale applications, the capillary-based Sanger sequencing technology [4–7] has been mostly replaced by the NGS technologies for large and/or genome scale applications. The first NGS platform became available in 2004 [8]. The new platform, which was developed based on pyrosequencing chemistry by 454 Life Sciences, enabled massively parallel generation of sequencing data [9,10], and hence drastically increased the throughput of the sequencing data production while reducing the cost. As the NGS technologies continue to improve and newer platforms became available [11,12], the cost and the efficiency of DNA sequencing continue to improve (Fig. 2.1). As of the beginning of 2015, a draft human genome using the Illumina HiSeq X 10 [13] platform can be obtained for approximately $1000 and in less than 2 weeks.

To illustrate the nature of the reductions in DNA sequencing cost, the graph in Fig. 2.1 shows hypothetical data reflecting Moore's law, which describes a long-term trend in the computer hardware industry that involves the doubling of "compute power" every 2 years. Technology improvements that "keep up" with Moore's law are widely regarded to be doing exceedingly well, making it useful for comparison. Note the use of a logarithmic scale on the Y axis and the sudden and profound out-pacing of Moore's law beginning in January 2008. The latter represents the time when the sequencing centers transitioned from Sanger-based (dideoxy chain termination sequencing) to "second generation" (or next generation) DNA sequencing technologies.

COMPUTATIONAL AND INFORMATION TECHNOLOGIES

Although the development of NGS technologies dramatically improved the efficiency and cost of genomic data generation, it created new challenges for computational assembly and analysis of the genomic data.

The process for genome sequencing involves shearing the genomic DNA of an organism into a large number of small fragments, each of which are then sequenced separately, because the sequencing technologies are only capable of reading the sequence length of up to several hundred or thousand nucleotide base pairs (a human genome has 3 billion bases) [13]. The resulting DNA sequence reads are reconstructed into the original genome using computer programs called *genome assemblers* or are mapped to the reference genome. The genome assembly process is analogous to solving a jigsaw puzzle. The smaller the puzzle pieces, the more similar the puzzle pieces and the more pieces there are to assemble, and the more challenging it is to assemble the whole puzzle correctly and rapidly. Sophisticated computational algorithms have been developed over the past years to

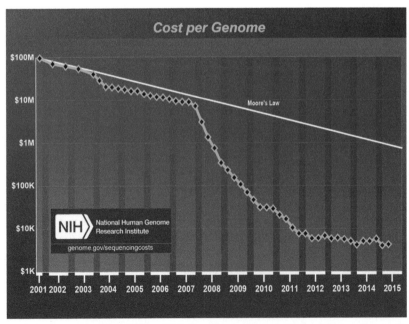

FIGURE 2.1   Cost per human genome from 2001 to 2015. *Based on data from the National Human Genome Research Institute.*

overcome these challenges. These genome assemblers are capable of reconstructing large mammalian genomes, including the human genome [14,15]. However, the development and widespread use of the NGS technologies posed new challenges to the computational scientists [16,17], mainly as a result of the large data volume and the characteristics of the new NGS data, as well as the emerging novel applications. A single NGS platform can provide data throughput similar to what was previously only achievable at a large genome center. Currently, the highest throughput NGS system is capable of generating up to 1.8 trillion bases and/or 6 billion sequencing reads per instrument run in 3 days at a cost of less than $1000 per human genome [13]. However, these sequence reads are only 2–300 base pairs long, which are much shorter than the thousand–base pair reads generated by the previously used capillary-based Sanger sequencing technology [4–7]. They also contain more sequencing errors, many of which are new types of errors specific to the technology platforms and the sequencing chemistries. Moreover, new applications of DNA sequencing in biological research, such as the human microbiome initiative [18], which applies the large-scale sequencing technology and requires computational analysis of entire microbial communities (metagenomics), demands the development of entirely new genome assembly and analysis software. In fact, analysis of the sequencing data rather than the generation of the data is the biggest bottleneck of the genomic-based projects because the analysis process is often complex with multiple steps, programs, and databases involved.

A large number of tools and software have been developed to meet some of the NGS data analysis challenges. These include tools for mapping and assembly [21], sequence alignments [20], algorithms for single nucleotide polymorphism (SNP) and genotype calling [19,21], or detection of structural variations (SVs) and copy number variations (CNVs) [22], Chromatin immunoprecipitation sequencing (ChIP-seq) and RNA-seq studies [23], and metagenomic analysis [24], to name a few. Even though most of these tools are readily accessible to scientists conducting genomic research, selecting the appropriate tools and being able to utilize them effectively for a specific application is no frivolous task, especially for untrained and inexperienced users, because a majority of these tools are not standardized or well integrated. Therefore strong bioinformatics resources and expertise are essential for successful execution of genomic projects of any scale.

Processing, analysis, management, and storage of the massive genomic data demand vast, efficient, and complex computational and data storage infrastructure and capacity. For effective collaborations and data sharing, data standardization or harmonization is necessary. Furthermore, when personal genomic data and clinical information are involved, special considerations need to be given to data integrity, confidentiality, and security, as well as

data-sharing and retention policies. Academic institutes engaging in genomic research have continued to invest a large amount of capital in purchasing, building, and managing computational and storage capacities to meet the demand; however, with the sequencing data production increasing on a logarithmic scale and computational power and storage density improving on a linear scale (see Fig. 2.1), it has become increasingly challenging financially and operationally for the information technology (IT) enterprise to keep pace with the sequence production and analysis in majority of the academic institutes. The IT industry has begun to offer cloud-based computing [25] and storage solutions as well as data management tools specifically to address infrastructure demands for genomic sciences (Google, https://cloud.google.com/genomics/; Amazon, http://aws.amazon.com/health/genomics). Various alliances and consortia (Global Alliance for Genomics and Health, https://genomicsandhealth.org/; International Pan-Cancer Analysis of Whole Genome, http://cancergenome.nih.gov/newsevents/events/AACR2015-PCAWG) have also been established to create common frameworks of harmonized approaches to enable the responsible, voluntary, and secure sharing of genomic and clinical data across institutions and internationally. Much more work still needs to be done to effectively analyze and manage the explosive amount of genomic data.

APPLICATIONS OF GENOMIC TECHNOLOGIES

Whereas the utility and comparative effectiveness of the genomic approaches for diagnostics, development of therapeutics, and treatment decisions are still being evaluated and validated for many applications, it is widely accepted that genomic medicine has the potential to revolutionize the entire healthcare system as a result of its predictive, personalized, preventive, and participatory nature [26]. With the advanced genomic and computational technologies, our understanding of the genomes' functionalities and molecular pathways of diseases will expand progressively and hence greatly enhance our ability in discovering and developing better and more precise diagnostics and therapeutics for diseases, as well as ways for early diagnostics and intervention. Healthcare providers will have knowledge of an individual patient's genome and microbiome and be able to use the information to devise the most effective treatment options and to cure disease and extend patient lives. Individuals will have their own genomic profiles and an understanding of how these factors affect their health and disease states and be able to use the knowledge to tailor their lifestyles, including diet, behaviors, and exercise routings, to manage their health and lives, to slow down the aging process, and to prevent or delay the onset of chronic diseases.

THE MICROBIOME AND HUMAN HEALTH

A microbiome is the ecological community of commensal, symbiotic, and pathogenic microorganisms that literally share our body space [18]. Various studies have linked the microbial communities' composition and function to health and disease states in different organs and tissues within the human body. Comprehensive survey and measurements of microbiomes combined with clinical information could yield valuable information regarding the human health condition and propensity for diseases that are linked to the microbial ecology of human bodies. With more in-depth and accurate knowledge of the human microbiome, new diagnostics and therapeutics for various acute, chronic, localized, and systemic human diseases could also be developed [27]. The following sections illustrate several examples of utilizing knowledge pertaining to the microbiome in disease prevention and wellness management.

The Pediatric Microbiome

A number of microbiome studies in pediatric populations has shown that a baby's diet [27] and antibiotic treatment have a marked effect on the microbial composition in the baby's gastrointestinal (GI) tract [28,29]. Moreover, the pediatric microbiome has been linked to the development of early-onset type 1diabetes mellitus. Researchers have found, in addition to genetic predisposition and other factors, significant differences in the gut microbiomes of children who develop type 1 diabetes mellitus and children who do not. The study also indicated that the presence of butyrate-producing bacteria appears to be especially important for maintaining a healthy gut and preventing type 1 diabetes mellitus [30]. This is an example of where a microbiome-based approach could potentially be taken to monitor for the development of a childhood disease that impacts millions of children worldwide and used in the formulation of preventive measures.

The Microbiome and Oral Health

A healthy oral bacterial consortium plays a key role in healthy tissue development and function of the oral cavity. Researchers have identified a number of bacteria that are associated with each other and with periodontitis [31]. The presence of certain bacteria combined with naturally existing commensal has been shown to induce or accelerate natural bone and tissue loss in oral disease [31]. Microbiome-based knowledge in these cases can be used to monitor oral health and develop cleaning routines and products for prevention of oral diseases.

The Microbiome and Gastrointestinal Health

Microbes play a rather complex role in human health and disease. Furthermore, most of these microbial organisms act as commensals rather than individually within and on human bodies. It has been suggested that the indigenous gut microbiome serves many vital functions in addition to its immense metabolic capacity. One of those functions could be to provide "colonization resistance," thereby protecting the human gut from being colonized by harmful bacteria. For example, lower gut microbiome diversity has been linked to low efficacy of antibiotic treatment for *Clostridium difficile* and recurrence of the infection. It has been demonstrated with a greater than 95% success rate that by restoring the microbiota diversity and therefore colonization resistance within the patients' gut, *C. difficile* can be successfully treated in patients with multiple recurrences [30]. This example illustrates the importance of gut microbiome diversity and its impact on human health. Diet or preventive treatment could be developed to assist in maintaining a high diversity in the gut microbiome.

THE 1000 GENOMES PROJECT AND STRUCTURAL VARIATIONS IN THE HUMAN GENOMES

One of the central goals in medical genetics and genomics is to understand the relationship between genotype and phenotype. SVs, including deletions, duplications, insertions, and inversions, account for most varying base pairs among individual human genomes. Numerous studies have indicated that SVs may play important roles in human health, including disease susceptibility and pathogenesis. The 1000 Genomes Project is the first project to sequence the genomes of a large number of people and to provide a comprehensive public catalog of human genetic variation, including SNPs, SVs, and their haplotype contexts [32]. Through 2015, this project has completed the studies of the pilot phase, phase 1, phase 2, and phase 3. The pilot phase was designed to develop and compare different strategies for genome-wide sequencing with high-throughput platforms. This study included low-coverage WGS of 179 individuals from four populations, high-coverage sequencing of two mother-father-child trios, and exon sequencing of 697 individuals from seven populations. The pilot phase reported approximately 15 million SNPs, 1 million short insertions and deletions (indels), and 20,000 SVs [33]. The phase 1 study included 1092 individuals from 14 populations and used a combination of low-coverage whole-genome and exome sequencing. This study further developed and optimized several algorithms for the analysis of sequencing data. The phase 1 study reported approximately 38 million SNPs, 1.4 million indels, and more than 14,000 larger deletions [34]. The data also demonstrated that individuals from different populations carry different rare and common variants. The phase 2 and 3 studies included 2504 individuals from 26 populations in Africa, East Asia, Europe, South Asia, and the Americas using a combination of low genome sequencing, deep exome sequencing, and dense microarray genotyping. The results from these two phases reported approximately 84.7 million SNPs, 3.6 million short indels, and 60,000 SVs [35,36]. The data from the 1000 Genomes Project will not only help us better understand the evolution of human populations, the genetic contribution to human phenotypes, and disease susceptibility, but also will serve as a global reference for human genetic variations and aid the interpretation and classification of genetic variants identified in clinical patients.

NONINVASIVE PRENATAL TESTING BY SEQUENCING OF CELL-FREE FETAL DNA IN THE MATERNAL BLOOD

The identification of fetal DNA in the cell-free DNA (cfDNA) in the maternal blood and the development of noninvasive prenatal testing (NIPT) by using massively parallel sequencing of cfDNA have revolutionized prenatal screening and diagnosis [37–41]. Previous studies have demonstrated that the fetal DNA in the maternal blood is mainly derived from the placenta and is cleared from the maternal blood within hours after delivery. NIPT has been recommended by the American College of Obstetricians and Gynecologists (ACOG) and the American College of Medical Genetics and Genomics (ACMGG) for those with any of the following conditions: (1) maternal age 35 years or older at delivery, (2) ultrasonographic findings with an increased risk of aneuploidy, (3) history of a prior pregnancy with a trisomy, (4) positive first or second trimester test result for aneuploidy, and (5) parental balanced robertsonian translocation with increased risk of fetal trisomy 13 or 21 [40]. NIPT has been validated for screening of trisomies 13, 18, and 21. However, it does not replace the diagnostic accuracy and precision of the conventional invasive diagnostic procedures, chorionic villus sampling and amniocentesis. Moreover, a negative test result does not ensure an unaffected pregnancy, whereas a positive test result should be confirmed by the invasive methods. Since its recommendation for clinical application by ACOG and ACMGG in 2012, NIPT has become the largest and fastest growing clinical test in genomic medicine. Several million tests have been performed globally. Compared with serum screening, NIPT has better sensitivities, specificities, positive and negative predictive values, and significantly lower false positive rates [41]. Since 2013, several large-scale clinical studies in general obstetrical populations have demonstrated that NIPT performs better than both serum biochemistry and the first trimester combined test, with significantly higher positive predictive values and lower false positive rates [41–43]. Since its incorporation into prenatal screening for aneuploidy, NIPT has had a major clinical impact on patient care worldwide and has resulted in about a 70% decrease in the number of conventional invasive diagnostic procedures in the United States [41–44]. Recently, NIPT has

been expanded to screen sex chromosome aneuploidies and microdeletions/microduplications in addition to trisomies 13, 18, and 21 [45].

One of the big challenges associated with the NIPT is the discordant results between NIPT and the fetal karyotype [43–47]. For an individual pregnant woman, not understanding the underlying basis for the discordant test results can cause emotional distress and may even be life threatening, if there is a tumor that is shedding DNA sequences into the maternal plasma that could be detected with appropriate evaluation [48]. There is also the possibility that, because of limited understanding of the underlying biology and the potential causes of false-positive results, chromosomally normal fetuses will be unnecessarily terminated. One study showed that 6% of women terminated their pregnancies for a positive NIPT result without obtaining a diagnostic fetal karyotype [41]. In addition, determining the relative incidence and consequences of confined placental mosaicism may allow identification of fetuses at risk for prenatal growth restriction and lead to improved prenatal care [45–47]. Further studies are expected to produce evidence-based guidelines for the clinical follow-up care in cases in which different types of false-positive NIPT results are found. The results of these studies will improve prenatal care and counseling by ascertaining the underlying biology of discordant NIPT results and producing specific sets of clinical recommendations. It is conceivable that NIPT will eventually become the primary method for screening aneuploidy and other chromosomal aberrations in all pregnant women as the associated costs decrease over time [49].

PRENATAL DIAGNOSIS BY WHOLE GENOME SEQUENCING OF "JUMPING LIBRARIES"

Precise mapping and localization of individual genes that are involved in the cytogenetically balanced chromosomal rearrangements may help improve the prediction of phenotypic outcomes and inform postnatal medical care. Routine karyotyping can detect apparently balanced rearrangements with a resolution larger than 5 Mb, but it cannot show whether any genes are affected in the regions involved in the chromosomal rearrangements. Whereas NIPT has become an important assay for prenatal screening of chromosomal aneuploidies, it has not yet been established and validated for detection of cytogenetically balanced chromosomal rearrangements. Talkowski et al. have developed an approach involving the WGS of the "jumping libraries" constructed from the fetal genomic DNA extracted from amniotic fluid cell cultures of a patient with a balanced de novo translocation and multiple fetal anomalies [50]. In this study, they successfully identified the disruption of the chromodomain helicase DNA binding protein 7 gene (CHD7), a gene known to be associated with the coloboma of the eye, heart anomaly, choanal atresia, retardation, and genital and ear anomalies (CHARGE) syndrome. This technology platform has demonstrated a great potential for prenatal diagnosis of patients with cytogenetically balanced chromosomal rearrangements.

POSTNATAL DIAGNOSIS

Individual genetic disorders are rare, but collectively, they are common in the pediatric population. Thousands of monogenic diseases have been reported and these disorders are often undifferentiated at birth [51]. In addition, genetic alternations have also been suggested to play important roles in a variety of congenital birth defects such as congenital diaphragmatic hernia, congenital heart defect, neural tube defect, and congenital hearing loss (approximately 50% of cases of hearing loss are attributed to genetic causes), as well as other complex disorders, such as autism spectrum disorders (ASDs) [52,53]. As of yet, clinical testing is not available for majority of these genetic disorders.

Whole exome sequencing (WES) and WGS have been successfully used for identification of genetic alternations in a variety of diseases [54–56]. In 2011, Worthey et al. performed WES for a 15-month-old child with intractable inflammatory bowel disease (IBD) without a definitive diagnosis, and subsequently identified a novel missense mutation in the X-linked inhibitor of apoptosis gene [57]. This mutation resulted in the substitution of a tyrosine for a highly conserved cysteine. Additional functional studies are consistent with loss of normal X-linked inhibitor of apoptosis protein function in apoptosis. As a result, this patient was diagnosed as having an X-linked inhibitor of apoptosis deficiency. According to the recommended treatment for X-linked inhibitor of apoptosis deficiency, the child received allogeneic hematopoietic stem cell transplantation and had no recurrence of IBD after 42 days posttransplantation. Dinwiddie et al. has also successfully used WES for the diagnosis of infantile onset IBD [58]. In this study, two siblings under the age of 1 year suffered from severe infantile-onset IBD, failure to thrive, skin rash, and perirectal abscesses refractory to medical management. They performed WES for the siblings and identified the compound heterozygous mutations in *IL10RA* in both of the patients. The siblings subsequently received hematopoietic stem cell transplantation and showed marked clinical improvement of all IBD-associated clinical symptoms. WGS has also been successfully used for molecular diagnosis of pediatric disorders. Brainbridge et al. performed WGS for a 14-year-old fraternal twin pair diagnosed with dopa-responsive dystonia and identified compound heterozygous mutations in the sepiapterin reductase (*SPR*) gene [59]. Previous studies have demonstrated that disruption of *SPR* causes a decrease in tetrahydrobiopterin, a cofactor required for the synthesis of dopamine and serotonin.

Indeed, supplementation of L-dopa with a serotonin precursor, 5-hydroxytryptophan, resulted in marked clinical improvements in both siblings. In neonatal intensive care units (NICUs), molecular diagnosis of newborn disorders must be quick in order to provide information for making relevant clinical decisions. Saunders et al. have developed a rapid WGS platform for identification of genetic alternations in newborns [60]. This platform has integrated the WGS work-flow into an automated bioinformatics analysis pipeline to make the differential diagnosis of genetic disorders in approximately 50 hours. They tested this rapid WGS platform in both retrospective and prospective clinical settings. In the retrospective study, they were able to identify the pathogenic mutations in two children with known molecular diagnoses. In the prospective study, they identified potential pathogenic mutations associated with severe *GJB2*-related skin disease in one neonate and *BRAT1*-related lethal neonatal rigidity and multifocal seizure syndrome in another infant. These studies demonstrated that both WES and WGS could be used for clinical diagnosis of genetic disorders in newborns and pediatric patients and subsequently improve the differential diagnosis, treatment, and genetic counseling of neonatal and pediatric diseases.

ASDs are neurodevelopmental diseases characterized by difficulties or deficits in communication and social interactions [53]. Rates of ASD diagnosis have risen sharply since 2000, from approximately 1 in 150 children to 1 in 68 in 2014, according to the Centers for Disease Control and Prevention [53]. Diagnostic criteria cover a broad range of clinical symptoms, including restricted, repetitive, and stereotyped patterns of behavior, interests, and activities; impairment in communication; and impairment in social interactions [61]. ASD patients are often diagnosed with other neurological disorders such as epilepsy [61]. The underlying disease pathways for the majority of the ASD cases were unknown. Recent research has shown that ASD and related disorders can be associated with de novo or rare genetic variations including large chromosomal alterations or single nucleotide variants (SNVs) [62–66]. Diagnostic tools include array comparative genome hybridization (aCGH), which identifies CNVs, and NGS-based sequencing assay for identification of mutations in the target gene panels. In collaboration with the Connecticut Children's Medical Center in Hartford, Connecticut, the scientists at the Jackson Laboratory for Genomic Medicine have been developing an innovative WGS-based assay for ASD and other developmental disorders [67,68]. The data from the pilot study on ASD patients demonstrated that this assay detected all the CNVs observed by aCGH and all the SNVs found by the gene-panel sequencing test, suggesting that this assay may replace aCGH and gene panel-based sequencing tests, and become a novel clinical test for ASD and other neurological disorders.

GENOME SEQUENCING IN NEWBORN HEALTHCARE

To further explore and expand the application of genome sequencing in newborn healthcare, the Eunice Kennedy Shriver National Institute of Child Health and Human Development and the National Human Genome Research Institute (NHGRI) at the National Institutes of Health (NIH), initiated and awarded four pilot projects in 2013 with $25 million over 5 years under the Genome Sequencing and Newborn Screening Disorders research program [69]. The objectives of these projects are to address the technical, clinical, and ethical aspects of genome sequencing in the newborn period and to help us better understand how genome sequencing information may improve health and prevent disease in infants and children. Dr. Robert Green's team at the Brigham and Women's Hospital in Boston will develop and test new methods for using information obtained from genome sequencing in the care of newborns. It will test a new approach to newborn screening, in which genomic data is available as a resource for parents and doctors throughout infancy and childhood to inform healthcare. In this study, a genetic counselor will provide the genome sequencing information and newborn screening results to the families. Parents will then be asked about the impact of receiving genome sequencing results and if the information was useful to them. Researchers will try to determine if the parents respond to receiving the genome sequencing results differently if their newborns are sick and if they respond differently to receiving genome sequencing results as compared with current newborn screening results. Investigators will also develop a process for reporting results of genome sequencing to the newborns' doctors and investigate how they act on these results. This team will recruit 240 healthy babies and 240 sick ones from NICUs and will perform WES for half of each group of infants randomly to determine whether those data alone can improve children's health. Dr. Stephen Kingsmore's team at the Children's Mercy Hospital in Kansas City will focus on the rapid genetic diagnosis of newborn disorders in the NICU through the use of genome sequencing. They will examine the benefits and risks of using rapid genome sequencing technology in the NICU population. They also aim to reduce the turnaround time for conducting and receiving genome sequencing results to approximately 50 hours, which is comparable to other newborn screening tests. The researchers will test if their methods increase the number of diagnoses or decrease the time it takes to reach a diagnosis in NICU newborns. They will also study whether genome sequencing changes the clinical care of newborns in the NICU. Additionally, the investigators will try to determine whether parents' perception of the benefits and risks associated with the results of sequencing change over time. Over the next 5 years, this group will sequence the genomes of 500 sick babies from the Children's Mercy Hospital NICU and compare the infants' clinical outcomes with those of 500 NICU babies who are diagnosed using conventional genetic and metabolic tests. The researchers will assess whether rapid sequencing allows babies to avoid unnecessary tests and unhelpful treatments, and whether it helps parents to make decisions about care when the child is diagnosed as having a fatal disease. Even when an infant does die, a genome sequence and diagnosis can provide closure to parents and give more information about the genetic conditions they carry. The so-called "rapid sequencing" technique requires that several specialists each perform one step of the process from the blood draw to the final diagnosis as quickly as possible. The group collects DNA from both of the parents and the baby to quickly identify mutations in the child's genome, then sequences the DNA and uses custom software to target specific parts of the genome on the basis of their symptoms. After making a gene-based diagnosis and delivering relevant information to the baby's physician, the group stores the sequence data anonymously in a secure database for use in future studies. Dr. Robert Nussbaum's team at the University of California in San Francisco will explore the potential of WES as a method of newborn screening for disorders currently screened for and others that are not currently screened for, but where newborns may benefit from screening. The researchers will examine whether the additional information from the WES may lead to improved care and treatment. Additionally, the researchers will explore parents' interest in receiving information beyond that typically available from newborn screening tests. This team will sequence exomes from 1400 blood spots collected from infants at birth to determine whether this information is useful for diagnosis. This team will also develop a participant protection framework for conducting genome sequencing during infancy and explore legal issues related to using genome analysis in newborn screening programs. Dr. Cynthia Powell's team at the University of North Carolina in Chapel Hill will identify and overcome the challenges that must be met in order to implement genome sequencing technology in a diverse newborn population. The researchers will sequence genomes of 400 healthy infants and infants with known conditions such as phenylketonuria, cystic fibrosis, or other metabolic disorders. Their goal is to help identify the best ways to return results to doctors and parents. The investigators will explore the ethical, legal, and social issues involved in helping doctors and parents make informed decisions, and develop best practices for returning results to parents after testing. They will also develop a tool to help parents understand the sequencing results and examine extra challenges that doctors may face as this new technology is used. This study will place a special emphasis on including multicultural families. Taken together, these studies have the potential to provide public health benefit for newborns and research-based information for policy makers.

GENOME SEQUENCING IN OTHER RESEARCH AREAS

The US NHGRI Clinical Sequencing Exploratory Research program and the National Cancer Institute (NCI) have awarded additional $27 million research grants to four multidisciplinary research teams to explore the use of genome sequencing in healthcare [70]. The research areas of these projects include genome sequencing for couples with reproductive risks, genome sequencing for children with developmental delays, and cancer genomics. Dr. Katrina Goddard's team at the Seattle Children's Research Institute will examine the use of WGS in informing couples about their potential carrier status for genetic disease. They will compare women and their partners who receive preconception genetic testing to those who receive WGS in addition to the testing. This project will enroll 380 people in the trial and look for genetic mutations for about 100 rare conditions. Couples with mutations that put children at risk for a condition will work with a genetic counselor and complete surveys to help researchers develop useful approaches to presenting information to patients. Approximately 1–3% of children worldwide are born with genetic disorders that lead to developmental or intellectual delays or disabilities. Dr. Richard Myers' group at the Hudson-Alpha Institute for Biotechnology in Huntsville will sequence the genomes of nearly 500 children with developmental delays and other disabilities, along with their parents, in the hopes of discovering genomic alterations behind such disorders. The researchers hope to uncover gene alterations that are common to more than one condition and gain insights to whether certain mutations cause milder or more severe cases of some conditions. The scientists plan to provide information on genetic differences to study participants and families and use questionnaires and interviews to better understand the impact of genomic testing results on families. Dr. Arul Chinnaiyan's team at the University of Michigan in Ann Arbor will sequence the genomes of 500 patients with advanced sarcoma or other rare cancers to discover new information about genomic alterations for personalized therapies. Few clinical trials have been conducted in most rare cancers, and scientists would like to know more about the genetic changes of these diseases. Investigators also plan to evaluate the patient consent process, and the delivery and use of genome sequencing results. Dr. Gail Jarvik's team at the University of Washington in Seattle will lead the coordinating center responsible for pulling together all of the scientific teams, helping to organize studies, interpreting study results, and helping groups focus on common goals. The coordinating center team members bring their own expertise in clinical genetics, genome sequencing, bioinformatics and ethics.

THE CANCER GENOME ATLAS AND CANCER GENOMICS

Cancer is a genetic disease. In particular, cancer is driven by genetic changes in the oncogenes, tumor suppressor genes, and DNA repair genes. These genes are involved in normal cell growth and division, as well as the repair of damaged DNA. Cells with certain genetic alternations in these genes may divide and grow in an uncontrolled manner and become cancerous. Genetic alternations in other regions of human genome may also play important roles in tumorigenesis. Therefore identification and characterization of genetic alternations in the cancer genomes are important for better understanding the pathogenesis, prevention, and treatment of cancers. In order to systematically and comprehensively identify genetic alternations associated with cancers, US NCI and NHGRI launched a 3-year pilot study in 2006 with an investment of $100 million on cancer genome sequencing and related studies [The Cancer Genome Atlas (TCGA)] [71]. This pilot project demonstrated the feasibility of the creation of an atlas of genetic changes for specific cancer types. It also showed that multiple national and international teams could work together on distinct but related projects and share their research resources and data. Since then, the US NIH has continued to support TCGA to collect and characterize more than 20 additional tumor types. As a result, TCGA has built up huge databases of cancer genomes that are available for all researchers worldwide. More importantly, some of the genetic alternations identified and validated by TCGA and other research groups have been widely used for diagnosis, prognosis, target therapy, and risk assessment of cancers.

CANCER DIAGNOSIS

The findings from TCGA and other cancer genome sequencing projects have laid a solid foundation for cancer diagnosis through molecular profiling of genetic alternations in cancers [71]. Detection of genetic alternations in cancer genomes can be performed either directly on the tumor tissues obtained from biopsy or surgical removal or on the tumor-derived cfDNA in cancer patient plasma.

A number of genomic profiling tests based on sequencing of targeted gene panels for cancer diagnostics have been developed, validated, and provided as clinical services [72–74]. These gene panels typically consist of clinically relevant cancer genes or transcripts that are oncogenes, tumor suppressors, genes with known drug targets, genes that are commonly mutated in cancer, and genes that are believed to become future targets for therapies. The tests aim to provide comprehensive molecular information by examining these genes or transcripts for cancer diagnoses and patient-specific treatment recommendations. For instance, The Jackson Laboratory for Genomic Medicine (Farmington, Conn.)

offers the JAX Cancer Treatment Profile test for predicting drug responses of solid tumors [72]. Foundation Medicine (Cambridge, Mass.) offers FundationOne and Fundation-One Heme for genomic profiling of solid tumors and hematological malignancies, respectively [73]. Genomic Health (Redwood City, Calif.) offers the Oncotype DX Breast Cancer Assay for predicting chemotherapy benefit and the likelihood of distant breast cancer recurrence, the Oncotype DX Colon Cancer Assay for providing a quantitative recurrence risk assessment in stage II and stage III colon cancers, and the Oncotype DX Prostate Cancer Assay for providing a more precise and accurate assessment of risk based on individual tumor biology [74]. All these profiling tests are performed directly on tumor tissues.

Liquid biopsy is a noninvasive molecular profiling of genetic alternations on the tumor-associated cfDNA in the cancer patient plasma [75–81]. Tumor-associated cfDNA was released into the blood through various physiological and/or pathological events including apoptosis, necrosis, and secretion. This assay can detect both mutations and CNVs of oncogenes and tumor suppressor genes such as *KRAS*, *APC*, *p53*, *PICK3A*, *BRAF*, *EGFR*, and *ERBB2* (*HER2/Neu*) [75–81]. It can also detect the methylation of tumor-associated genes and loss of heterozygosity, as well as the tumor-related viral genes [eg, for Epstein-Barr virus (EBV) and human papilloma virus]. Liquid biopsy has been successfully used to detect residual disease after surgery or radiotherapy, monitor disease progression and recurrence, and monitor drug response and resistance for a variety of cancers. Leary et al. performed WGS analysis of cfDNA from colorectal and breast cancer patients as well as healthy individuals to identify tumor-derived chromosomal alterations [75]. Amplification of *ERBB2* and *CDK6* (a cell-cycle regulatory gene) were detected in the cancer patients but not in the healthy individuals. The chromosomal aberrations observed in the cfDNA were found to be similar to those of the resected tumors from the same patients. Because almost all human cancers harbor some type of chromosomal aberrations, this approach may become a powerful method for cancer diagnosis. Forshew et al. detected *TP53*, *EGFR*, and other cancer-related gene mutations by deep sequencing of cfDNA in the plasma samples of advanced ovarian cancer and metastatic breast cancer patients and identified the origin of metastatic relapse in a patient with multiple primary tumors [76]. Spindler et al. used a quantitative polymerase method to assess the *KRAS* and *BRAF* mutations in the cfDNA from the patients with metastatic colorectal cancer in relation to the treatment with cetuximab and irinotecan [77]. The data indicated that *KRAS* mutations detected in the cfDNA are strongly correlated with the clinical outcomes; ie, high levels of *KRAS* mutations indicated a poor outcome. Jones et al. examined the cfDNA in lymphoma patients and observed that EBV-DNA had a strong correlation with the clinical/radiological response in EBV-associated lymphoma, but not in EBV-negative lymphoma, indicating that liquid biopsy can be used to monitor disease progress and treatment response in patients with EBV-associated lymphoma [78]. Dawson et al. performed sequencing on cfDNA to identify somatic genomic alterations in women with metastatic breast cancer and compared the sequencing results with the levels of cancer antigen 15-3, number of circulating tumor cells, and the radiographical imaging of tumors [79]. Somatic genomic alterations were identified in the cfDNA in 97% (29/30) of the patients examined in this study, and they appeared to be better correlated with the changes in tumor burden and response to the treatment than other assays. Murtaza et al. performed exome sequencing on the plasma cfDNA from patients with advanced breast, ovarian, or lung cancers to monitor the genomic evolution of metastatic cancers in response to therapy over 1–2 years [80]. The data indicated that exome sequencing analysis of tumor related cfDNA could become an alternative approach to identify mutations associated with acquired drug resistance (eg, *PIK3CA*, *EGFR*, *RB1*, *MED1*, and *GAS56*) in the treatment of advanced cancers. Heitzer et al. performed WGS analysis of plasma cfDNA from castration-resistant prostate cancer (CRPC) and castration-sensitive prostate cancer (CSPC) patients [81]. They identified multiple copy number aberrations including losses in 8p and gains in 8q. They also detected amplifications of androgen receptor in CRPC patients but not in patients with CSPC. Moreover, they identified *TMPRSS2-ERG* rearrangement in these cases. This study demonstrated that the liquid biopsy may become a powerful approach to profile genomic aberrations in prostate cancer patients.

PRECISION MEDICINE

Precision medicine (PM) (see chapter: The provision of medical and health genetics and genomics in the developing world) is an approach for disease prevention and treatment that takes into consideration individual variability in genes, environment, and lifestyle for each person [82].

Development of Targeted Therapeutics for Genetic Disorders

The efficiency of the traditional drug development process has reached a plateau in the past 50 years, with the number of new drug approvals remaining steady, while spending on biomedical research and development has tripled since 1990. This not a financially sustainable trajectory for the businesses and stakeholders involved. The advancement in genomic technologies affords new mechanisms to introducing greater efficiency and knowledge into the drug development process by enabling a more targeted approach for therapeutics discovery, identification, screening, validation, and trail testing, all of which help lower the costs and increase the success rate of drug discovery and development. Pharmacogenomics in particular applies genomic technologies and available genomic information, such as genomic profiling by sequencing, statistical genetics, and gene expression analysis to clinical evaluation and development of new therapeutic agents.

It is undisputable that our understanding of diseases and disease pathways has increased significantly through acquisition and examination of genomic information and integration of the genomic information with other biological data and clinical data. Genomic-based approaches are also being incorporated into drug discovery and development processes. Large-scale systematic genomic approaches are being applied to accelerate the identification of markers that are potentially drug-able targets, involved in drug metabolism, or in the disease pathways. A number of new targeted therapeutics has resulted from these efforts. A landmark example is ivacaftor (Kalydeco) for treating cystic fibrosis (CF). Ivacaftor is the first drug to address the underlying genomic cause rather than the symptoms of CF. It was approved by the US Food and Drug Administration (FDA) in a very short time [83]. These examples demonstrate that skillful application of genomic science allowed researchers to understand at a molecular level the reasons why a protein fails to function, to discover and develop a medicine specifically to improve its function, and to use the results of a genetic test to select the right patients for the drug.

Development of Targeted Therapeutics for Cancers

As described previously, a large number of genomic alternations have been identified in the cancer genomes by TCGA and other cancer research projects around the world. These findings have dramatically advanced our understanding of cancer biology and the development of more effective ways to diagnose, treat, and prevent cancer. A number of new drugs have been successfully developed and used for treatment of cancer patients by targeting the specific genetic alternations. For example, the drug imatinib (Gleevec) was used for treatment of chronic myelogenous leukemia, GI stromal tumor, and acute lymphocytic leukemia with *BCR-ABL* fusion gene. The drug trastuzumab (Herceptin) was used for treatment of breast cancer and gastric cancers with *ERBB2* (*HER2/neu*) amplification/overexpression. The drug crizotinib (Xalkori) was used for treatment of non–small cell lung cancer and neuroblastoma with *EML4-ALK* fusion or *ALK* missense mutations. The drug vemurafenib (Zelboraf) was used for treatment of melanoma with *BRAF V600E* mutation. The drugs gefitinib (Iressa) and erlotinib (Tarceva) were used for treatment of cancers with *EGFR* missense mutations [84–88]. We are expecting more and more specific target therapies for cancers to be developed in future.

Genomic Profiling and Counseling

Genomic profiling can be used to define patient populations genetically to improve outcomes by predicting individual responses to drugs. This enables the disease treatments that are tailored to genetic and specific individual markers through the selection of medications and dosages that are optimized for individual patients. Hence the safety and efficacy of therapeutic treatment for patients can be greatly improved. Much like the targeted approach for the development of drugs, significant achievements in treatment safety and efficacy are possible with an interdisciplinary approach that includes genetic, technological, and therapeutic measures. For widespread dissemination of the personalized treatment option approach, there are still several issues that are not completely resolved. These issues include the broad demonstration of clinical utilities of genomic profiling, integration of genomic information with clinical information and patient records, comparative effectiveness, cost, reimbursement policy, financial model, insurance coverage, and business practices. Because the environmental factors play a substantial role in complex or multifactorial diseases, genomic profiling alone is likely not a sufficient predictor for risk assessment. However, genomic profiling combined with longitudinal clinical and family information could be used for risk stratification. It has been shown that whole genome profiling coupled with clinical measures in immunological, metabolic, cardiovascular, musculoskeletal repertory, and mental health disciplines can provide risk assessment for type 2 diabetes mellitus, coronary artery disease, hypertriglyceridemia, hypertension, and obesity [54]. By combining the genomic profiles with clinical and family historical data and identifying the concordant and discordant areas, an individual's risk for subclasses of disease can be more accurately assessed. The information can then be used to develop personalized health action plans for disease prevention and wellness management in consultation with a healthcare provider.

The Precision Medicine Initiative

The advancements in genomic technologies and computational sciences have become the main driver for precision medicine [89]. On January 31, 2015, President Obama announced a new precision medicine initiative [90], requesting an investment of $215 million in 2016 alone. The initiative called for a specific allocation of funds, including: $130 million to create a population-scale study and develop an integrated database, $70 million to the NCI for accelerating the development of genetically-defined therapies, $10 million to the FDA to support regulatory modernization, and $5 million to the Office of the National Coordinator for facilitating clinical data interoperability. The President's initiative speaks to a long-term vision, which is the necessity to integrate genomic and other data types (including clinical data) to drive genomic science toward clinical utilities and relevance. Also outlined is the need to address regulatory and policy issues to enable genomic advancements to have a true impact on the healthcare system.

Furthermore, as we contemplate the possibilities and tremendous improvements in clinical care and medicine that will be brought about by genomic advancements, we must also consider the social and economic impacts of this venture. As genomic medicine becomes more precise in identifying genetic subsets of the population benefitting from a particular therapy, the market for these specialized therapies becomes smaller and thus potentially more costly to make available to patients. Ironically, the main obstacle to the realization of a personalized genomic medicine paradigm no longer rests solely with the cost and complexity of the science and technology itself, but rather its implementation into the current healthcare system. For initial adaptation and integration of personalized genomic medicine into the healthcare system, well-designed and effectively executed genomic education and training programs for healthcare providers are necessary. Significant shifts in federal and private healthcare and insurance reimbursement policies will also be required for it to gain entrance into the diagnostic and therapeutic markets. Added to these challenges, there will be ethical and regulatory debates and considerations for individual equality, privacy, and safety protection. The integration of genomic information into the practice of medicine could induce significant changes in lifestyle choices and social behaviors (as modifiers of genomics). Furthermore, it seems probable that new business and financial models will be needed to ensure sustainability across healthcare, drug development, and biomedical research.

POLICY AND REGULATORY ISSUES

The FDA has been building infrastructure to support personalized medicine since shortly after the completed sequencing of the first human genome. Personalized medicine goes beyond genomic medicine. However, advancements in genomic technologies have been the major driver in personalized medicine efforts. The FDA's effort in building infrastructure to support personalized medicine has intensified over the past 7 years because the genomic tools have become more powerful and gained widespread use. It has created a center for bioinformatics, functional genomics, and structural genomics; a genomics and targeted therapy group within its office of clinical pharmacology; a genomics evaluation team for safety; and an FDA genomics working group. These groups are established to focus on scientific, technological, and clinical and regulatory challenges related to development and management of genomic-based drugs, biological substances, and medical devices, as well as genomic information management. Much work still needs to be done and policy must be developed

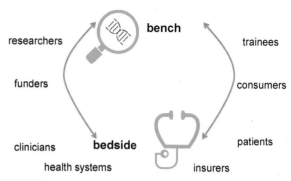

FIGURE 2.2   The future model for genomic and precision medicine.

to properly regulate genomic based medicine (Fig 2.2). In 2008, the Genetic Information Nondiscrimination Act was signed into law, which prohibits employers from requesting genetic information or making hiring, firing, and promotion decision based an individual's genetic profile [91]. The act also prohibits the insurance companies from requiring genetic tests or using genetic information for coverage decision or to set premiums or deductibles. Ethical questions related to the consideration of an individual's health need to be addressed and guidelines must be established for the use of genomics in clinical research and clinical medicine.

GENOMIC EDUCATION

Genomic science as a discipline has penetrated well into both academic and industry with genomic explorations occurring across many fields such as biomedical research, biotechnology, pharmaceuticals, agriculture, bioenergy production, synthetic biology, environmental science, computational science, and information technology. Genetics have also been applied to health and wellness management for more than 2 decades. However, personalized genomic medicine or precision medicine is a relatively new concept [89] for the healthcare industry and the vast majority of the stakeholders within the system. For successful adaptation and integration of personalized genomic medicine into the healthcare system, a number of countries have developed some genomic education and training programs for healthcare providers and the public [92–99].

REFERENCES

[1] International Human Genome Sequencing Consortium. Initial sequencing and analysis of the human genome. Nature 2001;409:860–921.

[2] Venter JC, Adams MC, Myers EW, Li PW, Mural RJ, Sutton GG, et al. The sequence of the human genome. Science 2001;291: 1304–51.

[3] National Research Council (US) Committee on A Framework for Developing a New Taxonomy of Disease. Toward precision medicine: building a knowledge network for biomedical research and a new taxonomy of disease. Washington (DC): National Academies Press (US); 2011.

[4] Sanger F, Coulson AR. A rapid method for determining sequences in DNA by primed synthesis with DNA polymerase. J Mol Biol 1975;94:441–8.

[5] Sanger F, Nicklen S, Coulson AR. DNA sequencing with chain-terminating inhibitors. Proc Natl Acad Sci USA 1977;74:5463–7.

[6] Smith LM, Fung S, Hunkapiller MW, Hunkapiller TJ, Hood LE. The synthesis of oligonucleotides containing an aliphatic amino group at the 5′ terminus: synthesis of fluorescent DNA primers for use in DNA sequence analysis. Nucleic Acids Res 1985;13: 2399–412.

[7] Smith LM, Sanders JZ, Kaiser RJ, Hughes P, Dodd C, Connell CR, et al. Fluorescence detection in automated DNA sequence analysis. Nature 1986;32:674–9.

[8] Margulies M, Egholm M, Altman WE, Attiya S, Bader JS, Bemben LA, et al. Genome sequencing in microfabricated high-density pico-litre reactors. Nature 2005;437:376–80.

[9] Ronaghi M, Karamohamed S, Pettersson B, Uhlén M, Nyrén P. Real-time DNA sequencing using detection of pyrophosphate release. Anal Biochem 1996;242:84–9.

[10] Bentley DR, Balasubramanian S, Swerdlow HP, Smith GP, Milton J, Brown CG, et al. Accurate whole human genome sequencing using reversible terminator chemistry. Nature 2008;456:53–9.

[11] Rusk N. Torrents of sequence. Nat Meth 2011;8:44.

[12] http://ghr.nlm.nih.gov/handbook/basics/dna.

[13] http://www.illumina.com/systems/hiseq-x-sequencing-system.html.

[14] Myers EW, Sutton GG, Delcher AL, Dew IM, Fasulo DP, Flanigan MJ, et al. A whole-genome assembly of *Drosophila*. Science 2000;287:2196–204.

[15] Jaffe DB, Butler J, Gnerre S, Mauceli E, Lindblad-Toh K, Mesirov JP, et al. Whole-genome sequence assembly for mammalian genomes: arachne 2. Genome Res 2003;13:91–6.

[16] Pop M, Salzberg SL. Bioinformatics challenges of new sequencing technology. Trends Genet 2008;24:142–9.

[17] Mardis ER. The impact of next-generation sequencing technology on genetics. Trends Genet 2008;24:133–41.

[18] Turnbaugh PJ, Ley RE, Hamady M, Fraser-Liggett CM, Knight R, Gordon JI. The human microbiome project. Nature 2007;449: 804–10.

[19] Bao S, Jiang R, Kwan W, Wang B, Ma X, Song YQ. Evaluation of next-generation sequencing software in mapping and assembly. J Hum Genet 2011;56:406–14.

[20] Li H, Homer N. A survey of sequence alignment algorithms for next-generation sequencing. Briefings Bioinf 2010;11:473–83.

[21] Nielsen R, Paul JS, Albrechtsen A, Sung YS. Genotype and SNP calling from next-generation sequencing data. Nat Rev Genet 2011;12:443–51.

[22] Koboldt DC, Larson DE, Chen K, Ding L, Wilson RK. Massively parallel sequencing approaches for characterization of structural variation. Methods Mol Biol 2012;838:369–84.

[23] Pepke S, Wold B, Mortazavi A. Computation for ChIP-seq and RNA-seq studies. Nat Methods 2009;6:S22–32.

[24] Thomas T, Gilbert J, Meyer F. Metagenomics: a guide from sampling to data analysis. Microb Inf Exp 2012;2:3.

[25] The NIST definition of cloud computing, http://csrc.nist.gov/publications/nistpubs/800-145/SP800-145.pdf. National Institute of Standards and Technology.

[26] Hood L, Friend SH. Predictive, personalized, preventive, participatory (P4) cancer medicine. Nat Rev Clin Oncol 2011;8:184–7.

[27] Pflughoeft KJ, Versalovic J. Human microbiome in health and disease. Annu Rev Pathol Mech Dis 2011;7:99–122.

[28] Palmer C, Bik EM, DiGiulio DB, Relman DA, Brown PO. Development of the human infant intestinal microbiota. PLoS Biol 2007;5(7):e177.

[29] Koenig JE, Spor A, Scalfone N, Fricker AD, Stombaugh J, Knight R, et al. Succession of microbial consortia in the developing infant gut microbiome. PNAS 2011;108(Suppl. 1):4578–85.

[30] Brown CT, Davis-Richardson AG, Giongo A, Gano KA, Crabb DB, Mukherjee N, et al. Gut microbiome metagenomics analysis suggests a functional model for the development of autoimmunity for type 1 diabetes. PLoS One 2011;6(10):e25792. pmid: 20668239.

[31] Socransky SS, Haffajee AD, Cugini MA, Smith C, Kent Jr RL. Microbial complexes in subgingival plaque. J Clin Periodontol 1998;25:134–44.

[32] http://www.1000genomes.org/.

[33] The 1000 Genomes Project Consortium. A map of human genome variation from population-scale sequencing. Nature 2010;467: 1061–73.

[34] The 1000 Genomes Project Consortium. An integrated map of genetic variation from 1092 human genomes. Nature 2012;491:56–65.

[35] The 1000 Genomes Project Consortium. An integrated map of structural variation in 2504 human genomes. Nature 2016;526(7571): 75–81.

[36] The 1000 Genomes Project Consortium. A global reference for human genetic variation. Nature 2016;526(7571):68–74.

[37] Lo YM, Corbetta N, Chamberlain PF, Rai V, Sargent IL, Redman CW, et al. Presence of fetal DNA in maternal plasma and serum. Lancet 1997;350:485–7.

[38] Ehrich M, Deciu C, Zwiefelhofer T, Tynan JA, Cagasan L, Tim R, et al. Noninvasive detection of fetal trisomy 21 by sequencing of DNA in maternal blood: a study in a clinical setting. Am J Obstet Gynecol 2011;204:205. e1–11.

[39] Palomaki GE, Kloza EM, Lambert-Messerlian GM, Haddow JE, Neveux LM, Ehrich M, et al. DNA sequencing of maternal plasma to detect down syndrome: an international clinical validation study. Genet Med 2011;13:913–20.

[40] The American College of Obstetricians and Gynecologists Committee on Genetics, The Society for Maternal-Fetal Medicine Publications Committee. Noninvasive prenatal testing for fetal aneuploidy. Committee Opinion No. 545. American College of Obstetricians and Gynecologists. Obstet Gynecol 2012;120:1532–4.

[41] Bianchi DW, Parker RL, Wentworth J, Madankumar R, Saffer C, Das AF, et al. DNA sequencing versus standard prenatal aneuploidy screening. N Engl J Med 2014;370:799–808.

[42] Taglauer ES, Wilkins-Haug L, Bianchi DW. Cell-free fetal DNA in the maternal circulation as an indication of placental health and disease. Placenta 2014;35(Suppl):S64–8.

[43] Bianchi DW, Wilkins-Haug L. Integration of noninvasive DNA testing for aneuploidy into prenatal care: what has happened since the rubber met the road? Clin Chem 2014;60:78–87.

[44] Kagan KO, Eiben B, Kozlowski P. Combined first trimester screening and cell-free fetal DNA: "next generation screening". Ultraschall Med 2014;35:229–36.

[45] Srinivasan A, Bianchi DW, Huang H, Sehnert AJ, Rava RP. Noninvasive detection of fetal sub-chromosome abnormalities via deep sequencing of maternal plasma. Am J Hum Genet 2013;92:167–76.

[46] Mao J, Wang T, Wang BJ, Liu YH, Li H, Zhang J, et al. Confined placental origin of the circulating cell free fetal DNA revealed by a discordant non-invasive prenatal test result in a trisomy 18 pregnancy. Clin Chim Acta 2014;433:190–3.

[47] Wang Y, Chen Y, Tian F, Zhang J, Song Z, Wu Y, et al. Maternal mosaicism is a significant contributor to discordant sex chromosomal aneuploidies associated with noninvasive prenatal testing. Clin Chem 2014;60:251–9.

[48] Bianchi DW, Chudova D, Sehnert AJ, Bhatt S, Murray K, Prosen TL, et al. Noninvasive prenatal testing and incidental detection of occult maternal malignancies. JAMA 2015;314:162–9.

[49] Benn P, Curnow KJ, Chapman S, Michalopoulos SN, Hornberger J, Rabinowitz M. An economic analysis of cell-free DNA non-invasive prenatal testing in the US general pregnancy population. PLoS One 2015;10:e0132313.

[50] Talkowski ME, Ordulu Z, Pillalamarri V, Benson CB, Blumenthal I, Connolly S, et al. Clinical diagnosis by whole-genome sequencing of a prenatal sample. N Engl J Med 2012;367:2226–32.

[51] http://www.ncbi.nlm.nih.gov/omim.

[52] http://www.cdc.gov/ncbddd/birthdefects/data.html.

[53] http://www.cdc.gov/ncbddd/autism/index.html.

[54] Patel CJ, Sivadas A, Tabassum R, Preeprem T, Zhao J, Arafat D, et al. Whole genome sequencing in support of wellness and health maintenance. Genome Med 2013;5:58.

[55] Thompson R, Drew CJ, Thomas RH. Next generation sequencing in the clinical domain: clinical advantages, practical, and ethical challenges. Adv Protein Chem Struct Biol 2012;89:27–63.

[56] Bick D, Dimmock D. Whole exome and whole genome sequencing. Curr Opin Pediatr 2011;23:594–600.

[57] Worthey EA, Mayer AN, Syverson GD, Helbling D, Bonacci BB, Decker B, et al. Making a definitive diagnosis: successful clinical application of whole exome sequencing in a child with intractable inflammatory bowel disease. Genet Med 2011;13:255–62.

[58] Dinwiddie DL, Bracken JM, Bass JA, Christenson K, Soden SE, Saunders CJ, et al. Molecular diagnosis of infantile onset inflammatory bowel disease by exome sequencing. Genomics 2013;102:442–7.

[59] Bainbridge MN, Wiszniewski W, Murdock DR, Friedman J, Gonzaga-Jauregui C, Newsham I, et al. Whole-genome sequencing for optimized patient management. Sci Transl Med 2011;3:87re3.

[60] Saunders CJ, Miller NA, Soden SE, Dinwiddie DL, Noll A, Alnadi NA, et al. Rapid whole-genome sequencing for genetic disease diagnosis in neonatal intensive care units. Sci Transl Med 2012;4:154ra135.

[61] Carter MT, Scherer SW. Autism spectrum disorder in the genetics clinic: a review. Clin Genet 2013;83:399–407.

[62] Jiang YH, Yuen RK, Jin X, Wang M, Chen N, Wu X, et al. Detection of clinically relevant genetic variants in autism spectrum disorder by whole-genome sequencing. Am J Hum Genet 2013;93:249–63.

[63] Rosti RO, Sadek AA, Vaux KK, Gleeson JG. The genetic landscape of autism spectrum disorders. Dev Med Child Neurol 2014;56:12–8.

[64] Schaefer GB, Mendelsohn NJ, Professional Practice and Guidelines Committee. Clinical genetics evaluation in identifying the etiology of autism spectrum disorders: 2013 guideline revisions. Genet Med 2013;15:399–407.

[65] Shen Y, Dies KA, Holm IA, Bridgemohan C, Sobeih MM, Caronna EB, et al. Clinical genetic testing for patients with autism spectrum disorders. Pediatrics 2010;125: 2009-1684.

[66] Weiss LA, Shen Y, Korn JM, Arking DE, Miller DT, Fossdal R, et al. Association between microdeletion and microduplication at 16p11.2 and autism. N Engl J Med 2008;358:667–75.

[67] Li G, Fullwood MJ, Xu H, Mulawadi FH, Velkov S, Vega V, et al. ChIA-PET tool for comprehensive chromatin interaction analysis with paired-end tag sequencing. Genome Biol 2010;11:R22.

[68] Yao F, Ariyaratne PN, Hillmer AM, Lee WH, Li G, Teo AS, et al. Long span DNA paired-end-tag (DNA-PET) sequencing strategy for the interrogation of genomic structural mutations and fusion-point-guided reconstruction of amplicons. PLoS One 2012;7:e46152.

[69] http://www.nih.gov/news/health/sep2013/nhgri-04.htm.

[70] http://www.genome.gov/27546194.

[71] http://cancergenome.nih.gov/.

[72] http://www.jax.org/ct/clinical/jax-cancer-treatment-profile.html.

[73] http://www.foundationmedicine.com/.

[74] http://www.genomichealth.com/en-US/OncotypeDX.aspx.

[75] Leary RJ, Sausen M, Kinde I, Papadopoulos N, Carpten JD, Craig D, et al. Detection of chromosomal alterations in the circulation of cancer patients with whole-genome sequencing. Sci Transl Med 2012;4:162ra154.

[76] Forshew T, Murtaza M, Parkinson C, Gale D, Tsui DW, Kaper F, et al. Noninvasive identification and monitoring of cancer mutations by targeted deep sequencing of plasma DNA. Sci Transl Med 2012;4:136ra68.

[77] Spindler KL, Pallisgaard N, Vogelius I, Jakobsen A. Quantitative cell-free DNA, *KRAS,* and *BRAF* mutations in plasma from patients with metastatic colorectal cancer during treatment with cetuximab and irinotecan. Clin Cancer Res 2012;18:1177–85.

[78] Jones K, Nourse JP, Keane C, Crooks P, Gottlieb D, Ritchie DS, et al. Tumor-specific but not nonspecific cell-free circulating DNA can be used to monitor disease response in lymphoma. Am J Hematol 2012;87:258–65.

[79] Dawson SJ, Tsui DW, Murtaza M, Biggs H, Rueda OM, Chin SF, et al. Analysis of circulating tumor DNA to monitor metastatic breast cancer. N Engl J Med 2013;368:1199–209. http://www.ncbi.nlm.nih.gov/pubmed/23484797-comments.

[80] Murtaza M, Dawson SJ, Tsui DW, Gale D, Forshew T, Piskorz AM, et al. Non-invasive analysis of acquired resistance to cancer therapy by sequencing of plasma DNA. Nature 2013;497:108–12. http://www.ncbi.nlm.nih.gov/pubmed/23563269-comments.

[81] Heitzer E, Ulz P, Belic J, Gutschi S, Quehenberger F, Fischereder K, et al. Tumor-associated copy number changes in the circulation of patients with prostate cancer identified through whole-genome sequencing. Genome Med 2013;5:30.

[82] http://www.nih.gov/precisionmedicine/.

[83] http://www.fda.gov/NewsEvents/Newsroom/PressAnnouncements/ucm289633.htm.

[84] http://www.fda.gov/Drugs/DrugSafety/PostmarketDrugSafetyInformationforPatientsandProviders/ucm110502.htm.

[85] http://www.fda.gov/Drugs/DevelopmentApprovalProcess/HowDrugsareDevelopedandApproved/ApprovalApplications/TherapeuticBiologicApplications/ucm080591.htm.

[86] http://www.fda.gov/Drugs/InformationOnDrugs/ApprovedDrugs/ucm376058.htm#.

[87] http://www.fda.gov/AboutFDA/CentersOffices/OfficeofMedicalProductsandTobacco/CDER/ucm268301.htm.

[88] http://www.fda.gov/Drugs/InformationOnDrugs/ApprovedDrugs/ucm454692.htm.

[89] https://www.whitehouse.gov/the-press-office/2015/01/30/fact-sheet-president-obama-s-precision-medicine-initiative.

[90] Economic impact of the human genome project – battelle. August 2013.

[91] http://www.eeoc.gov/laws/statutes/gina.cfm.

[92] http://www.nature.com/gim/journal/v13/n3/full/gim9201139a.html.

[93] https://www.genome.gov/Education/.

[94] http://www.abmgg.org/pages/resources_meetings.shtml.

[95] http://www.cdc.gov/genomics/public/features/educational.htm.

[96] http://www.geneticseducation.nhs.uk/genomics-in-health.

[97] http://www.geneticseducation.nhs.uk/.

[98] http://www.genomicseducation.hee.nhs.uk/.

[99] http://english.big.cas.cn/et/iet/.

Chapter 3

Genomic Databases, Access Review, and Data Access Committees

M. Shabani[1], B.M. Knoppers[2], P. Borry[1]

[1]University of Leuven, Leuven, Belgium; [2]McGill University, Montreal, QC, Canada

Chapter Outline

Introduction 29
Underlying Principles, Policies, and Guidelines 30
Examples of Controlled-Access Databases
and Pertinent Data Access Committees 30
  European Genome-phenome Archive 30
  Database of Genotypes and Phenotypes 31
Ethical and Legal Challenges 32
  Consent 32
  Oversight Mechanisms on Downstream
  Uses and Enforceability 33

Data Producers and Data Users' Interests
  and Benefits 33
Conclusion 34
Acknowledgments 34
References 34

INTRODUCTION

In recent years, sharing data and results from publicly funded genomic studies with a broader range of users has gained momentum [1]. Policy makers and funding organizations endorse genomic data sharing in order to optimize use of public funds and to increase the statistical power of databases. Yet beyond national approaches, given the legal barriers related to the transborder exchange of samples, sharing data resulting from analyses of these samples is necessary but difficult [2].

In order to further facilitate genomic data-sharing practices, public databases have been established to host and distribute annonymized results of studies investigating genotype–phenotype interactions [3]. The database of Genotypes and Phenotypes (dbGaP) [4] and the European Genome-Phenome Archive (EGA) [5], for example, contain mainly clinical information, genomic characterization data, and genomic data. Depending on the type of data and privacy policies, users' access to these databases is managed either under open-access mechanisms or via controlled access through central or local committees [6]. Malaria Genomic Epidemiology Network (MalariaGEN), the Wellcome Trust Case Control Consortium, the International Cancer Genome Consortium (ICGC), and the Wellcome

Trust Sanger Institute exemplify research consortia or institutes that share their results through open/controlled databases that are centrally accessible.

Various international and national principles and policy statements support data sharing. The Fort Lauderdale Agreement, among others, sets out a system of tripartite responsibility for sharing data from large-scale biological research projects. Funding agencies, resource producers and resource users are assigned "a unique and critical role to play in ensuring the growth and development of the community resource system" [7]. This system underscored the importance of developing institution-or consortium-wide data access arrangements in order to facilitate data-sharing practices in a way that meets both ethical principles and legal rules. The Organisation for Economic Co-operation and Development in its *Principles and Guidelines for Access to Research Data from Public Funding* defines access arrangements as "the regulatory, policy and procedural framework established by research institutions, research funding agencies and other partners involved, to determine the conditions of access to and use of research data" [8].

Data access committees (DACs) are an integral component of data access arrangements in controlled-access databases. DACs are responsible for reviewing, approving,

Medical and Health Genomics. http://dx.doi.org/10.1016/B978-0-12-420196-5.00003-4

or disapproving access requests from potential users for a variety of controlled-access genomic databases [9,10]. DACs, however, function in different ways. As Lowrance illustrates, "some of these groups are formally constituted, have terms of reference and hold regular meetings. Others, are casual, rarely meeting but existing to be consulted from time to time by the custodian and in a position to address serious problems should any arise" [11].

The DAC governance approach to data sharing is to "develop a new tier of oversight in addition to research ethics committee approval" [12]. DAC policies, composition, and infrastructure should be aligned with the oversight tasks to accomplish this goal. Although several aspects of genomic data-sharing practices have received much attention to date, the ability of DACs to accomplish their duties and responsibilities regarding data access governance in research is understudied [13]. In order to bridge this gap, this chapter first sheds light on a number of current controlled-access genomic databases and the structure, duties, and responsibilities of the pertinent DACs pursuant to underlying principles of genomic data sharing. The current DAC practices within various genomic data producing studies, institutions, and consortia are described from information available on their websites and relevant publications. Lastly, this chapter investigates the challenges associated with access review in the context of such databases and offers suggestions for improvements.

UNDERLYING PRINCIPLES, POLICIES, AND GUIDELINES

The overarching principles of genomic data sharing are enshrined in international policies and guidelines. The Bermuda Principles endorsed the rapid public release of primary genomic sequence data in order to "encourage research and development, and to maximize its benefit to society" [14]. A 2002 statement on human genetic databases from the Human Genome Organization subsequently reinforced the notion that genomic databases were to be considered "public goods." In 2003, the Fort Lauderdale Agreement [7] elaborated several aspects of data sharing, followed by the International Summit on Proteomics Data Release (2008) [15] and the Toronto Statement on Prepublication Data Sharing [16]. Funders for health research issued a joint statement in 2011 [17] in order to promote "greater access to and use of data" in ways that are equitable, ethical, and efficient. Other national documents such as the UK Research Council's Principles on Data Policy again considered publicly funded research data as "a public good, produced in the public interest, which should be made openly available with as few restrictions as possible in a timely and responsible manner" [18].

Likewise, the United Nations Educational, Scientific and Cultural Organization's International Declaration on Human Genetic Data (2003) adopted a "fair access"

approach toward genetic data sharing stipulated in Article 18 (a): "States should regulate, in accordance with their domestic law and international agreements, the cross-border flow of human genetic data, human proteomic data and biological samples so as to foster international medical and scientific cooperation and ensure fair access to these data. Such a system should seek to ensure that the receiving party provides adequate protection in accordance with the principles set out in this Declaration." The principle of "fair access" resonated in the framework of the UK DNA Banking Network, a consortium that promotes fair access policies with respect to data subjects, producers, users, and the producers' and users' institutions [19]. Fair access arrangements therefore depend on: privacy and confidentiality of data subjects; exploration and attribution rights of the data producers; transparency, accessibility, and availability of databases for users; and intellectual property rights of the original institutions.

Whereas the aforementioned documents illuminate the overarching principles of data sharing in research, it remains the responsibility of funding or research institutions to detail the access arrangements accordingly. The Expert Advisory Group on Data Access, which was established in 2012 by the four major funding organizations in the United Kingdom, is an example of a national advisory board that endeavors to provide funders with strategic advice on "… the emerging scientific, legal and ethical issues associated with data access for human genetics research and cohort studies." In the same vein, the Genomic Data Sharing Policy issued by the US National Institutes of Health (NIH) in August 2014 provides the data-sharing and access arrangement policies applicable for NIH-sponsored genomic research studies [20]. Admittedly, such guidelines and policies issued by funding organizations will only concern their grantees; that is, not necessarily all data producers, but they remain influential nonetheless.

EXAMPLES OF CONTROLLED-ACCESS DATABASES AND PERTINENT DATA ACCESS COMMITTEES

European Genome-phenome Archive

The EGA is a database of all types of "sequence and genotype experiments, including case control, population, and family studies, hosted at the European Bioinformatics Institute" [21]. According to the EGA website: "The EGA will serve as a permanent archive that will archive several levels of data including the raw data (which could, for example, be re-analysed in the future by other algorithms) as well as the genotype calls provided by the submitters."

Data submitters via EGA maintain control over the downstream uses of datasets via DACs located in the original study or consortium. To date, 212 DACs are listed on the EGA website. These DACs are charged with a number of similar tasks. First, they ascertain whether the proposed research uses are in conformity with the original purpose of the data collection. For instance, the Data Access Committee Office (DACO) of the ICGC is required to "verify the conformity of users' projects with the goals and policies of ICGC, including, but not limited to, policies concerning the purpose and relevance of the research, the protection of participants, and the security of participants' data." In doing so, DACs usually refrain from reviewing the scientific quality of the proposals and focus instead on the consistency of the proposals with the original consent forms or ethics approvals [22]. According to the assessment criteria of the Wellcome Trust Case Control Consortium Data Access Committee (CDAC), "CDAC does NOT attempt to peer review the scientific quality of proposals. However, it does ask for a brief summary of the research to be undertaken, in order to judge whether it falls within the scope of the consents." The DACs do not conduct an ethical evaluation of the proposed projects beyond ensuring consistency of the consent forms. Rather, this task remains the duty of ethics committees [23]. In cases where research purposes seem challenging from ethical and legal standpoints, however, some DACs have considered the possibility of further evaluations. For instance, the Wellcome Trust Case Control CDAC "...considers that grossly inadequate research is ethically questionable, and reserves the right to refer back for clarification those requests that do not appear to attain even a minimal standard of competence" [23]. Others see this as the responsibility of the data user's institution. The ICGC reports, for example, "...[International Data Access Committee (IDAC)] opinion has always been that DACO itself is not constituted as an ethics review committee and should not evaluate users' consent forms or research protocols. Essentially, IDAC and DACO rely on the local ethics processes of the data users without imposing another layer of ethics review requirements on them" [22].

Second, controlled-access databases are accessible only to qualified or bonafide researchers. DACs are therefore responsible for verifying the eligibility of the potential users. Nonetheless, criteria for a qualified researcher are rarely articulated. DACs determine eligibility mainly on the basis of user's self-reported information concerning their affiliation and scientific merits in the field, possibly as proof of seniority and expertise. For the Genome of the Netherlands Project (GoNL), for instance, a qualified researcher is referred to "a senior investigator who is employed or legitimately affiliated with an academic, non-profit or government institution and who has a track record in the field" [24].

A third responsibility for DAC is oversight on data use. DACs often utilize a legally binding instrument, often called a *data access agreement*, signed between the DAC and the user's home institution. Data access agreements involve numerous stipulations to which data users must adhere. The data access agreement of the GoNL, for example, elaborates

the terms of data access, including issues related to publication, data ownership, data return, intellectual property rights, data disposal, ethics approval, confidentiality, and commercialization [24]. The corresponding publication policies are also illustrated either within the data access agreements or in a separate document. They outline provisions concerning data use, such as acknowledging the data producers and funders in resulting publications. Additionally DACs can reserve a right to audit data users' practices. The audit ensures data users respect the terms of use: "The DACO of ICGC reserves the right to audit users' IT security documents, if warranted by the circumstances" [22].

There is limited information available to evaluate the expertise of members and the construction of the listed DACs, particularly in smaller or single studies. Consortia, on the other hand, often have a two-tier access assessment. The framework of UK10K project testifies to this, whereby two bodies oversee data-access requests: the management committee (MC), composed of representatives from each of the cohorts and disease groups; and the Wellcome Trust Data Access Committee. Whereas the MC assesses the validity of the request and consistency of the data requests given usage constraints, the Wellcome Trust DAC delivers the final approval for access requests [25]. In other consortia, a steering committee can offer consultation on the scientific relevance, suitability, or feasibility prior to a DAC's evaluation, or provide broader policy level observations in controversial cases. For instance, members from the MalariaGEN project with relevant expertise were appointed to the IDAC to facilitate data access review at the consortium level. Additionally, "[to] complement the IDAC, ... a broader consultative group involving partner institutions, ethical review bodies, and funding agencies [could] receive regular reports of the uses made of the released data and may be asked to consider issues of policy from time to time by IDAC" [26]. Likewise, the ICGC experience has been that the IDAC functions as an independent oversight committee that monitors the DACO activities, as well as advising on complex access requests [27].

Database of Genotypes and Phenotypes

The dbGaP is designed to archive and distribute the results of studies that have investigated the interaction of genotype and phenotype. After publication of a study revealing reidentifiability of aggregate data in 2008 [26], all participants' genotype data were moved to the controlled-access databases of dbGaP. The governance of dbGaP exemplifies a centralized approach to managing data access requests. Within this database, 16 DACs "review requests for consistency with any data use limitations and approve, disapprove or return requests for revision," except for large studies in which a local DAC leads the access review [28,29]. According to the dbGaP website, NIH DACs "review and approve or disapprove all requests from the research community for access to datasets within the database of Genotypes and Phenotypes (dbGaP). Decisions to grant access are made based on whether the request conforms to the specifications within the Genomic Data Sharing Policy and program specific requirements or procedures (if any). In particular, all data uses proposed for dbGaP data must be consistent with the data use limitations proscribed for the dataset by the submitting institution and identified on the public Website for dbGaP." Indeed, data submitters should determine the data use limitations.

The DACs of the NIH include a range of member expertise. The DAC of the Genetic Association Information Network is composed of the senior staff of the NIH with experience in the diseases under study and in genetics, epidemiology, bioethics, and human subjects' concerns. The DAC of the Extramural National Cancer Institute (NCI) appointed 10 NCI senior staff members with appropriate scientific, bioethics, and human subject research expertise for a 2-year term.

In order to grant access to the NIH datasets, the involved parties [namely the principal investigator requesting access to the genomic study dataset (the "approved user"), his or her home institution, and the relevant NIH institute or center] sign a data use certification agreement. The agreement stipulates research use and responsibilities of requesters and approved users such as nonidentification, nontransferability, data security, and data release reporting. Although seemingly straightforward, a recent study by Simpson et al. suggests that such procedures may not be well-defined across all institutions [27].

The NIH Genomic Data Sharing Policy envisages oversight for data use as "reporting research progress using controlled-access datasets through annual access renewal requests or project close-out reports" [20]. For transparency, the approved data uses are listed on the dbGaP website. According to Paltoo et al., "Sixty-nine percent (12,391/17,746) of requests were approved as of 1 December 2013. The most common reason for disapproving requests was inconsistency between the proposed research and the data use limitations of the requested data set" [29].

ETHICAL AND LEGAL CHALLENGES

Consent

Informing research participants about the associated risks, benefits, research goals, and withdrawal options is a core element of ethical research with human subjects. Traditionally, informed consent has been recognized as an appropriate mechanism to address the informational needs of research participants and to ensure the voluntary nature of research participation. However, the subsequent use of genomic data originally obtained for a single project challenges the adequacy of the traditional, paper-based, one-off consent [30].

Indeed, the consistency of further data uses with the original consent should be assessed in order to share data with external researchers. DACs must assess this consistency by considering the existing consent forms or ethics committees approvals. In the absence of an adequate consent for further uses, the subsequent use of genomic data may fail to respect the wishes of individuals and ultimately undermine public trust in research [31].

Interpreting the content of consent forms can be challenging for DACs. For instance, the scope of data-access limitations by commercial parties may be questioned if investigators fail to elaborate it in consent forms. Guidelines or institutional policies could be instrumental in illustrating the scope of such commercial involvement. Regardless, research participants and the concerns of the public should be taken into consideration when drafting such policies and guidelines. Empirical studies demonstrate that the general public may object to commercial parties' access to databases given the presumed incongruence between the goals of private and public bodies [32]. The restrictive data-sharing polices often adopted by private companies are seen as at odds with the ultimate goal of broad data-sharing polices intended to promote public good [33].

Furthermore, focusing only on individual consent for research data use may fail to take group-based considerations into account. The unique nature of genomic data and potential implications of study results for the identified communities or subpopulations should also be factored into decision making about data sharing. When these concerns occur, a representative of the group in question could be made an ad hoc member of the access committees and group-based concerns via public consultations could be elicited. Public commentary in response to a draft of the new NIH Policy for Genomic Data Sharing reflects this concern and the need to adequately respond to it: "Tribal groups expressed concerns about the ability of DACs to represent tribal preferences in the review of requests for tribal data." The final policy has been modified accordingly to "reference explicitly that tribal law, in addition to other factors...should be considered in assessing the secondary use of some genomic data" [20].

Representation of concerned parties in the DACs is epitomized in the unique case of the HeLa Genome Data Access Working Group, where two members are appointed to represent the Henrietta Lacks family in processing access requests to the HeLa genomic sequence data [34,35]. "NIH has taken this extraordinary step because full sequence data can reveal certain heritable aspects of Henrietta Lacks' germline DNA, and can thus be used to draw inferences of possible, if uncertain, significance about her descendants" [36]. Although the HeLa Data Access Working group is a unique case, it draws attentions to the general implications of genomic data sharing for family members and inclusion of relevant stakeholders' values in the course of genomic data sharing and access.

Oversight Mechanisms on Downstream Uses and Enforceability

Downstream uses of personal genomic data trigger concerns regarding data privacy and confidentiality. The unique features of genomic databases and the complexities of deidentification raise concerns about the effectiveness of methods [37] to mitigate privacy concerns and eliminate the need for rigorous oversight. In recent years, reidentifiability of genomic databases has resulted in adoption of further data protection and oversight mechanisms when sharing genomic data is desired [38–40].

In principle, oversight over downstream uses of databases is a responsibility of DACs. The tenets of protecting human subjects and personal data in biomedical research underscore the necessity for such oversight. To this end, some DACs require submission of resulting publications and periodic reports in order to ensure control over data uses. Users' compliance with certain standards has also been required to ensure that effective security protection measures are in place, including physical protections, administrative discipline, and cyber security [48]. DACs, however, are limited in their tools and resources to assess users' compliance with those standards. Downstream data use is often distanced from the original sources of the data, making it difficult for DACs to monitor the practices and identify potential violations. To address this concern, it has been suggested that instead of allowing users to download data into their local systems, "…all analysis activity … be executed on the trusted mediator's computing resources and logged" [41], as a means of monitoring data use. This alternative approach is in its infancy, and it remains to be seen how it would evolve in accord with the data sharing oversight requirements.

Moreover, various steps of genomic data sharing and access could become the subject of an ethics and access oversight by the involved parties. The broad nature of data sharing and use through controlled-access public databases, however, results in heterogeneous oversight approaches [42,43]. Studies investigating the awareness and practices of ethics committees reveal variations in ethics review requirements for secondary data uses and limited understanding of such requirements in data submission and sharing [44]. Whereas some of the identified inconsistencies have been rooted in this regulatory variety, unclear guidelines and limited technical expertise among committee members exacerbates "the complexity of current data-handling systems" [45].

Indeed the lack of coherent and consistent access oversight across the data sharing pipeline can subsequently result in duplicate efforts or leave some aspects out of

any oversight in the view of broad genomic data sharing. Although collaboration between access and ethics committees has been seen as a potential solution to address these oversight challenges [46], identifying the pertinent risks and potential protective mechanisms to steer oversight across both committees is vital. As Rothstein highlights, "The increased scale of research and new computer technologies demand a more nuanced assessment of the risks and benefits of research using a range of de-identified information and biological materials" [47]. Notably, informational risks associated with sharing of genomic data draw on ethical imperatives to protection of human subjects and personal data. This may require adopting an appropriate approach in ethics and access review that is attentive to such informational risks. Recent ongoing revisions to Recommendation (2006) 4 of the Committee of Ministers of the Council of Europe concerning research using biological materials suggest: "review procedures may be adapted to the nature of the research and the extent to which the persons from whom biological materials have been removed could be identified from these biological materials" [18].

Data Producers and Data Users' Interests and Benefits

Sharing results of genomic studies with a broader range of researchers may conflict with the interests of the data producers. Primary investigators may perceive data sharing as being at odds with their professional interests in terms of acknowledging their efforts in primary data production through publication [48]. Data access arrangements should consider these competing interests and ensure that such concerns will not adversely influence data-sharing practices. Including data release policies and imposing publication embargoes, whether within data access agreements or in a separate document, may resolve some of these concerns [49]. Enforcing such provisions is not clear cut [50]. It requires a joint effort by journals, funders, and access committees to monitor the practices and identify violations. To date, only a few cases of disrespect of publication policies such as publication embargoes have been reported [51]. This indeed highlights the role of journal editors or reviewers to identify infringements in using genomic research data [52].

Also, provisions about data ownership and potential intellectual property rights are discussed in some data access agreements. Likewise, a "collaboration" requirement can be embedded in some data access arrangements, particularly when original investigators see an overlap between the proposed use and their own research interests. Although these provisions may remedy the concerns of data producers, it might be unnecessary for the ultimate goals of data sharing. Thorough attention should be paid to drafting access arrangements that speak to the interests of both data producers and data users.

Fairness of the access review procedure is an important goal of DACs. To this end, assessment criteria should be delineated and readily available in a transparent and efficient manner. As Piwowar et al. note "equity depends on local adherence to formal guidelines, otherwise decision-making may appear ad-hoc and opaque" [53]. Fortin et al. also underscore the importance of providing a "well-defined mandate and decision-making process" as a requirement to "establish and secure the committee's authority and independence" [9].

CONCLUSION

Governance of genomic data sharing via controlled-access databases entails adopting adequate mechanisms on the basis of underlying ethical and legal principles. Data protection and respecting informational self-determination of the data subjects underpin the policies and regulatory regimes concerning oversight of genomic data sharing. DACs play a critical role in implementing access arrangements that are framed by overarching principles of genomic data sharing. However, the functionality of DACs in the face of fast-growing genomic data sharing policies needs to be closely scrutinized [13]. Ensuring that DACs are adequately equipped with tools and necessary expertise to conduct data-use reviews and maintain access standards is challenging. Data users adhere to various security standards depending on their country of origin. Privacy rules could require data producers and users to meet different standards that permit data sharing.

Harmonizing the access arrangements and data-sharing policies adopted across the world is key to successful data sharing. International and European initiatives have identified roadblocks in this regard, and have been engaged in finding solutions to facilitate access to research data. The Global Alliance for Genomics and Health exemplifies this in their *Framework for Responsible Sharing of Genomic and Health-Related Data*, which supports an international agenda for data sharing [54]. Despite ethical and legal differences between jurisdictions, the *Framework* and policies of the Global Alliance may serve to harmonize approaches toward data sharing and mitigate the regulatory incongruity that currently impedes data sharing internationally.

ACKNOWLEDGMENTS

This work is kindly supported by the Interfaculty Council for Development Co-operation (IRO) of the University of Leuven, the EUCelLEX project by the European Commission FP7 program (grant agreement 601806), the FWO (Flanders-Québec) project, and the Ministère de l'Économie, de l'Innovation et des Exportations du Québec, PSR-SIIRI-850 (Canada).

REFERENCES

[1] Johnson AD, Leslie R, O'Donnell CJ. Temporal trends in results availability from genome-wide association studies. PLoS Genet 2011;7(9):e1002269.
[2] Moraia LB, Kaye J, Tasse AM, Knoppers BM, Mitchell C, Soini S, et al. A comparative analysis of the requirements for the use of data in biobanks based in Finland, Germany, the Netherlands, Norway and the United Kingdom. Med Law Int 2015. http://dx.doi.org/10.1177/0968533215571956.
[3] Church DM, Lappalainen I, Sneddon TP, Hinton J, Maguire M, Lopez J, et al. Public data archives for genomic structural variation. Nat Genet 2010;42(10):813–4.
[4] dbGaP. http://www.ncbi.nlm.nih.gov/gap; [accessed on 13.05.15].
[5] European Genome-Phenome Archive. https://www.ebi.ac.uk/ega/home; [accessed on 14.05.15].
[6] O'Brien SJ. Stewardship of human biospecimens, DNA, genotype, and clinical data in the GWAS era. Annu Rev Genomics Hum Genet 2009;10:193–209.
[7] Report of a meeting organized by the Wellcome Trust and held on 14–15 January 2003 at Fort Lauderdale USA. Sharing data from large-scale biological research projects: a system of tripartite responsibility; 2003.
[8] Principles and guidelines for access to research data from public funding. OECD; 2007.
[9] Fortin S, Pathmasiri S, Grintuch R, Deschenes M. Access arrangements for biobanks: a fine line between facilitating and hindering collaboration. Public Health Genomics 2010;14(2):104–14.
[10] Kaye J, Hawkins N. Data sharing policy design for consortia: challenges for sustainability. Genome Med 2014;6(4).
[11] Lowrance WW. Access to collections of data and materials for health research A report to the Medical Research Council and the Wellcome Trust London. MRC and Wellcome Trust; 2006.
[12] Kaye J. From single biobanks to international networks: developing e-governance. Hum Genet 2011;130(3):377–82.
[13] Shabani M, Knoppers BM, Borry P. From the principles of genomic data sharing to the practices of data access committees. EMBO Mol Med 2015;7(5):507–9. http://dx.doi.org/10.15252/emmm.201405002.
[14] Bermuda principles. In: International strategy meeting on human genome sequencing; 1996.
[15] Rodriguez H, Snyder M, Uhlen M, Andrews P, Beavis R, Borchers C, et al. Recommendations from the 2008 International Summit on Proteomics data release and sharing policy: the Amsterdam principles. J Proteome Res 2009;8(7):3689–92.
[16] Birney E, Hudson TJ, Green ED, Gunter C, Eddy S, Rogers J, et al. Prepublication data sharing. Nature 2009;461(7261):168–70.
[17] Walport M, Brest P. Sharing research data to improve public health. Lancet 2011;377(9765):537–9.
[18] Research Councils UK. Common principles on data policy. 2011. http://www.rcuk.ac.uk/research/datapolicy/. [accessed on 13.05.15].
[19] Yuille M, Dixon K, Platt A, Pullum S, Lewis D, Hall A, et al. The UK DNA banking network: a fair access biobank. Cell Tissue Bank 2010;11(3):241–51.
[20] NIH. Genomic data sharing policy. 2014. http://gds.nih.gov/03policy2.html. [accessed on 13.05.15].
[21] European Genome-Phenome Archive. https://www.ebi.ac.uk/ega/about; [accessed on 13.05.15].

[22] Joly Y, Dove ES, Knoppers BM, Bobrow M, Chalmers D. Data sharing in the post-genomic world: the experience of the International Cancer Genome Consortium (ICGC) Data Access Compliance Office (DACO). PLoS Comput Biol 2012;8(7):e1002549.

[23] Wellcome Trust Case Control Consortium WTCCC: Access to genotype data. http://www.wtccc.org.uk/info/access_to_data_samples.html; [accessed on 15.05.15].

[24] Genome of the Netherlands. Data access agreement. http://www.nlgenome.nl/wp-content/uploads/2015/02/14-09-2011_GoNL-DATA_ACCESS_POLICY_CONDITIONS_FINAL_MTA_ONLY.pdf; [accessed on 15.05.15].

[25] Muddyman D, Smee C, Griffin H, Kaye J. Implementing a successful data-management framework: the UK10K managed access model. Genome Med 2013;5(11):1–9.

[26] Parker M, Bull SJ, de Vries J, Agbenyega T, Doumbo OK, Kwiatkowski DP. Ethical data release in genome-wide association studies in developing countries. PLoS Med 2009;6(11):e1000143.

[27] International Cancer Genome Consortium (ICGC). Informed consent, access and ethical oversight. https://icgc.org/icgc/goals-structure-policies-guidelines/e1-informed-consent-access-and-ethical-oversight; [accessed on 15.05.15].

[28] Walker L, Starks H, West KM, Fullerton SM. DbGaP data access requests: a call for greater transparency. Sci Transl Med 2011;3(113):113cm34.

[29] Paltoo DN, Rodriguez LL, Feolo M, Gillanders E, Ramos EM, Rutter JL, et al. Data use under the NIH GWAS data sharing policy and future directions. Nat Genet 2014;46(9):934–8.

[30] Hayden EC. Informed consent: a broken contract. Nature 2012; (7403).

[31] Budin-Ljosne I, Tasse AM, Knoppers BM, Harris JR. Bridging consent: from toll bridges to lift bridges? BMC Med Genomics 2011;4:69.

[32] Caulfield T, Burningham S, Joly Y, Master Z, Shabani M, Borry P, et al. A review of the key issues associated with the commercialization of biobanks. J Law Biosci 2014;1(1):94–110.

[33] Shabani M, Bezuidenhout L, Borry P. Attitudes of research participants and the general public towards genomic data sharing: a systematic literature review. Expert Rev Mol Diagn 2014;14(8):1053–65.

[34] Hudson KL, Collins FS. Biospecimen policy: family matters. Nature 2013;500(7461):141–2.

[35] NIH. HeLa Genome Data Access Working Group. http://acd.od.nih.gov/hlgda.htm; [accessed on 13.05.15].

[36] National Institutes of Health Advisory Committee to the Director. HeLa Genome Data Access Working Group (Charge). http://acd.od.nih.gov/HeLaGenomeDataAccessWG-charge.PDF; [accessed on 13.05.15].

[37] Rodriguez LL, Brooks LD, Greenberg JH, Green ED. The complexities of genomic identifiability. Science 2013;339(6117):275–6.

[38] Gymrek M, McGuire AL, Golan D, Halperin E, Erlich Y. Identifying personal genomes by surname inference. Science 2013;339 (6117):321–4.

[39] Homer N, Szelinger S, Redman M, Duggan D, Tembe W, Muehling J, et al. Resolving individuals contributing trace amounts of DNA to highly complex mixtures using high-density SNP genotyping microarrays. PLoS Genet 2008;4(8):e1000167.

[40] Church G, Heeney C, Hawkins N, de Vries J, Boddington P, Kaye J, et al. Public access to genome-wide data: five views on balancing research with privacy and protection. PLoS Genet 2009;5(10):e1000665.

[41] Erlich Y, Williams JB, Glazer D, Yocum K, Farahany N, Olson M, et al. Redefining genomic privacy: trust and empowerment. PLoS Biol 2014;12(11):e1001983.

[42] Simpson CL, Goldenberg AJ, Culverhouse R, Daley D, Igo RP, Jarvik GP, et al. Practical barriers and ethical challenges in genetic data sharing. Int J Environ Res Public Health 2014;11(8):8383–98.

[43] Groisman IJ, Egalite N, Godard B. Consenting for current genetic research: is Canadian practice adequate? BMC Med Ethics 2014;15(80).

[44] Lemke AA, Smith ME, Wolf WA, Trinidad SB, GRRIP Consortium. Broad data sharing in genetic research: views of institutional review board professionals. IRB 2011;33(3).

[45] Lowrance WW. Privacy, confidentiality, and health research. Cambridge University Press; 2012.

[46] Dove ES, Knoppers BM, Zawati MH. An ethics safe harbor for international genomics research. Genome Med 2013;5(99).

[47] Rothstein MA. Is deidentification sufficient to protect health privacy in research? Am J Bioeth 2010;10(9):3–11.

[48] Kaye J, Heeney C, Hawkins N, de Vries J, Boddington P. Data sharing in genomics: re-shaping scientific practice. Nat Rev Genet 2009;10(5):331–5.

[49] Knoppers BM, Chisholm RL, Kaye J, Cox D, Thorogood A, Burton P, et al. A P3G generic access agreement for population genomic studies. Nat Biotechnol 2013;31(5):384–5.

[50] Contreras JL. NIH's genomic data sharing policy: timing and tradeoffs. Trends Genet 2015;31(2):55–7.

[51] Holden C. Paper retracted following genome data breach. Science 2009;325(5947):1486–7.

[52] Nanda S, Kowalczuk MK. Unpublished genomic data: how to share? BMC Genomics 2014;15(1):5.

[53] Piwowar HA, Becich MJ, Bilofsky H, Crowley RS. Towards a data sharing culture: recommendations for leadership from academic health centers. PLoS Med 2008;5(9):e183.

[54] Hayden EC. Geneticists push for global data-sharing. Nature 2013;498(7452):16.

Chapter 4

Diagnostic Genomics and Clinical Bioinformatics

A. Haworth, H. Savage, N. Lench

Congenica Ltd., Hinxton, United Kingdom

Chapter Outline

Introduction	**37**	Reference Genome	43
Current Clinical Practice: Diagnosis	37	Interpretation of DNA Sequence Variants	44
Chromosomal Microarrays	37	Protein-Altering Variants	45
Single Gene Testing	38	Variants Causing Aberrant Messenger RNA Splicing	45
Gene Panel Testing	38	Nucleotide Conservation	45
Exome Sequencing	39	Inheritance and Mutation Type	45
Whole Genome Sequencing	39	Segregation Analysis	45
Diagnostic Testing Strategy	39	Mutational Mechanism and Variant Spectrum	46
Inherited Cardiac Conditions	40	De Novo Mutations	46
Hypertrophic Cardiomyopathy	40	Functional Evidence	46
Aortopathies	40	Databases and Literature	46
Intellectual Disability	41	Literature Databases	47
Deciphering Developmental Disorders	41	Online Mendelian Inheritance in Man	47
The UK100K Genomes Project: Large-Scale Implementation of		Incidental Findings	47
Whole Genome Sequencing	**42**	Data Sharing	48
Clinical Bioinformatics	42	**Conclusion**	**48**
Clinical Phenotyping	42	**References**	**48**
Bioinformatics Pipelines	43		

INTRODUCTION

Current Clinical Practice: Diagnosis

Historically, genetic testing has been segregated into cytogenetics and molecular genetics. Cytogenetics provides a genome-wide analysis capable of identifying major structural abnormalities such as chromosome trisomies, translocations, inversions, duplications, and microinsertions and microdeletions. Over the past few years, the introduction of chromosome microarrays has significantly improved resolution such that copy number variants (CNVs) can now be routinely detected. Molecular genetic testing has focused on single gene analysis, primarily using Sanger DNA sequencing. However, with the advent of next generation sequencing (NGS) technologies, we are beginning to see the convergence of these hitherto distinct disciplines. For many reasons it is preferable to perform a single test that has an improved diagnostic yield and a shorter reporting time.

Chromosomal Microarrays

Chromosomal microarray (CMA) testing is now part of standard clinical practice for patients with developmental delay, intellectual disability (ID), and multiple congenital anomalies with/without epilepsy. Both the UK Association for Clinical Cytogenetics and the American College of Medical Genetics and Genomics (ACMG) have produced best practice guidelines for the use of CMAs [1,2].

A CMA is a genome-wide assay that is capable of detecting chromosomal imbalances; eg, duplications and deletions at a resolution significantly higher than traditional

Medical and Health Genomics. http://dx.doi.org/10.1016/B978-0-12-420196-5.00004-6

chromosomal karyotyping. The level of resolution is dependent on the microarray design; arrays consist of many thousands of unique, short, single-stranded nucleic acid molecules (oligonucleotides) immobilized to a solid substrate such as a glass slide. The oligonucleotides are distributed across the genome to ensure even coverage. This is often referred to as a *backbone design*; additional oligonucleotides ("content") can then be targeted to known regions of microdeletions and microinsertions, as well as in a gene-specific or exon-specific manner. The total number of oligonucleotide probes contained on an array is referred to as the *array density*. Using an array with a density of 135,000 probes, it is possible to achieve a resolution in the region of 50–250 kb DNA in contrast with conventional karyotyping, which is only able to achieve a resolution of approximately 10 Mbp. However, microarrays do have limitations and cannot detect balanced rearrangements or inversions.

Another important feature of array design is the inclusion of single nucleotide polymorphism (SNP) probes. These are oligonucleotides that are able to discriminate SNPs at known, defined regions in the human genome. These provide the added advantage of being able to detect extended regions of homozygosity (often indicative of consanguinity), as well as uniparental disomy (UPD) and germline mosaicism.

There are many different permutations for CMA design; however, most certified genetic testing laboratories use a standard commercial design; custom array designs are generally used in research projects. The testing laboratory is normally able to provide referrers with appropriate advice as required.

Inevitably, the choice of the CMA used is more often than not governed by financial considerations, with higher density arrays costing significantly more. Within the UK Genetic Testing Network [3], National Health Service regional genetics laboratories use a range of array densities (60k, 135k, and 750k) with and without SNP content.

Single Gene Testing

Currently the majority of genetic testing for single gene disorders is performed by a combination of Sanger DNA sequencing to identify small causative mutations involving a few nucleotides and methodologies such as multiplex ligation-dependent probe amplification (MLPA) to identify larger pathogenic variants caused by intragenic insertions or deletions.

Although Sanger DNA sequencing is considered to be the gold standard (clinical sensitivity and specificity are high), it has some notable disadvantages, the most significant of which are cost and scalability. Typically Sanger DNA sequencing involves the serial amplification of individual coding and splice-site regions of each gene associated with the patient phenotype. In this way, it is possible

to achieve 100% coverage and importantly have confidence in the reporting of negative (no mutation detected) findings, for single nucleotide variants (SNVs) and small insertions and deletions (indels).

Whereas in many diseases, SNVs and indels represent the major mutational load, in others, large-scale rearrangements undetectable by routine Sanger DNA sequencing represent a significant contribution to the mutational spectrum. For example, the most common type of mutation in Duchenne muscular dystrophy (DMD) involves the deletion or duplication of large sections of the *DMD* gene, and in others such as hereditary breast cancer (*BRCA1* and *BRCA2*), it represents a lower (~10%) but clinically significant group of patients. Indeed, it is estimated that approximately 10% of disease-causing variants involve hundreds to thousands of nucleotides. Therefore in these cases, genetic testing requires both Sanger DNA sequencing and the use of additional strategies such as MLPA and high-density microarrays.

This approach has been particularly successful for patients with single gene disorders with a well-characterized clinical phenotype such as cystic fibrosis, sickle cell disease, and phenylketonuria. However, many inherited disorders display significant genetic heterogeneity with mutations in many different genes contributing to an apparent single clinical phenotype, for example, inherited neuropathies and hypertrophic cardiomyopathy (HCM). Additionally, many patients arrive at the clinic with complex phenotypes, without an obvious candidate gene for testing. In such cases, the required sequential testing of each candidate gene is slow, costly, and inefficient, resulting in patients waiting for months, even years, for a molecularly confirmed diagnosis.

The arrival of NGS technologies has transformed genetic research and is in the process of doing the same for genetic testing of patients. NGS allows massively parallel sequencing of thousands of genes simultaneously and thus overcomes many of the constraints of Sanger DNA sequencing. The use of clinical grade NGS analysis for patient diagnosis is now widespread and has been adopted into routine practice by many laboratories worldwide, in particular the use of gene panels and exome sequencing.

Nevertheless it is important to note that some disorders, such as repeat expansion disorders (eg, fragile X syndrome, myotonic dystrophy, Huntington disease) and imprinting disorders (eg, Angelman syndrome, Beckwith-Wiedemann syndrome) remain refractory to both traditional and NGS technologies owing to the nature of the mutational mechanisms involved.

Gene Panel Testing

The analysis of a collection of genes implicated in a given heterogeneous phenotype is referred to as a *gene panel*. The size of a gene panel may range from two to three genes to in excess of 1000 genes. Gene panels can be real or virtual; in a real gene panel only the genes making up the panel are physically isolated and sequenced. Virtual gene panels are generated bioinformatically and applied to the analysis of DNA sequence data generated from a clinical exome, whole exome, or whole genome. The attraction of using a gene panel (real or virtual) is the ability to limit analysis to genes known to be associated with a given phenotype. This reduces the risk of encountering secondary (non-pertinent) findings not related to the patient's phenotype. Virtual gene panels provide the flexibility to add or remove genes analyzed without the need for further laboratory investigation; for example, after publication demonstrating association of a new gene with an established phenotype.

However, the range of gene panels available for testing can often be confusing for the non-specialist, an issue which is further exacerbated by variation in both panel content and design between different genetic testing service providers.

Exome Sequencing

The exome is defined as the proportion of the genome (~1–2%) that encodes for functional proteins. The Genome Reference Consortium (GRC) [4] genome assembly release GRCh38 has a total of 20,300 annotated coding genes. A number of commercially available whole exome kits are available that selectively capture these coding regions; alternatively, it is possible to capture a smaller subset of genes that have an established association with disease. This is referred to as a *clinical* or *medical exome* and typically the gene list is derived from sources such as Online Mendelian Inheritance in Man (OMIM) [5] and the Human Gene Mutation Database (HGMD) [6]. For example, the TruSight One sequencing panel (Illumina Inc., San Diego, USA) contains 4813 genes associated with known clinical phenotypes.

The key advantages of using a smaller, selective design include cost and the ability to restrict analysis to clinically relevant genes. Interpreting variation in "unknown" genes is challenging and is likely to result in an extended list of variants of unknown significance (VUSs) within a clinical report. Also, it is possible to exclude the capture and sequencing of particular genes; for example, severe adult-onset disease genes in pediatric patients.

However, the drawback of using a restricted gene list is the inability to extend analysis to the entire exome. This is often referred to as "opening up the exome" and is where the boundaries between diagnostics and research discovery become blurred. Clinicians and researchers appreciate the ability to review results on a genome-wide basis and interrogate gene sequences that are currently not associated with a given clinical phenotype. As new genes become associated with clinical phenotypes, it is helpful to be able to reanalyze patients' DNA sequence data, although it is necessary to ensure that appropriate consents are in place.

An increasingly favored approach is to perform whole exome sequencing and restrict analysis to specific genes using bioinformatic filtering. In this way it is possible to create virtual gene panels; genes can be easily added and removed as well as blacklisted such that analyzed data is never available for certain genes (eg, *BRCA1*). A first-line test can be performed using a defined set of genes based on the patient's clinical phenotype; if negative, a second-line (or *reflex*) test could involve opening up the exome to analyze an extended set of genes or alternatively all coding genes.

Whole Genome Sequencing

Whole genome sequencing (WGS) offers the ability to interrogate the entire DNA sequence of the genome without the need to use selective capture techniques to isolate specific regions of DNA. Conceptually, this approach is very appealing and will enable the identification of additional classes of mutation that are refractory to detection by exome sequencing. These include the identification of large structural rearrangements, balanced translocations, uniparental isodisomy, and mosaicism. WGS also offers the opportunity to interrogate noncoding regions of DNA and identify functionally important sequence variants that influence gene expression. Removing the need to capture sequences removes selection bias so that coverage across sequences is more uniform. The main obstacles to the uptake of WGS include cost and dealing with the enormous amount of data produced. Moving, analyzing, interpreting, and storing large amounts of genetic data have significant resource and cost implications, many of which are currently beyond the majority of routine diagnostic laboratories. As the cost of sequencing continues to decrease and experience is gained in data analysis and interpretation, we can anticipate that WGS will be the method of choice for the clinical diagnosis of rare genetic disorders.

Diagnostic Testing Strategy

From a technical point of view, parent–offspring (trio) analysis is the preferred method for the diagnosis of patients with complex clinical phenotypes. However, a number of key factors will influence a clinician's decision to order such a test. Foremost will be cost; at present, DNA sequence analysis of family trios (exomes or genomes) is prohibitively expensive for use in routine clinical practice. Also, the referrer needs to be clear about the purpose of the test. Is it *only* to provide an unequivocal diagnosis for the patient or should it also be used to provide additional genetic information that will (1) help guide treatment and patient management and (2) uncover susceptibility genes and risk markers (that may or may not be clinically actionable) unrelated to the patient's phenotype? If the latter is the case, then it is possible to make the argument that it is more cost-effective to generate comprehensive genetic data in a single assay, eg, WGS, and reanalyze the sequence on an *as-needed* basis, rather than undertake multiple sequential tests over the lifetime of an individual.

Two clinical areas where gene panel testing and whole exome and whole genome analysis have already had a major impact are inherited cardiac conditions and ID associated with multiple congenital anomalies.

Inherited Cardiac Conditions

Inherited cardiac conditions (ICCs) are a group of monogenic disorders primarily affecting the heart, its conducting system, or vasculature. In some cases, the first indication that an individual has an ICC is sudden cardiac death, often in adolescence or early adulthood. Therefore in deceased individuals it is important to ascertain the molecular cause of death and subsequently test and identify at-risk relatives in order to provide surveillance (eg, echocardiograph) and health-management options. The main categories of ICCs include arrhythmias, cardiomyopathies, and aortopathies; collectively there are approximately 250 known genes associated with these conditions. These include genes that code for large structural proteins such as actins and myosins, as well as titin, which is the largest gene yet characterized, containing 363 exons and the single largest known exon (17,106 bp). To provide sequential, single gene testing for ICCs is unrealistic and as a consequence, this has precluded the routine characterization and diagnosis of patients and the identification of at-risk individuals. However, the advent of NGS has transformed the ability to provide comprehensive gene panel–based tests for ICCs.

Hypertrophic Cardiomyopathy

HCM is a heterogeneous, autosomal-dominant genetic disorder that affects up to 1 in 500 individuals in Europe and the United States [7]. The disorder is compounded by incomplete penetrance and variable expressivity. It is also important to identify at-risk individuals as early as possible, because the risk of sudden cardiac death is greater in adolescence. In around 50% of cases, HCM is caused by mutations in genes coding for sarcomere or sarcomere-related genes. HCM cases also carry rare variants in genes coding for desmosomal, ion-channel, and other proteins implicated in inherited heart disease, but their relevance to disease expression is unknown.

A UK study by Lopes et al. [8] investigated a cohort of 223 patients with HCM; 69 predicted pathogenic mutations were identified, of which 83% occurred in sarcomeric protein genes (*ACT1, MYBPC3, MYH7, MYL2, MYL3, TNNI3, TNNT2, TPM1*) and 17% in Z-disc–related and calcium-handling protein genes (*ANK2, CASQ2, PLN, SCN5A*). It

is possible that rare variants in sarcomere genes and also in non-sarcomere genes implicated in other forms of inherited cardiac disorders modify the clinical characteristics and severity of HCM. Therefore it may be prudent to only offer extended gene panel testing in HCM patients with atypical clinical phenotypes.

Alfares et al. [9] investigated the change in test sensitivity for HCM diagnosis as a result of gene panel content increasing from 10 genes to 51 genes over 10 years. The study cohort comprised 2912 individuals selected for HCM. All subjects were sequenced for at least five sarcomeric protein genes: *MYBPC3, MYH7, TNNI3, TNNT2,* and *TPM1*. Interestingly, 83% of all pathogenic/likely pathogenic mutations were present in two genes: *MYBPC3* and *MYH7*; a further 5% were caused by mutations in *TNNI3* and 4% in *TNNT2*. Therefore 92% of all pathogenic/likely pathogenic mutations occurred in five genes. Broadening genetic testing to more comprehensive panels identified a limited number of additional mutations, with only one being in a non-sarcomeric protein gene.

Aortopathies

Aortopathies are a group of disorders characterized by aneurysms and dilation of the aorta that ultimately can lead to catastrophic rupture of the blood vessels. These disorders may be syndromic or non-syndromic, familial, or sporadic. The more common disorders include Marfan syndrome (MFS; OMIM 154700), Loeys-Dietz syndrome (LDS; OMIM 609192), and Ehlers-Danlos syndrome (130,050).

MFS (prevalence ~1/5000) is an autosomal-dominant disorder characterized by the involvement of the cardiovascular, ocular, and skeletal systems; approximately 90% of MFS is caused by mutations in the Fibrillin-1 gene (*FBN1*) located on chromosome 15q21.1 and encoding a large multidomain extracellular matrix protein. Seventy-five percent of MFS patients have a family history and 25% of MFS cases represent de novo mutations. Two thirds of mutations that cause MFS are missense and usually involve cysteine residues or residues that are important for calcium binding; indel mutations account for approximately 13% of all mutations. Nonsense mutations are responsible for 10% and splicing mutations account for a further 10–15% of the mutation spectrum. Gene rearrangements have been reported but are thought to be rare. *FBN1* is a large gene covering over 230 kb of genomic DNA and consists of 65 exons. There is a high degree of intrafamilial and interfamilial variation in the clinical symptoms displayed by patients. *FBN1* mutations can cause other diseases related to MFS (type 1 fibrillinopathies). Clinical diagnosis of MFS is based on the Ghent nosology [10]. There is a greater chance that a mutation will be detected in a patient if that individual fulfils these criteria. If no mutations are detected in *FBN1* when screened for a diagnostic test, this does not exclude a diagnosis of MFS. Similarly the *TGFBR1* and *TGFBR2* genes have now been implicated in a MFS-like phenotype.

LDS is characterized by vascular and skeletal manifestations; approximately 95% of patients with LDS have aortic root dilation. LDS is caused by mutations in the *SMAD3*, *TGFBR1*, *TGFBR2*, and *TGFB2* genes. LDS has an autosomal dominant mode of inheritance, with many more cases being caused by *TGFBR2* mutations compared with *TGFBR1* (75% compared with 25%, respectively). A high proportion of mutations detected in patients arise de novo with no previous family history.

Wooderchak-Donahue et al. [11] assessed the clinical utility of a NGS assay comprising a panel of 10 genes associated with Marfan and Marfanlike syndromes with aortopathy. A cohort of 175 patients was analyzed: 18 individuals (~10%) tested positive for a pathogenic mutation and 32 had VUSs. Pathogenic mutations were identified in *ACTA2*, *FBN1*, *SMAD3*, and *TGFBR1*, whereas VUSs were present across all 10 genes. Seventy-two percent of pathogenic mutations were present in *FBN1*; however, the majority of cases included for testing had a clinical or differential diagnosis of MFS. Phenotypically, mutation-positive patients exhibited a higher incidence of aortic dilation, dissection, and rupture. The current diagnostic strategy for patients with an MFS phenotype is *FBN1* sequence analysis followed by deletion/duplication analysis, with sensitivities of ~70–93% and ~7% respectively. If mutation negative, patients are screened for mutations in *TGFBR1* and *TGFBR2*. Clearly, a single assay is preferable, providing increased sensitivity and shorter reporting times.

Moving forward, whole exome or whole genome analysis will be required to identify mutations in patients that currently test negative using gene panels for ICCs.

Intellectual Disability

Exome and genome sequencing has been applied with great success to the diagnosis of patients with ID. As previously described, the majority of patients with ID undergo genome-wide investigation using CMAs. However, a number of studies have now demonstrated the utility of exome and genome sequencing in "CMA-negative" subjects by significantly increasing diagnostic yield from ~10% to ~50% (the precise figure is dependent on a number of factors mainly related to patient ascertainment). These studies have also increased our understanding of the mechanisms of disease, providing evidence of a high proportion of de novo mutations and novel genes responsible for ID, as well as providing the opportunity of novel therapies for ID through the identification of key biological pathways and molecular targets.

One of the first large-scale studies using whole exome sequencing in family trios with probands with ID (IQ < 50) was performed by de Ligt et al. [12]. A total of 100 trios were investigated; on average, 24,324 variants were identified per patient and subsequently prioritized based on de novo occurrence. After Sanger DNA sequencing, 79 de novo mutations were confirmed in 53 patients, of which 12 occurred in known ID genes. A diagnostic yield of 16% was achieved. This study further demonstrated the importance of de novo mutations in the etiology of ID. In a subsequent study, de Ligt et al. [13] used whole exome sequencing to identify CNVs as well as SNVs and indels, offering the possibility of using WES as a first-line test for patients with ID, developmental delay, and multiple congenital abnormalities. WES is not as sensitive as high-resolution microarrays; however, it was possible to identify 88% of rare coding CNVs containing three or more exons.

The effectiveness of WGS was investigated in 50 ID family trios [14]. Individuals were sequenced to an average genome-wide coverage of 80-fold, revealing an average of 4.4 million SNVs and 276 CNVs per genome. Analysis focused initially on de novo mutations; after confirmation by Sanger DNA sequencing, an overrepresentation of de novo substitution mutations with a bias toward loss-of-function mutations in coding regions of known ID genes was observed. A conclusive diagnosis was achieved in 21/50 patients (42% yield).

Deciphering Developmental Disorders

The UK Deciphering Developmental Disorders (DDD) research program performed genome-wide microarray and whole exome sequencing on 1133 family trios (mother, father, and affected offspring) with undiagnosed developmental disorders [15]. High-density microarray analysis (Agilent 1M array comparative genomic hybridization and Illumina 800K SNP array) was used to identify CNVs in the proband and whole exome sequencing to identify SNVs, small indels, and CNVs in the coding regions of the genome. Family trios enable: (1) the identification of de novo mutations in the proband and (2) the identity of the parent of origin for inherited genomic variants. The latter is particularly important for comparing the penetrance of phenotype as well as being important for recurrence risk estimates and reproductive counseling.

A major feature of the DDD study was the incorporation of standardized and structured clinical data using the Human Phenotype Ontology (HPO) classification system [16]. Across the cohort, 1435 unique HPO terms were used to describe clinical presentations. Automated variant filtering was performed and excluded common and nonfunctional variants. Potentially pathogenic variants were identified in known disease genes contained in the Developmental Disorders Genotype to Phenotype database. Likely diagnostic DNA sequence variants were identified in 311 probands, representing a diagnostic yield of 27%. De novo variants had by far the highest predictive value, accounting for 65%

of diagnoses and demonstrating the value of trio analysis. Further analysis identified 12 novel genes associated with developmental disorders; clustering of missense mutations in six of these genes suggests that normal development is being perturbed by activating or dominant-negative mutations [17]. These additional findings increased the overall diagnostic yield of the study to 31%.

Whole exome sequencing data has also enabled the development of novel analytical methods to detect other mutational mechanisms such as UPD and somatic mosaicism; UPD detection was performed on 1057 trios from the DDD study and six validated events identified [18]; likely pathogenic structural mosaic abnormalities were detected in 12/1303 children with developmental disorders [19].

THE UK100K GENOMES PROJECT: LARGE-SCALE IMPLEMENTATION OF WHOLE GENOME SEQUENCING

The UK100K Genomes Project [20] is a ground-breaking initiative aimed at sequencing the genomes of rare disease patients and cancer patients. Approximately 15,000 family trios with rare disease will be recruited into the study covering a diverse range of clinical phenotypes. WGS to an average depth of 30-fold will be performed and automated variant calling and analysis employed to identify pathogenic mutations from multiple classes (ie, SNVs, indels, CNVs, UPD, and mosaicism). A key advantage of WGS is the removal of sequence capture bias that is a feature of targeted selection methods. At comparable sequencing depth, WGS outperforms WES in terms of covered coding regions. At higher sequencing depth (95×–160×), WES successfully captures 95% of the coding regions with a minimal coverage of 20×, compared with 98% for WGS at 87-fold coverage [21]. As the price of WGS decreases, more genomes will be sequenced, thus improving our understanding of the contribution of noncoding sequence variation to disease pathology. This will include the analysis of promoter and enhancer regions, the effect of alternative splicing and transcript diversity and epigenetic control.

For the foreseeable future, clinical diagnosis using genetics will be a mixture of single gene testing, gene panel testing (real and virtual), exome sequencing (clinical and whole exome), and WGS. Cost, assay sensitivity and specificity, clinical utility, and test reporting times are considerations that will influence the referring clinician's choice of test.

Clinical Bioinformatics

A new discipline of clinical bioinformatics is emerging that brings together bioinformatics, genotype, and phenotype data and interpretation and decision-making. This will lead to personalized treatment and care based on a patient's genotype.

It is now widely accepted that even within apparently homogeneous single-gene disorders, there can be a high level of phenotypic and genetic heterogeneity. Over 2000 different potentially pathogenic mutations have been reported in the cystic fibrosis (*CFTR*)gene; however, proving the deleterious functional consequences of these mutations is difficult. There is also widespread heterogeneity in clinical symptoms with variability in severity, rate of disease progression, and organ involvement [22]. Therefore it is becoming increasingly important to gather comprehensive clinical information (deep phenotyping) in order to define precisely phenotype–genotype correlations, identify genetic risk modifiers, and manage the treatment and care of patients appropriately [23]. The use of structured nomenclatures enables the sharing and comparison of standardized clinical terms. Subjects recruited to the UK100K Genomes Project require a phenotypic core minimum dataset that includes medical terms, clinical tests, and medication history.

Clinical Phenotyping

A phenotype can be defined as a list of traits exhibited by an individual. Clinical phenotypes in medicine tend to relate to specific features that show some level of deviation from an accepted standard. Phenotyping of a patient is arguably the most critical step in allowing a clinician to develop a diagnostic hypothesis, drawing together information from a number of sources: verbal or documented patient medical history, physical examination, observation, and collation of previous test results. The formation of a diagnostic hypothesis should provide the basis from which to initiate and direct further investigation and to confirm or refute a clinical diagnosis.

In general, phenotype information tends to be poorly communicated and collated, and expressed in non-standardized terms. As there are multiple different ways to describe a set of features, correctly and accurately linking vast amounts of genotype information generated through exome and genome sequencing, with an individual's phenotype data is very challenging. In order to provide some guidance and work toward a level of standardization the HPO was established. The HPO [16] aims to provide a standardized vocabulary of phenotypic abnormalities encountered in human disease and provides a framework from which clinicians can compile patient phenotype information. These terms are not only more easily transferable between patients, but reduce the risk of misphenotyping an individual by removing a level of objective interpretation, allowing an accurate and comprehensive phenotype to be described. This detailed patient phenotype is known as a *deep phenotype*.

Deep phenotyping of patients has become increasingly important as genetic testing moves away from screening a single gene known to cause a single, well-characterized disorder to screening panels of genes associated with multiple diseases with overlapping phenotypes. More recently, exome and genome sequencing are driving the shift toward the need for deep phenotyping, as an increasing number of variants with unknown clinical significance are identified. It is particularly challenging for clinical scientists to interpret the relevance of a variant in a gene associated with disease A in a patient primarily referred for disease B without having access to detailed information about the patient's phenotype.

As clinical sharing of genome, exome, and variant data becomes normal practice, identifying patients with overlapping phenotypes and variants in the same, as yet uncharacterized, gene allows collaboration to identify and describe new, rare syndromes. Excluding deep phenotype information, or using locally preferred descriptives for a particular feature, makes new genotype–phenotype links almost impossible to identify, ultimately leading to patients and their families being denied a diagnosis and without access to any accompanying downstream support. Deep phenotype information can and should be integrated into NGS analysis pipelines to allow more thorough and efficient review of evidence relating to the possible pathogenicity of a variant. Indeed, some bioinformatics software packages have integrated the HPO database into their pipeline, allowing users to select standardized terms when providing a list of the clinical features seen in their patients. In the near future, it is envisaged that there will be a move toward analytical software providing a suggested diagnosis based on the provided HPO terms before the formation of a diagnostic testing strategy. This would allow the application of a tailored virtual gene panel containing only a select set of relevant genes, making analysis faster, as there will be fewer variants to review, and will help to minimize the detection of incidental findings. Additionally, a properly curated, anonymized genotype–phenotype database accessible to research and industry users is likely to lead to the improved ability to stratify patients for the most suitable treatments, drug doses, or clinical trials based on their genetic results and phenotypic information.

Bioinformatics Pipelines

The analysis of NGS data requires a complex set of algorithms to map, align, perform quality control (QC), annotate variants, filter variants, and display the outputs; together these steps form the bioinformatics pipeline. Essentially, the bioinformatics pipeline is a set of instructions needed to transform a raw DNA sequence data file into a list of DNA sequence variants for the user to interpret and report. The increasing uptake and implementation of NGS has resulted in a plethora of bioinformatics tools that have been developed to analyze sequence data, with over 570 tools published within a span of only 2 years [24]. These methods include programs for assessing the quality of short-read data: Burrows-Wheeler Aligner (BWA) [25] and Bowtie2 [26] for sequence alignment and Genome Analysis Toolkit (GATK) [27] for detecting single nucleotide variations and small indels. Many of these tools are constantly being improved and updated, with new versions released on a frequent basis (eg, BWA and GATK). Despite these large number of tools, the bottleneck in a typical clinical diagnostics laboratory remains in the bioinformatics analysis phase owing to lack of access to informatics expertise. Therefore providing a robust and rigorously tested pipeline as part of a commercial software offering is an attractive option for many laboratories.

Each NGS platform outputs a specific file type containing the raw sequence data; the initial step of the pipeline trims the data to remove any ancillary sequence used for target selection (eg, primer and adapter sequences) or sample unique identifiers (DNA barcodes). The data is then QC assessed to determine the quality of each nucleotide in the read; any poor quality reads will be excluded at this point and no further analysis performed. Quality scores relate to the confidence that a nucleotide detected at a particular position is a true reflection of the sequence in that sample.

Reads achieving minimum QC requirements are aligned to the reference genome sequence; this is dependent on the type of test being performed and might be the whole genome sequence or predefined regions of interest. Duplicate reads are detected and "stacked" to generate a total depth of coverage at each particular base; if there are 100 duplicate reads covering a single nucleotide, it is said to have a read depth of 100-fold. In order to detect small insertions (gains) and deletions (losses), the reads are then realigned using a different algorithm tailored to account for longer mismatches in the nucleotide sequence.

Variant calling identifies any differences between the aligned nucleotide sequence and the reference sequence. The variants are then annotated according to their genomic location and the type of change that has occurred; for example, a deletion of two nucleotides (A and G) at complementary DNA positions 68–69 in the *BRCA1* gene would be annotated c.68_69delAG according to Human Genome Variation Society nomenclature guidelines [28]. At this stage, all variants are called regardless of whether they are considered to be benign or pathogenic in the patient. A number of different algorithms are available to perform variant calling and each can be tailored to the platform and specific sequencing application used.

Reference Genome

In order to identify and describe changes in DNA, a standardized representative sequence derived from the genomes of multiple individuals in a normal population is required to act as a genetic baseline for comparison. This artificially assembled sequence is known as the *reference genome.* The aim of a reference genome is to provide a consensus sequence to which patient data can be compared; however, it has significant limitations; the pooling of sequence information from only a small number of participants results in a biased representation of allelic diversity in the global population. As WGS becomes more commonplace, the volume of data available from which to generate an artificial sequence increases exponentially, allowing the current sequence to be improved and updated to improve accuracy. It is the role of the GRC, a collaboration of scientists primarily based at The Wellcome Trust Sanger Institute, The Genome Institute, The European Bioinformatics Institute, and the National Center for Biotechnology Information, to maintain and update these reference sequences. The current reference genome is GRCh38 [4].

As each update of the reference genome includes slight changes in the sequence, reference transcripts are assigned version numbers to allow users to accurately record the sequence used to identify variants in a clinical sample; for example, a commonly used reference sequence for *BRCA1* (a gene associated with hereditary breast and ovarian cancer) is NM_007294.3, where the 0.3 indicates that this is revision 3 of this reference transcript. To complicate matters further, many genes have multiple reference transcripts, varying in content and length, which could all be used for variant calling; *BRCA1* also has reference transcripts NM_007297.3, NM_007298.3, NM_007299.3, and NM_007300.3.

Clearly this is a complex situation; interpretation and clinical reporting require expert review from highly trained specialists. This expertise has been built upon specific genes and/or specific clinical phenotypes and this has also led to the development of the Locus Reference Genomic (LRG) sequences. LRGs are manually curated, stable reference sequences used for the reporting of variants within genes of known clinical significance. Provision of these stable sequences reduces the risk of variant annotation errors and aligns with the aims of the clinical sequencing nomenclature currently under development by the Global Alliance for Genomics and Health (GA4GH) [29]. Migration to this set of stable reference transcripts for diagnostic variant annotation and reporting will facilitate interlaboratory data sharing and aid variant interpretation for the benefit of patients and healthcare providers.

Interpretation of DNA Sequence Variants

The results of genomic testing generated via high throughput sequence technologies are increasingly being used in medical decision making to inform clinical recommendations and interventions such as recruitment to screening programs, prophylactic mastectomy, and prenatal diagnosis. Furthermore, whole genome analysis has been used for diagnosis in an acute clinical setting for newborns on high dependency [30]. Therefore the potential for harm caused by incorrect variant interpretation is clearly substantial.

The strategies used for variant discovery and interpretation in a clinical setting vary slightly depending on context. Currently, examination of a restricted number of known genes via gene panel analysis is often the test of choice because of high clinical sensitivity and specificity. However, exome analysis or even whole genome analysis may be more suitable in other instances. In each scenario, a similar variant interpretation pathway is used, but analysis of exomes and whole genomes have the additional complication of not only associating a variant with a phenotype, but on occasion additional evidence is also required to associate a gene itself with causality.

In an effort to systematically standardize how variants are categorized and improve reporting of clinically actionable variants, several professional bodies, such as the ACMG, the European Society of Human Genetics, and the British Society of Genomic Medicine have published guidelines in an attempt to standardize how variant interpretation and clinical reporting are performed [31–33]. Such guidance outlines several broad categories of resource that can assist in variant interpretation and suggests describing variants via a five-tier categorization of variant consequence (from pathogenic to benign) and subsequent contribution to patient phenotype (from full contribution to phenotype to no contribution). The five-tier categorization has been widely adopted by diagnostic laboratories both in the United States and Europe.

Once prefiltered, there are several lines of evidence and prediction that can be utilized to systematically assess pathogenicity of variants for clinical interpretation. Most of the tools available are particularly relevant to interpretation of variants located within coding exons and known splice regions because this is the part of the genome about which most is known. Tools for variant interpretation outside these regions are in their infancy and generally such variants are not currently included in clinical reporting.

In general, a significant number of variants detected in patients are relatively straightforward to interpret, because they are clearly pathogenic and causative of the patient phenotype. Generally this is because of the type of variant (those that ablate/terminate protein production) and/ or is supported by a large body of clinical and functional evidence; for example, the common p.Phe508del in-frame pathogenic variant in the *CFTR* gene.

For other variants, interpretation relies on the sum of several lines of evidence, details of which are outlined in summary. Until recently, each line of evidence was gained via independent interrogation of disparate resources, which in the clinical context is unsustainable because each decision on variant significance needs to be transparent and auditable. In response to this clinical need, a number of commercial clinical decision support tools have been developed, including Sapientia (Congenica Ltd, Hinxton, UK), which allows systematic variant interpretation via a single web interface.

Protein-Altering Variants

Missense variants in human disease genes are relatively common and are often difficult to interpret. Nucleotide variants that alter an amino acid residue may result in disease via either amino acid substitution, altering protein function or via disruption of splicing via the specific nucleotide change. Protein-altering effects will be discussed here and putative splicing effects will be discussed in the section on splicing.

A large number of publically available tools exist to enable variant interpretation.

- **SIFT**: evolutionary conservation across species. In silico predictive tools exist that assess whether an amino acid is highly conserved across different diverse species and use algorithms to predict whether substitution of the amino acid is likely to have a deleterious effect on the protein and result in disease [34].
- **PFAM, UniProt, InterPro**: location and context within the protein. These tools and resources display and predict, among other considerations, whether a variant [35–37]:
  - affects a functionally important part of the protein
  - affects correct processing of the protein
  - affects transport and localization of the protein
  - affects the structural integrity of the protein.
- **SIFT, PolyPHEN2, Mutation Taster, Align GVGD**: biochemical consequence of the amino acid change (in isolation and in the context of evolutionary conservation) [38–40].

Evaluation of the clinical utility of these tools has been assessed using known pathogenic missense mutations, [41] with most algorithms reliably predicting pathogenicity for known disease variants. However, such prediction tools tend to overpredict pathogenicity and are not as reliable at predicting missense variants with milder consequences. Current best practice guidelines recommend using three separate missense prediction tools for clinical variant interpretation [33].

Variants Causing Aberrant Messenger RNA Splicing

Disruption of normal splicing can result in loss of a functional gene product or, less commonly, result in the absence of large sections of the translated protein via exon skipping as typified by some mutations causal of Becker muscular dystrophy. Interpretation of variants affecting the highly conserved canonical dinucleotides either side of each exon is relatively straightforward, because they are highly likely to result in aberrant splicing and generally are clinically

reported as such. Variants affecting other sequences are more difficult to interpret. As with missense prediction, there are multiple tools to predict: (1) loss of normal splicing or (2) creation of new cryptic splice sites within intronic or exonic sequences. Tools such as MaxEntScan [42], Human Splice Finder [43], and NNsplice [44] use a number of different algorithms to predict motifs within the genome that when mutated may affect normal splicing [45,46]. Prediction of consequences of variants affecting near exon nucleotides is clinically more significant, and prediction of deep intronic variants is less accurate.

Nucleotide Conservation

In addition to conservation of amino acid sequence within proteins, there is evidence of conservation of nucleotide sequence within genes (across species) at functionally significant sites. A number of tools such as Genomic Evolutionary Rate Profiling can be used to provide a measure of the level of conservation [47].

All in silico prediction tools have their own strengths and weaknesses and so it is recommended to use multiple software predictions to collate an overall interpretation of the functional consequences of a sequence variant.

Inheritance and Mutation Type

In autosomal recessive conditions establishing the phase of two variants in a gene, ie, whether a variant occurs in *cis* (on the same chromosome) or in *trans* (on the opposite homologous chromosome), can be important for assessing pathogenicity. If one known pathogenic variant and one variant of unknown pathogenicity is detected in an individual, establishing whether the second variant is in *trans* can be considered to be supportive, or at least consistent, with pathogenicity. If the two variants are shown to be in *cis* this reduces the likelihood that the second variant is contributing to pathogenicity.

If two variants are of uncertain pathogenicity are detected, establishment of inheritance in *trans* does not necessarily provide additional evidence to support pathogenicity of either variant; conversely, inheritance in *cis* is supportive of the likelihood that at least one of the variants is not pathogenic. Most commonly used mainstream sequencing techniques and platforms, such as Illumina and Life Technologies, are not able to confidently determine phase. Therefore phase is usually established by testing parental samples.

Segregation Analysis

The segregation of a variant with a disease phenotype within a family is commonly cited as evidence to support pathogenicity. However, in reality the segregation of a variant with disease is evidence of linkage of the gene locus to the disorder but is not direct evidence of pathogenicity of the variant itself. It is possible that the variant in question is in linkage disequilibrium with the true pathogenic variant. In order to provide robust evidence from segregation analyses, it is important to factor in additional information such as age-related penetrance, variable expressivity, and phenocopy rates to statistically model the likelihood of pathogenicity. Conversely, lack of segregation of a variant in a family provides strong evidence against pathogenicity (again accounting for factors previously described).

Mutational Mechanism and Variant Spectrum

The mutational mechanism (eg, haploinsufficiency, loss of function, gain of function) and the known inheritance pattern of the disorder are important considerations in interpretation of likely pathogenicity and contribution to phenotype; for example, a single heterozygous nonsense mutation would be predicted to be pathogenic and contributing to phenotype in a dominant disease with haploinsufficiency as a mechanism, but would not be expected to contribute to phenotype to a recessive disorder in the absence of another pathogenic variant in *trans*.

Many genes have a defined spectrum of mutations that reflect the mechanism of action of the disorder; for example gain-of-function missense mutations within a restricted number of codons of the FGFR3 gene are causative of achondroplasia and thus it is highly unlikely that a nonsense mutation affecting the same codons would result in an identical phenotype [48]. Conversely protein-ablating mutations (eg, nonsense, frameshift) are the primary type of mutation in the choroideremia gene that causes X-linked recessive retinal dystrophy [49] and therefore novel missense variants in this gene may be considered to be unlikely to be pathogenic. Additionally genes vary significantly in size, the degree of variation, and mutation rate, and such factors must be considered during variant interpretation.

De Novo Mutations

New, or de novo, mutations have long been known to cause genetic diseases and in some instances, such as *DMD* mutations in DMD and *SCN1A* mutations in Dravet syndrome, they form a significant part of the mutational load [50,51]. One hypothesis is that de novo mutations also provide a mechanism by which reproductively lethal diseases, often early-onset, remain common in the population. The DDD study revealed that over 60% of mutations in this patient cohort assigned as pathogenic were de novo, reinforcing the hypothesis that they are a common cause of rare, early-onset neurodevelopmental disorders.

Although the presence of a de novo mutation in an affected offspring of two unaffected parents is compelling evidence for pathogenicity, other factors must be considered. The mutation type (eg, missense, nonsense, frameshift), disease mechanism (eg, gain of function, loss of function), and association with patient phenotype are central to interpretation of its impact. Thus the presence of a de novo mutation in a disease gene associated with the relevant phenotype is suggestive of pathogenicity, but that alone does not represent sufficient proof for causality. The contribution of de novo mutations to rare and common disease has become clearer with the use of family-based exomes and genome analysis studies [52,53]. As the number of such studies increases, the number of de novo mutations found to be responsible for sporadic cases of rare Mendelian diseases is increasing, and trio analysis for whole exome studies have been successful in identifying new disease genes; for example, the identification of the EZH2 gene underlying Weaver syndrome [54].

Functional Evidence

Experimental approaches devised to investigate the impact of a variant on gene function can provide compelling evidence for pathogenicity, particularly when the gene and disease mechanism are well elucidated; for example, effects of variants on CLCN1 ion channel function [55] and the use of minigene assays to analyze gene splicing [56]. However, it is not often possible to carry out functional assays in a clinical diagnostic environment and so such evidence should be critically assessed from literature sources.

Databases and Literature

Variant databases generally conform to one of two types: population or disease specific. Generally the presence of a variant at high incidence (>1%) in a population database is indicative that it is unlikely to be pathogenic, especially where the rate is higher than indicated by disease prevalence. In reality, this is simplistic and will not apply in all cases; for example, the most common mutations in the CFTR gene, which cause cystic fibrosis, are present in Northern European populations at a rate of approximately 3–4%.

Conversely, the presence of a variant at very low incidence in a population database but at significantly higher incidence in a disease cohort is supportive of causality.

Population databases can be very powerful tools for variant interpretation because of the large amount of data captured and collated within them. Several large international studies have contributed significantly to this knowledge base, such as the Single Nucleotide Polymorphism database [57], the Exome Aggregation Consortium (variants from 62,000 exomes) [58], the Exome Variant Server (variants from exomes from several large cohorts of individuals of African American or European ancestry) [59], and the 1000 Genomes Project (variants found during genomic and targeted sequencing of 26 populations) [60]. Most studies contributing to population databases cover cohorts of ethnically defined individuals and aim to exclude related individuals. However, all will contain individual data from patients with disease phenotypes (genetic and nongenetic), either as a result of project design or by virtue of variable expressivity and age-related disease onset/penetrance. Therefore population databases do not represent a repository of nondisease variation; rather they represent a picture of all/most variations.

Unlike population-specific databases most disease-specific databases have developed organically from research and clinical groups involved in analysis of very specific phenotypic patient cohorts. Often, but not always, the group responsible for the creation and curation are experts in the field and have a wealth of experience and knowledge to bring to variant interpretation in a restricted number of genes and phenotypes. In addition, disease-specific databases often contain literature references supporting variant interpretation. As such, disease-specific databases can be a valuable repository of information.

However, common problems exist with even the largest, most commonly used disease-specific databases, such as The Breast Cancer Information Core [61]. Issues for disease-specific databases include a lack of standardization of data quality, data assessment, and nomenclature. Data are often out of date and disease-specific databases often contain variants that are incorrectly classified. All these issues make it difficult to automatically mine and collate data for large-scale projects.

The Human Variome Project [62] has endeavored to mitigate some of the problems with disease-specific databases and establish publically available standardized curated disease-specific data repositories, often using the Leiden Open Variation Database Platform [63]. The project has met with some success, such as the establishment of the International Society for Gastrointestinal Hereditary Tumors database [64].

Literature Databases

In addition to literature-based disease-specific databases, there are several commercially available variant databases that provide a curated resource of literature related to all known monogenic genetic disorders. The advantage of such resources is that the data is up to date, maintained, and standardized, and can be easily mined. Some clinical decision support software packages display commercially available variant databases in their interface to assist in variant interpretation. HGMD [6] and CentoMD (Centogene AG, Rostock, Germany) are examples of well-known commercially available resources.

Online Mendelian Inheritance in Man

OMIM is a comprehensive authoritative and continuously updated catalog of human genes and genetic disorders and

traits that is freely available. It has a particular focus on the molecular relationship between genetic variation and phenotype and is thus considered to be a phenotypic companion to the Human Genome Project. However, whereas OMIM outlines genotype/phenotype correlation and collates a number of other fields such as gene function, cytogenetic location, genomic coordinates, gene structure, and mapping, it does not attempt to capture all variations in a gene.

Incidental Findings

Although NGS offers unrivalled potential to discover variants contributory to human disease and health, it will inevitably uncover variants within patients that though not contributory to the presenting disease phenotype, are likely to be clinically significant for the individual and their families. Such "incidental" or "secondary" findings present an ethical conundrum for the healthcare and patient communities alike.

In an attempt to quantitate the likely scale of incidental findings and the efficacy of using WGS for clinical analysis, Dewey et al. [65] performed whole genome analysis of 12 healthy volunteer adults without known history of genetic disease. The study reported two to six personal disease risk findings in each normal participant; one or more of which was categorized as disease causing in five individuals.

Although incidental findings are not new to medicine, the impact of those uncovered during genomic analysis affects not only the patient but also the family. Given the potential scale and impact of whole exome and exome sequencing in uncovering such incidental findings, some in the genetics community, including the European Society of Human Genetics have advised caution in the use of whole genome and whole exome sequencing for clinical use [66,67].

Central to the dilemma around incidental findings is not only whether and how to reveal such findings to patients but also whether to actively look for their presence in key genes. Several studies exploring patient attitudes to disclosure of incidental findings have indicated that patients hold diverse views, with many expressing "a right to know" and others "a right not to know" [68–71].

In an attempt to address these issues, the ACMG published recommendations [72] that all laboratories performing whole exome or WGS, for any clinical indication, should in addition analyze the sequence of 56 specific genes; so-called "looked for" findings. Pathogenic mutation in the 56 genes in question result in the high likelihood of severe disease that is preventable if identified before disease symptoms occur. The list includes, but is not restricted to, high penetrant genes involved in cancer and cardiomyopathy predisposition. The updated recommendations from the ACMG [31] include a clear statement that this analysis is undertaken only after appropriate consent has been obtained and the patient has a right to opt out.

In the United Kingdom the landmark 100K Genomes Project has gained ethical approval for the screening of a more restricted list of "looked for" findings in genes with potential severe impact on patient health or on the health of their offspring.

The recommendation to actively seek to analyze genes unrelated to the reason for referral has met with considerable debate, and is still under active review by professional bodies. And although an in-depth discussion on the ethical aspects of genomic screening is beyond the remit of this chapter and beyond the expertise of the authors, it is fundamental to the use of genomic screening in healthcare.

Data Sharing

As more DNA sequence and detailed clinical phenotype data are produced, the development of guidelines for data release and standardizing methods for data sharing are becoming increasingly important. A number of data sharing initiatives have been established, including GA4GH [29]. The mission of the GA4GH is to "accelerate progress in human health by helping to establish a common framework of harmonized approaches to enable effective and responsible sharing of genomic and clinical data, and by catalyzing data sharing projects that drive and demonstrate the value of data sharing". The organization has over 300 members representing the academic, healthcare, and commercial sectors. Agreement on standard file formats and the development of application program interfaces (APIs) will improve interoperability, facilitate the deposition of genotype and phenotype data into relevant databases, and build in QC mechanisms. Meta-analysis across large datasets is dependent on the input of standardized data.

Information on specific genotypes and phenotypes is already shared between many diagnostic services; however, through programs such as the DDD study and initiatives such as ClinGen [73], a more formal information exchange and alerting service is possible. This will be particularly important for establishing pathogenicity of novel DNA sequence variants in rare disease patients. A cornerstone of assigning pathogenicity to a particular variant is that it has been reported previously in a patient with an overlapping or identical phenotype. For patients with extremely rare phenotypes, it is often very difficult to locate patients, in some instances there may only be a handful of reported cases worldwide. Initiatives such as Matchmaker Exchange [74] are addressing this problem; this program involves a large and growing number of teams and projects working toward a federated platform (exchange) to facilitate the matching of cases with similar phenotypic and genotypic profiles (matchmaking) through standardized APIs and procedural conventions in line with the goals of the GA4GH.

CONCLUSION

Rapid advances in exome and genome sequencing technologies are enabling a transformational change in the way inherited rare diseases are diagnosed. Our understanding of the underlying molecular pathology of these diseases will increase rapidly over the next few years and provide a major positive impact on patient management and treatment. Tailored therapies based on genotype (precision medicine) will become standard treatments, providing increased drug efficacy. Healthcare professionals will be required to have the skills and knowledge to interpret, report, and relay genetic information to patients and fellow healthcare specialists. Developing an appropriately skilled workforce through high-quality education and training will be key to realizing the potential of genomic medicine and translation into routine practice.

REFERENCES

[1] Association for Clinical Cytogenetics. Professional guidelines for clinical cytogenetics constitutional chromosomal microarray best practice guidelines. 2011. v.2.00. [Online] Available from: http://www.acgs.uk.com/media/765587/acc_array_bp_dec2011_2.00.pdf.

[2] South ST, et al. ACMG standards and guidelines for constitutional cytogenomic microarray analysis including postnatal and prenatal applications: revision, 2013. Genet Med 2013;15(11):901–9.

[3] UK Genetic Testing Network. [Online] http://ukgtn.nhs.uk.

[4] The Genome Reference Consortium. [Online] http://www.ncbi.nlm.nih.gov/projects/genome/assembly/grc/.

[5] Online Mendelian Inheritance in Man (OMIM). [Online] http://www.omim.org.

[6] Human Gene Mutation Database (HGMD). [Online] http://www.hgmd.cf.ac.uk.

[7] Elliott PM, et al. 2014 ESC guidelines on diagnosis and management of hypertrophic cardiomyopathy. Eur Heart J 2014;35:2733–79.

[8] Lopes LR, et al. Genetic complexity in hypertrophic cardiomyopathy revealed by high throughput sequencing. J Med Genet 2013;50:228–39.

[9] Alfares AA, et al. Results of clinical genetic testing of 2912 probands with hypertrophic cardiomyopathy: expanded panels offer limited additional sensitivity. Genet Med 2015. Available from: http://dx.doi.org/10.1038/gim.2014.205.

[10] Loeys BL, et al. The revised Ghent nosology for the Marfan syndrome. J Med Genet 2010;47(7):476–85.

[11] Wooderchak-Donahue W, et al. Clinical utility of a next generation sequencing panel assay for Marfan and Marfan-like syndromes featuring aortopathy. Am J Med Genet 2015:1–11. Part A 9999A.

[12] de Ligt J, et al. Diagnostic exome sequencing in persons with severe intellectual disability. N Engl J Med 2012;367(20):1921–9.

[13] de Ligt J, et al. Detection of clinically relevant copy number variants with whole-exome sequencing. Hum Mutat 2013;34(10):1439–48.

[14] Gilissen C, et al. Genome sequencing identifies major causes of severe intellectual disability. Nature 2014;511(7509):344–7.

[15] Wright CF, et al. Genetic diagnosis of developmental disorders in the DDD study: a scalable analysis of genome-wide research data. Lancet 2015;385(9975):1305–14.

[16] The Human Phenotype Ontology. [Online] http://www.human-phenotype-ontology.org.

[17] Deciphering Developmental Disorders Study. Large-scale discovery of novel genetic causes of developmental disorders. Nature March 12, 2015;519(7542):223–8.

[18] King DA, et al. A novel method for detecting uniparental disomy from trio genotypes identifies a significant excess in children with developmental disorders. Genome Res 2014;24(4):673–87.

[19] King DA, et al. Mosaic structural variation in children with developmental disorders. Hum Mol Genet 2015;24(10):2733–45.

[20] The UK100K Genomes Project. [Online] http://www.genomicsengland.co.uk/the-100000-genomes-project/.

[21] Lelieveld SH, et al. Comparison of exome and genome sequencing technologies for the complete capture of protein coding regions. Hum Mutat 2015. http://dx.doi.org/10.1002/humu.22813.

[22] Sosnay PR, et al. Defining the disease liability of variants in the cystic fibrosis transmembrane conductance regulator gene. Nat Genet 2013;45:1160–7.

[23] Cutting GR. Cystic fibrosis genetics: from molecular understanding to clinical application. Nat Rev Genet 2015;16(1):45–56.

[24] Li J-W, et al. The SEQanswers Wiki: a Wiki database of tools for high-throughput sequencing analysis. Nucleic Acids Res 2012;40:D1313–7.

[25] Li H, et al. Fast and accurate short read alignment with Burrows-Wheeler transform. Bioinformatics 2009;25:1754–60.

[26] Langmead B, et al. Fast gapped-read alignment with Bowtie 2. Nat Methods 2012;9:357–9.

[27] McKenna A, et al. The Genome Analysis Toolkit: a MapReduce framework for analyzing next-generation DNA sequencing data. Genome Res 2010;20:1297–303.

[28] The Human Genome Variation Society. [Online] http://www.hgvs.org.

[29] The Global Alliance for Genomics and Health. [Online] http://genomicsandhealth.org.

[30] Saunders CJ, et al. Rapid whole-genome sequencing for genetic disease diagnosis in neonatal intensive care units. Sci Transl Med 2012;4(154):154ra135.

[31] Richards S, et al. Standards and guidelines for the interpretation of sequence variants: a joint consensus recommendation of the American College of Medical Genetics and Genomics and the Association for Molecular Pathology. Genet Med 2015;17(5):405–23.

[32] Claustres, et al. Recommendations for reporting results of diagnostic genetic testing (biochemical, cytogenetic and molecular genetic). Eur J Hum Genet 2014;22:160–70.

[33] Ellard S. et al., ACGS Practice guidelines for targeted next generation sequencing analysis and interpretation. [Online] Available from: http://www.acgs.uk.com/media/774807/bpg_for_targeted_next_generation_sequencing_may_2014_final.pdf.

[34] SIFT. [Online] http://sift.jcvi.org.

[35] Finn RD, et al. The PFAM protein families database. Nucl Acids Res 2014;42(Database Issue):D222–30.

[36] The UniProt Consortium. UniProt: a hub for protein information. Nucl Acids Res 2015;43:D204–12.

[37] Mitchell A, et al. The InterPro protein families database: the classification resource after 15 years. Nucl Acids Res 2015;43(D1):D213–21.

[38] Adzhubei IA, et al. A method and server for predicting damaging missense mutations. Nat Methods 2010;7(4):248–9.

[39] Schwarz JM, et al. MutationTaster2: mutation prediction for the deep-sequencing age. Nat Methods 2014;11(4):361–2.

[40] Mathe E, et al. Computational approaches for predicting the biological effect of p53 missense mutations: a comparison of three sequence analysis based methods. Nucl Acids Res 2006;34(5):1317–25.

[41] Thursberg J, et al. Performance of mutation pathogenicity prediction methods on missense variants. Hum Mutat 2011;32(4):358–68.

[42] Yeo G, et al. Maximum entropy modeling of short sequence motifs with applications to RNA splicing signals. J Comput Biol 2004;11(2–3):377–94.

[43] Desmet FO, et al. Human Splicing Finder: an online bioinformatics tool to predict splicing signals. Nucl Acids Res 2009;37(9):e67.

[44] Reese MG, et al. Improved splice site detection in Genie. J Comp Biol 1997;4(3):311–23.

[45] Hellen B. Splice site tools: a comparative analysis report. 2009. [Online] Available from: http://www.ngrl.org.uk/Manchester/sites/default/files/publications/Informatics/NGRL_Splice_Site_Tools_Analysis_2009.pdf.

[46] Jian X, et al. In silico prediction of splice-altering single nucleotide variants in the human genome. Nucl Acids Res 2014. http://dx.doi.org/10.1093/nar/gku1206.

[47] Davydov E.V. et al. Identifying a high fraction of the human genome to be under selective constraint using GERP++. 6(12)PLoS Comput Biol. 2010: e1001025.

[48] Heuertz S, et al. Novel FGFR3 mutations creating cysteine residues in the extracellular domain of the receptor cause achondroplasia or severe forms of hypochondroplasia. Eur J Hum Genet 2006;14:1240–7.

[49] Ramsden SC, et al. A clinical molecular genetic service for United Kingdom families with choroideraemia. Eur J Med Genet 2013;56(8):432–8.

[50] Lee T, et al. Differences in carrier frequency between mothers of Duchenne and Becker muscular dystrophy patients. J Hum Genet 2014;59:46–50.

[51] Claes L, et al. De novo SCN1A mutations are a major cause of severe myoclonic epilepsy of infancy. Hum Mutat 2003;21(6):615–21.

[52] Veltman JA, et al. De novo mutations in human genetic disease. Nat Rev Gen 2012;13:565–75.

[53] Ku C-S, et al. A new era in the discovery of de novo mutations underlying human genetic disease. Hum Genomics 2012;6:27.

[54] Gibson WT, et al. Mutations in EZH2 cause Weaver syndrome. Am J Hum Genet 2012;90(1):110–8.

[55] Simpson BJ, et al. Characterization of three myotonia-associated mutations of the CLCN1 chloride channel gene via heterologous expression. Hum Mutat 2004;24(2):185.

[56] Gaildrat P, et al. Use of splicing reporter minigene assay to evaluate the effect on splicing of unclassified genetic variants. Methods Mol Biol 2010;653:249–57.

[57] dbSNP. [Online] http://www.ncbi.nlm.nih.gov/snp.

[58] The Exome Aggregation Consortium (ExAC). [Online] http://exac.broadinstitute.org.

[59] The Exome Variant Server. [Online] http://evs.gs.washington.edu/EVS/.

[60] 1000 Genomes Project. [Online] http://www.1000genomes.org.

[61] Breast Cancer Information Core (BIC) Database. [Online] http://lgdfm3.ncifcrf.gov/bic/BIC.html.

[62] The Human Variome Project. [Online] http://www.humanvariomeproject.org.

[63] Leiden Open Variation Database. [Online] http://www.lovd.nl.

[64] Thompson BA, et al. Application of a 5-tiered scheme for standardized classification of 2360 unique mismatch repair gene variants in the InSiGHT locus-specific database. Nat Genet 2014;46(2):107–15.

[65] Dewey FE, et al. Clinical interpretation and implications of whole-genome sequencing. JAMA 2014;311(10):1035–45.

[66] Feero WG. Clinical application of whole-genome sequencing: proceed with care. JAMA 2014;311(10):1017–9.

[67] van El CG, et al. Whole-genome sequencing in health care: recommendations of the European Society of Human Genetics. Eur J Hum Genet 2013;21(Suppl. 1):S1–5.

[68] Wolf SM, et al. Managing incidental findings in human subjects research: analysis and recommendations. J Law Med Ethics 2008;36(2):219–48.

[69] Wolf SM, et al. Patient autonomy and incidental findings in clinical genomics. Science 2013;340(6136):1049–50.

[70] Tabor HK, et al. Pathogenic variants for Mendelian and complex traits in exomes of 6517 European and African Americans: implications for the return of incidental results. Am J Hum Genet 2014;95(2):183–93.

[71] Clift CE, et al. Patients' views on incidental findings from clinical exome sequencing. Appl Transl Genomics 2015;4:38–43.

[72] Green RC, et al. ACMG recommendations for reporting of incidental findings in clinical exome and genome sequencing. Genet Med 2013;15(7):565–74.

[73] Rehm H, et al. ClinGen: the clinical genome resource. N Engl J Med 2015;372(23):2235–42.

[74] Matchmaker Exchange. [Online] http://www.matchmakerexchange.org.

Chapter 5

Epigenetics and Epigenomics in Human Health and Disease

R. Festenstein

Imperial College, London, United Kingdom

Chapter Outline

Introduction 51
Epigenotype and Regulation of Gene Expression 51
Epigenotypes and Human Disease 53
  Trinucleotide Repeats and Human Disease 54
  Friedreich Ataxia 54
    Frataxin Gene Repression in Friedreich Ataxia 54
  Myotonic Dystrophy 55
  Fragile X Syndrome 56
  Facioscapulohumeral Dystrophy 57
Immunodeficiency, Centromeric Region Instability, and Facial Anomalies Syndrome: Genetic Mutations Causing Methylation 59
  Rett Syndrome 60
  α-Thalassemia X-Linked Mental Retardation Syndrome 61
Epigenetic Mechanisms in Cancer 62
Conclusion 63
Acknowledgment 63
References 63

INTRODUCTION

The different cellular phenotypes that compose multicellular organisms are generated by the expression of housekeeping and cell-type–specific genes and repression of inappropriate ones. The pattern of gene expression that defines a cell type is termed the *epigenotype,* which is established and maintained by "epigenetic" mechanisms able to govern gene expression regardless of the underlying genetic code [1]. Genomic imprinting, where genes are expressed from only one of the inherited parental alleles, represents a classical example of epigenetic gene regulation; memory of the expression state, presumably established during gametic meiosis, is thus transmitted to the zygote, maintained throughout embryonic and postembryonic development, and reestablished during gametogenesis of the newly formed organism in a sex-specific manner [2,3]. It follows that epigenetic "plasticity" would enable pluripotent stem cells to give rise to a variety of epigenotypes. Conversely, *induced pluripotent stem cell (iPSC)* technology allows one to artificially stimulate epigenetic reprogramming of differentiated adult cells and hence lead to their pluripotency [4–6].

EPIGENOTYPE AND REGULATION OF GENE EXPRESSION

Cells can acquire an epigenotype by modulating the availability of *trans*-acting factors to regulatory *cis*-acting genetic elements that specify gene activity or inactivity. Such availability can be controlled by the manner in which DNA is packaged as chromatin inside the nucleus. Silent genes may thus be packaged in "condensed" chromatin such as heterochromatin. Conversely, active genes may be packaged in "open" chromatin, termed *euchromatin.*

Chromatin is formed when approximately 147 base pairs of DNA associate with the histone octamer (two histone H3-H4 dimers plus two histone H2A-H2B dimers) to form the nucleosome [7]. Linker histones of the H1 class associate with the DNA between single nucleosomes, facilitating a higher level of organization, the so-called "solenoid" helical fibers (30-nm fibers) [8,9]. Although a detailed image of in vivo chromatin fibers has not been fully elucidated yet, it is generally accepted that higher order folding (condensation) and unfolding (decondensation) are conceived as functionally relevant chromatin states [10].

Centromeric/pericentromeric heterochromatic regions are gathered together in the mammalian interphase nucleus as distinct domains known as *centromeric clusters* [11]. These clusters have characteristic biochemical features that help define heterochromatin: abundant repetitive DNA (satellite DNA) [12], replication in mid to late S phase [13], histone hypoacetylation, [14] and specific methylation at lysine 9 of histone H3, the latter placed primarily by the histone methyltransferase *Suv39h* [15]. These modifications, together with an RNA component [16,17] and possibly the

Medical and Health Genomics. http://dx.doi.org/10.1016/B978-0-12-420196-5.00005-8

RNA interference machinery [18–20], maintain the presence of heterochromatin protein 1 (*HP1*), a component of constitutive heterochromatin [21].

Previously regarded as a static, condensed structure, heterochromatin is now known to be a highly dynamic structure. Fluorescence recovery after photobleaching experiments using green fluorescent protein-tagged *HP1* (HP1-GFP) revealed that *HP1* is highly mobile as both heterochromatin and euchromatin [22–24]. The dynamic exchange of *HP1* indicates the "windows of opportunity" for the binding of additional factors and suggests that gene regulation in heterochromatin results from a dynamic competition between regulatory factors [25].

Processes exist to allow changes in accessibility at both the chromatin fiber and the core DNA. Covalent modifications in DNA (cytosine methylation or hydroxymethylation) and histones (lysine acetylation, methylation or ubiquitination, serine and threonine phosphorylation, and arginine methylation) modify the interaction between histones, DNA, and chromatin binding factors, [26] whereas nucleosome remodeling factors modify core histone and DNA accessibility [27]. It is very likely that these biochemical processes are regulated by an "epigenetic code" written as modifications in DNA and histones, and "read" by factors that specifically recognize single or combined modifications [28–30]. For example, histone acetylation, recognized by proteins with bromodomains [31], as well as histone H3 lysine 4 (*H3K4*) methylation, recognized by *WDR5*-containing *H3K4* methyltransferase complexes [32] are generally associated with open chromatin and gene expression [14,33,34]. On the other hand, histone H3 lysine 9 (*H3K9*) or 27 (*H3K27*) methylation, recognized by the chromodomains of *HP1* and polycomb proteins, respectively [35,36], and DNA methylation, recognized by methyl binding domain (MBD) proteins [37], associate with condensed chromatin and gene repression [38–40]. Recently, it has also been uncovered that DNA can be hydroxymethylated preferentially at cytosines of CpG residues via 10-11 translocation enzymes [41]. The exact function of this mark has not been resolved yet; however, it is thought to be an intermediary step between unmethylated CpGs and methylated ones [42]. In line with this, 5-hydroxymethylcytosine residues were associated mainly with transcriptionally active euchromatin and polycomb silencing of formerly euchromatic promoters, where it possibly allows a rapid release of the repressed state in poised genes [42,43].

Methylated DNA, through MBD proteins, may target histone deacetylases (*HDACs*) [44,45] and *H3K9* methyltransferases (eg, *Suv39h*) to specific loci [46]. Conversely, histone *H3K9* methylation may bring about DNA methylation (by association with DNA de novo methyltransferases) [47] as well as *HP1* [48,49]. Therefore both DNA and histone modification machineries may cross-talk to generate condensed chromatin (Fig. 5.1A and B). DNA methylation and *H3K9* methylation may, however, function independently [49,50] and thus confer different degrees of epigenetic plasticity. DNA methylation, thought to be more stable throughout meiosis and mitosis, may provide long-term (repressive) epigenetic memory, whereas histone modifications leave a more labile epigenetic mark.

Many aspects of chromatin explained so far have been evaluated using the chromatin immunoprecipitation (ChIP) assay, which involves the immunoprecipitation of protein-bound DNA using specific antibodies followed by polymerase chain reaction (PCR) [51,52]. Until recently, results were only obtained in limited contexts. Importantly, though, development of modern genomics techniques allowed high-throughput sequencing of the genomic ChIP-DNA and thereby global examination of chromatin [53]. In line with this, the human Encyclopedia of DNA Elements (ENCODE) project aimed to reveal a detailed picture (eg, histone modifications, transcription factor binding) of how a human genome functions [54–59]. In addition, genome-wide analyses on identical twins have furthered our understanding of epigenetic differences [60,61].

Where genes reside in the nucleus also provides a mechanism for epigenetic gene regulation. Genes may dynamically relocate to associate with nuclear structures rich in transcription factors required for expression [62], or be developmentally inactivated by their relocation to heterochromatic pericentromeric clusters [63]. In the phenomenon of position effect variegation (PEV), genes are repressed in a proportion of the cells when abnormally juxtaposed to heterochromatin [64–67]. That gene expression is affected by placement of genes near heterochromatin has focused attention on the associations between genes and their native *cis*-acting regulatory elements. DNA sequences, such as locus control regions [68] or boundary elements/insulators [69], are thought to regulate gene expression by establishing "permissive" chromatin domains [70] or by facilitating nuclear relocation and/or regulating interactions between gene promoters and enhancer or silencer elements [69]. Imprint control regions, typically found as DNA differentially methylated regions in imprinted gene clusters, are conceived as sites from which chromatin structures are propagated bidirectionally to control the expression of genes within imprinted domains [71]. Notably, recent advances now allow scientists to study the three-dimensional (3D) organization of the genome via chromosome-conformation-capture (3C) approaches, which are based on the restriction enzyme digestion of cross-linked DNA and its subsequent ligation followed by PCR [72]. Within the human ENCODE project, a global long-range interaction map of gene promoters

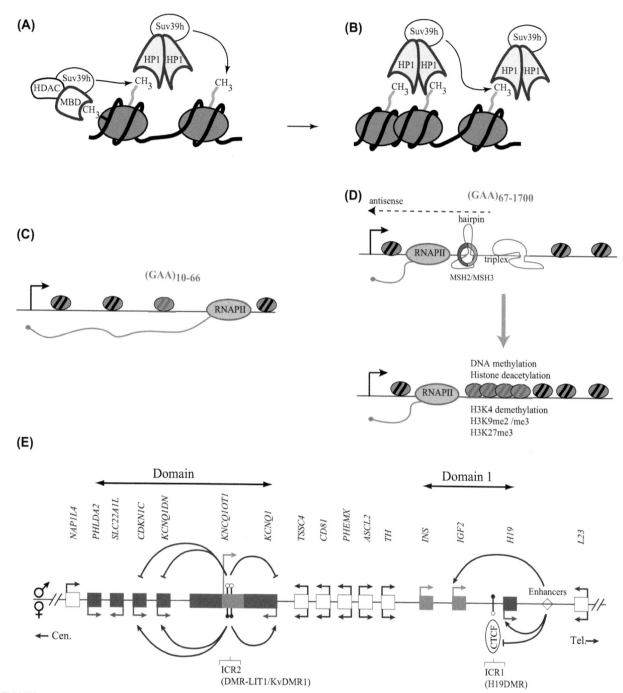

FIGURE 5.1 (A) Gene expression regulation of Sav39a-wild copy. (B) Silencing of Sav39a by methylation. (C) Full frataxin (FXN) expression loose chromatin. Wild-type. (D) Silencing of FXN expression condensed chromatin, Friedrich ataxia (FRDA). (E) Epigenetic multi-gene expression regulation in Beckwith-Weidemann syndrome. *Cen.,* centromere; *Tel.,* telomore.

was revealed thanks to the fusion of 3C technique with high-throughput sequencing [73].

EPIGENOTYPES AND HUMAN DISEASE

Disease states with an epigenetic basis are classified into those where changes in chromatin structure at the deregulated gene(s) result from mutations in DNA sequences in the same chromosome (ie, in *cis*) and those where genetic mutations affect the genes that encode for factors that establish or maintain chromatin structures (ie, in *trans*). Table 11.4 (see Chapter 11) illustrates disease states with a confirmed or possible epigenetic basis, some of which are discussed later in this chapter.

Importantly, epigenetic mechanisms appear to play a key role in the development of numerous different types of cancer, and given their potential reversible nature, offer exciting therapeutic targets. It has been also suggested that epigenetic mechanisms might differ between genders based on sex chromosome complements and the male determining gene "*SRY*" [74,75]. These may explain different predisposition rates toward disease among different genders.

Trinucleotide Repeats and Human Disease

Several human diseases are associated with expansion of untranslated trinucleotide repeats and their molecular pathogenesis may be mediated by effects on chromatin packaging of nearby genes [76]. These include Friedreich ataxia (FRDA), myotonic dystrophy, and fragile X syndrome (FXS). Deletion of repetitive DNA may deregulate nearby genes via an epigenetic effect in facioscapulohumeral dystrophy. Nearby genes may also be deregulated by genetic mutations in imprint centers that control the expression of genes within imprinted gene domains.

Friedreich Ataxia

FRDA was described by Nicholaus Friedreich in 1863. It is the most common of the hereditary ataxias with a prevalence of one in 50,000. FRDA is an autosomal recessive disease with age of onset usually before 25 years. Progressive ataxia, cardiomyopathy, and associated diabetes are the core features [77]. The most common genetic abnormality is a homozygous expanded GAA trinucleotide repeat in the first intron of the frataxin *(FXN)* gene located on chromosome 9q [78]. How this expansion is exactly triggered remains elusive yet; however, mechanisms in relation to DNA repair, replication, or transcription were suggested to be important in this phenomenon [79]. The *FXN* gene encodes the mitochondrial protein frataxin, which is involved in the biogenesis of iron–sulphur clusters and is therefore vital for iron homeostasis [80]. The correlation of the severity of certain clinical features and age of onset with the shorter of the two expanded repeats [81–85] may be explained by frataxin expression being inversely proportional to the length of the expanded GAA repeat, which is particularly true for smaller expansions [86]. Therefore residual expression from the allele with the shorter expansion may be important in modulating disease severity. Interestingly, atypical patients have GAA repeats of similar length to patients with more classical features [81,87,88]. Factors such as environment, modifier genes, and somatic mosaicism may play a role in such phenotypic variation.

Frataxin Gene Repression in Friedreich Ataxia

Abnormally expanded GAA repeats within the *FXN* gene impair frataxin expression. This impairment could be a direct result of physical blockage effects caused by unusual

conformations of DNA adapted by GAA triplets on the elongation of transcription (see Fig. 5.1C and D). The most common such structure is hairpin DNA resulting by unusual hydrogen bonding between G·G, G·A and A·A, RRY (purine:purine:pyrimidine) triplexes, which is also known as "Sticky" DNA [89]. Consistent with the idea of transcriptional blockage effects, in vitro transcription assays based on ribonuclease protection and northern blots revealed a transcriptional elongation defect in the presence of expanded GAA repeats [90–96]. Moreover, ChIP assays performed on lymphoblastoid cell lines using antibodies against initiating and elongating forms of *RNAPII*, as well as histone marks associated with transcriptional elongation (ie, *H3K36* and *H3K79* methylation) also underlined a deficit in transcriptional elongation on the *FXN* gene [97–99]. Whether expanded GAA triplets affect the initiation of *RNAPII* at the *FXN* gene promoter, however, remains elusive.

Importantly, expanded GAA repeats are also associated with the heterochromatinization of the *FXN* gene. The first line of evidence regarding this phenomenon came from an experiment in transgenic mice. Here, A (GAA)$_{200}$ repeat expansion was linked to a human *CD2 (hCD2)* reporter. The direct inhibitory transcriptional effect of GAA repeats on DNA transcription was excluded as the GAA repeat was linked to the 3′ untranscribed region of the *hCD2* transgene. In this transgenic mouse model, the *hCD2* reporter gene alone is sensitive to juxtaposition to constitutively tightly packaged DNA (heterochromatin; eg, centromeres), and results in variegated hCD2 expression on T cells, or PEV [66]. In PEV, rather than gene expression being silenced in all cells, a proportion of cells become silenced with the remaining continuing to express. Linking a GAA repeat expansion to the *hCD2* transgene also resulted in PEV [100] but, importantly, this occurred even when the transgene was situated in regions of the chromosome that are usually loosely packaged in euchromatin, suggesting that GAA repeats induced chromatin condensation and heterochromatin formation. In T cells where *hCD2* was silenced, deoxyribonuclease I hypersensitive site analysis showed condensation of chromatin packaging at the promoter of the gene. This silencing was also modified by altering the dosage of *HP1β*, an important component of heterochromatin [100].

In recent years, it has been shown that a similar heterochromatinization also takes place at the pathologically silenced *FXN* gene (see Fig. 5.1C and D). DNA methylation levels in lymphoblastoid cells, peripheral blood mononuclear cells (PBMCs), [101–103] and nerve tissues in FRDA mouse models [104] were found to be higher on the *FXN* gene in the presence of expanded GAA repeats. This result was also accompanied by an increase in heterochromatic histone marks (ie, *H3K9* dimethylation and trimethylation and *H3K27* trimethylation) and an overall decrease in histone acetylation in the flanking regions of expanded GAA repeats in patient-derived lymphoblastoid cells/PBMCs

[97–99,103,105], fibroblasts [106], and FRDA mouse models [104]. Indeed the pattern of heterochromatic marks on pathologically silenced alleles supports the hypothesis that heterochromatin is spreading bidirectionally from the GAA repeats causing the silencing of the gene [89].

How exactly heterochromatin is triggered in FRDA is an important question that still remains to be answered. Interestingly, unusual DNA conformations adopted by expanded GAA repeats were shown to be recognized by the cell's mismatch repair mechanism. *MSH2/MSH3* dimers were shown to be attracted by expanded GAA repeats (see Fig. 5.1D) in various FRDA models including iPSCs derived from patients ([107–111]). Some studies suggest heterochromatinization as a protective response against faulty transcription, which may be caused by DNA damage [112–117]. Moreover, De Biase et al. (2009) reported increased antisense transcription in pathological *FXN* alleles and suggest this as a potential trigger for the heterochromatinization. However, it still remains unclear whether antisense transcription may lead to heterochromatic silencing in mammals [118]. De Biase et al. (2009) also described a *CTCF* binding site near the promoter of *FXN*. *CTCF* is a chromatin insulator protein associated with enhancer blocking or chromatin insulation activities [119,120]. The effect of *CTCF* on *FXN* has not been clearly characterized yet; however, the study of De Biase et al. reports a depletion of *CTCF* binding on silenced alleles. Interestingly, knockdown of *CTCF* in healthy and patient fibroblasts resulted in increased antisense transcription in the promoter of the gene. This may imply that *CTCF* may have an inhibitory function against the spreading of heterochromatin nucleated by GAA repeats.

One of the challenges in the FRDA field is limited access to primary nerve tissue, which is predominantly affected in this disease. Most of the studies so far presented results obtained from Epstein-Barr virus–transformed lymphoblastoid cells derived from patients. Other sources of research were fibroblasts and primary PBMCs, as well as mouse models, which only exhibit a mild disease phenotype [121]. Notably, the arrival of iPSC technology has allowed scientists to differentiate neuronlike cells using fibroblasts obtained from FRDA patients [111,122].

A lot has been resolved about the pathological silencing of the *FXN* gene since 2005. Importantly, uncovering different aspects of heterochromatin brought up the possibility of treating FRDA with *HDAC* inhibitors, which have the potential to reduce histone deacetylation and thereby a subsequent methylation. Indeed, synthetically derived *HDAC* inhibitor BML-210 and its derivatives were shown to upregulate *FXN* expression significantly in FRDA cells [105,123–127]. Moreover, a recent study from our group showed that an *HDAC* class III (Sirtuin) inhibitor nicotinamide also upregulates *FXN* in vitro, ex vivo, and in vivo (mouse model) (Personal communication). Clinical trials for nicotinamide and BML-210 derivatives have recently

provided proof of concept that *HDAC* inhibition can partially restore expression in FRDA patients [128,129]. Undoubtedly, identification of specific epigenetic modifiers responsible for the silencing of *FXN* will help scientists further develop such radical therapeutic strategies, which specifically address the primary cause of this currently incurable disease rather than its symptoms.

Myotonic Dystrophy

Myotonic dystrophy is an autosomal dominant disease that is the most common adult form of muscular dystrophy. It is a multisystem neuromuscular disorder characterized by clinical manifestations including myotonia (skeletal muscle hyperexcitability), progressive muscular dystrophy, cataracts, cardiac conduction defects, cognitive deficits, and endocrine anomalies.

Two forms of the disease with similar features are caused by different microsatellite expansions in two different gene loci. Myotonic dystrophy type 1 (DM1) is caused by a CTG repeat expansion located in the 3′ untranslated region (UTR) of the *DMPK* gene on chromosome 19q, [130–132] whereas myotonic dystrophy type 2 (DM2) is caused by expansion of a CCTG repeat in the intron one of the zinc finger 9 *(ZNF9)* gene on chromosome 3q21 [133]. Expanded CTG or CCTG repeats are highly unstable in both germline and somatic tissues [133,134] and the length of the repeats are correlated with the severity of symptoms and the earlier disease onset [135,136].

Although DM1 is now thought to be largely mediated by an "RNA gain of function mechanism" in which CUG repeats containing RNA transcripts cause sequestration of muscle blind protein ligand *(MBNL)* and increase *CUGBP/* Elav-like family member 1 *(CELF1)* protein activities [137–140], creating a spliceopathy, the CTG expansion has also been found to affect chromatin packaging of DNA. For example, in fibroblasts from myotonic dystrophy patients, a CTG expansion in the *DMPK* gene is associated with condensation of chromatin as indicated by nuclease resistance at the *SIX5* enhancer present in the 3′ region of the *DMPK* gene, rendering it inaccessible to transcription factors and causing downregulation of SIX5 expression [141,142]. CTG repeats also efficiently recruit nucleosomes, the basic structural element of chromatin [143,144]. CTG expansions also behave in a similar fashion to pericentromeric heterochromatin, causing gene silencing and chromatin condensation in the *hCD2* transgenic mouse model [100]. Some features of myotonic dystrophy (particularly cataract formation as seen in *six5* knockout mice [145]) may be secondary to *deregulation* of the *six5* gene located near the CTG repeat [145–147]; this deregulation may be secondary to CTG repeat–induced chromatin condensation. Evidence supporting this notion has emerged as loss of *CTCF* binding to the regions flanking the CTG repeat expansion has been shown in the DM1 locus, which might be associated with DNA hypermethylation of these regions [148,149]. Spreading of heterochromatin at the CTG-expanded allele was indicated by the enrichment of histone *H3K9* methylation and HP1γ recruitment where antisense transcription of *DMPK* was activated. This might lead to a wider dysregulation of the messenger RNA (mRNA) and protein amount in DM1-affected cells [148]. The exact mechanisms of CTG repeat-mediated chromatin remodeling are still uncertain; however, the repeat itself strongly stimulates nucleosome formation [144]. In addition, it has been suggested that DNA binding proteins that recognize DNA triplet repeats [150–152] contribute to the epigenetic changes seen at the *DMPK* locus.

Other diseases where DNA repeats are found in UTRs, like spinocerebellar ataxia type 10 [153] and DM2 [133], may share similar so-called "epigenetic" molecular pathogenic mechanisms.

Fragile X Syndrome

FXS (OMIM #300624) is the most common inherited form of mental retardation [154] and is one of the best characterized forms of autism spectrum disorder [155]. It is an X-linked dominant disorder characterized by variable penetrance. The name *fragile X* is derived from the observation that the cytogenetic band Xq27.3, where the causative fragile X mental retardation 1 (*FMR1*) gene resides, is a fragile site in individuals carrying the full mutation (FM) [156].

Because *FMR1* is located on the X chromosome, the degree of cognitive disability is more severe in males, who possess only one X chromosome. Females (with two X chromosomes) manifest a less severe phenotype than males, which is correlated to the extent of X inactivation on the abnormal chromosome [157]. FXS has an estimated prevalence of 1/5000 in males and 1/10,000 in females [158].

The main disease manifestations are moderate to severe intellectual disability, autistic features, seizures, hypersensitivity to sensory stimuli, attention deficit, hyperactivity, motor incoordination, growth abnormalities, sleep disturbances, connective tissue dysplasia, craniofacial abnormalities, and macroorchidism [159].

The syndrome is a trinucleotide repeat disorder caused by the expansion of the triplet CGG in the 5′ UTR of the *FMR1* gene [160]. A CGG expansion greater than 200 units results in hypermethylation at CpG sites at the *FMR1* promoter region; this is responsible for the silencing of *FMR1* and the subsequent loss of the protein it codes for, the fragile X mental retardation protein *(FMRP)* [156]. *FMRP* is an RNA binding protein that is able to bind to several mRNAs, including its own [161], and is believed to be involved in the transportation of these target mRNAs throughout neuronal dendrites and in the inhibition of their translation upon stimulation of the metabotropic glutamate receptor 5

at the synapse [162]. It is ubiquitously expressed until day 14 of embryonic development, after which its expression becomes restricted to the brain (specifically in neurons) and the gonads [163]. Loss of *FMRP* is thought to be critical owing to the important role played in neuronal function. Abnormal dendritic spines were observed in both FXS patients and in *FMR1* knockout mice, supporting *FMRP* involvement in synaptic maturation [164]. More generally, the absence of *FMRP* appears to lead to a global increase in brain protein synthesis [165] and several studies in human patients are testing whether this observation could be a biomarker of the disease.

If cells from FXS patients are treated with a DNA methylation inhibitor [5-aza-2-deoxycitidine (5-aza DC)], the level of CpG methylation decreases and FMR1 expression is reactivated [166–170], suggesting that DNA methylation is the major factor causing *FMR1* silencing, rather than the triplet expansion itself. DNA methylation has been to shown to also cause local histone deacetylation creating another mechanism of transcriptional silencing [171]. Treatment of the same lymphoblastoid fragile X cells with trichostatin A (an *HDAC* inhibitor) also resulted in moderate transcriptional activation of *FMR1*, suggesting that in FXS, histones are deacetylated at the inactive promoter. Further, treatment of fragile X cells with *HDAC* inhibitors and 5-aza DC synergistically activated transcription [172].

The exact mechanisms behind CGG expansion and the consequent alteration in *FMR1* transcription and translation are still not fully understood. The generation of stable cell lines harboring the *FMR1* 5′ UTR with varying CGG repeat lengths targeted to the correct gene locus have proven to be a useful model for studying FXS. The promoter with variable (CGG)$_n$ length has been fused to the coding sequence of a reporter gene. The construct has shown that a full-mutation CGG repeat length inhibits reporter gene expression, whereas a premutation CGG repeat did not affect reporter gene expression. Therefore this model could be a better tool to elucidate the molecular mechanisms of *FMR1* deregulation in FXS [173].

Transcriptional silencing of the *FMR1* gene as a result of hypermethylation of CpG islands and the consequent loss of *FMRP* expression is still considered to be the major cause of the disease [156]. However, a mouse model of FXS with mice carrying long CGG repeats of nearly 230 units show high levels of the *FMR1* mRNA, although low levels of the *FMRP* protein, arguing against a purely transcriptional deficit. Promoters in these mice do not show the abnormal methylation previously described, which suggests that modeling FXS in mice requires more genetic manipulation, and that perhaps the threshold number of repeats in mice might be higher than the level observed in human subjects [174].

In addition, within humans there are cases of males expressing *FMR1* to some extent despite carrying the FM. Mosaicism of the gene promoter methylation pattern is thought to underlie this, allowing some transcription to occur. The presence of three types of mosaicism was tested in cells derived from male expressing patients, and the data suggests that intercell mosaicism in DNA methylation patterns might explain the presence of *FMR1* mRNA in some FXS-affected individuals [175].

The CGG repeat region is unstable and repeat length can vary in unaffected individuals from 6 to 55 repeats. The instability of the region can result in an expansion of this region upon maternal transmission to the next generation. A $(CGG)_n$ range between 55 and 200 is referred to as the *premutation (PM)* [176]. It was previously thought that carriers of the PM simply had a higher risk of developing FXS upon transmission of the PM allele to the next generation. However, it has become evident that certain carriers show autistic features or anxiety disorders [177–179]. Nearly 20% of female PM carriers manifest premature ovarian failure (cessation of menstruation at or before 40 years of age) [180]. PM carriers are also at risk of developing a progressive neurodegenerative disorder called *fragile X–associated tremor/ataxia syndrome (FXTAS)* [181]. The PM expansion results in a two–eight-fold upregulation of *FMR1* mRNA, but the *FMRP* translation is inhibited [182–184]. It is believed that a toxic RNA gain of function of the CGG-expanded *FMR1* mRNA is responsible for FXTAS [154,185,186].

Recently the use of human embryonic stem cells derived from genetically diagnosed preimplantation embryos has allowed scientists to look at the temporal order of the events that lead to the *FMR1* silencing. This system has shown that transcriptional downregulation of *FMR1* and some other chromatin modifications occur before DNA methylation and that they might be responsible for the initiation of *FMR1* inactivation. In fact, undifferentiated human embryonic stem cells express FMR1 and their DNA is unmethylated despite the presence of an FM. But DNA methylation and other histone modifications occur upon differentiation of these cells, establishing epigenetic silencing of *FMR1* and leading to intellectual impairment [174].

More recently, human-induced pluripotent stem cells from FXS patients and from healthy controls were used as an FXS model. This model looked at the *FMRP* protein in early neurodevelopment before synaptogenesis and from it, a direct link has been identified between the epigenetic modifications of the *FMR1* gene, the consequent loss of *FMRP* expression, and aberrant neuronal differentiation [187].

In addition to methylation, hydroxymethylation increasingly is thought to play a role in FXS pathophysiology. 5-Hydroxymethylcytosine (5-hmC) is a modified form of cytosine thought to play an important role in neurodevelopment. For example, genome-wide DNA hydroxymethylation studies have revealed a positive correlation between 5-hmC levels and cerebellum development [188]. This modified cytosine is enriched in exons and UTRs of protein coding genes, but it is absent on introns and intergenic regions. Different regions of DNA also show different hydroxymethylation patterns depending on the stage of development. Several mRNA targets of *FMRP* are differentially enriched in hydroxymethylated regions during development and these hydroxymethylation patterns are disrupted in several autism genes. Consequently the disruption of 5-hmC–mediated epigenetic regulation might contribute to the pathogenesis of FXS [188].

Facioscapulohumeral Dystrophy

Facioscapulohumeral dystrophy (FSHD) is the third most common inherited muscular dystrophy, inherited in an autosomal dominant fashion and affecting approximately one in 20,000 individuals worldwide [189]. The condition usually manifests clinically by the second decade, beginning with progressive weakness of the facial, scapular, and humeral muscles, and later involving the abdominal musculature and lower limbs [189]. It is primarily a disease of skeletal muscle, but retinal telangiectasia, sensorineural deafness, cardiac arrhythmias, and mental retardation may also be complications [190,191]. Its molecular genetic pathogenesis is probably mediated by epigenetic mechanisms.

Contraction of a tandem array of 3.3 kilobase (kb) *D4Z4* repeats lying in the subtelomeric region of chromosome 4q is associated with FSHD [192,193]. Ninety-five percent of FSHD cases are caused by this deletion, and contraction to a threshold of less than 11 repeats results in disease [193,194]. There are problems in correlating genotype and phenotype in FSHD, owing to the large variability in symptom severity, even in patients from the same affected family [195]; however, generally the shorter the remaining *D4Z4* tandem array, the earlier the onset and more severe the disease [194,196]. Each *D4Z4* repeat unit of the array contains several GC-rich sequences and also an open reading frame (ORF) of a double homeobox transcription factor designated *DUX4* [197]. Several sequences similar to this *D4Z4* array are found throughout the human genome, with the array at 10q26 almost completely identical to the one at 4q35 [198]. Large contractions of this other *D4Z4* array, found in 10% of the normal population, are not associated with FSHD, confirming 4q35 as the causative genome region [193]. Importantly, haploinsufficiency of the entire chromosome 4q subtelomeric region, including the *D4Z4* tandem repeats and nearby genes, does not cause FSHD, [199] suggesting a gain of function underlying the pathogenesis of this disease.

The genetic basis of FSHD is further complicated by the fact that additional 4q variants have been described, for example 4qA, 4qB, and 4qC. Of these, FSHD appears to be only associated with the 4qA variant [200–202]. A particular simple sequence–length polymorphic site proximal to the 4q array is found in three haplotypes, which, if a large

D4Z4 repeat contraction is also present, seems to be specifically associated with the FSHD phenotype [203,204]. It therefore appears that the *D4Z4* repeat contraction will only cause disease when found on a certain genetic background.

The way in which this repeat contraction causes disease is probably epigenetically mediated. It was hypothesized initially that in healthy patients, *D4Z4* repeat tracts are heterochromatinized and that variable spreading of heterochromatin silences nearby genes. Loss of these repeats may produce a more open chromatin configuration, resulting in inappropriate derepression of these nearby genes [205]. However, early ChIP analysis looking at H4 acetylation of the FSHD locus in human lymphoid cells as well as human–rodent somatic cell hybrids suggested that regions close to the *D4Z4* repeats and of two nearby genes (*FRG1* and *ANT1*) showed histones acetylated more in the pattern of euchromatin than heterochromatin [206]. Subsequent ChIP analysis has looked at other histone modifications, both repressive such as trimethylation at lysines 9 and 27 on histone 3 (*H3K9me3* and *H3K27me3*) as well as markers of transcriptional activation such as dimethylation at lysine 4 on histone 3 (*H3K4me2*). This has in fact shown that in healthy controls, *D4Z4* arrays display both heterochromatic as well as euchromatic regions, whereas in FSHD patients, there are significantly reduced levels of *H3K9me3* with unaltered levels of *H3K27me3* and *H3K4me2* [207], indicating a relative reduction in repressive modifications, a more open chromatin configuration as a consequence, and perhaps gene upregulation.

Along with histone modifications, DNA methylation patterns also reflect chromatin configuration. Heterochromatic regions are usually hypermethylated and normal *D4Z4* arrays reflect this. Contracted *D4Z4* arrays, however, display hypomethylation [208], and, rather akin to phenotype, the level of this hypomethylation correlates with repeat length: the shorter the array, the lower the level of methylation [209]. However, hypomethylation is also seen in some unusual asymptomatic individuals who carry both the contracted *D4Z4* array as well as a permissive haplotype, perhaps suggesting that hypomethylation is not sufficient for disease onset [208]. In addition, more profound levels of hypomethylation are seen at *D4Z4* repeats in the immunodeficiency, centromeric region instability, and facial anomalies (ICF) syndrome [208], which is phenotypically entirely different to FSHD. In the same syndrome, *H3K9me3* enrichment is normal at *D4Z4* [207], suggesting that it is the loss of *H3K9me3* that is crucial in FSHD, rather than any change in DNA methylation.

The previous discussion indicates transcriptional upregulation is key to FSHD pathogenesis, and given the clearly demonstrated association with the chromosome 4q35 region, a number of candidate genes found in that area have been investigated as potentially causative. These include *FRG1*, *FRG2*, *ANT1*, and, more recently, *DUX4c* and *DUX4*.

FRG1 in particular seemed a promising candidate, because functional studies have suggested a role in muscle development [210]. Early work suggested that *FRG1*, *FRG2*, and *ANT1* seemed to be upregulated in muscle cells from FSHD patients in a manner specific to FSHD (not found in other hereditary myopathies) and also to muscle (not replicated in patient lymphocytes) [211]. However, subsequent studies failed to fully corroborate these findings, being unable to replicate any increase in *FRG1* or *ANT1* expression [206,212–214] and showed that *FRG2* was only upregulated in FSHD myoblasts and not mature myocytes [215]. In fact, studies in certain FSHD families have largely ruled out these genes as causative for the disease. In one family, the disease-associated allele showed a large deletion in the *D4Z4* repeat array including both the *DUX4c* and *FRG2* genes, arguing against a role for them in causing disease [216]. Similarly, in another family, the pathological allele was actually found on chromosome 10q26, where the short repeat array had been extended by a *D4Z4* fragment translocated from 4q35. This translocated fragment included part of the distal *D4Z4* fragment, allowing stable *DUX4* mRNA expression, but did not include *FRG1*, *ANT1*, or *DUX4c*, again arguing against a role for these genes [203] and suggesting that *DUX4* is the key gene involved in FSHD.

Early work looking at *DUX4* was greatly hampered by the difficulty in detecting specific mRNA transcripts. This, coupled with a lack of an identifiable polyadenylation site led to the theory that *DUX4* was actually a nonfunctional pseudogene [217]. The major breakthrough arose with the identification of a mature mRNA transcript containing the *DUX4* ORF, with reverse transcription PCR [218]. This transcript was shown to originate from the distal *D4Z4* unit and extend to an adjacent region conferring the polyadenylated tail, termed *pLAM1* [218]. There is evidence that DUX4 expression at high levels in muscle has numerous effects, including inhibition of differentiation [219] and induction of genes involved in muscle atrophy, apoptosis, and cell death, implying that *DUX4* overexpression could be responsible for FSHD [220,221]. However, *DUX4* mRNA abundance is very low; it was not always detectable in FSHD muscle biopsies and only around 1/1000 FSHD myoblasts express *DUX4* in culture [222]. The same authors have shown that the low abundance in muscle actually reflects a small number of nuclei expressing abundant amounts of *DUX4* and that *DUX4* is highly expressed in human testis and germline cells [222]. This has led to a developmental model being proposed for FSHD, whereby in normal individuals, *DUX4* is expressed in early development and is heterochromatically silenced in mature tissues. In FSHD, there is a defect in this silencing mechanism, which leads to occasional escape from repression in muscle cells, with consequent DUX4 expression and cell death [222,223].

There have been great advances in our understanding of the epigenetic mechanisms underlying FSHD recently.

Outstanding questions concern the exact function of *DUX4* and how expression in so few cells in adult tissue can lead to a progressive myopathy.

Immunodeficiency, Centromeric Region Instability, and Facial Anomalies Syndrome: Genetic Mutations Causing Methylation

ICF syndrome is an extremely rare autosomal recessive disorder in which patients display immunodeficiency, centromere instability (association, breakage, and stretching of the pericentromeric heterochromatin of chromosomes 1, 9, and 16), and facial anomalies. Mental retardation and developmental delay are also observed [224]. The disease maps to chromosome 20 and the gene responsible is the DNA de novo methyltransferase *DNMT3B* [225]. Mutations in both the catalytic domain responsible for methyltransferase activity and the aminoterminal end of the protein, which is likely to be responsible for its targeting to pericentromeric sequences [226,227], are present in ICF syndrome patients [228,229]. Even though deletion of the *Dnmt3b* catalytic domain in mice results in perinatal lethality [230], missense mutations in this domain in ICF syndrome patients probably impair rather than completely abolish enzymatic activity [228]. Indeed, the majority of ICF syndrome patients have missense mutations in the catalytic domain of *DNMT3B* and no patients have been found to be homozygous for nonsense alleles [231]. This has been further confirmed in a mouse model of the disease where mutant alleles carrying missense mutations as in ICF syndrome were constructed. Unlike total knockout, which is embryonically lethal, these mice are alive at birth and display higher methylation levels suggestive of a residual level of *DNMT3B* activity as hypothesized in ICF syndrome patients [232].

Around 40% of ICF syndrome patients do not have mutations in *DNMT3B,* yet have exactly the same clinical phenotype; this group is denoted *ICF syndrome type 2* [233,234]. Early hypotheses for this subgroup of patients suggested alternative catalytically inactive splice variants of *DNMT3B* might be overexpressed, a phenomenon noticed in human hepatocarcinogenesis with overexpression of the splice variant *DNMT3B4* [235]. However, the *DNMT3B3* isoform remains the most abundant isoform in ICF syndrome type 2 patients, arguing against this alternative splicing mechanism [233]. Recent work has attempted to find another candidate gene that might be responsible for ICF syndrome type 2. Homozygosity mapping and whole exome sequencing of patients with ICF syndrome type 2 has identified a mutation in exon 3 of *ZBTB24* [236], and another mutation in the same gene has been found in another pedigree [237]. This gene is thought to be involved in B-cell differentiation [236], and it is mutations in this gene that are thought to be responsible for ICF syndrome type 2. This gene also appears to

have a hypermethylated promoter in ICF syndrome type 1 patients, thus *ZBTB24* dysfunction might be a common mechanism to account for the similar phenotypes seen in both classes of patient [238].

Reduced DNA methylation levels in ICF cells are not global but primarily observed in specific DNA sequences at specific loci. As expected from the sites of chromosomal instability, there is hypomethylation of satellites 2 and 3 of pericentromeric heterochromatin but only on chromosomes 1, 16, and occasionally chromosome 9. At these sites, a number of aberrant chromatin rearrangements are seen [239], and these cytological abnormalities are seen primarily in lymphocytes, although they have also been noted in other cell types in ICF syndrome patients [240]. Such ICF-specific cytogenetic abnormalities are also seen in similar rates in normal lymphoblastoid cell lines and lymphocytes that are treated with inhibitors of DNA methylation [241,242], again confirming a direct link between DNA methylation abnormalities and chromosomal instability as seen in ICF syndrome. Undermethylation of sequences is seen in other locations such as at Yqh in males and within the inactive X chromosome in females [240,243], but these changes are not likely to be biologically significant, because sex-specific differences in disease severity are not observed [244]. A whole genome scan to identify sequences that are consistently hypomethylated in lymphoblasts from ICF syndrome patients compared with controls demonstrated a methylation deficit on only a small proportion of the genome, in particular two types of repeats, one of which was the *D4Z4* repeats implicated in FSHD (see Table 5.1 and previous section) [245]. Further work has raised the possibility that in addition to the hypomethylation seen in pericentromeric heterochromatin mentioned, there may be hypermethylation in other regions within genes and promoters, suggesting that it may be the overall methylation pattern in ICF syndrome that is perturbed rather than hypomethylation specifically [238].

Why should particular sequences have altered methylation states in ICF cells? One possibility is that these sequences are present at the chromosomal sites where the *DNMT3B* enzyme is normally localized. The subcellular localization of endogenous human *DNMT3B* has not yet been reported, but exogenously tagged murine *Dnmt3b* colocalizes with pericentric heterochromatin in some cell types [226]. The domain necessary for targeting *DNMT3B* to heterochromatin has not been determined, but it is likely to be at the aminoterminus where there are two domains commonly found in other chromatin-associated proteins. One of these is a plant homeodomain (PHD) finger protein similar to that found in *ATRX*, a chromatin remodeling protein that is concentrated at sites of heterochromatin and repetitive DNA sequences [246]. There is also hypomethylation of specific repetitive DNA sequences in patients with α-thalassemia X-linked mental retardation (ATR-X) (but hypermethylation of other sequences). However, there is no chromosome instability reported at the sites of hypomethylation in ATR-X cells, and little phenotypic overlap between ATR-X and ICF syndromes is observed (see α-Thalassemia X-Linked Mental Retardation Syndrome). The other domain that may be involved in targeting *DNMT3B* is a conserved proline and tryptophan (PWWP) domain, which binds DNA, [247] and based on similarity to tudor and chromodomains [248] may also recognize methylated proteins. A homozygous missense mutation in the PWWP domain of *DNMT3B* has been identified in siblings with ICF syndrome, resulting in a serine to proline change that may have a profound consequence on the mutant protein's structure [229]. In addition, the PWWP domain of murine *Dnmt3b* has been shown to bind DNA nonspecifically and to be required for targeting DNA methyltransferase activity to murine pericentric sequences [227].

Presumably DNA hypomethylation in ICF syndrome leads to *deregulation* of genes that perturb craniofacial, cerebral, and immunological development. Microarray analysis has been used to identify genes with significantly altered mRNA levels in ICF lymphoblastoid cell lines as compared with controls [249]. Many of the genes identified have known roles in immune function in both B and T cell lineages that could account for the immunodeficiency consistently manifested in ICF syndrome. No alteration of DNA methylation was detected at the promoters of any of the deregulated genes tested [249] consistent with the findings of the whole-genome scan [245], arguing against a direct *cis* effect of a methylation abnormality. Furthermore, none of these genes are located on chromosomes 1, 9, or 16. This raises the question of how hypomethylation of specific repetitive DNA sequences in ICF syndrome patients can lead to altered expression of genes located at distant genomic sites. One possibility is that the hypomethylation of satellite DNA alters their heterochromatin properties, and that it is the physical association of deregulated genes with these domains in the nucleus that is aberrant in ICF cells [250]. Silenced genes in human B and T lymphocytes have been shown to colocalize with domains of pericentromeric heterochromatin and to relocate away form these domains upon gene activation [251]. Centromere distribution within the nucleus of lymphoid cells has been shown to vary with different stages of differentiation, suggesting that the distribution of heterochromatin can influence gene expression in *trans* [252,253]. The 3D organization of intranuclear pericentric heterochromatin has been shown to be abnormal in ICF syndrome patient lymphoblastoid B cells, at least for chromosome 1, and treatment with a demethylating agent partially induces this heterochromatic remodeling [254,255]. Further work is needed to see if particular genes also show a altered spatial distributions in ICF cells in parallel to heterochromatic changes.

Alternatively, the loss of DNA methylation at large arrays of satellite repeats may release or recruit protein complexes and affect the balance of regulatory complexes throughout the genome. Interestingly, ICF lymphoblastoid cell lines showed altered binding patterns of *HP1* with the formation of large foci containing *HP1* and components of promyelocytic (PML) nuclear bodies that colocalize with chromosome 1qh and 16qh DNA sequences [256]. The altered pattern, however, was only observed during the G2 phase of the cell cycle and not in fibroblasts. These results suggest that binding of *HP1* is not dependent on DNA methylation (because the abnormal *HP1* distribution only occurs at one stage in the cell cycle) and also indicate that cell type–specific defects in the timing of heterochromatin packaging may be a major determinant of chromosomal abnormalities and gene deregulation [256]. The aggregation of such chromatin proteins in ICF syndrome may simply be a result of undercondensation of heterochromatin at those specific loci, but also may have an effect in *trans* on gene expression elsewhere in the genome [244].

Finally, in addition to hypomethylation and any changes in transcription, a recent study has shown that DNA replication itself is altered in ICF syndrome, either as a result of altered transcription of genes involved in replication, or perhaps because of alteration of chromatin structure affecting the access of the replication machinery [257]. Given one of the key features of ICF syndrome is the chromosomal instability at certain locations, a parallel is seen here with other conditions in which DNA replication defects resulting in chromosomal instability, such as certain cancers or cancer syndromes [257].

Since the mutation responsible for ICF syndrome was first described, it has been clear that hypomethylation plays a key role in this disorder. However, many questions remain unanswered. It is not clear why only certain regions of the genome are hypomethylated in ICF syndrome, and although it is plausible that reduced methylation could lead to derepression of normally silenced genes, specific target genes are yet to be confirmed.

Rett Syndrome

Rett syndrome is a severe mental retardation syndrome characterized by loss of intellectual functioning, motor skills and communicative abilities, microcephaly, development of stereotypic hand movements, respiratory abnormalities, seizures, scoliosis, growth defects, and hypotonia. The syndrome is characterized by normal early infant development, followed by developmental regression from around 6–18 months of age, and then stabilization of phenotype, which may persist for the lifetime of the patient. Patients normally survive into adulthood but require intensive support [258].

The vast majority of cases (90–95%) of Rett syndrome result from sporadic de novo [259] mutations in the *MECP2* gene, which is located on chromosome X (Xq28) and encodes nuclear methyl-CpG binding protein 2 [260]. To date, more than 600 pathogenic *MECP2* mutations, including nonsense, missense, and frameshift have been described [261].

Males carrying comparable *MECP2* mutations present with infantile encephalopathy and rarely survive beyond 2 years [262], and so Rett syndrome patients are usually heterozygous females. After random X inactivation, around half of the cells express the wild-type allele and the other half a mutated *MECP2*. Cases of symptom-free female carriers of *MECP2* mutations are very rare where skewing of X chromosome inactivation prevents expression of the mutant allele [262].

MeCP2 is a member of the family of MBD proteins, a group of transcriptional repressors [263,264]. The functional domains of MeCP2 consist of an aminoterminal MBD, a transcriptional repressor domain (TRD) and a C-terminal domain (CTD). The MBD enables binding to methylated CpG dinucleotides, preferentially adjacent to A/T sequences [265]. The TRD interacts with corepressor complexes, such as Sin3a, to recruit histone deacetylases and heterochromatic proteins, resulting in chromatin condensation and transcriptional repression [44]. The CTD has been shown to interact with splicing factors. [266] MeCP2 has been shown to suppress transcription of repetitive elements, thus controlling transcriptional noise resulting from transcription of these sequences within the cell [267].

As well as a role in transcriptional repression, recent studies have identified MeCP2 involvement in transcriptional activation via interaction with CREB [268], chromatin compaction, and nucleosome clustering, [269,270] as well as RNA processing [271].

The *MeCP2* gene is expressed in various tissues, with the highest level in postnatal neurons and lowest level in glial cells [272,273]. Two major splice variants, *MeCP2_e2* (*MECP2β* or *MECP2A* form) and *MeCP2_e1* (*MECP2α* or *MECP2B* form), have been characterized, the latter more abundantly expressed in somatic tissues, including brain [274,275].

Much recent work in Rett syndrome has focused on *MeCP2* knockdown mice. Deletion of *MeCP2* in mice results in a remarkably similar phenotype to that observed in Rett syndrome [276,277]. Male *MeCP2*-null mice develop normally until around 6 weeks of age, after which there is a period of rapid regression resulting in reduced spontaneous movement, clumsy gait, irregular breathing, hind limb clasping and tremors, leading to death by 20 weeks. In contrast, heterozygous female mice show delayed onset of symptoms (4–12 months) with stabilization of the phenotype later on. Neurobiological examination of *MeCP2*-null mice has revealed reduced synaptic plasticity, reduced synaptic connectivity, altered network excitability, and overall reduced neuron size. Interestingly, the activation of the *MECP2* gene in *MeCP2*-null mice after the onset of Rett syndrome symptoms was discovered to rescue disease phenotype in both males and females [278–280]. This reversibility suggests an essential role for *MeCP2* in the maintenance of normal neurological function in the developed brain, and questions its involvement in early stage brain development. In addition, *MeCP2* inactivation at

different time points in postnatal mouse brain cells results in the appearance of Rett-like phenotypes, further supporting this hypothesis [281]. Another interesting finding is that *MECP2* has different functions based on different posttranslational modifications [282]. To date, two phosphorylation modifications have been studied, linking Ser421 phosphorylation exclusively to neuronal activity [283].

To fully understand the role of the *MeCP2* in brain function, it is essential to identify downstream target genes of *MeCP2*. Analysis of the genomic distribution of *MECP2* has revealed affinity for both methylated as well as unmethylated DNA regions [284]. An interesting target for *MeCP2* is the promoter site of the brain-derived neurotrophic factor (*BDNF*) in resting neurons. Recent evidence suggests in vivo interaction of *MeCP2* and *BDNF* and a role for the latter in the Rett syndrome phenotype. Overexpression of *BDNF* in mice has been shown to reverse the Rett-like phenotype, in contrast with knockout, which worsens symptoms [285]. In a different experiment, ChIP analysis has identified *MeCP2* binding within the mouse chromosome 6 imprinted domain where *Dlx5* and its nonimprinted neighbor *Dlx6* are present. *Dlx5* and *Dlx6* encode proteins that regulate neurotransmitter production [286], and were shown be upregulated twofold in *MeCP2*-null mice brains with accompanying changes in histone modifications and lack of chromatin loop formation [287].

Around 5–10% of individuals clinically diagnosed with Rett syndrome do not appear to have mutations in the *MECP2* gene. In addition, *MECP2* mutations have been found in other disorders, including neonatal-onset encephalopathy, autism, patients exhibiting Angelman phenotype, nonsyndromic X-linked mental retardation and PPM-X (psychosis, pyramidal signs, and macroorchidism-X) syndrome. Recently, mutations in cyclin-dependent kinase 5 (*CDKL5*), which is shown to directly interact with *MeCP2*, and netrin G1, have been found in patients with a very similar phenotype to Rett syndrome [288–290].

Recent work has confirmed the importance of epigenetic mechanisms in Rett syndrome pathophysiology, but specific downstream targets are still to be determined. Perhaps most significantly, the accepted hypothesis that Rett syndrome is purely a disorder of neurodevelopment has been recently called into question, with reversal of phenotype in adult mouse models of disease.

α-Thalassemia X-Linked Mental Retardation Syndrome

The ATR-X syndrome is another example of genetic mutation in a factor involved in chromatin organization affecting disease loci in *trans*. The *ATRX* gene at chromosome Xq13.3 encodes for a *SNF2*-like chromatin remodeling helicase. Functional domains of *ATRX* include a PHD zinc finger–like motif at its aminoterminus (homologous to the PHD motifs in *DNMT3A* and *DNMT3B*), an adjacent coil–coil motif termed *ATRX-DNMT3-DNMT3L (ADD)* and a helicase domain at its carboxyterminus [291]. Mutations in *ATRX* are clustered in these domains and are thought to impair its nuclear localization, protein–protein interactions, or chromatin remodeling functions [246,292–294].

Affected individuals have low levels of α-globin subunits, which favor the formation of unstable β-globin tetramers that precipitate within erythrocytes causing varying degrees of hemolysis and splenomegaly. Affected males have relatively severe mental retardation together with facial and skeletal abnormalities, urogenital abnormalities and microcephaly, whereas heterozygous females are usually asymptomatic because of a skewed pattern of X-chromosome inactivation preferentially silencing the mutated allele [295–297].

ATRX may be involved in gene activation, suggested by the reduced expression of the α-globin locus in ATR-X syndrome. This, however, does not explain the additional phenotypic traits observed and presumably deregulation of many other loci gives rise to the complex phenotype. The observation that diverse DNA methylation defects (hypermethylation at *DYZ2* Y-chromosome repeats and hypomethylation of ribosomal DNA) are present in disease [298] indicates *ATRX* is able to regulate chromatin structure at several distinct loci. Direct (stimulatory) effects of *ATRX* upon loci are also indicated by its association with the transcriptional regulatory death-domain associated protein 6 *(DAXX)*. The *DAXX–ATRX* containing complex, levels of which are reduced in ATR-X syndrome patient cell lines, is able to remodel nucleosomes in vitro [299], and patient mutations in *ATRX* cause a reduction in this function [294]. Furthermore, *ATRX* associates with PML bodies, which are thought to function as regulatory (activator) factor reservoirs in the nucleus [299,300].

On the other hand, chromatin remodeling by *ATRX* may facilitate chromatin condensation and gene silencing. The ADD (ATRX-DNMT3-DNMT3L, ADDATRX) domain of *ATRX* has been shown to bind directly to histone *H3K9me3,* a hallmark of pericentric chromatin, and disease-causing mutations impair this association [301]. *ATRX* was also found to associate with the histone methyltransferase enhancer of zeste homologue 2 *(EZH2)*. *EZH2* is part of the polycomb group repressor complexes (PRCs) that methylate histone *H3K27* (PRC2) and histone *H1K26* (PRC3) [303]. *H3K27* methylation is a mark recognized by the chromodomain of the polycomb protein (contained within the PRC1 complex) implicated in chromatin condensation and developmentally regulated gene silencing [304–306]. Methylation at *H1K26* may also be involved in chromatin condensation, because *HP1* has been shown to specifically interact with this mark [307]. These interactions reveal potential new pathways of gene regulation where the *HP1* and the polycomb group silencing pathways may be synergistic. For example, a novel polycomb group complex, PRC4, was recently shown to arise and be upregulated upon oncogenic transformation [308]. PRC4 has special activity toward histone *H1bK26,* particular native complex subunit

isoforms (Eed2) and the presence of histone deacetylases. Excess of PRC4 might lead to increased lysine 26 methylation and subsequent recognition by *HP1*. Dependent on the genomic context and associated complexes, *HP1* may either silence or derepress the affected loci [309]. Furthermore, *ATRX* and *HP1* have been shown to colocalize by immunofluorescence [246] and to directly interact [310–312]. These interactions of *ATRX* with other molecular chaperones may generate regulatory complexes that directly affect chromatin condensation at pericentric sites.

Mutations in *ATRX* are common at the PHD-zinc finger motif, which is likely to be responsible for the targeting of *ATRX* to pericentromeric heterochromatin [246]. Inadequate targeting of catalytically active *ATRX* may therefore result in ectopic binding at loci deregulated in disease. Additionally, owing to the predominant localization of *ATRX* to pericentromeric heterochromatin, *ATRX* may be indirectly regulating additional loci by modulating the nature of gene association with pericentromeric heterochromatin. Interestingly, different patients with identical mutations in the *ATRX* gene exhibit significant phenotypic variation [291,313]. Recent ChIP sequencing data has suggested that *ATRX* is found associated with variable number tandem repeats (VNTRs), and given the size of these VNTRs vary between individuals, this may explain some of this variation [314].

Acquired forms of α-thalassemia myelodysplastic syndrome with mutations in *ATRX* exhibit a much more severe form of α-thalassemia compared with inherited forms even when the mutation itself is identical, again implicating epigenetic mechanisms in causing severity [315]. The previous discussions demonstrate that *ATRX* is clearly a chromain modifier acting in *trans*, particularly at pericentric heterochromatin, with the potential for both gene activation or silencing. However, the downstream targets and the exact nature of modifying complexes formed are yet to be elucidated.

EPIGENETIC MECHANISMS IN CANCER

Cancer is a complex multifactorial disease [316]. Genetic mutations may promote tumorigenesis by hampering tumor suppressor activity or by hyperactivation of oncogenic pathways (see Chapter 20). The balance between these activities promotes gene expression programs that fuel neoplastic metabolic states. How the genome responds to these signals in terms of transcriptional output is thus central to the generation and/or maintenance of the cancer phenotype.

Access to genetic information and subsequent transcriptional readout are modulated by epigenetic factors: DNA methylation, chromatin histone modifications, noncoding RNA, and higher order chromatin structures. The epigenome constitutes a "buffering" system that regulates gene expression noise. This buffering system is grossly disrupted in cancer, leading to increased noise and heterogeneity of gene expression [317].

The cancer methylome is characterized by a global reduction in DNA methylation that enhances genomic instability, and by focal gains of methylation in particular at promoters enriched in CpG dinucleotides (CpG islands). DNA hypermethylation of promoter CpG islands commonly associates with repression of the associated genes, more often than not of tumors suppressor loci [318]. Nevertheless, the function of DNA methylation is now understood to be dependent on where exactly it is located; it has been shown to also associate with active genes [319].

It is becoming increasingly clear that an interplay exists between DNA methylation and histone modification [320]. In colonic tumors, increased DNA methylation is common at promoters, marked by both histone *H3K4* and *H3K27* trimethylation in embryonic stem cells [321]. This bivalency for histone H3 modification is therefore part of an instructive process that guides the DNA methylation machinery to the affected loci. Indeed, polycomb group complexes that place *H3K27* methylation are known to interact with the de novo DNA methyltransferases [322].

Long noncoding RNA (>200 bp) and small noncoding RNA (~22 bp) such as microRNAs are novel functional elements capable of regulating gene expression and seem to be involved in tumorigenesis [323,324]. For example, a long noncoding RNA that is antisense to the *CDKN2B* (p15) tumor suppressor locus can regulate the chromatin and DNA methylation of the p15 locus [325] and a similar effect in *cis* was seen at the *CDKN1A* (p21) locus [326]. Effects in *trans*, where the noncoding RNA has an effect on a chromosome other than the one where it originates, have been described where large intergenic noncoding RNAs appear to target chromatin-modifying machineries to affect the expression of distant loci [327]. MicroRNAs are very powerful regulatory elements [328] grossly deregulated in cancers [324] and thus a very attractive target for cancer therapy [329].

Effects at a distance are also brought about by chromatin looping, where higher order chromatin architectures enable the interaction of physically distant regulatory elements, enhancer–promoter interaction, for example, or place loci within nuclear–chromosomal territories that facilitate or inhibit gene activity. Higher order architectures such as those mediated by *CTCF* binding sites and cohesin are thought to be part of networks of long-range interactions that control developmentally regulated transcriptional programs [330]. Because *CTCF* binding is sensitive to DNA methylation, further refinement of the role that long-distance interactions play in tumorigenesis is needed.

Contrary to genetic mutation, epigenetic mutation/modification is potentially reversible and thus a very attractive target for therapeutic intervention. Inhibitors of DNA methylation and histone deacetylation are already in use for some types of tumors [320,329] and extensive research is underway to identify further epigenetic targets and to understand how the epigenetic modifications interrelate such that combinatorial therapies can be designed. Together with the technical advances that make genome-wide profiling of epigenetic modifications in individuals possible, the future of epigenetic therapy in cancer is very promising and very exciting.

CONCLUSION

It is clear that previously mysterious aspects of gene regulation that can be grouped under the terms *epigenetic* and *epigenomic* are finally yielding to molecular biology approaches and have revealed a new level of genome organization and regulation. Already, the rapid increase in our understanding of the control of gene expression patterns has helped to unravel complex molecular mechanisms underpinning the complex phenotypes in many mendelian uncommon genetic conditions. In addition, *epigenetics* and *epigenomics* investigations have led to potentially powerful new therapeutic avenues for an ever-increasing number of human diseases, and a great variety of human cancers, many of which are currently incurable.

ACKNOWLEDGMENT

Funding received from Imperial College Biomedical Research Center (NIHR), the MRC (UK), and Wellcome Trust.

REFERENCES

[1] Holliday R. DNA methylation and epigenotypes. Biochemistry (Mosc) 2005;70:500–4. Published online Epub May.

[2] Reik W, Walter J. Genomic imprinting: parental influence on the genome. Nat Rev Genet 2001;2:21–32. Published online Epub Jan.

[3] Ferguson-Smith AC. Genomic imprinting: the emergence of an epigenetic paradigm. Nat Rev Genet 2011;12:565–75. http://dx.doi.org/10.1038/nrg3032. Published online Epub Aug.

[4] Hochedlinger K, Plath K. Epigenetic reprogramming and induced pluripotency. Development 2009;136:509–23. http://dx.doi.org/10.1242/dev.020867. Published online Epub Feb.

[5] Takahashi K, Tanabe K, Ohnuki M, Narita M, Ichisaka T, Tomoda K, et al. Induction of pluripotent stem cells from adult human fibroblasts by defined factors. Cell 2007;131:861–72. http://dx.doi.org/10.1016/j.cell.2007.11.019. Published online Epub Nov 30, S0092-8674(07)01471-7 [pii].

[6] Takahashi K, Yamanaka S. Induction of pluripotent stem cells from mouse embryonic and adult fibroblast cultures by defined factors. Cell 2006;126:663–76. http://dx.doi.org/10.1016/j.cell.2006.07.024. Published online Epub Aug 25, S0092-8674(06)00976-7 [pii].

[7] Kornberg RD, Lorch Y. Twenty-five years of the nucleosome: fundamental particle of the eukaryote chromosome. Cell 1999;98:285–94. Published online Epub Aug 6.

[8] Thoma F, Koller T, Klug A. Involvement of histone *H1* in the organization of the nucleosome and of the salt-dependent superstructures of chromatin. J Cell Biol 1979;83:403–27. Published online Epub Nov.

[9] Rattner JB, Hamkalo BA. Nucleosome packing in interphase chromatin. J Cell Biol 1979;81:453–7. Published online Epub May.

[10] Tremethick DJ. Higher-order structures of chromatin: the elusive 30 nm fiber. Cell 2007;128:651–4. http://dx.doi.org/10.1016/j.cell.2007.02.008. Published online Epub Feb 23.

[11] Hsu TC, Cooper JE, Mace Jr ML, Brinkley BR. Arrangement of centromeres in mouse cells. Chromosoma 1971;34:73–87.

[12] Joseph A, Mitchell AR, Miller OJ. The organization of the mouse satellite DNA at centromeres. Exp Cell Res 1989;183:494–500. Published online Epub Aug.

[13] Guenatri M, Bailly D, Maison C, Almouzni G. Mouse centric and pericentric satellite repeats form distinct functional heterochromatin. J Cell Biol 2004;166:493–505. Published online Epub Aug 16.

[14] Jeppesen P, Mitchell A, Turner B, Perry P. Antibodies to defined histone epitopes reveal variations in chromatin conformation and under acetylation of centric heterochromatin in human metaphase chromosomes. Chromosoma 1992;101:322–32. Published online Epub Mar.

[15] Peters AH, O'Carroll D, Scherthan H, Mechtler K, Sauer S, Schofer C, et al. Loss of the *Suv39h* histone methyltransferases impairs mammalian heterochromatin and genome stability. Cell 2001;107:323–37. Published online Epub Nov 2.

[16] Maison C, Bailly D, Peters AH, Quivy JP, Roche D, Taddei A, et al. Higher-order structure in pericentric heterochromatin involves a distinct pattern of histone modification and an RNA component. Nat Genet 2002;30:329–34.

[17] Maison C, Bailly D, Roche D, Montes de Oca R, Probst AV, Vassias I, et al. Sumoylation promotes de novo targeting of *HP1alpha* to pericentric heterochromatin. Nat Genet 2011;43:220–7. http://dx.doi.org/10.1038/ng.765. Published online Epub Mar, ng.765 [pii].

[18] Schramke V, Sheedy DM, Denli AM, Bonila C, Ekwall K, Hannon GJ, et al. RNA-interference-directed chromatin modification coupled to RNA polymerase II transcription. Nature 2005;435:1275–9. Published online Epub Jun 30.

[19] Fukagawa T, Nogami M, Yoshikawa M, Ikeno M, Okazaki T, Takami Y, et al. *DICER* is essential for formation of the heterochromatin structure in vertebrate cells. Nat Cell Biol 2004;6:784–91. Published online Epub Aug.

[20] Pal-Bhadra M, Leibovitch BA, Gandhi SG, Rao M, Bhadra U, Birchler JA, et al. Heterochromatic silencing and *HP1* localization in *Drosophila* are dependent on the RNAi machinery. Science 2004;303:669–72. Published online Epub Jan 30.

[21] Eissenberg JC, Elgin SC. The *HP1* protein family: getting a grip on chromatin. Curr Opin Genet Dev 2000;10:204–10. Published online Epub Apr.

[22] Festenstein R, Pagakis SN, Hiragami K, Lyon D, Verreault A, Sekkali B, et al. Modulation of heterochromatin protein 1 dynamics in primary mammalian cells. Science 2003;299:719–21. Published online Epub Jan 31.

[23] Cheutin T, McNairn AJ, Jenuwein T, Gilbert DM, Singh PB, Misteli T. Maintenance of stable heterochromatin domains by dynamic *HP1* binding. Science 2003;299:721–5. Published online Epub Jan 31.

[24] Schmiedeberg L, Weisshart K, Diekmann S, Zu Hoerste GM, Hemmerich P. High- and low-mobility populations of *HP1* in heterochromatin of mammalian cells. Mol Biol Cell 2004;15:2819–33. Published online Epub Jun.

[25] Dillon N, Festenstein R. Unravelling heterochromatin: competition between positive and negative factors regulates accessibility. Trends Genet 2002;18:252–8. Published online Epub May.

[26] Bannister AJ, Kouzarides T. Regulation of chromatin by histone modifications. Cell Res 2011;21:381–95. http://dx.doi.org/10.1038/cr.2011.22. Published online Epub Mar, cr201122 [pii].

[27] Hargreaves DC, Crabtree GR. ATP-dependent chromatin remodeling: genetics, genomics and mechanisms. Cell Res 2011;21:396–420. http://dx.doi.org/10.1038/cr.2011.32. Published online Epub Mar, cr201132 [pii].

[28] Turner BM. Histone acetylation and an epigenetic code. Bioessays 2000;22:836–45. Published online Epub Sep.

[29] Strahl BD, Allis CD. The language of covalent histone modifications. Nature 2000;403:41–5. Published online Epub Jan 6.

[30] Narlikar GJ, Fan HY, Kingston RE. Cooperation between complexes that regulate chromatin structure and transcription. Cell 2002;108:475–87. Published online Epub Feb 22.

[31] de la Cruz X, Lois S, Sanchez-Molina S, Martinez-Balbas MA. Do protein motifs read the histone code?. Bioessays 2005;27:164–75. Published online Epub Feb.

[32] Wysocka J, Swigut T, Milne TA, Dou Y, Zhang X, Burlingame AL, et al. *WDR5* associates with histone H3 methylated at K4 and is essential for *H3 K4* methylation and vertebrate development. Cell 2005;121:859–72. Published online Epub Jun 17.

[33] Bernstein BE, Humphrey EL, Erlich RL, Schneider R, Bouman P, Liu JS, et al. Methylation of histone H3 Lys 4 in coding regions of active genes. Proc Natl Acad Sci USA 2002;99:8695–700. Published online Epub Jun 25.

[34] Schubeler D, MacAlpine DM, Scalzo D, Wirbelauer C, Kooperberg C, van Leeuwen F, et al. The histone modification pattern of active genes revealed through genome-wide chromatin analysis of a higher eukaryote. Genes Dev 2004;18:1263–71. Published online Epub Jun 1.

[35] Lachner M, O'Carroll D, Rea S, Mechtler K, Jenuwein T. Methylation of histone H3 lysine 9 creates a binding site for *HP1* proteins. Nature 2001;410:116–20.

[36] Fischle W, Wang Y, Jacobs SA, Kim Y, Allis CD, Khorasanizadeh S. Molecular basis for the discrimination of repressive methyl-lysine marks in histone H3 by polycomb and *HP1* chromodomains. Genes Dev 2003;17:1870–81. Published online Epub Aug 1.

[37] Hermann A, Gowher H, Jeltsch A. Biochemistry and biology of mammalian DNA methyltransferases. Cell Mol Life Sci 2004;61:2571–87. Published online Epub Oct.

[38] Bird AP, Wolffe AP. Methylation-induced repression: belts, braces, and chromatin. Cell 1999;99:451–4. Published online Epub Nov 24.

[39] Lachner M, Jenuwein T. The many faces of histone lysine methylation. Curr Opin Cell Biol 2002;14:286–98. Published online Epub Jun.

[40] Dellino GI, Schwartz YB, Farkas G, McCabe D, Elgin SC, Pirrotta V. Polycomb silencing blocks transcription initiation. Mol Cell 2004;13:887–93. Published online Epub Mar 26.

[41] Tahiliani M, Koh KP, Shen Y, Pastor WA, Bandukwala H, Brudno Y, et al. Conversion of 5-methylcytosine to 5-hydroxymethylcytosine in mammalian DNA by MLL partner *TET1*. Science 2009;324:930–5. http://dx.doi.org/10.1126/science.1170116. Published online Epub May 15.

[42] Branco MR, Ficz G, Reik W. Uncovering the role of 5-hydroxymethylcytosine in the epigenome. Nat Rev Genet 2012;13:7–13. http://dx.doi.org/10.1038/nrg3080. Published online Epub Jan.

[43] Williams K, Christensen J, Pedersen MT, Johansen JV, Cloos PA, Rappsilber J, et al. *TET1* and hydroxymethylcytosine in transcription and DNA methylation fidelity. Nature 2011;473:343–8. http://dx.doi.org/10.1038/nature10066. Published online Epub May 19.

[44] Nan X, Ng HH, Johnson CA, Laherty CD, Turner BM, Eisenman RN, et al. Transcriptional repression by the methyl-CpG-binding protein *MeCP2* involves a histone deacetylase complex. Nature 1998;393:386–9. http://dx.doi.org/10.1038/30764. Published online Epub May 28.

[45] Ng HH, Zhang Y, Hendrich B, Johnson CA, Turner BM, Erdjument-Bromage H, et al. *MBD2* is a transcriptional repressor belonging to the *MeCP1* histone deacetylase complex. Nat Genet 1999;23:58–61. Published online Epub Sep.

[46] Fuks F, Hurd PJ, Wolf D, Nan X, Bird AP, Kouzarides T. The methyl-CpG-binding protein *MeCP2* links DNA methylation to histone methylation. J Biol Chem 2003;278:4035–40. http://dx.doi.org/10.1074/jbc.M210256200. Published online Epub Feb 7.

[47] Fuks F, Hurd PJ, Deplus R, Kouzarides T. The DNA methyltransferases associate with *HP1* and the *SUV39H1* histone methyltransferase. Nucleic Acids Res 2003;31:2305–12. Published online Epub May 1.

[48] Aagaard L, Laible G, Selenko P, Schmid M, Dorn R, Schotta G, et al. Functional mammalian homologues of the *Drosophila* PEV-modifier Su(var)3-9 encode centromere-associated proteins which complex with the heterochromatin component *M31*. Embo J 1999;18:1923–38. Published online Epub Apr 1.

[49] Lehnertz B, Ueda Y, Derijck AA, Braunschweig U, Perez-Burgos L, Kubicek S, et al. *Suv39h*-mediated histone H3 lysine 9 methylation directs DNA methylation to major satellite repeats at pericentric heterochromatin. Curr Biol 2003;13:1192–200. Published online Epub Jul 15.

[50] Lewis A, Mitsuya K, Umlauf D, Smith P, Dean W, Walter J, et al. Imprinting on distal chromosome 7 in the placenta involves repressive histone methylation independent of DNA methylation. Nat Genet 2004;36:1291–5. Published online Epub Dec.

[51] Ebralidse KK, Hebbes TR, Clayton AL, Thorne AW, Crane-Robinson C. Nucleosomal structure at hyperacetylated loci probed in nuclei by DNA-histone crosslinking. Nucleic Acids Res 1993;21:4734–8. Published online Epub Oct 11.

[52] Orlando V, Strutt H, Paro R. Analysis of chromatin structure by in vivo formaldehyde cross-linking. Methods 1997;11:205–14. http://dx.doi.org/10.1006/meth.1996.0407. Published online Epub Feb, S1046-2023(96)90407-7 [pii].

[53] Shendure J, Ji H. Next-generation DNA sequencing. Nat Biotechnol 2008;26:1135–45. http://dx.doi.org/10.1038/nbt1486. Published online Epub Oct.

[54] Arvey A, Agius P, Noble WS, Leslie C. Sequence and chromatin determinants of cell-type–specific transcription factor binding. Genome Res 2012;22:1723–34. http://dx.doi.org/10.1101/gr.127712.111. Published online Epub Sep.

[55] Consortium EP, Dunham I, Kundaje A, Aldred SF, Collins PJ, Davis CA, et al. An integrated encyclopedia of DNA elements in the human genome. Nature 2012;489:57–74. http://dx.doi.org/10.1038/nature11247. Published online Epub Sep 6.

[56] Djebali S, Davis CA, Merkel A, Dobin A, Lassmann T, Mortazavi A, et al. Landscape of transcription in human cells. Nature 2012;489:101–8. http://dx.doi.org/10.1038/nature11233. Published online Epub Sep 6.

[57] Harrow J, Frankish A, Gonzalez JM, Tapanari E, Diekhans M, Kokocinski F, et al. GENCODE: the reference human genome annotation for the ENCODE Project. Genome Res 2012;22:1760–74. http://dx.doi.org/10.1101/gr.135350.111. Published online Epub Sep.

[58] Thurman RE, Rynes E, Humbert R, Vierstra J, Maurano MT, Haugen E, et al. The accessible chromatin landscape of the human genome. Nature 2012;489:75–82. Published online Epub Sep 6 (10.1038/nature11232).

[59] Wang J, Zhuang J, Iyer S, Lin X, Whitfield TW, Greven MC, et al. Sequence features and chromatin structure around the genomic regions bound by 119 human transcription factors. Genome Res 2012;22:1798–812. http://dx.doi.org/10.1101/gr.139105.112. Published online Epub Sep.

[60] Wong CC, Caspi A, Williams B, Craig IW, Houts R, Ambler A, et al. A longitudinal study of epigenetic variation in twins. Epigenetics Official J DNA Methylation Soc 2010;5:516–26. Published online Epub Aug 16.

[61] Bell JT, Spector TD. A twin approach to unraveling epigenetics. Trends Genet 2011;27:116–25. http://dx.doi.org/10.1016/j.tig.2010.12.005. Published online Epub Mar.

[62] Osborne CS, Chakalova L, Brown KE, Carter D, Horton A, Debrand E, et al. Active genes dynamically colocalize to shared sites of ongoing transcription. Nat Genet 2004;36:1065–71. Published online Epub Oct.

[63] Brown K, Guest S, Smale S, Hahm K, Merkenschlager M, Fisher A. Association of transcriptionally silent genes with Ikaros complexes at centromeric heterochromatin. Cell 1997;91:845–54.

[64] Locke J, Kotarski MA, Tartof KD. Dosage-dependent modifiers of position effect variegation in *Drosophila* and a mass action model that explains their effect. Genetics 1988;120:181–98. Published online Epub Sep.

[65] Cattanach BM. Position effect variegation in the mouse. Genet Res 1974;23:291–306. Published online Epub Jun.

[66] Festenstein R, Tolaini M, Corbella P, Mamalaki C, Parrington J, Fox M, et al. Locus control region function and heterochromatin-induced position effect variegation. Science 1996;271:1123–5. Published online Epub Feb 23.

[67] Henikoff S. Dosage-dependent modification of position-effect variegation in *Drosophila*. Bioessays 1996;18:401–9. Published online Epub May.

[68] Grosveld F. Activation by locus control regions?. Curr Opin Genet Dev 1999;9:152–7. Published online Epub Apr.

[69] West AG, Fraser P. Remote control of gene transcription. Hum Mol Genet 2005;14(Spec No 1):R101–11. Published online Epub Apr 15.

[70] Festenstein R, Kioussis D. Locus control regions and epigenetic chromatin modifiers. Curr Opin Genet Dev 2000;10:199–203. Published online Epub Apr.

[71] Soejima H, Wagstaff J. Imprinting centers, chromatin structure, and disease. J Cell Biochem 2005;95:226–33. Published online Epub May 15.

[72] Gavrilov A, Eivazova E, Priozhkova I, Lipinski M, Razin S, Vassetzky Y. Chromosome conformation capture (from 3C to 5C) and its ChIP-based modification. Methods Mol Biol 2009;567:171–88. http://dx.doi.org/10.1007/978-1-60327-414-2_12.

[73] Sanyal A, Lajoie BR, Jain G, Dekker J. The long-range interaction landscape of gene promoters. Nature 2012;489:109–13. http://dx.doi.org/10.1038/nature11279. Published online Epub Sep 6.

[74] Wijchers PJ, Yandim C, Panousopoulou E, Ahmad M, Harker N, Saveliev A, et al. Sexual dimorphism in mammalian autosomal gene regulation is determined not only by *Sry* but by sex chromosome complement as well. Dev Cell 2010;19:477–84. http://dx.doi.org/10.1016/j.devcel.2010.08.005. Published online Epub Sep.

[75] Wijchers PJ, Festenstein RJ. Epigenetic regulation of autosomal gene expression by sex chromosomes. Trends Genet 2011;27:132–40. http://dx.doi.org/10.1016/j.tig.2011.01.004. Published online Epub Apr.

[76] Nageshwaran S, Festenstein R. Epigenetics and triplet-repeat neurological diseases. Front Neurology 2015. http://dx.doi.org/10.3389/fneur.2015.00262. Published online Epub 21/12/2015.

[77] Harding AE. Friedreich's ataxia: a clinical and genetic study of 90 families with an analysis of early diagnostic criteria and intrafamilial clustering of clinical features. Brain 1981;104:589–620.

[78] Campuzano V, Montermini L, Molto MD, Pianese L, Cossee M, Cavalcanti F, et al. Friedreich's ataxia: autosomal recessive disease caused by an intronic GAA triplet repeat expansion. Science 1996;271:1423–7.

[79] Mirkin SM. Expandable DNA repeats and human disease. Nature 2007;447:932–40. http://dx.doi.org/10.1038/nature05977. Published online Epub Jun 21, nature05977 [pii].

[80] Pandolfo M, Pastore A. The pathogenesis of Friedreich ataxia and the structure and function of frataxin. J Neurol 2009;256(Suppl 1):9–17. http://dx.doi.org/10.1007/s00415-009-1003-2. Published online Epub Mar.

[81] Montermini L, Richter A, Morgan K, Justice CM, Julien D, Castellotti B, et al. Phenotypic variability in Friedreich ataxia: role of the associated GAA triplet repeat expansion. Ann Neurol 1997;41:675–82.

[82] Monros E, Molto MD, Martinez F, Canizares J, Blanca J, Vilchez JJ, et al. Phenotype correlation and intergenerational dynamics of the Friedreich ataxia GAA trinucleotide repeat. Am J Hum Genet 1997;61:101–10. Published online Epub Jul, S0002-9297(07)64282-7 [pii].

[83] Filla A, De Michele G, Cavalcanti F, Pianese L, Monticelli A, Campanella G, et al. The relationship between trinucleotide (GAA) repeat length and clinical features in Friedreich ataxia. Am J Hum Genet 1996;59:554–60. Published online Epub Sep.

[84] Durr A, Cossee M, Agid Y, Campuzano V, Mignard C, Penet C, et al. Clinical and genetic abnormalities in patients with Friedreich's ataxia [see comments]. N Engl J Med 1996;335:1169–75.

[85] Lamont PJ, Davis MB, Wood NW. Identification and sizing of the GAA trinucleotide repeat expansion of Friedreich's ataxia in 56 patients: clinical and genetic correlates. Brain 1997;120:673–80.

[86] Campuzano V, Montermini L, Lutz Y, Cova L, Hindelang C, Jiralerspong S, et al. Frataxin is reduced in Friedreich ataxia patients and is associated with mitochondrial membranes. Hum Mol Genet 1997;6:1771–80. Published online Epub Oct.

[87] Geschwind DH, Perlman S, Grody WW, Telatar M, Montermini L, Pandolfo M, et al. Friedreich's ataxia GAA repeat expansion in patients with recessive or sporadic ataxia. Neurology 1997;49:1004–9.

[88] Gellera C, Pareyson D, Castellotti B, Mazzucchelli F, Zappacosta B, Pandolfo M, et al. Very late onset Friedreich's ataxia without cardiomyopathy is associated with limited GAA expansion in the X25 gene. Neurology 1997;49:1153–5.

[89] Kumari D, Usdin K. Is Friedreich ataxia an epigenetic disorder?. Clin Epigenetics 2012;4(2):1868–7083. http://dx.doi.org/10.1186/1868-7083-4-2. -4-2 [pii].

[90] Bidichandani SI, Ashizawa T, Patel PI. The GAA triplet-repeat expansion in Friedreich ataxia interferes with transcription and may be associated with an unusual DNA structure. Am J Hum Genet 1998;62:111–21. http://dx.doi.org/10.1086/301680. Published online Epub Jan (S0002-9297(07)60128-1 [pii].

[91] Grabczyk E, Usdin K. Alleviating transcript insufficiency caused by Friedreich's ataxia triplet repeats. Nucleic Acids Res 2000;28:4930–7. Published online Epub Dec 15.

[92] Grabczyk E, Usdin K. The GAA*TTC triplet repeat expanded in Friedreich's ataxia impedes transcription elongation by T7 RNA polymerase in a length and supercoil dependent manner. Nucleic Acids Res 2000;28:2815–22. Published online Epub Jul 15.

[93] Krasilnikova MM, Kireeva ML, Petrovic V, Knijnikova N, Kashlev M, Mirkin SM. Effects of Friedreich's ataxia (GAA)n*(TTC) n repeats on RNA synthesis and stability. Nucleic Acids Res 2007;35:1075–84. http://dx.doi.org/10.1093/nar/gkl1140. gkl1140 [pii].

[94] Sakamoto N, Chastain PD, Parniewski P, Ohshima K, Pandolfo M, Griffith JD, et al. Sticky DNA: self-association properties of long GAA.TTC repeats in R.R.Y triplex structures from Friedreich's ataxia. Mol Cell 1999;3:465–75. Published online Epub Apr, S1097-2765(00)80474-8 [pii].

[95] Ohshima K, Montermini L, Wells RD, Pandolfo M. Inhibitory effects of expanded GAA.TTC triplet repeats from intron I of the Friedreich ataxia gene on transcription and replication in vivo. J Biol Chem 1998;273:14588–95. Published online Epub Jun 5.

[96] Sakamoto N, Ohshima K, Montermini L, Pandolfo M, Wells RD. Sticky DNA, a self-associated complex formed at long GAA*TTC repeats in intron 1 of the frataxin gene, inhibits transcription. J Biol Chem 2001;276:27171–7. http://dx.doi.org/10.1074/jbc.M101879200. Published online Epub Jul 20, M101879200 [pii].

[97] Kim E, Napierala M, Dent SY. Hyperexpansion of GAA repeats affects post-initiation steps of FXN transcription in Friedreich's ataxia. Nucleic Acids Res 2011;39:8366–77. http://dx.doi.org/10.1093/nar/gkr542. Published online Epub Oct 1, gkr542 [pii].

[98] Kumari D, Biacsi RE, Usdin K. Repeat expansion affects both transcription initiation and elongation in Friedreich ataxia cells. J Biol Chem 2011;286:4209–15. Published online Epub Feb 11.

[99] Punga T, Buhler M. Long intronic GAA repeats causing Friedreich ataxia impede transcription elongation. EMBO Mol Med 2010;2:120–9. Published online Epub Apr.

[100] Saveliev A, Everett C, Sharpe T, Webster Z, Festenstein R. DNA triplet repeats mediate heterochromatin-protein-1-sensitive variegated gene silencing. Nature 2003;422:909–13. http://dx.doi.org/10.1038/nature01596. Published online Epub Apr 24, nature01596 [pii].

[101] Castaldo I, Pinelli M, Monticelli A, Acquaviva F, Giacchetti M, Filla A, et al. DNA methylation in intron 1 of the frataxin gene is related to GAA repeat length and age of onset in Friedreich ataxia patients. J Med Genet 2008;45:808–12. http://dx.doi.org/10.1136/jmg.2008.058594. Published online Epub Dec, jmg.2008.058594 [pii].

[102] Evans-Galea MV, Carrodus N, Rowley SM, Corben LA, Tai G, Saffery R, et al. FXN methylation predicts expression and clinical outcome in Friedreich ataxia. Ann Neurol 2012;71:487–97. http://dx.doi.org/10.1002/ana.22671. Published online Epub Apr.

[103] Greene E, Mahishi L, Entezam A, Kumari D, Usdin K. Repeat-induced epigenetic changes in intron 1 of the frataxin gene and its consequences in Friedreich ataxia. Nucleic Acids Res 2007;35:3383–90. http://dx.doi.org/10.1093/nar/gkm271. gkm271 [pii].

[104] Al-Mahdawi S, Pinto RM, Ismail O, Varshney D, Lymperi S, Sandi C, et al. The Friedreich ataxia GAA repeat expansion mutation induces comparable epigenetic changes in human and transgenic mouse brain and heart tissues. Hum Mol Genet 2008;17:735–46. http://dx.doi.org/10.1093/hmg/ddm346. Published online Epub Mar 1, ddm346 [pii].

[105] Herman D, Jenssen K, Burnett R, Soragni E, Perlman SL, Gottesfeld JM. Histone deacetylase inhibitors reverse gene silencing in Friedreich's ataxia. Nat Chem Biol 2006;2:551–8. http://dx.doi.org/10.1038/nchembio815. Published online Epub Oct, nchembio815 [pii].

[106] De Biase I, Chutake YK, Rindler PM, Bidichandani SI. Epigenetic silencing in Friedreich ataxia is associated with depletion of *CTCF* (*CCCTC*-binding factor) and antisense transcription. Plos One 2009;4:e7914. http://dx.doi.org/10.1371/journal.pone.0007914.

[107] Bourn RL, Rindler PM, Pollard LM, Bidichandani SI. *E. Coli* mismatch repair acts downstream of replication fork stalling to stabilize the expanded (GAA.TTC)(n) sequence. Mutat Res 2009;661:71–7. http://dx.doi.org/10.1016/j.mrfmmm.2008.11.003. Published online Epub Feb 10, S0027-5107(08)00281-9 [pii].

[108] Du J, Campau E, Soragni E, Ku S, Puckett JW, Dervan PB, et al. Role of mismatch repair enzymes in GAA.TTC triplet-repeat expansion in Friedreich ataxia induced pluripotent stem cells. J Biol Chem 2012;287:29861–72. http://dx.doi.org/10.1074/jbc.M112.391961. Published online Epub Aug 24, M112.391961 [pii].

[109] Ezzatizadeh V, Pinto RM, Sandi C, Sandi M, Al-Mahdawi S, Te Riele H, et al. The mismatch repair system protects against intergenerational GAA repeat instability in a Friedreich ataxia mouse model. Neurobiol Dis 2012;46:165–71. http://dx.doi.org/10.1016/j.nbd.2012.01.002. Published online Epub Apr, S0969-9961(12)00017-4 [pii].

[110] Kim HM, Narayanan V, Mieczkowski PA, Petes TD, Krasilnikova MM, Mirkin SM, et al. Chromosome fragility at GAA tracts in yeast depends on repeat orientation and requires mismatch repair. EMBO J 2008;27:2896–906. http://dx.doi.org/10.1038/emboj.2008.205. Published online Epub Nov 5, emboj2008205 [pii].

[111] Ku S, Soragni E, Campau E, Thomas EA, Altun G, Laurent LC, et al. Friedreich's ataxia induced pluripotent stem cells model intergenerational GAATTC triplet repeat instability. Cell Stem Cell 2010;7:631–7. http://dx.doi.org/10.1016/j.stem.2010.09.014. Published online Epub Nov 5.

[112] Ayoub N, Jeyasekharan AD, Bernal JA, Venkitaraman AR. *HP1*-beta mobilization promotes chromatin changes that initiate the DNA damage response. Nature 2008;453:682–6. http://dx.doi.org/10.1038/nature06875. Published online Epub May 29, nature06875 [pii].

[113] Goodarzi AA, Noon AT, Deckbar D, Ziv Y, Shiloh Y, Lobrich M, et al. ATM signaling facilitates repair of DNA double-strand breaks associated with heterochromatin. Mol Cell 2008;31:167–77. http://dx.doi.org/10.1016/j.molcel.2008.05.017. Published online Epub Jul 25, S1097-2765(08)00395-X [pii].

[114] Luijsterburg MS, Dinant C, Lans H, Stap J, Wiernasz E, Lagerwerf S, et al. Heterochromatin protein 1 is recruited to various types of DNA damage. J Cell Biol 2009;185:577–86. http://dx.doi.org/10.1083/jcb.200810035. Published online Epub May 18, jcb.200810035 [pii].

[115] Peng JC, Karpen GH. Heterochromatic genome stability requires regulators of histone *H3 K9* methylation. PLoS Genet 2009;5:e1000435. http://dx.doi.org/10.1371/journal.pgen.1000435. Published online Epub Mar.

[116] Sun Y, Jiang X, Xu Y, Ayrapetov MK, Moreau LA, Whetstine JR, et al. Histone H3 methylation links DNA damage detection to activation of the tumour suppressor *Tip60*. Nat Cell Biol 2009;11:1376–82. http://dx.doi.org/10.1038/ncb1982. Published online Epub Nov, ncb1982 [pii].

[117] Zarebski M, Wiernasz E, Dobrucki JW. Recruitment of heterochromatin protein 1 to DNA repair sites. Cytom A 2009;75:619–25. http://dx.doi.org/10.1002/cyto.a.20734. Published online Epub Jul.

[118] Djupedal I, Ekwall K. Epigenetics: heterochromatin meets RNAi. Cell Res 2009;19:282–95. http://dx.doi.org/10.1038/cr.2009.13. Published online Epub Mar, cr200913 [pii].

[119] Barkess G, West AG. Chromatin insulator elements: establishing barriers to set heterochromatin boundaries. Epigenomics 2012;4:67–80. http://dx.doi.org/10.2217/epi.11.112. Published online Epub Feb.

[120] Bushey AM, Dorman ER, Corces VG. Chromatin insulators: regulatory mechanisms and epigenetic inheritance. Mol Cell 2008;32:1–9. http://dx.doi.org/10.1016/j.molcel.2008.08.017. Published online Epub Oct 10, S1097-2765(08)00605-9 [pii].

[121] Puccio H. Multicellular models of Friedreich ataxia. J Neurol 2009;256(Suppl. 1):18–24. http://dx.doi.org/10.1007/s00415-009-1004-1. Published online Epub Mar.

[122] Liu J, Verma PJ, Evans-Galea MV, Delatycki MB, Michalska A, Leung J, et al. Generation of induced pluripotent stem cell lines from Friedreich ataxia patients. Stem Cell Rev 2011;7:703–13. http://dx.doi.org/10.1007/s12015-010-9210-x. Published online Epub Sep.

[123] Chou CJ, Herman D, Gottesfeld JM. Pimelic diphenylamide 106 is a slow, tight-binding inhibitor of class I histone deacetylases. J Biol Chem 2008;283:35402–9. http://dx.doi.org/10.1074/jbc.M807045200. Published online Epub Dec 19, M807045200 [pii].

[124] Rai M, Soragni E, Chou CJ, Barnes G, Jones S, Rusche JR, et al. Two new pimelic diphenylamide *HDAC* inhibitors induce sustained frataxin upregulation in cells from Friedreich's ataxia patients and in a mouse model. PLoS One 2010;5:e8825. http://dx.doi.org/10.1371/journal.pone.0008825.

[125] Rai M, Soragni E, Jenssen K, Burnett R, Herman D, Coppola G, et al. *HDAC* inhibitors correct frataxin deficiency in a Friedreich ataxia mouse model. PLoS one 2008;3:e1958.

[126] Sandi C, Pinto RM, Al-Mahdawi S, Ezzatizadeh V, Barnes G, Jones S, et al. Prolonged treatment with pimelic o-aminobenzamide *HDAC* inhibitors ameliorates the disease phenotype of a Friedreich ataxia mouse model. Neurobiol Dis 2011;42:496–505. http://dx.doi.org/10.1016/j.nbd.2011.02.016. Published online Epub Jun, S0969-9961(11)00070-2 [pii].

[127] Xu C, Soragni E, Chou CJ, Herman D, Plasterer HL, Rusche JR, et al. Chemical probes identify a role for histone deacetylase 3 in Friedreich's ataxia gene silencing. Chem Biol 2009;16:980–9. http://dx.doi.org/10.1016/j.chembiol.2009.07.010. Published online Epub Sep 25.

[128] Libri V, Yandim C, Athanasopoulos S, Loyse N, Natisvili T, Law PP, et al. Epigenetic and neurological effects and safety of high-dose nicotinamide in patients with Friedreich's ataxia: an exploratory, open-label, dose-escalation study. Lancet 2014. http://dx.doi.org/10.1016/S0140–6736(14)60382-2. Published online Epub Apr 30.

[129] Soragni E, Miao W, Iudicello M, Jacoby D, De Mercanti S, Clerico M, et al. Epigenetic therapy for Friedreich ataxia. Ann Neurology 2014;76:489–508. http://dx.doi.org/10.1002/ana.24260. Published online Epub Oct.

[130] Fu YH, Pizzuti A, Fenwick Jr RG, King J, Rajnarayan S, Dunne PW, et al. An unstable triplet repeat in a gene related to myotonic muscular dystrophy. Science 1992;255:1256–8. Published online Epub Mar 6.

[131] Mahadevan M, Tsilfidis C, Sabourin L, Shutler G, Amemiya C, Jansen G, et al. Myotonic dystrophy mutation: an unstable CTG repeat in the 3' untranslated region of the gene. Science 1992;255:1253–5.

[132] Brook JD, McCurrach ME, Harley HG, Buckler AJ, Church D, Aburatani H, et al. Molecular basis of myotonic dystrophy: expansion of a trinucleotide (CTG) repeat at the 3' end of a transcript encoding a protein kinase family member. Cell 1992;69:385. Published online Epub Apr 17.

[133] Liquori CL, Ricker K, Moseley ML, Jacobsen JF, Kress W, Naylor SL, et al. Myotonic dystrophy type 2 caused by a CCTG expansion in intron 1 of *ZNF9*. Science 2001;293:864–7. Published online Epub Aug 3 (10.1126/science.1062125293/5531/864 [pii]).

[134] Gomes-Pereira M, Monckton DG. Chemical modifiers of unstable expanded simple sequence repeats: what goes up, could come down. Mutat Res 2006;598:15–34. http://dx.doi.org/10.1016/j.mrfmmm.2006.01.011. Published online Epub Jun 25, S0027-5107(06)00054-6 [pii].

[135] Ashizawa T, Sarkar PS. Myotonic dystrophy types 1 and 2. Handb Clin Neurol 2011;101:193–237. http://dx.doi.org/10.1016/B978-0-08-045031-5.00015-3. B978-0-08-045031-5.00015-3 [pii].

[136] Lee JE, Cooper TA. Pathogenic mechanisms of myotonic dystrophy. Biochem Soc Trans 2009;37:1281–6. http://dx.doi.org/10.1042/BST0371281. Published online Epub Dec, BST0371281 [pii].

[137] Mankodi A, Lin X, Blaxall BC, Swanson MS, Thornton CA. Nuclear RNA foci in the heart in myotonic dystrophy. Circ Res 2005;97:1152–5. http://dx.doi.org/10.1161/01.RES.0000193598.89753.e3. Published online Epub Nov 25, 01.RES.0000193598.89753.e3 [pii].

[138] Jiang H, Mankodi A, Swanson MS, Moxley RT, Thornton CA. Myotonic dystrophy type 1 is associated with nuclear foci of mutant RNA, sequestration of muscleblind proteins and deregulated alternative splicing in neurons. Hum Mol Genet 2004;13:3079–88. Published online Epub Dec 15.

[139] Lin X, Miller JW, Mankodi A, Kanadia RN, Yuan Y, Moxley RT, et al. Failure of *MBNL1*-dependent post-natal splicing transitions in myotonic dystrophy. Hum Mol Genet 2006;15:2087–97. http://dx.doi.org/10.1093/hmg/ddl132. Published online Epub Jul 1, ddl132 [pii].

[140] Kuyumcu-Martinez NM, Wang GS, Cooper TA. Increased steady-state levels of *CUGBP1* in myotonic dystrophy 1 are due to *PKC*-mediated hyperphosphorylation. Mol Cell 2007;28:68–78. http://dx.doi.org/10.1016/j.molcel.2007.07.027. Published online Epub Oct 12, S1097-2765(07)00544-8 [pii].

[141] Otten AD, Tapscott SJ. Triplet repeat expansion in myotonic dystrophy alters the adjacent chromatin structure. Proc Natl Acad Sci USA 1995;92:5465–9.

[142] Klesert TR, Otten AD, Bird TD, Tapscott SJ. Trinucleotide repeat expansion at the myotonic dystrophy locus reduces expression of *DMAHP*. Nat Genet 1997;16:402–6.

[143] Wang YH, Griffith J. Expanded CTG triplet blocks from the myotonic dystrophy gene create the strongest known natural nucleosome positioning elements. Genomics 1995;25:570–3. Published online Epub Jan 20.

[144] Wang YH, Amirhaeri S, Kang S, Wells RD, Griffith JD. Preferential nucleosome assembly at DNA triplet repeats from the myotonic dystrophy gene. Science 1994;265:669–71. Published online Epub Jul.

[145] Klesert TR, Cho DH, Clark JI, Maylie J, Adelman J, Snider L, et al. Mice deficient in *Six5* develop cataracts: implications for myotonic dystrophy. Nat Genet 2000;25:105–9. http://dx.doi.org/10.1038/75490. Published online Epub May.

[146] Inukai A, Doyu M, Kato T, Liang Y, Kuru S, Yamamoto M, et al. Reduced expression of *DMAHP/SIX5* gene in myotonic dystrophy muscle. Muscle Nerve 2000;23:1421–6.

[147] Sarkar PS, Appukuttan B, Han J, Ito Y, Ai C, Tsai W, et al. Heterozygous loss of *Six5* in mice is sufficient to cause ocular cataracts. Nat Genet 2000;25:110–4.

[148] Cho DH, Thienes CP, Mahoney SE, Analau E, Filippova GN, Tapscott SJ. Antisense transcription and heterochromatin at the *DM1* CTG repeats are constrained by *CTCF*. Mol Cell 2005;20:483–9. http://dx.doi.org/10.1016/j.molcel.2005.09.002. Published online Epub Nov 11.

[149] Lopez Castel A, Nakamori M, Tome S, Chitayat D, Gourdon G, Thornton CA, et al. Pearson, Expanded CTG repeat demarcates a boundary for abnormal CpG methylation in myotonic dystrophy patient tissues. Hum Mol Genet 2011;20:1–15. http://dx.doi.org/10.1093/hmg/ddq427. Published online Epub Jan 1.

[150] Maueler W, Bassili G, Hardt C, Keyl HG, Epplen JT. A complex containing at least one zinc dependent HeLa nuclear protein binds to the intronic (GAA)(n) block of the frataxin gene. Gene 2001;270:131–43. Published online Epub May 30, S0378111901004875 [pii].

[151] Naumann F, Remus R, Schmitz B, Doerfler W. Gene structure and expression of the 5'-(CGG)(n)-3'-binding protein *(CGGBP1)*. Genomics 2004;83:106–18. Published online Epub Jan, S088875430300212X [pii].

[152] Timchenko LT, Timchenko NA, Caskey CT, Roberts R. Novel proteins with binding specificity for DNA CTG repeats and RNA CUG repeats: implications for myotonic dystrophy. Hum Mol Genet 1996;5:115–21. Published online Epub Jan, 5w0268 [pii].

[153] Matsuura T, Yamagata T, Burgess DL, Rasmussen A, Grewal RP, Watase K, et al. Large expansion of the ATTCT pentanucleotide repeat in spinocerebellar ataxia type 10. Nat Genet 2000;26:191–4.

[154] Bardoni B, Davidovic L, Bensaid M, Khandjian EW. The fragile X syndrome: exploring its molecular basis and seeking a treatment. Expert Rev Mol Med 2006;8:1–16. http://dx.doi.org/10.1017/S1462399406010751.

[155] Penagarikano O, Mulle JG, Warren ST. The pathophysiology of fragile x syndrome. Annu Rev Genomics Hum Genet 2007;8:109–29. http://dx.doi.org/10.1146/annurev.genom.8.080706.092249.

[156] Verkerk AJ, Pieretti M, Sutcliffe JS, Fu YH, Kuhl DP, Pizzuti A, et al. Identification of a gene *(FMR-1)* containing a CGG repeat coincident with a breakpoint cluster region exhibiting length variation in fragile X syndrome. Cell 1991;65:905–14. Published online Epub May, 0092-8674(91)90397-H [pii].

[157] Merenstein SA, Sobesky WE, Taylor AK, Riddle JE, Tran HX, Hagerman RJ. Molecular-clinical correlations in males with an expanded *FMR1* mutation. Am J Med Genet 1996;64:388–94. Published online Epub Aug (10.1002/(SICI)1096-8628(19960809)64:2<388::AID-AJMG31>3.0.CO;2-9).

[158] Bourgeois JA, Coffey SM, Rivera SM, Hessl D, Gane LW, Tassone F, et al. A review of fragile X premutation disorders: expanding the psychiatric perspective. J Clin Psychiatry 2009;70:852–62. http://dx.doi.org/10.4088/JCP.08m04476. Published online Epub Jun.

[159] Bhakar AL, Dölen G, Bear MF. The pathophysiology of fragile X (and what it teaches us about synapses). Annu Rev Neurosci 2012;35:417–43. http://dx.doi.org/10.1146/annurev-neuro-060909-153138.

[160] Oberle I, Rousseau F, Heitz D, Kretz C, Devys D, Hanauer A, et al. Instability of a 550-base pair DNA segment and abnormal methylation in fragile X syndrome. Science 1991;252:1097–102. http://dx.doi.org/10.1126/science.252.5009.1097. Published online Epub May.

[161] Ashley CT, Wilkinson KD, Reines D, Warren ST. *FMR1* protein: conserved *RNP* family domains and selective RNA binding. Science 1993;262:563–6. Published online Epub Oct.

[162] Gantois I, Pop AS, de Esch CE, Buijsen RA, Pooters T, Gomez-Mancilla B, et al. Chronic administration of AFQ056/mavoglurant restores social behaviour in *Fmr1* knockout mice. Behav Brain Res 2013;239:72–9. http://dx.doi.org/10.1016/j.bbr.2012.10.059. Published online Epub Feb.

[163] Bakker CE, de Diego Otero Y, Bontekoe C, Raghoe P, Luteijn T, Hoogeveen AT, et al. Immunocytochemical and biochemical characterization of *FMRP, FXR1P,* and *FXR2P* in the mouse. Exp Cell Res 2000;258:162–70. http://dx.doi.org/10.1006/excr.2000.4932. Published online Epub Jul.

[164] Zhao X, Pak C, Smrt RD, Jin P. Epigenetics and neural developmental disorders: Washington DC, September 18 and 19, 2006. Epigenetics 2007;2:126–34. Published online Epub 2007 Apr-Jun.

[165] Qin M, Kang J, Burlin TV, Jiang C, Smith CB. Postadolescent changes in regional cerebral protein synthesis: an in vivo study in the *FMR1* null mouse. J Neurosci 2005;25:5087–95. http://dx.doi.org/10.1523/JNEUROSCI.0093-05.2005. Published online Epub May.

[166] Chiurazzi P, Pomponi MG, Willemsen R, Oostra BA, Neri G. In vitro reactivation of the *FMR1* gene involved in fragile X syndrome. Hum Mol Genet 1998;7:109–13. Published online Epub Jan.

[167] Chiurazzi P, Pomponi MG, Pietrobono R, Bakker CE, Neri G, Oostra BA. Synergistic effect of histone hyperacetylation and DNA demethylation in the reactivation of the *FMR1* gene. Hum Mol Genet 1999;8:2317–23. Published online Epub Nov.

[168] Coffee B, Zhang F, Warren ST, Reines D. Acetylated histones are associated with *FMR1* in normal but not fragile X syndrome cells. Nat Genet 1999;22:98–101. http://dx.doi.org/10.1038/8807. Published online Epub May.

[169] Coffee B, Zhang F, Ceman S, Warren ST, Reines D. Histone modifications depict an aberrantly heterochromatinized *FMR1* gene in fragile X syndrome. Am J Hum Genet 2002;71:923–32. http://dx.doi.org/10.1086/342931. Published online Epub Oct.

[170] Tabolacci E, Pietrobono R, Moscato U, Oostra BA, Chiurazzi P, Neri G. Differential epigenetic modifications in the *FMR1* gene of the fragile X syndrome after reactivating pharmacological treatments. Eur J Hum Genet 2005;13:641–8. http://dx.doi.org/10.1038/sj.ejhg.5201393. Published online Epub May.

[171] Eden S, Hashimshony T, Keshet I, Cedar H, Thorne AW. DNA methylation models histone acetylation. Nature 1998;394:842. http://dx.doi.org/10.1038/29680. Published online Epub Aug.

[172] Neri G, Chiurazzi P. X-linked mental retardation. Adv Genet 1999;41:55–94.

[173] Sølvsten C, Nielsen AL. *FMR1* CGG repeat lengths mediate different regulation of reporter gene expression in comparative transient and locus specific integration assays. Gene 2011;486:15–22. http://dx.doi.org/10.1016/j.gene.2011.06.034. Published online Epub Oct.

[174] Ben-Yosef D, Malcov M, Eiges R. *PGD*-derived human embryonic stem cell lines as a powerful tool for the study of human genetic disorders. Mol Cell Endocrinol 2008;282:153–8. http://dx.doi.org/10.1016/j.mce.2007.11.010. Published online Epub Jan.

[175] Stöger R, Genereux DP, Hagerman RJ, Hagerman PJ, Tassone F, Laird CD. Testing the *FMR1* promoter for mosaicism in DNA methylation among CpG sites, strands, and cells in *FMR1*-expressing males with fragile X syndrome. PLoS One 2011;6:e23648. http://dx.doi.org/10.1371/journal.pone.0023648.

[176] Maddalena A, Richards CS, McGinniss MJ, Brothman A, Desnick RJ, Grier RE, et al. Technical standards and guidelines for fragile X: the first of a series of disease-specific supplements to the standards and guidelines for clinical genetics Laboratories of the American College of Medical Genetics. Quality Assurance Subcommittee of the Laboratory Practice Committee. Genet Med 2001;3:200–5. http://dx.doi.org/10.1097/00125817-200105000-00010. Published online Epub 2001 May-Jun.

[177] Hagerman RJ, Hagerman PJ. The fragile X premutation: into the phenotypic fold. Curr Opin Genet Dev 2002;12:278–83. Published online Epub Jun.

[178] Hagerman PJ, Hagerman RJ. The fragile-X premutation: a maturing perspective. Am J Hum Genet 2004;74:805–16. http://dx.doi.org/10.1086/386296. Published online Epub May.

[179] Bacalman S, Farzin F, Bourgeois JA, Cogswell J, Goodlin-Jones BL, Gane LW, et al. Psychiatric phenotype of the fragile X-associated tremor/ataxia syndrome (FXTAS) in males: newly described fronto-subcortical dementia. J Clin Psychiatry 2006;67:87–94. Published online Epub Jan.

[180] Sherman SL. Premature ovarian failure among fragile X premutation carriers: parent-of-origin effect?. Am J Hum Genet 2000;67:11–3. http://dx.doi.org/10.1086/302985. Published online Epub Jul.

[181] Jacquemont S, Hagerman RJ, Leehey M, Grigsby J, Zhang L, Brunberg JA, et al. Fragile X premutation tremor/ataxia syndrome: molecular, clinical, and neuroimaging correlates. Am J Hum Genet 2003;72:869–78. http://dx.doi.org/10.1086/374321. Published online Epub Apr.

[182] Tassone F, Hagerman RJ, Taylor AK, Gane LW, Godfrey TE, Hagerman PJ. Elevated levels of *FMR1* mRNA in carrier males: a new mechanism of involvement in the fragile-X syndrome. Am J Hum Genet 2000;66:6–15. http://dx.doi.org/10.1086/302720. Published online Epub Jan.

[183] Tassone F, Beilina A, Carosi C, Albertosi S, Bagni C, Li L, et al. Elevated *FMR1* mRNA in premutation carriers is due to increased transcription. RNA 2007;13:555–62. http://dx.doi.org/10.1261/rna.280807. Published online Epub Apr.

[184] Kenneson A, Zhang F, Hagedorn CH, Warren ST. Reduced *FMRP* and increased *FMR1* transcription is proportionally associated with CGG repeat number in intermediate-length and premutation carriers. Hum Mol Genet 2001;10:1449–54. Published online Epub Jul.

[185] Hagerman RJ, Hagerman PJ. Fragile X syndrome: a model of gene-brain-behavior relationships. Mol Genet Metab 2001;74:89–97. http://dx.doi.org/10.1006/mgme.2001.3225. Published online Epub 2001 Sep-Oct.

[186] Hoem G, Raske CR, Garcia-Arocena D, Tassone F, Sanchez E, Ludwig AL, et al. CGG-repeat length threshold for *FMR1* RNA pathogenesis in a cellular model for FXTAS. Hum Mol Genet 2011;20:2161–70. http://dx.doi.org/10.1093/hmg/ddr101. Published online Epub Jun.

[187] Sheridan SD, Theriault KM, Reis SA, Zhou F, Madison JM, Daheron L, et al. Epigenetic characterization of the *FMR1* gene and aberrant neurodevelopment in human induced pluripotent stem cell models of fragile X syndrome. PLoS One 2011;6:e26203. http://dx.doi.org/10.1371/journal.pone.0026203.

[188] Wang T, Pan Q, Lin L, Szulwach KE, Song CX, He C, et al. Genome-wide DNA hydroxymethylation changes are associated with neurodevelopmental genes in the developing human cerebellum. Hum Mol Genet 2012;21:5500–10. http://dx.doi.org/10.1093/hmg/dds394. Published online Epub Dec.

[189] Padberg G. In: U.M., C.D.N., editors. Facioscapulohumeral muscular dystrophy: clinical medicine and molecular cell biology. Abingdon: (Garland Science/BIOS Scientific; 2004. p. 41–54.

[190] Tawil R, Figlewicz DA, Griggs RC, Weiffenbach B. Facioscapulohumeral dystrophy: a distinct regional myopathy with a novel molecular pathogenesis. FSH Consortium. Ann Neurol 1998;43:279–82. Published online Epub Mar.

[191] Tawil R. Facioscapulohumeral muscular dystrophy. Curr Neurol Neurosci Rep 2004;4:51–4. Published online Epub Jan.

[192] van Deutekom JC, Wijmenga C, van Tienhoven EA, Gruter AM, Hewitt JE, Padberg GW, et al. FSHD-associated DNA rearrangements are due to deletions of integral copies of a 3.2 kb tandemly repeated unit. Hum Mol Genet 1993;2:2037–42. Published online Epub Dec.

[193] Wijmenga C, Hewitt JE, Sandkuijl LA, Clark LN, Wright TJ, Dauwerse HG, et al. Chromosome 4q DNA rearrangements associated with facioscapulohumeral muscular dystrophy. Nat Genet 1992;2:26–30. Published online Epub Sep.

[194] Lunt PW. In: 44th ENMC International Workshop. Facioscapulohumeral muscular dystrophy: molecular studies 19-21 July 1996, Naarden, The Netherlands. Neuromuscul Disord, ;8. 1998. p. 126–30. Published online Epub Apr.

[195] Richards M, Coppée F, Thomas N, Belayew A, Upadhyaya M. Facioscapulohumeral muscular dystrophy (FSHD): an enigma unravelled?. Hum Genet 2012;131:325–40. http://dx.doi.org/10.1007/s00439-011-1100-z. Published online Epub Mar.

[196] Tawil R, Forrester J, Griggs RC, Mendell J, Kissel J, McDermott M, et al. Evidence for anticipation and association of deletion size with severity in facioscapulohumeral muscular dystrophy. The FSH-DY Group. Ann Neurol 1996;39:744–8. Published online Epub Jun.

[197] Gabriels J, Beckers MC, Ding H, De Vriese A, Plaisance S, van der Maarel SM, et al. Nucleotide sequence of the partially deleted *D4Z4* locus in a patient with FSHD identifies a putative gene within each 3.3 kb element. Gene 1999;236:25–32. Published online Epub Aug 5.

[198] Lyle R, Wright TJ, Clark LN, Hewitt JE. The FSHD-associated repeat, *D4Z4*, is a member of a dispersed family of homeobox-containing repeats, subsets of which are clustered on the short arms of the acrocentric chromosomes. Genomics 1995;28:389–97. http://dx.doi.org/10.1006/geno.1995.1166. Published online Epub Aug.

[199] Tupler R, Berardinelli A, Barbierato L, Frants R, Hewitt JE, Lanzi G, et al. Monosomy of distal 4q does not cause facioscapulohumeral muscular dystrophy. J Med Genet 1996;33:366–70. Published online Epub May.

[200] Lemmers RJ, de Kievit P, Sandkuijl L, Padberg GW, van Ommen GJ, Frants RR, et al. Facioscapulohumeral muscular dystrophy is uniquely associated with one of the two variants of the 4q subtelomere. Nat Genet 2002;32:235–6. http://dx.doi.org/10.1038/ng999. Published online Epub Oct.

[201] Lemmers RJ, van der Vliet PJ, van der Gaag KJ, Zuniga S, Frants RR, de Knijff P, et al. Worldwide population analysis of the 4q and 10q subtelomeres identifies only four discrete interchromosomal sequence transfers in human evolution. Am J Hum Genet 2010;86: 364–77. http://dx.doi.org/10.1016/j.ajhg.2010.01.035. Published online Epub Mar.

[202] Thomas NS, Wiseman K, Spurlock G, MacDonald M, Ustek D, Upadhyaya M. A large patient study confirming that facioscapulohumeral muscular dystrophy (FSHD) disease expression is almost exclusively associated with an FSHD locus located on a 4qA-defined 4qter subtelomere. J Med Genet 2007;44:215–8. http://dx.doi.org/10.1136/jmg.2006.042804. Published online Epub Mar.

[203] Lemmers RJ, van der Vliet PJ, Klooster R, Sacconi S, Camaño P, Dauwerse JG, et al. A unifying genetic model for facioscapulohumeral muscular dystrophy. Science 2010;329:1650–3. http://dx.doi.org/10.1126/science.1189044. Published online Epub Sep.

[204] Spurlock G, Jim HP, Upadhyaya M. Confirmation that the specific *SSLP* microsatellite allele 4qA161 segregates with fascioscapulohumeral muscular dystrophy (FSHD) in a cohort of multiplex and simplex FSHD families. Muscle Nerve 2010;42:820–1. http://dx.doi.org/10.1002/mus.21766. Published online Epub Nov.

[205] Fisher J, Upadhyaya M. Molecular genetics of facioscapulohumeral muscular dystrophy (FSHD). Neuromuscular Disorders. NMD 1997;7:55–62. Published online Epub Jan.

[206] Jiang G, Yang F, van Overveld PG, Vedanarayanan V, van der Maarel S, Ehrlich M. Testing the position-effect variegation hypothesis for facioscapulohumeral muscular dystrophy by analysis of histone modification and gene expression in subtelomeric 4q. Hum Mol Genet 2003;12:2909–21. http://dx.doi.org/10.1093/hmg/ddg323. Published online Epub Nov.

[207] Zeng W, de Greef JC, Chen YY, Chien R, Kong X, Gregson HC, et al. Specific loss of histone H3 lysine 9 trimethylation and *HP1gamma*/cohesin binding at *D4Z4* repeats is associated with facioscapulohumeral muscular dystrophy (FSHD). PLoS Genet 2009;5:e1000559. http://dx.doi.org/10.1371/journal.pgen.1000559. Published online Epub Jul.

[208] van Overveld PG, Lemmers RJ, Sandkuijl LA, Enthoven L, Winokur ST, Bakels F, et al. Hypomethylation of *D4Z4* in 4q-linked and non-4q-linked facioscapulohumeral muscular dystrophy. Nat Genet 2003;35:315–7. http://dx.doi.org/10.1038/ng1262. Published online Epub Dec.

[209] van Overveld PG, Enthov L, Ricci E, Rossi M, Felicetti L, Jeanpierre M, et al. Variable hypomethylation of D4Z4 in facioscapulohumeral muscular dystrophy. Ann Neurol 2005;58:569–76. http://dx.doi.org/10.1002/ana.20625. Published online Epub Oct.

[210] Gabellini D, D'Antona G, Moggio M, Prelle A, Zecca C, Adami R, et al. Facioscapulohumeral muscular dystrophy in mice overexpressing *Frg1*. Nature 2006;439:973–7. Published online Epub Feb 23.

[211] Gabellini D, Green MR, Tupler R. Inappropriate gene activation in FSHD: a repressor complex binds a chromosomal repeat deleted in dystrophic muscle. Cell 2002;110:339–48. Published online Epub Aug.

[212] Klooster R, Straasheijm K, Shah B, Sowden J, Frants R, Thornton C, et al. Comprehensive expression analysis of FSHD candidate genes at the mRNA and protein level. Eur J Hum Genet 2009;17:1615–24. http://dx.doi.org/10.1038/ejhg.2009.62. Published online Epub Dec.

[213] Arashiro P, Eisenberg I, Kho AT, Cerqueira AM, Canovas M, Silva HC, et al. Transcriptional regulation differs in affected facioscapulohumeral muscular dystrophy patients compared to asymptomatic related carriers. Proc Natl Acad Sci U. S. A 2009;106:6220–5. http://dx.doi.org/10.1073/pnas.0901573106. Published online Epub Apr.

[214] Masny PS, Chan OY, de Greef JC, Bengtsson U, Ehrlich M, Tawil R, et al. Analysis of allele-specific RNA transcription in FSHD by RNA-DNA FISH in single myonuclei. Eur J Hum Genet 2010;18:448–56. http://dx.doi.org/10.1038/ejhg.2009.183. Published online Epub Apr.

[215] Rijkers T, Deidda G, van Koningsbruggen S, van Geel M, Lemmers RJ, van Deutekom JC, et al. *Frg2*, an FSHD candidate gene, is transcriptionally upregulated in differentiating primary myoblast cultures of FSHD patients. J Med Genet 2004;41:826–36. http://dx.doi.org/10.1136/jmg.2004.019364. Published online Epub Nov.

[216] Deak KL, Lemmers RJ, Stajich JM, Klooster R, Tawil R, Frants RR, et al. Genotype-phenotype study in an FSHD family with a proximal deletion encompassing p13E-11 and D4Z4. Neurology 2007;68:578–82. http://dx.doi.org/10.1212/01.wnl.0000254991.21818.f3. Published online Epub Feb.

[217] Yip DJ, Picketts DJ. Increasing D4Z4 repeat copy number compromises C2C12 myoblast differentiation. FEBS Lett 2003;537:133–8. Published online Epub Feb.

[218] Dixit M, Ansseau E, Tassin A, Winokur S, Shi R, Qian H, et al. DUX4, a candidate gene of facioscapulohumeral muscular dystrophy, encodes a transcriptional activator of PITX1. Proc Natl Acad Sci U. S. A 2007;104:18157–62. http://dx.doi.org/10.1073/pnas.0708659104. Published online Epub Nov.

[219] Geng LN, Yao Z, Snider L, Fong AP, Cech JN, Young JM, et al. DUX4 activates germline genes, retroelements, and immune mediators: implications for facioscapulohumeral dystrophy. Dev Cell 2012;22:38–51. http://dx.doi.org/10.1016/j.devcel.2011.11.013. Published online Epub Jan.

[220] Kowaljow V, Marcowycz A, Ansseau E, Conde CB, Sauvage S, Mattéotti C, et al. The DUX4 gene at the FSHD1A locus encodes a pro-apoptotic protein. Neuromuscul Disord 2007;17:611–23. http://dx.doi.org/10.1016/j.nmd.2007.04.002. Published online Epub Aug.

[221] Vanderplanck C, Ansseau E, Charron S, Stricwant N, Tassin A, Laoudj-Chenivesse D, et al. The FSHD atrophic myotube phenotype is caused by DUX4 expression. PLoS One 2011;6:e26820. http://dx.doi.org/10.1371/journal.pone.0026820.

[222] Snider L, Geng LN, Lemmers RJ, Kyba M, Ware CB, Nelson AM, et al. Facioscapulohumeral dystrophy: incomplete suppression of a retrotransposed gene. PLoS Genet 2010;6:e1001181. http://dx.doi.org/10.1371/journal.pgen.1001181. Published online Epub Oct.

[223] van der Maarel SM, Tawil R, Tapscott SJ. Facioscapulohumeral muscular dystrophy and DUX4: breaking the silence. Trends Mol Med 2011;17:252–8. Published online Epub May.

[224] Smeets DF, Moog U, Weemaes CM, Vaes-Peeters G, Merkx GF, Niehof JP, et al. ICF syndrome: a new case and review of the literature. Hum Genet 1994;94:240–6. Published online Epub Sep.

[225] Hansen RS, Wijmenga C, Luo P, Stanek AM, Canfield TK, Weemaes CM, et al. The DNMT3B DNA methyltransferase gene is mutated in the ICF immunodeficiency syndrome. Proc Natl Acad Sci U. S. A 1999;96:14412–7. Published online Epub Dec.

[226] Bachman KE, Rountree MR, Baylin SB. Dnmt3a and Dnmt3b are transcriptional repressors that exhibit unique localization properties to heterochromatin. J Biol Chem 2001;276:32282–7. Published online Epub Aug 24.

[227] Chen T, Tsujimoto N, Li E. The PWWP domain of Dnmt3a and Dnmt3b is required for directing DNA methylation to the major satellite repeats at pericentric heterochromatin. Mol Cell Biol 2004;24:9048–58. Published online Epub Oct.

[228] Wijmenga C, Hansen RS, Gimelli G, Bjorck EJ, Davies EG, Valentine D, et al. Genetic variation in ICF syndrome: evidence for genetic heterogeneity. Hum Mutat 2000;16:509–17. Published online Epub Dec.

[229] Shirohzu H, Kubota T, Kumazawa A, Sado T, Chijiwa T, Inagaki K, et al. Three novel DNMT3B mutations in Japanese patients with ICF syndrome. Am J Med Genet 2002;112:31–7. Published online Epub Sep. 15.

[230] Okano M, Bell DW, Haber DA, Li E. DNA methyltransferases Dnmt3a and Dnmt3b are essential for de novo methylation and mammalian development. Cell 1999;99:247–57. Published online Epub Oct 29.

[231] Matarazzo MR, De Bonis ML, Vacca M, Della Ragione F, D'Esposito M. Lessons from two human chromatin diseases: ICF syndrome and Rett syndrome. Int J Biochem Cell Biol 2009;41:117–26. http://dx.doi.org/10.1016/j.biocel.2008.07.026. Published online Epub Jan.

[232] Ueda Y, Okano M, Williams C, Chen T, Georgopoulos K, Li E. Roles for Dnmt3b in mammalian development: a mouse model for the ICF syndrome. Development 2006;133:1183–92. http://dx.doi.org/10.1242/dev.02293. Published online Epub Mar.

[233] Jiang YL, Rigolet M, Bourc'his D, Nigon F, Bokesoy I, Fryns JP, et al. DNMT3B mutations and DNA methylation defect define two types of ICF syndrome. Hum Mutat 2005;25:56–63. http://dx.doi.org/10.1002/humu.20113. Published online Epub Jan.

[234] Hagleitner MM, Lankester A, Maraschio P, Hultén M, Fryns JP, Schuetz C, et al. Clinical spectrum of immunodeficiency, centromeric instability and facial dysmorphism (ICF syndrome). J Med Genet 2008;45:93–9. http://dx.doi.org/10.1136/jmg.2007.053397. Published online Epub Feb.

[235] Saito Y, Kanai Y, Sakamoto M, Saito H, Ishii H, Hirohashi S. Overexpression of a splice variant of DNA methyltransferase 3b, DNMT3b4, associated with DNA hypomethylation on pericentromeric satellite regions during human hepatocarcinogenesis. Proc Natl Acad Sci U. S. A 2002;99:10060–5. Published online Epub Jul 23.

[236] de Greef JC, Wang J, Balog J, den Dunnen JT, Frants RR, Straasheijm KR, et al. Mutations in ZBTB24 are associated with immunodeficiency, centromeric instability, and facial anomalies syndrome type 2. Am J Hum Genet 2011;88:796–804. http://dx.doi.org/10.1016/j.ajhg.2011.04.018. Published online Epub Jun.

[237] Chouery E, Abou-Ghoch J, Corbani S, El Ali N, Korban R, Salem N, et al. A novel deletion in ZBTB24 in a Lebanese family with immunodeficiency, centromeric instability, and facial anomalies syndrome type 2. Clin Genet 2012;82:489–93. http://dx.doi.org/10.1111/j.1399-0004.2011.01783.x. Published online Epub Nov.

[238] Heyn H, Vidal E, Sayols S, Sanchez-Mut JV, Moran S, Medina I, et al. Whole-genome bisulfite DNA sequencing of a DNMT3B mutant patient. Epigenetics 2012;7:542–50. http://dx.doi.org/10.4161/epi.20523. Published online Epub Jun.

[239] Gisselsson D, Shao C, Tuck-Muller CM, Sogorovic S, Pålsson E, Smeets D, et al. Interphase chromosomal abnormalities and mitotic missegregation of hypomethylated sequences in ICF syndrome cells. Chromosoma 2005;114:118–26. http://dx.doi.org/10.1007/s00412-005-0343-7. Published online Epub Jul.

[240] Jeanpierre M, Turleau C, Aurias A, Prieur M, Ledeist F, Fischer A, et al. An embryonic-like methylation pattern of classical satellite DNA is observed in ICF syndrome. Hum Mol Genet 1993;2:731–5. Published online Epub Jun.

[241] Ji W, Hernandez R, Zhang XY, Qu GZ, Frady A, Varela M, et al. DNA demethylation and pericentromeric rearrangements of chromosome 1. Mutat Res 1997;379:33–41. Published online Epub Sep.

[242] Prada D, González R, Sánchez L, Castro C, Fabián E, Herrera LA. Satellite 2 demethylation induced by 5-azacytidine is associated with missegregation of chromosomes 1 and 16 in human somatic cells. Mutat Res 2012;729:100–5. http://dx.doi.org/10.1016/j.mrfmmm.2011.10.007. Published online Epub Jan.

[243] Miniou P, Jeanpierre M, Blanquet V, Sibella V, Bonneau D, Herbelin C, et al. Abnormal methylation pattern in constitutive and facultative (X inactive chromosome) heterochromatin of ICF patients. Hum Mol Genet 1994;3:2093–102. Published online Epub Dec.

[244] Ehrlich M, Sanchez C, Shao C, Nishiyama R, Kehrl J, Kuick R, et al. ICF, an immunodeficiency syndrome: DNA methyltransferase 3B involvement, chromosome anomalies, and gene dysregulation. Autoimmunity 2008;41:253–71. http://dx.doi.org/10.1080/08916930802024202. Published online Epub May.

[245] Kondo T, Bobek MP, Kuick R, Lamb B, Zhu X, Narayan A, et al. Whole-genome methylation scan in ICF syndrome: hypomethylation of non-satellite DNA repeats D4Z4 and NBL2. Hum Mol Genet 2000;9:597–604. Published online Epub Mar 1.

[246] McDowell TL, Gibbons RJ, Sutherland H, O'Rourke DM, Bickmore WA, Pombo A, et al. Localization of a putative transcriptional regulator (ATRX) at pericentromeric heterochromatin and the short arms of acrocentric chromosomes. Proc Natl Acad Sci U. S. A 1999;96:13983–8. Published online Epub Nov 23.

[247] Qiu C, Sawada K, Zhang X, Cheng X. The PWWP domain of mammalian DNA methyltransferase Dnmt3b defines a new family of DNA-binding folds. Nat Struct Biol 2002;9:217–24. Published online Epub Mar.

[248] Maurer-Stroh S, Dickens NJ, Hughes-Davies L, Kouzarides T, Eisenhaber F, Ponting CP. The tudor domain 'royal family': tudor, plant agenet, chromo, PWWP, and MBT domains. Trends Biochem Sci 2003;28:69–74. Published online Epub Feb.

[249] Ehrlich M, Buchanan KL, Tsien F, Jiang G, Sun B, Uicker W, et al. DNA methyltransferase 3B mutations linked to the ICF syndrome cause dysregulation of lymphogenesis genes. Hum Mol Genet 2001;10:2917–31. Published online Epub Dec 1.

[250] Bickmore WA, van der Maarel SM. Perturbations of chromatin structure in human genetic disease: recent advances. Hum Mol Genet 2003;12(Spec No 2):R207–13. Published online Epub Oct 15.

[251] Brown KE, Baxter J, Graf D, Merkenschlager M, Fisher AG. Dynamic repositioning of genes in the nucleus of lymphocytes preparing for cell division. Mol Cell 1999;3:207–17.

[252] Alcobia I, Quina AS, Neves H, Clode N, Parreira L. The spatial organization of centromeric heterochromatin during normal human lymphopoiesis: evidence for ontogenically determined spatial patterns. Exp Cell Res 2003;290:358–69. Published online Epub Nov.

[253] Gasser SM. Positions of potential: nuclear organization and gene expression. Cell 2001;104:639–42. Published online Epub Mar.

[254] Jefferson A, Colella S, Moralli D, Wilson N, Yusuf M, Gimelli G, et al. Altered intra-nuclear organisation of heterochromatin and genes in ICF syndrome. PLoS One 2010;5:e11364. http://dx.doi.org/10.1371/journal.pone.0011364.

[255] Dupont C, Guimiot F, Perrin L, Marey I, Smiljkovski D, Le Tessier D, et al. 3D position of pericentromeric heterochromatin within the nucleus of a patient with ICF syndrome. Clin Genet 2012;82:187–92. http://dx.doi.org/10.1111/j.1399-0004.2011.01697.x. Published online Epub Aug.

[256] Luciani JJ, Depetris D, Missirian C, Mignon-Ravix C, Metzler-Guillemain C, Megarbane A, et al. Subcellular distribution of HP1 proteins is altered in ICF syndrome. Eur J Hum Genet 2005;13:41–51. Published online Epub Jan.

[257] Lana E, Mégarbané A, Tourrière H, Sarda P, Lefranc G, Claustres M, et al. DNA replication is altered in immunodeficiency centromeric instability facial anomalies (ICF) cells carrying DNMT3B mutations. Eur J Hum Genet 2012;20:1044–50. http://dx.doi.org/10.1038/ejhg.2012.41. Published online Epub Oct.

[258] Neul JL, Kaufmann WE, Glaze DG, Christodoulou J, Clarke AJ, Bahi-Buisson N, et al. Rett syndrome: revised diagnostic criteria and nomenclature. Ann Neurol 2010;68:944–50. Published online Epub Dec.

[259] Girard M, Couvert P, Carrie A, Tardieu M, Chelly J, Beldjord C, et al. Parental origin of de novo MECP2 mutations in Rett syndrome. Eur J Hum Genet 2001;9:231–6. Published online Epub Mar.

[260] Amir RE, Van den Veyver IB, Wan M, Tran CQ, Francke U, Zoghbi HY. Rett syndrome is caused by mutations in X-linked MECP2, encoding methyl-CpG-binding protein 2. Nat Genet 1999;23:185–8. Published online Epub Oct.

[261] J. Christodoulou.

[262] Villard L, Kpebe A, Cardoso C, Chelly PJ, Tardieu PM, Fontes M. Two affected boys in a Rett syndrome family: clinical and molecular findings. Neurology 2000;55:1188–93. Published online Epub Oct.

[263] Klose RJ, Bird AP. Genomic DNA methylation: the mark and its mediators. Trends Biochem Sci 2006;31:89–97. Published online Epub Feb.

[264] Meehan RR, Lewis JD, Bird AP. Characterization of MeCP2, a vertebrate DNA binding protein with affinity for methylated DNA. Nucleic Acids Res 1992;20:5085–92. Published online Epub Oct 11.

[265] Klose RJ, Sarraf SA, Schmiedeberg L, McDermott SM, Stancheva I, Bird AP. DNA binding selectivity of MeCP2 due to a requirement for A/T sequences adjacent to methyl-CpG. Mol Cell 2005;19:667–78. Published online Epub Sep 2.

[266] Buschdorf JP, Stratling WH. A WW domain binding region in methyl-CpG-binding protein MeCP2: impact on Rett syndrome. J Mol Med Berl 2004;82:135–43. Published online Epub Feb.

[267] Muotri AR, Marchetto MC, Coufal NG, Oefner R, Yeo G, Nakashima K, et al. L1 retrotransposition in neurons is modulated by MeCP2. Nature 2010;468:443–6. Published online Epub Nov 18.

[268] Chahrour M, Jung SY, Shaw C, Zhou X, Wong ST, Qin J, et al. MeCP2, a key contributor to neurological disease, activates and represses transcription. Science 2008;320:1224–9. Published online Epub May 30.

[269] Nikitina T, Shi X, Ghosh RP, Horowitz-Scherer RA, Hansen JC, Woodcock CL. Multiple modes of interaction between the methylated DNA binding protein MeCP2 and chromatin. Mol Cell Biol 2007;27:864–77. Published online Epub Feb.

[270] Georgel PT, Horowitz-Scherer RA, Adkins N, Woodcock CL, Wade PA, Hansen JC. Chromatin compaction by human MeCP2: assembly of novel secondary chromatin structures in the absence of DNA methylation. J Biol Chem 2003;278:32181–8. Published online Epub Aug 22.

[271] Long SW, Ooi JY, Yau PM, Jones PL. A brain-derived MeCP2 complex supports a role for MeCP2 in RNA processing. Biosci Rep 2011;31:333–43. Published online Epub Oct 1.

[272] Shahbazian MD, Antalffy B, Armstrong DL, Zoghbi HY. Insight into Rett syndrome: MeCP2 levels display tissue- and cell-specific differences and correlate with neuronal maturation. Hum Mol Genet 2002;11:115–24. Published online Epub Jan 15.

[273] Ballas N, Lioy DT, Grunseich C, Mandel G. Non-cell autonomous influence of MeCP2-deficient glia on neuronal dendritic morphology. Nat Neurosci 2009;12:311–7. Published online Epub Mar.

[274] Kriaucionis S, Bird A. The major form of MeCP2 has a novel N-terminus generated by alternative splicing. Nucleic Acids Res 2004;32:1818–23.

[275] Mnatzakanian GN, Lohi H, Munteanu I, Alfred SE, Yamada T, MacLeod PJ, et al. A previously unidentified MECP2 open reading frame defines a new protein isoform relevant to Rett syndrome. Nat Genet 2004;36:339–41. Published online Epub Apr.

[276] Chen RZ, Akbarian S, Tudor M, Jaenisch R. Deficiency of methyl-CpG binding protein-2 in CNS neurons results in a Rett-like phenotype in mice. Nat Genet 2001;27:327–31. http://dx.doi.org/10.1038/85906. Published online Epub Mar.

[277] Guy J, Hendrich B, Holmes M, Martin JE, Bird A. A mouse *Mecp2*-null mutation causes neurological symptoms that mimic Rett syndrome. Nat Genet 2001;27:322–6. http://dx.doi.org/10.1038/85899. Published online Epub Mar.

[278] Guy J, Gan J, Selfridge J, Cobb S, Bird A. Reversal of neurological defects in a mouse model of Rett syndrome. Science 2007;315:1143–7. http://dx.doi.org/10.1126/science.1138389. Published online Epub Feb.

[279] Giacometti E, Luikenhuis S, Beard C, Jaenisch R. Partial rescue of *MeCP2* deficiency by postnatal activation of *MeCP2*. Proc Natl Acad Sci U. S. A 2007;104:1931–6. http://dx.doi.org/10.1073/pnas.0610593104. Published online Epub Feb.

[280] Robinson L, Guy J, McKay L, Brockett E, Spike RC, Selfridge J, et al. Morphological and functional reversal of phenotypes in a mouse model of Rett syndrome. Brain 2012;135:2699–710. http://dx.doi.org/10.1093/brain/aws096. Published online Epub Sep.

[281] Cheval H, Guy J, Merusi C, De Sousa D, Selfridge J, Bird A. Postnatal inactivation reveals enhanced requirement for *MeCP2* at distinct age windows. Hum Mol Genet 2012;21:3806–14. http://dx.doi.org/10.1093/hmg/dds208. Published online Epub Sep.

[282] Tao J, Hu K, Chang Q, Wu H, Sherman NE, Martinowich K, et al. Phosphorylation of *MeCP2* at serine 80 regulates its chromatin association and neurological function. Proc Natl Acad Sci U. S. A 2009;106:4882–7. http://dx.doi.org/10.1073/pnas.0811648106. Published online Epub Mar.

[283] Zhou Z, Hong EJ, Cohen S, Zhao WN, Ho HY, Schmidt L, et al. Brain-specific phosphorylation of *MeCP2* regulates activity-dependent *BDNF* transcription, dendritic growth, and spine maturation. Neuron 2006;52:255–69. http://dx.doi.org/10.1016/j.neuron.2006.09.037. Published online Epub Oct.

[284] Yasui DH, Peddada S, Bieda MC, Vallero RO, Hogart A, Nagarajan RP, et al. Integrated epigenomic analyses of neuronal *MeCP2* reveal a role for long-range interaction with active genes. Proc Natl Acad Sci U. S. A 2007;104:19416–21. http://dx.doi.org/10.1073/pnas.0707442104. Published online Epub Dec.

[285] Chang Q, Khare G, Dani V, Nelson S, Jaenisch R. The disease progression of *Mecp2* mutant mice is affected by the level of *BDNF* expression. Neuron 2006;49:341–8. http://dx.doi.org/10.1016/j.neuron.2005.12.027. Published online Epub Feb.

[286] Stühmer T, Puelles L, Ekker M, Rubenstein JL. Expression from a *Dlx* gene enhancer marks adult mouse cortical GABAergic neurons. Cereb Cortex 2002;12:75–85. Published online Epub Jan.

[287] Horike S, Cai S, Miyano M, Cheng JF, Kohwi-Shigematsu T. Loss of silent-chromatin looping and impaired imprinting of *DLX5* in Rett syndrome. Nat Genet 2005;37:31–40. http://dx.doi.org/10.1038/ng1491. Published online Epub Jan.

[288] Borg I, Freude K, Kübart S, Hoffmann K, Menzel C, Laccone F, et al. Disruption of Netrin G1 by a balanced chromosome translocation in a girl with Rett syndrome. Eur J Hum Genet 2005;13:921–7. http://dx.doi.org/10.1038/sj.ejhg.5201429. Published online Epub Aug.

[289] Evans JC, Archer HL, Colley JP, Ravn K, Nielsen JB, Kerr A, et al. Early onset seizures and Rett-like features associated with mutations in *CDKL5*. Eur J Hum Genet 2005;13:1113–20. http://dx.doi.org/10.1038/sj.ejhg.5201451. Published online Epub Oct.

[290] Tao J, Van Esch H, Hagedorn-Greiwe M, Hoffmann K, Moser B, Raynaud M, et al. Mutations in the X-linked cyclin-dependent kinase-like 5 *(CDKL5/STK9)* gene are associated with severe neurodevelopmental retardation. Am J Hum Genet 2004;75:1149–54. http://dx.doi.org/10.1086/426460. Published online Epub Dec.

[291] De La Fuente R, Baumann C, Viveiros MM. Role of *ATRX* in chromatin structure and function: implications for chromosome instability and human disease. Reproduction 2011;142:221–34. http://dx.doi.org/10.1530/REP-10-0380. Published online Epub Aug.

[292] Gibbons RJ, Bachoo S, Picketts DJ, Aftimos S, Asenbauer B, Bergoffen J, et al. Mutations in transcriptional regulator *ATRX* establish the functional significance of a PHD-like domain. Nat Genet 1997;17:146–8. Published online Epub Oct.

[293] Cardoso C, Lutz Y, Mignon C, Compe E, Depetris D, Mattei MG, et al. *ATR-X* mutations cause impaired nuclear location and altered DNA binding properties of the *XNP/ATR-X* protein. J Med Genet 2000;37:746–51. Published online Epub Oct.

[294] Tang J, Wu S, Liu H, Stratt R, Barak OG, Shiekhattar R, et al. A novel transcription regulatory complex containing death domain-associated protein and the ATR-X syndrome protein. J Biol Chem 2004;279:20369–77. Published online Epub May 7.

[295] Hendrich B, Bickmore W. Human diseases with underlying defects in chromatin structure and modification. Hum Mol Genet 2001;10:2233–42. Published online Epub Oct 1.

[296] Ausio J, Levin DB, De Amorim GV, Bakker S, Macleod PM. Syndromes of disordered chromatin remodeling. Clin Genet 2003;64:83–95. Published online Epub Aug.

[297] Gibbons RJ, Higgs DR. Molecular-clinical spectrum of the ATR-X syndrome. Am J Med Genet 2000;97:204–12. 10.1002/1096-8628(200023)97:3<204::AID-AJMG1038>3.0.CO;2-X).

[298] Gibbons RJ, McDowell TL, Raman S, O'Rourke DM, Garrick D, Ayyub H, et al. Mutations in *ATRX,* encoding a SWI/SNF-like protein, cause diverse changes in the pattern of DNA methylation. Nat Genet 2000;24:368–71. Published online Epub Apr.

[299] Xue Y, Gibbons R, Yan Z, Yang D, McDowell TL, Sechi S, et al. The ATRX syndrome protein forms a chromatin-remodeling complex with *Daxx* and localizes in promyelocytic leukemia nuclear bodies. Proc Natl Acad Sci USA 2003;100:10635–40. Published online Epub Sep 16.

[300] Wang J, Shiels C, Sasieni P, Wu PJ, Islam SA, Freemont PS, et al. Promyelocytic leukemia nuclear bodies associate with transcriptionally active genomic regions. J Cell Biol 2004;164:515–26. Published online Epub Feb 16.

[301] Dhayalan A, Tamas R, Bock I, Tattermusch A, Dimitrova E, Kudithipudi S, et al. The *ATRX*-ADD domain binds to H3 tail peptides and reads the combined methylation state of K4 and K9. Hum Mol Genet 2011;20:2195–203. http://dx.doi.org/10.1093/hmg/ddr107. Published online Epub Jun.

[302] Cardoso C, Timsit S, Villard L, Khrestchatisky M, Fontes M, Colleaux L. Specific interaction between the XNP/ATR-X gene product and the SET domain of the human *EZH2* protein. Hum Mol Genet 1998;7:679–84. Published online Epub Apr.

[303] Kuzmichev A, Jenuwein T, Tempst P, Reinberg D. Different *EZH2*-containing complexes target methylation of histone H1 or nucleosomal histone H3. Mol Cell 2004;14:183–93. Published online Epub Apr 23.

[304] Kuzmichev A, Nishioka K, Erdjument-Bromage H, Tempst P, Reinberg D. Histone methyltransferase activity associated with a human multiprotein complex containing the enhancer of zeste protein. Genes Dev 2002;16:2893–905. Published online Epub Nov 15.

[305] Cao R, Wang L, Wang H, Xia L, Erdjument-Bromage H, Tempst P, et al. Role of histone H3 lysine 27 methylation in polycomb-group silencing. Science 2002;298:1039–43. Published online Epub Nov 1.

[306] Czermin B, Melfi R, McCabe D, Seitz V, Imhof A, Pirrotta V. *Drosophila* enhancer of zeste/ESC complexes have a histone H3 methyltransferase activity that marks chromosomal polycomb sites. Cell 2002;111:185–96. Published online Epub Oct 18.

[307] Daujat S, Zeissler U, Waldmann T, Happel N, Schneider R. *HP1* binds specifically to Lys26-methylated histone H1.4, whereas simultaneous Ser27 phosphorylation blocks *HP1* binding. J Biol Chem 2005;280:38090–5. Published online Epub Nov 11.

[308] Kuzmichev A, Margueron R, Vaquero A, Preissner TS, Scher M, Kirmizis A, et al. Composition and histone substrates of polycomb repressive group complexes change during cellular differentiation. Proc Natl Acad Sci U. S. A 2005;102:1859–64. Published online Epub Feb 8.

[309] Hiragami K, Festenstein R. Heterochromatin protein 1: a pervasive controlling influence. Cell Mol Life Sci 2005;62(23):2711–26. Published online Epub Nov 2.

[310] Le Douarin B, Nielsen AL, Garnier JM, Ichinose H, Jeanmougin F, Losson R, et al. A possible involvement of *TIF1* alpha and *TIF1* beta in the epigenetic control of transcription by nuclear receptors. Embo J 1996;15:6701–15.

[311] Lechner MS, Schultz DC, Negorev D, Maul GG, Rauscher 3rd FJ. The mammalian heterochromatin protein 1 binds diverse nuclear proteins through a common motif that targets the chromoshadow domain. Biochem Biophys Res Commun 2005;331:929–37. Published online Epub Jun 17.

[312] Kourmouli N, Sun YM, van der Sar S, Singh PB, Brown JP. Epigenetic regulation of mammalian pericentric heterochromatin in vivo by *HP1*. Biochem Biophys Res Commun 2005;337(3):901–7. Published online Epub Sep 30.

[313] Gibbons RJ, Wada T, Fisher CA, Malik N, Mitson MJ, Steensma DP, et al. Mutations in the chromatin-associated protein *ATRX*. Hum Mutat 2008;29:796–802. http://dx.doi.org/10.1002/humu.20734. Published online Epub Jun.

[314] Law MJ, Lower KM, Voon HP, Hughes JR, Garrick D, Viprakasit V, et al. ATR-X syndrome protein targets tandem repeats and influences allele-specific expression in a size-dependent manner. Cell 2010;143:367–78. http://dx.doi.org/10.1016/j.cell.2010.09.023. Published online Epub Oct.

[315] Steensma DP, Gibbons RJ, Higgs DR. Acquired alpha-thalassemia in association with myelodysplastic syndrome and other hematologic malignancies. Blood 2005;105:443–52. http://dx.doi.org/10.1182/blood-2004-07-2792. Published online Epub Jan.

[316] Hanahan D, Weinberg RA. Hallmarks of cancer: the next generation. Cell 2011;144:646–74. http://dx.doi.org/10.1016/j.cell.2011.02.013. Published online Epub Mar 4, S0092-8674(11)00127-9 [pii].

[317] Pujadas E, Feinberg AP. Regulated noise in the epigenetic landscape of development and disease. Cell 2012;148:1123–31. http://dx.doi.org/10.1016/j.cell.2012.02.045. Published online Epub Mar 16, S0092-8674(12)00283-8 [pii].

[318] Kulis M, Esteller M. DNA methylation and cancer. Adv Genet 2010;70:27–56. http://dx.doi.org/10.1016/B978-0-12-380866-0.60002-2. B978-0-12-380866-0.60002-2 [pii].

[319] Jones PA. Functions of DNA methylation: islands, start sites, gene bodies and beyond. Nat Rev Genet 2012;13:484–92. http://dx.doi.org/10.1038/nrg3230. Published online Epub Jul, nrg3230 [pii].

[320] Baylin SB, Jones PA. A decade of exploring the cancer epigenome: biological and translational implications. Nat Rev Cancer 2011;11:726–34. http://dx.doi.org/10.1038/nrc3130. Published online Epub Oct, nrc3130 [pii].

[321] Ohm JE, McGarvey KM, Yu X, Cheng L, Schuebel KE, Cope L, et al. A stem cell-like chromatin pattern may predispose tumor suppressor genes to DNA hypermethylation and heritable silencing. Nat Genet 2007;39:237–42. http://dx.doi.org/10.1038/ng1972. Published online Epub Feb, ng1972 [pii].

[322] Vire E, Brenner C, Deplus R, Blanchon L, Fraga M, Didelot C, et al. The Polycomb group protein *EZH2* directly controls DNA methylation. Nature 2006;439:871–4. http://dx.doi.org/10.1038/nature04431. Published online Epub Feb 16, nature04431 [pii].

[323] Huarte M, Rinn JL. Large non-coding RNAs: missing links in cancer?. Hum Mol Genet 2010;19:R152–61. http://dx.doi.org/10.1093/hmg/ddq353. Published online Epub Oct 15, ddq353 [pii].

[324] Lovat F, Valeri N, Croce CM. MicroRNAs in the pathogenesis of cancer. Semin Oncol 2011;38:724–33. http://dx.doi.org/10.1053/j.seminoncol.2011.08.006. Published online Epub Dec, S0093-7754(11)00214-4 [pii].

[325] Yu W, Gius D, Onyango P, Muldoon-Jacobs K, Karp J, Feinberg AP, et al. Epigenetic silencing of tumour suppressor gene p15 by its antisense RNA. Nature 2008;451:202–6. http://dx.doi.org/10.1038/nature06468. Published online Epub Jan 10, nature06468 [pii].

[326] Morris KV, Santoso S, Turner AM, Pastori C, Hawkins PG. Bidirectional transcription directs both transcriptional gene activation and suppression in human cells. PLoS Genet 2008;4:e1000258. http://dx.doi.org/10.1371/journal.pgen.1000258. Published online Epub Nov.

[327] Khalil AM, Guttman M, Huarte M, Garber M, Raj A, Rivea Morales D, et al. Many human large intergenic noncoding RNAs associate with chromatin-modifying complexes and affect gene expression. Proc Natl Acad Sci U. S. A 2009;106:11667–72. Published online Epub Jul 14 (10.1073/pnas.09047151060904715106 [pii]).

[328] Baek D, Villen J, Shin C, Camargo FD, Gygi SP, Bartel DP. The impact of microRNAs on protein output. Nature 2008;455:64–71. http://dx.doi.org/10.1038/nature07242. Published online Epub Sep 4, nature07242 [pii].

[329] Heyn H, Esteller M. DNA methylation profiling in the clinic: applications and challenges. Nat Rev Genet 2012;13:679–92. http://dx.doi.org/10.1038/nrg3270. Published online Epub Oct, nrg3270 [pii].

[330] Merkenschlager M, Odom DT. *Ctcf* and cohesin: linking gene regulatory elements with their targets. Cell 2013;152:1285–97. http://dx.doi.org/10.1016/j.cell.2013.02.029. Published online Epub Mar 14, S0092-8674(13)00218-3 [pii].

Chapter 6

Mitochondrial Genomics: Emerging Paradigms and Challenges

V. Singh[1], P. Gupta[2,3], N.K. Rajput[2], A. Bhardwaj[2]

[1]Amity University, Noida, India; [2]Institute of Microbial Technology, Council of Scientific and Industrial Research, Chandigarh, India; [3]Bhaskaracharya College of Applied Sciences, University of Delhi, New Delhi, India

Chapter Outline

Introduction 75
Nongenetic Diagnostics for Mitochondrial Dysfunction 75
Genome Data Deluge 78
DNA Testing, Next Generation Sequencing, and Mitochondrial Disease Diagnosis 79
Next Generation Sequencing 80
Next Generation Sequencing–Based Diagnostic Assay Design for Mitochondrial Disorders 80
Translational Advancements 81
Semantic Data Standards for Community Collaboration 84
The Road Ahead 85
Conclusions 85
References 86

INTRODUCTION

Human mitochondrial DNA (mtDNA) is a 16,569-bp long circular genome and codes for 37 genes-13 protein-coding genes, 2 ribosomal RNAs, and 22 transfer RNAs (tRNAs). Human nuclear DNA (nDNA) encodes the larger set of over 1500 proteins crucial for mitochondrial function, including its replication, its repair, and sustenance of other key metabolic pathways [1]. The mtDNA codes for four respiratory complexes of the oxidative phosphorylation system. The displacement loop is the only noncoding segment of the mitochondrial genome. It comprises of the origin of replication of the H-strand and the promoters for L- and H-strand transcription [2] (Fig. 6.1). Mutations in either mtDNA, nDNA, or both may lead to mitochondrial dysfunction, which manifests in a broad spectrum of diseases affecting various tissues like brain, heart, liver, and skeletal muscles. The clinical symptoms of the disease depend on various factors like the cell type affected, heteroplasmy levels of the mutated DNA, and the spatial context of the mutation: mtDNA, nDNA, or both. The symptoms range from loss of motor control, muscle weakness, and cardiac disease to visual or hearing loss. Given the involvement of mitochondria in a large number of cellular pathways, it is extremely challenging to correlate the exact role of genome variation with the observed phenotype [3]. Besides, a majority of mitochondrial diseases are ascribed to mutations in the

nDNA. It is therefore imperative to scale up high-coverage sequencing of both mtDNA and nDNA in patients and controls to elucidate and catalog the quantum of variations that exist in the two, as also to unambiguously identify the disease-associated variations. Massive parallelization and high coverage, the two key features of next generation sequencing (NGS) technologies render them most suitable for mitochondrial disease diagnostics and prognostics. The Mitochondria Medicine Society recommends NGS as the first line of diagnostics for mitochondrial diseases [4].

This chapter elaborates on NGS methodologies in diagnosis of mitochondrial diseases, the current bottlenecks in mitochondrial disease research, and innovative ways to address these challenges.

NONGENETIC DIAGNOSTICS FOR MITOCHONDRIAL DYSFUNCTION

A major challenge in diagnosis of mitochondrial diseases is lack of sensitive and specific biomarkers. The biochemical testing recommended as part of the clinical diagnosis includes measurement of lactate and pyruvate in blood, urine, and spinal fluid. This method has high specificity ranging from 83% to 100%, but poor sensitivity (34–62%). Generally the blood lactate or pyruvate level (in cases where lactate is on higher end) is a better measure of mitochondrial

Medical and Health Genomics. http://dx.doi.org/10.1016/B978-0-12-420196-5.00006-X

Human Mitochondrial Genome: At a Glance

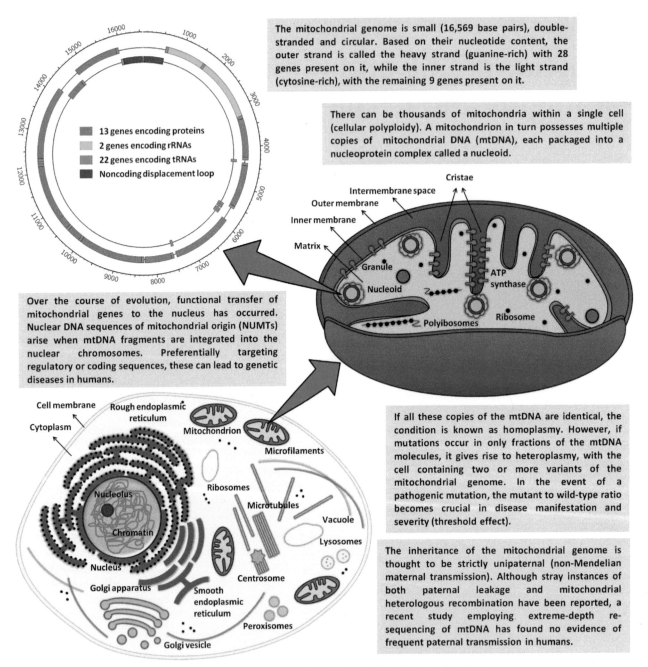

The mitochondrial genome is small (16,569 base pairs), double-stranded and circular. Based on their nucleotide content, the outer strand is called the heavy strand (guanine-rich) with 28 genes present on it, while the inner strand is the light strand (cytosine-rich), with the remaining 9 genes present on it.

13 genes encoding proteins
2 genes encoding rRNAs
22 genes encoding tRNAs
Noncoding displacement loop

There can be thousands of mitochondria within a single cell (cellular polyploidy). A mitochondrion in turn possesses multiple copies of mitochondrial DNA (mtDNA), each packaged into a nucleoprotein complex called a nucleoid.

Over the course of evolution, functional transfer of mitochondrial genes to the nucleus has occurred. Nuclear DNA sequences of mitochondrial origin (NUMTs) arise when mtDNA fragments are integrated into the nuclear chromosomes. Preferentially targeting regulatory or coding sequences, these can lead to genetic diseases in humans.

If all these copies of the mtDNA are identical, the condition is known as homoplasmy. However, if mutations occur in only fractions of the mtDNA molecules, it gives rise to heteroplasmy, with the cell containing two or more variants of the mitochondrial genome. In the event of a pathogenic mutation, the mutant to wild-type ratio becomes crucial in disease manifestation and severity (threshold effect).

The inheritance of the mitochondrial genome is thought to be strictly uniparental (non-Mendelian maternal transmission). Although stray instances of both paternal leakage and mitochondrial heterologous recombination have been reported, a recent study employing extreme-depth re-sequencing of mtDNA has found no evidence of frequent paternal transmission in humans.

FIGURE 6.1   The mitochondrion as seen in a eukaryotic cell and the mitochondrial genome.

dysfunction with 100% specificity. The sensitivity, however, is abysmally low at 31%, because increased lactate levels (>3 mmol/L) may also be a consequence of organic acidemias, other inborn errors of metabolism, toxins, tissue ischemia, or other conditions. On the other hand, pyruvate elevation is a better measure, with 75% sensitivity and approximately 87% specificity. Likewise, other biochemical tests like elevated levels of amino acids, malate, fumarate, creatine phosphokinase, and uric acid are also routinely performed wherever the symptoms suggest mitochondrial involvement. Of all these, specific quantification of 3-methylglutaconic acid in blood and urine is more reliable. Though a large number of tests are available and several scoring systems proposed and practiced to identify mitochondrial dysfunction, none of them are unambiguously affirmative (Table 6.1). Most of the biochemical diagnostic methods in practice today were developed before the advent of "-omics" era. It is only recently that evidence-based

TABLE 6.1 List of Nongenetic Diagnostic Tests for Mitochondrial Diseases

S. No.	Diagnostic	Pathological Basis
1.	**Biochemical testing of blood, cerebrospinal fluid (CSF), and urine**	Electron transport chain (ETC) dysfunction leads to a decrease in cellular adenosine triphosphate (ATP), resulting in the upregulation of cytosolic glycolysis; pyruvate production increases and pyruvate is either transaminated to alanine or converted to lactate. Hence lactic acidosis occurs. Pyruvate elevation indicates defects in enzymes pyruvate dehydrogenase and pyruvate carboxylase. Only when lactate levels are high, can the blood lactate/pyruvate ratio be reliably used to differentiate between ETC diseases and disorders of pyruvate metabolism. Amino acid levels are elevated because of respiratory chain dysfunction, altering the reduction–oxidation state. Urine amino acids used to assess mitochondrial disease associates renal tubulopathy. Total and free carnitine levels, along with acylcarnitine profiling and also some primary amino and organic acidemias, help in identification of primary or secondary fatty-acid oxidation defects. Creatine phosphokinase and uric acid elevations are caused by nucleic acid and nucleotide catabolism. Transaminases and albumin levels may help detect liver pathology resulting from mitochondrial DNA depletion and/or general liver dysfunction. Cerebral folate deficiency is diagnosed via measurements of 5-methylhydrofolate in the CSF.
a.	Blood	Lactate; pyruvate; amino acids like alanine, glycine, proline, and threonine; acylcarnitine; hematological abnormalities like aplastic, megaloblastic, and sideroblastic anemias; leukopenia; thrombocytopenia and pancytopenia; creatine phosphokinase and uric acid; transaminases; and albumin
b.	CSF	Lactate, pyruvate, amino acids, 5-methyltetrehydrofolate
c.	Urine	Amino acids and organic acids like malate, fumarate, 3-methylglutaconic acid, dicarboxylic acid, 2-oxoadipic acid and methylmalonic acid
2.	**Spectroscopy**	
a.	Near-infrared spectroscopy of hemoglobin	Oxygenated and deoxygenated hemoglobin emit different wavelengths. Normally, there is prompt deoxygenation during and prompt reoxygenation after exercise. In case of mitochondrial disorders, increased oxygenation during exercise is observed, with a rapid return to baseline thereafter (high delivery of oxygen via hemoglobin but reduced mitochondrial oxidative respiration).
b.	Phosphorous magnetic resonance spectroscopy	Five major peaks associated with energy metabolism: one for phosphocreatine, one for inorganic phosphate, and three for the phosphate groups of ATP. ATP utilization/ATP production is directly related to phosphocreatine/inorganic phosphate at a given pH. Normally, the phosphocreatine/inorganic phosphate ratio is high (low metabolic rate and high energy capacity at rest). In case of mitochondrial disorders, the phosphocreatine/inorganic phosphate ratio is low (low energy capacity at rest).
3.	**Light microscopy (muscle histology)**	
a.	Hematoxylin and eosin staining	For general muscle morphology
b.	Gomori trichrome stain	Subsarcolemmal accumulation of mitochondria visualized as ragged-red fibers
c.	Stain for succinate dehydrogenase (SDH)	Ragged-blue fibers
d.	Nicotinamide adenine dinucleotide hydrate (NADH)-tetrazolium reductase	Mitochondrial enzyme NADH-dehydrogenase activity

Continued

TABLE 6.1 List of Nongenetic Diagnostic Tests for Mitochondrial Diseases—cont'd

S. No.	Diagnostic	Pathological Basis
e.	Cytochrome oxidase (COX)	COX negative fibers
f.	SDH/COX staining	COX intermediate fibers
4.	**Ultrastructure/electron microscopy**	Increase in mitochondrial number or size, increased lipid, glycogen droplets, increased mitochondrial matrix, abnormal cristae, and paracrystalline inclusions
5.	**Nonspecific tissue pathology**	
a.	Hepatic	Steatosis, micronodular cirrhosis, cholestasis, bile-duct proliferation, fibrosis, apoptosis, and mitochondrial ultrastructural abnormalities
b	Cardiac	Ultrastructural mitochondrial abnormalities
c.	Neural	Focal necrosis, loss of cortical neurons, reactive astrogliosis, astrocytic inclusions, demyelination, spongiosis, endothelial proliferation, or capillary hyperplasia
6.	**Brain imaging**	
a.	Computed tomography	Punctuate calcifications particularly in the basal ganglia or cerebellum
b.	Magnetic resonance imaging	High T_2 signal in the basal ganglia, brainstem, or cerebellum; high cortical T_2 signal representing "stroke-like" lesions, edema, atrophy, or abnormal myelination
c.	Proton magnetic resonance imaging	Semiquantitative estimate of brain metabolites like lactate, creatine, and N-acetylaspartate
d.	Diffusion tensor imaging	Detection and quantification of major white matter tracts
7.	**ETC enzymes**	
a.	Polarographic assays	Fresh muscle tissue; overall estimate of ETC activity. However, increase in mitochondrial cellular polyploidy may lead to an overall increase in ETC activity, masking a deficit.
b.	Spectrophotometric assays	Frozen or fresh muscle tissue; activity of individual complexes in the ETC measured. The absolute activity of each complex shows high variability, but a constant ratio is maintained between the ETC enzymes in all tissues. Complex activity is also compared with an enzyme not directly involved in the ETC, citrate synthase, to detect general enzyme deficiency.
c.	Fibroblast ETC assay	Study of ETC enzyme activity, but abnormalities are evident only in the presence of persistent lactic acidosis.
d.	Western blots and gel electrophoresis	Measurement of various protein components with complexes and supercomplexes

None of these diagnostic methods can confirm or rule out mitochondrial diseases independently. Combinations thereof, as embodied in the sets of diagnostic criteria such as the Walker criteria, the modified Walker criteria, Nijmegen Center for Mitochondrial Disorders scoring system and the Mitochondrial Disease Criteria, are of greater utility, because corroborative evidence helps in reaching a more conclusive diagnosis.

clinical protocol has been proposed for diagnostics involving NGS as the key component [4].

GENOME DATA DELUGE

The first human DNA to be completely sequenced was the human mtDNA (16,569 bp) in 1981 [5,6]. Since then, more than 30,700 complete or near complete mtDNA sequences have been deposited in GenBank (National Institutes of Health, United States), primarily from evolutionary and migration studies, occasionally from disease association studies, and rarely from forensic applications. There has been a similar and, in fact, more rigorous efforts toward

sequencing the nDNA of multiple individuals. Most of this data has been generated in past 10 years, propelled by advances in sequencing technologies (Fig. 6.2). More recently, Human Longevity (San Diego, CA) launched an initiative to sequence half a million to one million human genomes per year in the world's largest sequencing effort to further the understanding of genotype–phenotype association (http://www.genengnews.com/gen-news-highlights/venter-s-new-goal-world-s-largest-sequencing-operation/81249577/).

Over the years, a large number of web-based resources have also been developed on various aspects of mitochondrial diseases, most of them focusing on the data from

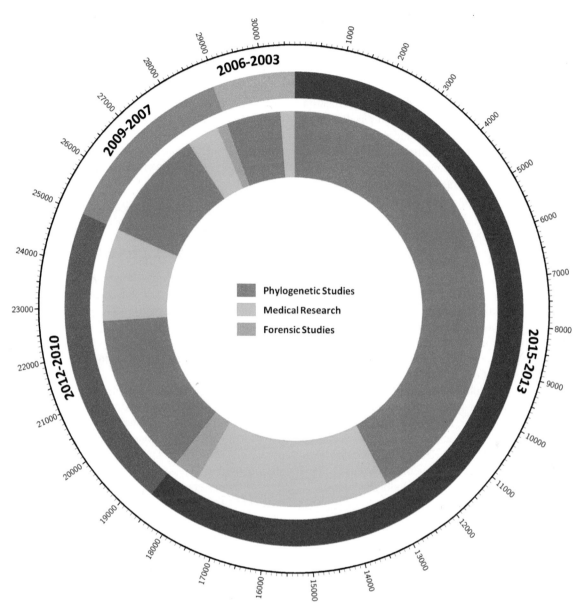

FIGURE 6.2 Representation of the overall distribution of around 30,000 mitochondrial genomes sequenced since 2003. The outer circle denotes the number of mitochondrial genomes. The second circle depicts the year-wise distribution of these mitochondrial genomes. It can be seen that more than 50% of genomes have been sequenced only since 2013. The innermost circle again divides these year-wise bins into three categories, namely, phylogenetic studies, medical research and forensic studies.

mtDNA (MITOMAP, MitoLSDB, MitoCarta, MitoMiner, MitoBreak, HmtDB, MitoDB, MtSNPScore, Mit-*o*-Matic, MseqDR) [7–17].

DNA TESTING, NEXT GENERATION SEQUENCING, AND MITOCHONDRIAL DISEASE DIAGNOSIS

As mentioned, mitochondrial disorders represent an extremely heterogeneous group of rare complex diseases with respect to phenotype, clinical features, stage of onset, site and nature of causal mutation, and pattern of inheritance. Since the cells are heteroplasmic for mitochondria, in addition to identifying

the causal mutation [most commonly a single nucleotide polymorphism (SNP) or a deletion], it is also imperative to ascertain the mutation load. Apart from mutations in mtDNA, more than 1500 nuclear genes are known or hypothesized to contribute significantly to mitochondrial disorders [18]. The fact that the causal mutation may be in the mtDNA, nDNA, or both makes the task of accurate, effective, and expedient diagnosis even more arduous.

Until recently, DNA diagnosis of the mitochondrial disorder included screening of the known common disease-causing single nucleotide variants (SNVs) in mitochondrial and nDNA by polymerase chain reaction (PCR)–based assays, restriction fragment length polymorphism, or allele-specific oligonucleotide dot blot hybridization [19], and quantitative analysis of the levels of heteroplasmy by amplification-refractory mutation system–based quantitative PCR [20]. Large deletions in mtDNA were detected by Southern hybridization and quantification of heteroplasmy facilitated by array comparative genome hybridization [21,22]. In case of novel mutations or multiple deletions, complete sequencing of the mitochondrial genome using overlapping primers remained the only alternative. Until the advent of NGS technologies around 2002, Sanger sequencing of the complete mtDNA was the gold standard for identifying novel disease-causing mutations in mtDNA [21,22].

NEXT GENERATION SEQUENCING

NGS technologies refers to a group of sequencing technologies developed since the sequencing and assembly of the reference human genome in 2001 [23]. The human reference genome took 10 years and approximately $3 billion to come by, paradoxically exposing the success and limitations of Sanger sequencing technology in sequencing large, repeat-rich complex genomes. The NGS platforms were developed with the stated objective of cutting down the cost of human genome sequencing to $1000 and ensuring the assembly and whole genome sequence in a reasonable time frame.

The three most popular NGS platforms are Pyrosequencing (454, Roche), Reversible Terminator Sequencing (Solexa, Illumina) and SOLiD (Applied Biosystems). Though all of these differ in the sequencing chemistry, they have several common features [24], including:

1. Library preparation: random fragmentation of starting DNA and addition of custom adaptors at each end.
2. Library amplification: to improve the signal to noise ratio, each library fragment is clonally amplified in a local space.
3. Imaging-based sequencing techniques: all the three sequencing techniques are imaging based, recording either the bioluminescence or fluorescence in each cycle of nucleotide (oligonucleotide in SOLiD) addition.
4. Massively parallel sequencing: one of the defining features of all NGS technologies resulting in high coverage, high quality, high throughput data.
5. Short read lengths: the average read length for Sanger sequencer is about 900 bases. The read length in all NGS techniques is much less and highly variable: Pyrosequencing, 250 bases; Reversible Terminator Sequencing, 36 bases; and SOLiD, 35 bases. It is also important to mention here that improvisations in basic sequencing chemistry have resulted in increased read lengths over the years.

Although discussing the sequencing chemistry is out of the purview of this chapter, it is worth mentioning that Pyrosequencing and Reversible Terminator Sequencing are polymerase based; in contrast SOLiD utilizes unique ligase-based chemistry for sequencing [25]. It is also important to note that whereas Pyrosequencing suffers from inherent problems in correctly sequencing long homopolymer tracts, Illumina and SOLiD are more accurate. SOLiD, by virtue of its unique two-base coding, dual interrogation of each nucleotide, and direct inference of true polymorphism or sequencing error by comparison to a reference, is the most accurate [26]. However, Illumina is reasonably accurate and economically a more favorable option and hence the most preferred.

NEXT GENERATION SEQUENCING–BASED DIAGNOSTIC ASSAY DESIGN FOR MITOCHONDRIAL DISORDERS

As mentioned earlier, mitochondrial disorders are characterized by heterogeneity in clinical presentation with broad range of overlapping spectrum of symptoms confounding the diagnosis by biochemical assays. The causal mutation may be in the mtDNA, nDNA, or both. Unambiguous diagnosis therefore necessitates sequencing of the mtDNA as well as the nDNA.

Mitochondrial DNA is small (16.5 kb), repeats sparse, and mostly coding (introns and repeats absent) [1]. In contrast, the nDNA in humans is large (3.2 billion bases), repeat-rich (>50%) and mostly noncoding (genes account for <2% of the genome) [23]. Approximately 1500 nuclear genes are proposed to play a role in mitochondrial disorders [18]. Because most pathogenic mutations lie within exonic regions of the nDNA, a rationalized strategy for identifying the causative mutation(s) for mitochondrial disorders entails complete sequencing of the mitochondrial genome, and exome or targeted sequencing of the nuclear genome [27,28].

Various steps in NGS-based diagnostic assay include:

1. Target enrichment
   The target enrichment strategy depends on the properties of the target. The preferred strategy for mtDNA is PCR-based enrichment, using 24–36 pairs of overlapping primers (multiplexing) and normal PCR, or using two or three pairs of overlapping primer pairs and long-range PCR. However, using multiple primers for enrichment may result in uneven coverage in regions of overlap, necessitating normalization. Another major drawback of using multiple primers for enrichment is the increased probability of the SNP being in the primer binding site, interfering with amplification and, in some cases, leading to preferential amplification of nuclear homologues of mtDNA sequences. Using one pair of back-to-back primers to generate a single amplicon of the entire mitochondrial genome in a long-range PCR is likely to yield most accurate results and minimize experimental artifacts.

Capture in solution using DNA probes is preferred for target enrichment of nuclear genes. A comprehensive nuclear gene panel for mitochondrial disorders may include more than 100 genes. As of now, more than 1500 nuclear genes are proposed to have a role in mitochondrial dysfunction. Whereas PCR-based enrichment methods have limitations in scaling up, capture-based methods offer flexibility of scaling up to capture all genes or the whole genome.

2. Target library, amplification, and sequencing
   After the enrichment of the mtDNA and the candidate nuclear genes, a fragment library is generated and each fragment clonally amplified either by emulsion PCR or bridge PCR (depending on the choice of platform) and sequenced.
3. Alignment to reference and variant calling
   The reads generated by the NGS sequencer are aligned to the reference genome to facilitate assembly, variant calling, and identification of indels.
4. Estimating the functional significance of the variants and indels
   Once the SNVs have been called with confidence, the functional consequence of each SNV must be assessed to unequivocally identify the deleterious ones from a background of nondeleterious ones. An in silico combined evidence approach is generally followed to distinguish the signal from the background. Several databases are available to identify known disease-associated SNPs. For the nuclear genome these include:
   Human Genome Mutation Database (http://www.hgmd.cf.ac.uk)
   Database of Single Nucleotide Polymorphisms (http://www.ncbi.nlm.nih.gov/projects/SNP)
   Online Mendelian Inheritance in Man (http://omim.org/)
   Locus Specific Mutation Databases (http://grenada.lumc.nl/LSDB_list/lsdbs)
   Leiden Open Variation Database (http://www.lovd.nl/3.0/home).
   Likewise, known mitochondrial SNPs are cataloged at:
   MITOMAP (http://www.mitomap.org/MITOMAP)
   Human Mitochondrial Genome Database (http://www.mtdb.igp.uu.se/)
   Mamit-tRNA database (http://mamit-trna.u-strasbg.fr/human.asp).
   The novel variants may be prioritized as disease causing using the following algorithms:
   PolyPhen (Polymorphism Phenotyping) (http://coot.embl.de/PolyPhen/)
   SIFT (Sorting Intolerant From Tolerant) (http://blocks.fhcrc.org/sift/SIFT.html)
   Align GVGD (Grantham Variation Grantham Deviation) (http://agvgd.iarc.fr/agvgd_input.php/)
   Panther (http://www.pantherdb.org/tools/csnpScoreForm.jsp)
   PMut (Pathogenic Mutation Prediction) (http://mmb.pcb.ub.es/PMut/)
   MtSNPscore (http://ab-openlab.csir.res.in/snpscore/). Generally a combined evidence approach is followed to increase the confidence in the results.

Advantages of NGS-based diagnostic methods over conventional methods:

1. The entire mitochondrial genome and a panel of candidate nuclear genes can be interrogated simultaneously in a single run. Multiplexing of samples is also possible.
2. In case of mtDNA mutations, quantification of mutation load is extremely critical in result interpretation and genetic counseling. In Illumina, by multiplexing 12 samples per lane of the flow cell and 75 cycles of sequencing, a 20,000× coverage per base is achievable. At this coverage, heteroplasmies greater than 1.5% can be accurately predicted [29].
3. Aligning the reads to the reference genome and scanning for sharp drop in coverage readily allows detection and accurate mapping of deletions. A gradual drop in coverage, on the other hand, indicates multiple deletions.

NGS-based clinical targeted gene assay for the mitochondrial genome and 108 selected nuclear genes associated with mitochondrial disorders have already been designed to facilitate the analysis and understanding of nuclear and mitochondrial variations in mitochondrial diseases [30]. These emerging technologies offer an excellent opportunity to further dissect the molecular basis of disease manifestation [31a]. With an increasing number of individuals that may be genetically screened across different populations, excellent datasets may be available to explore the genetic basis of disease. The genomics data along with clinical and biochemical profiles may also be used to identify disease biomarkers with high sensitivity and specificity, which is a major challenge in diagnosing mitochondrial dysfunction.

TRANSLATIONAL ADVANCEMENTS

The United Mitochondrial Disease Foundation (UMDF) lists more than 40 diseases caused by mitochondrial dysfunction. Surprisingly, however, there are no known drugs for any of these diseases. A simple search in the clinical trials registry (https://clinicaltrials.gov/) shows that only 0.2% of clinical trials ever done, ongoing, terminated, or planned are for mitochondrial diseases. The reasons may include the prohibitively expensive nature of clinical trials, dearth of expert clinicians, lack of acceptable number of patients, and lack of disease biomarkers for appropriate cohort selection. Given that mitochondrial diseases are rare and present a broad spectrum of clinically heterogeneous symptoms, it is important that these issues be addressed by derisking research, developing platforms for sharing data on patients, making centralized patient registries, and creating new models for development of therapeutic interventions.

Globally, many attempts have been made to systematically address the problem of mitochondrial diseases by establishing patient communities, network projects, focused programs and consortia-based approaches (Table 6.2). The North American Mitochondrial Disease Consortium (NAMDC) (http://www.rarediseasesnetwork.org/namdc/) collects data on mitochondrial disease patients as a clinical patient registry and periodically updates the patients on mitochondrial diseases. NAMDC also helps researchers to identify and recruit patients for future studies. The NAMDC is supported by the National Institute of Neurological Disorders and Stroke, the Eunice Kennedy Shriver National Institute of Child Health and Human Development, the Office of Rare Disease Research (ORDR), and the National Center for Advancing Translational Sciences (NCATS). Additionally, Therapeutics of Rare and Neglected Diseases (TRND) (http://www.ncats.nih.gov/research/rare-diseases/trnd/trnd. html) is a program led by NCATS (http://www.ncats.nih. gov/) and supports the development of potential treatments for rare and neglected diseases to first-in-human trials. Preclinical studies, including medicinal chemistry optimization, drug metabolism and pharmacokinetics, toxicology formulation, and other studies required to file and Investigational New Drug Application for regulatory approvals are also supported by TRND. ORDR (http://www.orphadata.org/cgi-bin/inc/ordo_orphanet.inc.php) coordinates a large number of collaborative research efforts toward rare diseases, including support to institutes and centers, management of a patient registry, and maintenance of a human biospecimen repository, to name a few. The Rare Diseases Clinical Research Network (http://rarediseases.info.nih.gov/ research/pages/41/rare-diseases-clinical-research-network), from ORDR, is the umbrella organization for NAMDC and is focused on advancing medical research on rare diseases by facilitating collaboration, study enrollment, and data sharing. The NCATS Global Rare Diseases Patient Registry Data Repository (http://www.ncats.nih.gov/research/rare-diseases/grdr/grdr.html) manages the data generated as part of the various initiatives involving patient information. This is a web-based resource that aggregates deidentified patient information across many registries and specifies a globally unique identifier to each patient's data. This unique identifier allows for patient follow-up observation across different registries, diseases, studies, and countries, and also ensures that clinical information is mapped to biospecimen datasets. In addition, the Bridging Interventional Development Gaps (http://www.ncats.nih.gov/research/rare-diseases/bridgs/ bridgs.html) division of NCATS manages preclinical innovations toward the development of new therapeutics.

There are various support and advocacy groups like International Mito-Patients, which is a network of patient organizations involved in mitochondrial diseases; MitoAction, which is a support group providing education and advocacy to mitochondrial disease patients; and UMDF, which provides support along with promoting education and research for the diagnosis, treatment, and cure of mitochondrial disorders. The Children's Mitochondrial Disease Network provides support and information for all mitochondrial disorders. MitoCanada, a Canada-based support and awareness group for mitochondrial patients also supports the advancement of research in the field of mitochondrial diseases. The Mitochondrial Research Guild, an organization of Seattle Children's Hospital, works for raising awareness, promoting research, and improving medical care for mitochondrial disorder patients. There are various organizations that are working in the area of research for diagnosis, treatment, and management of mitochondrial disorders, such as NAMDC, which is a network of clinicians and clinical investigators that works toward the collection of information and also creating a repository of specimens, including DNA from patients with mitochondrial disorders. The Mitochondrial Medicine Society (MitoSOC) represents the international group of physicians, clinicians, and researchers who works toward the better diagnosis, management, and treatment of mitochondrial disorders. The Mitochondrial Research Society is an organization of scientists and physicians who work for the cure of mitochondrial diseases by promoting research on mitochondria, mitochondrial pathogenesis, and prevention, diagnosis, and treatment of the disease. The Rare Mitochondrial Disease Service for Adults and Children is a web resource that provides information on all aspects of mitochondrial disease along with the diagnostic and management options available to clinicians, researchers, and healthcare professionals. The Society for Mitochondria Research and Medicine, India (SMRM) is a nonprofit organization of scientists, clinicians, and academicians. The purpose of SMRM is to foster research on basic science of mitochondria, mitochondrial pathogenesis, and prevention, diagnosis, and treatment throughout India and abroad.

One of the most formidable challenges in the treatment of rare diseases is patient recruitment for clinical trials. To overcome this challenge, patient advocacy groups and web based recruitment sites like PatientsLikeMe are turning out to be major game changers. The crowd sourcing and crowd funding strategy is a cost- and time-effective viable option for patient recruitment. Crowd sourcing initiatives on social media not only generate awareness about rare diseases but may also secondarily be used to generate funds for supporting research activities in resource crunch settings. The amyotrophic lateral sclerosis Ice Bucket Challenge is a case in point, raising over $100 million within 3 months of going viral on Facebook and Twitter (http://en.wikipedia. org/wiki/Ice_Bucket_Challenge).

TABLE 6.2 List of Databases/Analysis Platforms, Advocacy Groups, Translational Platforms, and Data Exchange Groups Helpful in Mitochondrial Diseases Research

Databases/Analysis Platforms	Translational Platforms
Human Mitochondrial Database (HmtDB): an online web resource for population genetics and mitochondrial disease http://www.hmtdb.uniba.it/hmdb/	North American Mitochondrial Disease Consortium (NAMDC): collects data on mitochondrial disease patients as a clinical patient registry http://www.rarediseasesnetwork.org/namdc/index.htm
MitoBreak: an online resource of curated datasets of mtDNA breakpoints http://mitobreak.portugene.com/cgi-bin/Mitobreak_home.cgi	Therapeutics of Rare and Neglected Diseases (TRND): a program led by the National Center for Advancing Translational Sciences (NCATS) that supports the development of potential treatments for rare and neglected diseases to first-in-human trials http://www.ncats.nih.gov/research/rare-diseases/trnd/trnd.html
MitoDB: a database and analysis resource for mitochondrial disease http://www.mitodb.com/	
MitoCarta: a data inventory for genes for human and mouse encoding proteins with strong support for mitochondrial localization http://www.broadinstitute.org/pubs/MitoCarta/	Mitochondrial Medicine Society (MMS): a society for physicians, clinicians, and researchers working for the better diagnosis, management, and treatment of mitochondrial diseases http://mitosoc.org/
Mitochondrial Disease Sequence Data Resource (MseqDR): a consortium for data capturing, integration, visualization, and analysis of genomic data for mitochondrial diseases https://mseqdr.org/index.php	Rare Mitochondrial Disease Service for Adults and Children (RMDSAC): an information web resource for all aspects of mitochondrial disease http://www.mitochondrialncg.nhs.uk/
Mitochondrial Locus Specific Database (MitoLSDB): the most comprehensive web resource for mtDNA variations on the Leiden Open Variation Database () platform https://ab-openlab.csir.res.in/mitolsdb/home.php	Mitochondrial Research Guild (MRG): an organization of Seattle Children's Hospital that works to raise awareness, promote research, and improve quality of medical care http://www.nwmito-research.org/
MITOMAP: a database of human mitochondrial genome http://mitomap.org/MITOMAP	**Networking/Advocacy Platforms**
Mit-o-Matic: an analysis tool for finding clinical correlation of human mtDNA variations http://genome.igib.res.in/mitomatic/index.html	International Mito-Patients (IMP): a network of patient organization working on mitochondrial disease http://www.mitopatients.org/index.html
MtSNPscore: a combined evidence-based approach for prioritizing disease-associated variants http://ab-openlab.csir.res.in/snpscore/	MitoAction: a support group that provides education and advocacy for mitochondrial diseases patients http://www.mitoaction.org/
MitoVariome: a database of human mitochondrial DNA variation http://variome.kobic.re.kr/MitoVariome/index.jsp	United Mitochondrial Disease Foundation (UMDF): promotes research and education for the diagnosis, treatment, and cure of mitochondrial disorders http://www.umdf.org/
MitoMiner: an integrated data warehouse of proteomic data for mitochondria http://mitominer.mrc-mbu.cam.ac.uk/release-3.1/begin.do	Children's Mitochondrial Disease Network: an information and support group for patients with mitochondrial disorders http://www.emdn-mitonet.co.uk/index.htm

Data Exchange Groups

Global Alliance for Genomics and Health (GA4GH): a coalition for effective and responsible data sharing for advancement in human health through genomics medicine
http://genomicsandhealth.org/
Matchmaker exchange: a federated platform to facilitate matching similar phenotypic and genotypic profiles (matchmaking) through standardized application programming interfaces
http://matchmakerexchange.org/
Beacon project: a web service for comparing genomic data maintaining data privacy
http://ga4gh.org/#/beacon

mtDNA, Mitochondrial DNA.

Alternatively, the n-of-1 trial method may be used to evaluate new treatments. The ultimate goal of an n-of-1 trial is to determine the optimal or best intervention for an individual patient using objective data-driven criteria. The n-of-1 trial may be more relevant in rare diseases, because the probability that the patient may harbor personal mutations is more likely, thereby necessitating personalized medicine regime [31b].

SEMANTIC DATA STANDARDS FOR COMMUNITY COLLABORATION

The advancements in web technologies offer unprecedented opportunities for seamless collaboration, widespread access, and sharing of research data. These mediums of collaborative research facilitate data integration and exchange of ideas, which may expedite the process of bringing new therapeutic interventions and diagnostic methods to the market.

Various web-based platforms exist today that are expected to foster innovation in the area of drug discovery research. However, it is important that these platforms cater to the needs of different stakeholders, namely, researchers, clinicians, and patient groups, and technology innovations are needed to interface the data generated by these stakeholders for getting deeper insight into disease diagnosis, treatment options, adverse effects, and comorbidity profiles in context of the host genetics. Given the sheer size and variety of the data involved in the process, it is absolutely necessary to define standards for data exchange to ensure seamless data integration and analysis.

There are several research groups that are involved in building ontologies over years. Some of the popular ones include Gene Ontology [32], which unifies representation of gene, and gene products, Medical Subject Headings [33], a controlled vocabulary used for the purpose of indexing journal articles and books in the life sciences in PubMed. It is also used by the ClinicalTrials.gov registry to classify which diseases are studied by trials registered in ClinicalTrials.gov. The popular medical ontologies include Systematized Nomenclature of Medicine: Clinical Terms [34], the most comprehensive clinical terminology, and RxNORM [35], with normalized names for clinical drugs and links to many of the drug vocabularies generally used in pharmacy management and drug interaction software. There are other ontologies proposed for personalized therapeutics, namely, the case profile ontology [36], which offers a knowledge base for the personalization of patient conditions and intervention plans. Similarly, there are other examples like Human Phenotype Ontology (HPO), which deals with standardized vocabulary of phenotypic abnormalities encountered in human disease [37]. Disease Ontology, on the other hand, enables cross-references among disease concepts, genes contributing to disease, and the host of symptoms, findings, and signs [38]. The Unified Medical Language System (http://www.nlm.nih.gov/research/

umls/) is a terminology integration system to establish correspondence among terms from different terminologies for a given biomedical entity. A detailed description of various clinical and disease ontologies may be read at http://lhncbc.nlm.nih.gov/system/files/pub2011017.pdf. Similarly, there are attempts to develop ontology for genetic variation like the Phenotype and Genotype Object Model (PAGE-OM) [39] and VariO [40]. More recently, a binary fingerprint-based method is also reported that includes the features from both PAGE-OM and VariO in providing a comprehensive classification for the effect of single nucleotide changes [41]. These advancements are necessary to complement the diagnosis offered by NGS-based methods. Unlike limited data points from conventional clinical diagnosis methods, the NGS may identify several genetic variants. Interpreting the role of these variations and their individual or cumulative contribution to the clinical manifestation of the disease is the real challenge. Huge gaps in the functional annotation of the genome further complicate genotype-phenotype interpretation, eg, for over 100,000 disease-associated variants reported, there is no functional information available. Genome-wide studies have also reported that different mutations in the same gene lead to different disease phenotypes [42,43]. As our understanding of genome regulation and function expands, we must update existing ontologies or create new ones that allow representation of this enhanced knowledge base. These data resources will go a long way in exploring the unknown patterns in disease identification and monitoring therapeutic responses. The semantic network of these terms will also allow deciphering and interpreting novel patterns in understanding disease biology.

A recent study has reported a clinical phenotype-based gene prioritization method using semantic similarity estimated using HPO terms [44]. This study clearly highlights that semantic similarity based on phenotype descriptors can rank a causative gene accurately in a gene list relative to patient phenotype characteristics. This also points to the potential applications of using standards for data sharing for drug repurposing (identify new therapeutic indications for existing drugs), which is even more meaningful in case of mitochondrial diseases, where the conventional drug discovery and development efforts are scanty. In addition, correlation of the molecular understanding of disease etiology and its manifestation in clinical symptoms will also serve as a valuable resource for identification of potential biomarkers. Such a semantic-based platform will be amenable for plugging in data and new methods with increasing understanding of disease biology. The scientific challenges involved in establishing genotype–phenotype correlations may be addressed by integrating existing resources and facilitating community collaborations toward a systems level understanding of the disease biology. It is therefore imperative that these data along with clinical readout are shared systematically for building systems level models to understand disease manifestation (Fig. 6.3). These models will not only be useful for companion diagnostics but also for predicting disease

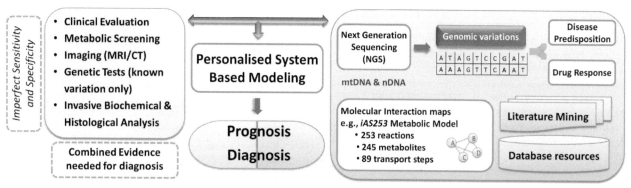

FIGURE 6.3 The current clinical evaluation methods for diagnosing mitochondrial diseases along with new NGS-based methods and system-based models could contribute to the development of personalized systems for treatment as well as diagnosis and prognosis. *MRI*, magneric resonance imaging; *CT*, computerized tomography; *mDNA*, mitochindrial DNA; *nDNA*, nuclear DNA.

comorbidities. However, this integration is possible only if the data sharing protocols are standardized and the data curated in globally acceptable ontology.

THE ROAD AHEAD

A global network of millions of genomes is vouched as the breakthrough technology with the power to advance the field of medicine (http://www.technologyreview.com/featuredstory/535016/internet-of-dna/?utm_campaign=newsletters&utm_source=newsletter-weekly-biomedicine&utm_medium=email&utm_content=20150224). Comparison of the genetic information across a large number of individuals with medical records is expected to deliver improved quantifiable measures for disease diagnosis. This is a formidable challenge, partly because of the technical reasons of moving petabytes of data across different laboratories, but especially because of the privacy issues surrounding patient identity. Both these issues must be addressed before the ever-increasing amount of genomic and clinical data piling up in laboratories and hospitals are utilized optimally. Initiatives like the MatchMaker Exchange are aiming to bring the genotype and phenotype data together on a common platform (http://matchmakerexchange.org/). Global Alliance for Genomics and Health (GA4GH) is an organization which specifies protocols, application program interface, and file formats for effective and responsible sharing of genomic and clinical data (http://genomicsandhealth.org/). The goal of GA4GH is to overcome challenges likes ethics and privacy involved with sharing of genomics data and to accelerate the potential of genomic medicine for advancement of human health. However, the issue of data privacy should be addressed using a patient-centric model where each patient decides on who can access their data and how it is used. This is only possible if the information generated on patient samples is made available to them in real-time.

Leading the route to advanced treatment options, Britain has given approval for mitochondrial donation, which allows mothers with mitochondrial diseases to have disease-free children (http://www.theguardian.com/politics/2015/feb/24/uk-house-of-lords-approves-conception-of-three-person-babies). In this methodology, in vitro fertilization (IVF) is used with biological material coming from three parents: mother, father (contributing 98.8% genetic material), and a female donor (contributing 0.2% genetic material). This three-parent IVF approval in the United Kingdom has received mixed opinions and the long-term implications of the same will be manifested and presented with time.

CONCLUSIONS

Unambiguous diagnosis, clinical prediction and assessment of long-term outcomes are often difficult for mitochondrial diseases because of heteroplasmy and allelic heterogeneity. Although inheritance pattern in classic mitochondrial disease is matrilineal, a majority of disorders of mitochondrial dysfunction are the result of specific pathogenic nDNA sequence alterations and mutations. The advent of NGS has revolutionized mitochondrial genomics. It marks a distinct break from the past, ensuring high-throughput, high-quality data in a very short time. Although NGS has its limitations in terms of short length and complicated assembly algorithms, luckily these are not as important with respect to mtDNA sequencing and targeted nDNA sequencing. It must also be mentioned that NGS is a rapidly evolving technique both with respect to sequencing chemistry and analysis algorithms. The high-throughput mtDNA and targeted nDNA data from NGS will facilitate cataloging the full range of disease-associated variations, improve genotype–phenotype correlation, and identify novel diagnostic variations. In synergy with the semantic web, improvised strategies for patient identification and recruitment, customized drug discovery strategies for rare diseases such as taking trials to the patient, and n-of-1 trials coupled with updated ontology will help in better genotype–phenotype correlation, developing better diagnostics and treatments/cures for mitochondrial diseases.

REFERENCES

[1] Kazak L, Reyes A, Holt IJ. Minimizing the damage: repair pathways keep mitochondrial DNA intact. Nat Rev Mol Cell Biol 2012;13(10):659–71.

[2] Iborra FJ, Kimura H, Cook PR. The functional organization of mitochondrial genomes in human cells. BMC Biol 2004;2:9.

[3] Tuppen HA, Blakely EL, Turnbull DM, Taylor RW. Mitochondrial DNA mutations and human disease. Biochim Biophys Acta 2010;1797(2):113–28.

[4] Parikh S, Goldstein A, Koenig MK, Scaglia F, Enns GM, Saneto R, et al. Diagnosis and management of mitochondrial disease: a consensus statement from the Mitochondrial Medicine Society. Genet Med 2014;17(9):689–701.

[5] Anderson S, Bankier AT, Barrell BG, de Bruijn MH, Coulson AR, Drouin J, et al. Sequence and organization of the human mitochondrial genome. Nature 1981;290(5806):457–65.

[6] Andrews RM, Kubacka I, Chinnery PF, Lightowlers RN, Turnbull DM, Howell N. Reanalysis and revision of the Cambridge reference sequence for human mitochondrial DNA. Nat Genet 1999;23(2):147.

[7] Lott MT, Leipzig JN, Derbeneva O, Xie HM, Chalkia D, Sarmady M, et al. mtDNA variation and analysis using MITOMAP and MITOMASTER. Curr Protoc Bioinform/Editoral board, Andreas D Baxevanis [et al]. 2013;1(123):1.23. 1–26.

[8] K S, Jalali S, Scaria V, Bhardwaj A. MitoLSDB: a comprehensive resource to study genotype to phenotype correlations in human mitochondrial DNA variations. PLoS One 2013;8(4):e60066.

[9] Pagliarini DJ, Calvo SE, Chang B, Sheth SA, Vafai SB, Ong SE, et al. A mitochondrial protein compendium elucidates complex I disease biology. Cell 2008;134(1):112–23.

[10] Smith AC, Blackshaw JA, Robinson AJ. MitoMiner: a data warehouse for mitochondrial proteomics data. Nucleic Acids Res 2012;40(Database issue):D1160–7.

[11] Smith AC, Robinson AJ. MitoMiner, an integrated database for the storage and analysis of mitochondrial proteomics data. Mol Cell Proteomics 2009;8(6):1324–37.

[12] Damas J, Carneiro J, Amorim A, Pereira F. MitoBreak: the mitochondrial DNA breakpoints database. Nucleic Acids Res 2014;42(Database issue):D1261–8.

[13] Rubino F, Piredda R, Calabrese FM, Simone D, Lang M, Calabrese C, et al. HmtDB, a genomic resource for mitochondrion-based human variability studies. Nucleic Acids Res 2012;40(Database issue):D1150–9.

[14] Scheibye-Knudsen M, Scheibye-Alsing K, Canugovi C, Croteau DL, Bohr VA. A novel diagnostic tool reveals mitochondrial pathology in human diseases and aging. Aging 2013;5(3):192–208.

[15] Bhardwaj A, Mukerji M, Sharma S, Paul J, Gokhale CS, Srivastava AK, et al. MtSNPscore: a combined evidence approach for assessing cumulative impact of mitochondrial variations in disease. BMC Bioinformatics 2009;10(Suppl. 8):S7.

[16] Vellarikkal SK, Dhiman H, Joshi K, Hasija Y, Sivasubbu S, Scaria V. mit-o-matic: a comprehensive computational pipeline for clinical evaluation of mitochondrial variations from next-generation sequencing datasets. Hum Mutat 2015;36(4):419–24.

[17] Falk MJ, Shen L, Gonzalez M, Leipzig J, Lott MT, Stassen AP, et al. Mitochondrial Disease Sequence Data Resource (MSeqDR): a global grass-roots consortium to facilitate deposition, curation, annotation, and integrated analysis of genomic data for the mitochondrial disease clinical and research communities. Mol Genet Metab 2014;114(3):388–96.

[18] Lopez MF, Kristal BS, Chernokalskaya E, Lazarev A, Shestopalov AI, Bogdanova A, et al. High-throughput profiling of the mitochondrial proteome using affinity fractionation and automation. Electrophoresis 2000;21(16):3427–40.

[19] Tang S, Halberg MC, Floyd KC, Wang J. Analysis of common mitochondrial DNA mutations by allele-specific oligonucleotide and Southern blot hybridization. Methods Mol Biol 2012;837:259–79.

[20] Venegas V, Halberg MC. Quantification of mtDNA mutation heteroplasmy (ARMS qPCR). Methods Mol Biol 2012;837:313–26.

[21] Chinault AC, Shaw CA, Brundage EK, Tang LY, Wong LJ. Application of dual-genome oligonucleotide array-based comparative genomic hybridization to the molecular diagnosis of mitochondrial DNA deletion and depletion syndromes. Genet Med 2009;11(7):518–26.

[22] Wong LJ, Dimmock D, Geraghty MT, Quan R, Lichter-Konecki U, Wang J, et al. Utility of oligonucleotide array-based comparative genomic hybridization for detection of target gene deletions. Clin Chem 2008;54(7):1141–8.

[23] Lander ES, Linton LM, Birren B, Nusbaum C, Zody MC, Baldwin J, et al. Initial sequencing and analysis of the human genome. Nature 2001;409(6822):860–921.

[24] Mardis ER. Next-generation DNA sequencing methods. Annu Rev Genomics Hum Genet 2008;9:387–402.

[25] van Dijk EL, Auger H, Jaszczyszyn Y, Thermes C. Ten years of next-generation sequencing technology. Trends Genet 2014;30(9):418–26.

[26] Buermans HP, den Dunnen JT. Next generation sequencing technology: advances and applications. Biochim Biophys Acta 2014;1842(10):1932–41.

[27] Calvo SE, Compton AG, Hershman SG, Lim SC, Lieber DS, Tucker EJ, et al. Molecular diagnosis of infantile mitochondrial disease with targeted next-generation sequencing. Sci Transl Med 2012;4(118):118–210.

[28] Wong LJ. Challenges of bringing next generation sequencing technologies to clinical molecular diagnostic laboratories. Neurotherapeutics 2013;10(2):262–72.

[29] Zhang W, Cui H, Wong LJ. Comprehensive one-step molecular analyses of mitochondrial genome by massively parallel sequencing. Clin Chem 2012;58(9):1322–31.

[30] Dames S, Chou LS, Xiao Y, Wayman T, Stocks J, Singleton M, et al. The development of next-generation sequencing assays for the mitochondrial genome and 108 nuclear genes associated with mitochondrial disorders. J Mol Diagn 2013;15(4):526–34.

[31] a. Sosa MX, Sivakumar IK, Maragh S, Veeramachaneni V, Hariharan R, Parulekar M, et al. Next-generation sequencing of human mitochondrial reference genomes uncovers high heteroplasmy frequency. PLoS Comput Biol 2012;8(10):e1002737.
b. Lillie EO, Patay B, Diamant J, Issell B, Topol EJ, Schork NJ. The n-of-1 clinical trial: the ultimate strategy for individualizing medicine? Per Med March 2011;8(2):161–73.

[32] Gene Ontology C. The gene ontology project in 2008. Nucleic Acids Res 2008;36(Database issue):D440–4.

[33] Rogers FB. Medical subject headings. Bull Med Libr Assoc 1963;51:114–6.

[34] Cote RA, Robboy S. Progress in medical information management: systematized nomenclature of medicine (SNOMED). JAMA 1980;243(8):756–62.

[35] Liu S, Wei M, Moore R, Ganesan V, Nelson S. RxNORM: prescription for electronic drug information exchange. IT Prof 2005;7(5):17–23.

[36] Riano D, Real F, Lopez-Vallverdu JA, Campana F, Ercolani S, Mecocci P, et al. An ontology-based personalization of health-care knowledge to support clinical decisions for chronically ill patients. J Biomed Inform 2012;45(3):429–46.

[37] Kohler S, Doelken SC, Mungall CJ, Bauer S, Firth HV, Bailleul-Forestier I, et al. The Human Phenotype Ontology project: linking molecular biology and disease through phenotype data. Nucleic Acids Res 2014;42(Database issue):D966–74.

[38] Schriml LM, Mitraka E. The Disease Ontology: fostering interoperability between biological and clinical human disease-related data. Mamm Genome 2015;26(9–10):584–9.

[39] Brookes AJ, Lehvaslaiho H, Muilu J, Shigemoto Y, Oroguchi T, Tomiki T, et al. The phenotype and genotype experiment object model (PAGE-OM): a robust data structure for information related to DNA variation. Hum Mutat 2009;30(6):968–77.

[40] Vihinen M. Variation ontology for annotation of variation effects and mechanisms. Genome Res 2014;24(2):356–64.

[41] Abinaya E, Narang P, Bhardwaj A. FROG: fingerprinting genomic variation ontology. PLoS One 2015;10(8):e0134693.

[42] Sahni N, Yi S, Taipale M, Fuxman Bass JI, Coulombe-Huntington J, Yang F, et al. Widespread macromolecular interaction perturbations in human genetic disorders. Cell 2015;161(3):647–60.

[43] Waszak SM, Delaneau O, Gschwind AR, Kilpinen H, Raghav SK, Witwicki RM, et al. Population variation and genetic control of modular chromatin architecture in humans. Cell 2015;162(5):1039–50.

[44] Masino AJ, Dechene ET, Dulik MC, Wilkens A, Spinner NB, Krantz ID, et al. Clinical phenotype-based gene prioritization: an initial study using semantic similarity and the human phenotype ontology. BMC Bioinformatics 2014;15:248.

Chapter 7

The Significance of Metabolomics in Human Health

D.F. Gomez-Casati, M. Grisolía, M.V. Busi
Universidad Nacional de Rosario, Rosario, Argentina

Chapter Outline

Introduction	89	Metabolomics in Environmental and Public Health	95
Metabolomics in Human Diseases	90	Human Nutrition	96
Biomarker Detection	93	Intestinal Metabolome	96
Group Discrimination (Diagnosis)	93	Conclusions	97
Inborn Metabolic Disorders	93	Acknowledgment	98
Cancer Metabolomics	94	References	98
Biomarkers for Psychiatric Diseases	94		

INTRODUCTION

Metabolites are the ultimate response of biological systems to genetic or environmental changes [1]. They are structurally and chemically heterogeneous and show dynamism in time and space. For this reason, the analytical procedures in their measurement presents technical and experimental challenges [2].

Metabolomics is the study of the complete, nonbiased, high-throughput analysis of a complex collection of metabolites present in a cell or tissue under a particular set of conditions, generating a biochemical profile called a *metabolome* [3]. This contributes to our understanding of the complex molecular interactions in biological systems [4]. It is worth recalling that there is a similar term, *metabonomics*, which is usually applied only to studies of human health, and usually compares profiles instead of identifying individual compounds. Metabolomics symbolizes the progression from large-scale analysis of RNA and proteins to metabolite at the systems level [5,6]. Because of the highly diverse molecular identity of metabolites, this task depends on sophisticated instrumentation such as mass spectrometry (MS) and nuclear magnetic resonance (NMR) spectroscopy [7]. As well as proteomic and transcriptomic techniques, the metabolomic determinations are low-expense, rapid, and automated techniques, allowing the possibility to analyze several biological samples. Thus metabolomic studies compared with genomics or proteomics alone are more suitable to reflect changes in a particular phenotype of an organism,

tissue, or cell [8,9]. Metabolomics could be used for the characterization of all measurable molecules in a sample (untargeted metabolomics), or for the analysis of a specific set of chemically characterized metabolites (targeted metabolomics) such as lipids, amino acids, nucleotides, or steroids [10].

MS is more sensitive than NMR. The first technique quantifies metabolites at picograms per milliliter level, whereas NMR detects metabolites in the order of micrograms per milliliter [10]. Nevertheless, NMR shows a remarkable reproducibility, making it a more robust analytical approach [11]. In addition, MS sensitivity is dependent on the type of compound to be analyzed and may be affected by matrix effects and absolute quantification by MS requires inner standards (eg, ribitol) whereas NMR samples do not need chemical derivatives to make them detectable [11]. This ease of sample preparation, in addition to the rapid analysis time of NMR techniques, allows a better adaptation of this approach to high-throughput fingerprinting. Furthermore, NMR provides a better analysis of highly polar metabolites including sugars and sugar phosphates, whereas the identification of those metabolites by MS techniques requires specific strategies using hydrophilic stationary phases in the metabolite resolution step [12].

However, one point to consider is that these techniques can still measure only a fraction of the total metabolome, suggesting that there are still problems associated with metabolite extraction and resolution rather than technique sensitivity [13]. For this reason, the combined use of NMR

Medical and Health Genomics. http://dx.doi.org/10.1016/B978-0-12-420196-5.00007-1

and MS could increase the number of detected metabolites, including the endogenous metabolites and those generated in response to environmental exposure [10].

NMR spectroscopy and MS technologies are continuously improved to increase sensitivity and spectral resolution of analytic assays on biological samples. Several strategies used in metabolomics are summarized in Table 7.1 [19–22], including metabolite target analysis, metabolite profiling, metabolomics, metabolite flux analysis, and metabolic fingerprinting (and footprinting) [3].

Metabolomics was first applied to the study of toxicology and pharmacology, inborn metabolic errors, and nutrition [9]. At present, there is a growing number of applications to characterize different human diseases (cancer, diabetes, and autoimmune, psychiatric, and coronary diseases) and human response to environmental stresses [3]. In this chapter, we attempt to introduce the reader to the most relevant medical and health applications of metabolomics. In consideration of this objective, the reader is presented with technical and analytical procedures supported with updated literature.

METABOLOMICS IN HUMAN DISEASES

The use of metabolomics in human health requires a detailed experimental design to improve reproducibility and sensitivity, allowing statistic signal/noise ratio optimization. Most metabolomic experimental designs use two approaches to balance the correct control of confounding factors: (1) the detailed evaluation of phenotype parameters, including gender, age, physiological status (for example, body mass index), pertinent pathological information, and medications used; (2) any cultural or environmental factors (eg, diet, lifestyle). However, because of the variable nature of endogenous metabolites in biological fluids, each metabonomic experiment would need to include a significant number of subjects (n) to ensure statistically significant outcomes.

The challenge in human metabolic characterization is the existence of many cell types located in topographically distinct locations in varied physiological states [23]. Therefore of all human samples (fluids or tissues) used for metabolomics methods, the urine and plasma are most representative of the general metabolic state [10]. Furthermore, these are preferable owing to relatively easy and noninvasive sampling.

The type of biological sample and storage conditions are also essential for sampling strategy [24,25]. For example, the NMR analysis of polar metabolites in some biological samples, such as urinary and tissue extracts, is straightforward. However, in blood plasma or sera, the relevant metabolites may be dispersed with lipoprotein particles, hindering the metabolite resolution unless separately extracted. Still, the metabolic profile of those samples could be increased using a range of pulse sequences (Carr-Purcell-Meiboom-Gill pulse sequence) to optimize the observation of slow tumbling metabolites [15]. Extensive reviews of sampling, analysis, and processing strategies for NMR and MS analyses have been carried out elsewhere [24–27].

For the comprehensive implementation of metabolomics, there is a need of a wide variety of databases (Table 7.2). For example, the KEGG PATHWAY database (http://www.genome.jp/kegg/pathway.html) [28] is a collection of metabolomics pathway maps containing the information of all the molecular interaction and reaction networks known so far. Compound-specific databases, such as the KEGG COMPOUND database (http://www.genome.jp/kegg/compound/), focus on providing detailed nomenclature, structural, or physicochemical data on small molecules of biological interest. Spectral databases such as the Madison Metabolic Consortium database (http://mmcd.nmrfam.wisc.edu/) [29] and the Golm Metabolome database (http://gmd.mpimp-golm.mpg.de/) from the Max Planck Institute for Molecular Plant Physiology [30] contain reference spectra for molecules along with spectral matching software.

TABLE 7.1 Comparison of Most Used Methods Applied in Metabolomic Research

Method	Use	Summary	References
NMR			
1D ^{1}H with presaturation	II, III, IV	Nondestructive, very reproducible, low sensitivity	[14]
CPMG	IV	Large molecules NMR signal suppressed, simpler spectra, nondestructive, very reproducible, low sensitivity	[15]
MS			
HPLC-MS	I, II, IV	High sensitivity, nonvolatile metabolites, expensive	[16]
GC-MS	II, III, IV	High sensitivity, only volatile compounds, inexpensive	[17]
Raman spectroscopy	IV	Low resolution, sensitivity, automated, high throughput	[18]

1D ^{1}H, one dimensional proton NMR; I, Targeted metabolomics; II, untargeted metabolomics; III, metabolite flux analysis; IV, metabolic fingerprinting; CPMG, Carr-Purcell-Meiboom-Gill pulse sequence; GC-MS, gas chromatography–mass spectrometry; HPLC-MS, high-performance liquid chromatography–mass spectrometry; MS, mass spectrometry; NMR, nuclear magnetic resonance.

TABLE 7.2 Metabolomic Resources

Database	Detail	URL
Drug Databases		
DrugBank	Combines detailed drug chemical data with drug target information	http://www.drugbank.ca/
Therapeutic Targets Database	Provides information about the known therapeutic targets and the corresponding drugs/ligands	http://bidd.nus.edu.sg/group/ttd/ttd.asp
PharmGKB	Pharmacogenomics knowledge resource that encompasses clinical information	http://www.pharmgkb.org/
Search Tool for Interactions of Chemicals (STITCH)	Database of known and predicted interactions between chemicals and proteins	http://stitch.embl.de/
SuperTarget	Drug-target relations database linked with pathways, ontologies, 3D structures and medical data	http://bioinf-apache.charite.de/supertarget_v2/
Metabolic Pathway Databases		
Small Molecule Pathway Database (SMPDB)	Database of metabolic, drug, and disease pathways	http://smpdb.ca/
Kyoto Encyclopedia of Genes and Genomes (KEGG)	Hyperlinked databases that contains metabolic pathways from a wide variety of organisms	http://www.genome.jp/kegg/
MetaCyc Metabolic Pathway Database	Nonredundant database of experimentally elucidated metabolic pathways from more than 2600 different organisms	http://metacyc.org/
HumanCyc: Encyclopedia of Human Genes and Metabolism	Encyclopedic database of the human metabolic pathways	http://humancyc.org/
BioCyc Database Collection	BioCyc is a collection of 5500 pathway/genome databases where each database describes the genome and metabolic pathways of a single organism. It also contains tools for understanding their data	http://biocyc.org/
Reactome	Reductionist database of biological pathways from several organisms, particularly *Homo sapiens*	http://www.reactome.org/
Compound or Compound-Specific Databases		
PubChem	Database of chemical structures of small organic molecules linked to their biological activities and NIH PubMed/Entrez information	https://pubchem.ncbi.nlm.nih.gov/
Chemical Entities of Biological Interest (ChEBI)	Dictionary of small molecules (natural or synthetic) including ontological classification	http://www.ebi.ac.uk/chebi/
ChemSpider	Chemical structure database of organic molecules	http://www.chemspider.com/
KEGG Glycan	Collection of experimentally determined glycan structures	http://www.genome.jp/kegg/glycan/
Toxin and Toxin Target Database (T3DB)	Database of common toxins and their associated toxin targets	http://www.t3db.ca/
In vivo/In silico Metabolite Database (IIMDB)	Database describing both known and computationally generated compounds	http://metabolomics.pharm.uconn.edu/iimdb/

Continued

TABLE 7.2 Metabolomic Resources—cont'd

Database	Detail	URL
Spectral Databases		
Biological Magnetic Resonance Data Bank (BMRB)	NMR spectral database, primarily for macromolecules	http://www.bmrb.wisc.edu/metabolomics/
Madison-Qingdao Metabolomics Consortium Database (MMCD)	Metabolite NMR and MS spectral database relating metabolites to their physical and chemical properties	http://mmcd.nmrfam.wisc.edu/
High Quality Mass Spectral Database (MassBank)	Database of high-resolution MS spectra of metabolites	http://www.massbank.jp/
Golm Metabolome Database	Mass spectral and retention time index libraries determined on several types of GC-MS instruments (quadrupole and time-of-flight)	http://gmd.mpimp-golm.mpg.de/
Metlin	Database of high-resolution metabolite MS/MS spectra and tandem MS experiments	http://metlin.scripps.edu/index.php
Platform for RIKEN Metabolomics (PRIMe)	Genomic and metabolomics data based on NMR spectroscopy, GC-MS, LC-MS and CE-MS	http://prime.psc.riken.jp/
The Armec Repository Project	Database of human metabolite FIE-MS/HPLC-ESI-MS spectra	http://www.armec.org/MetaboliteLibrary/
Birmingham Metabolite Library (BML-NMR)	Database of metabolite standards for NMR spectra	http://www.bml-nmr.org/
MetaboLights	Database for metabolomics experiments and derived information	http://www.ebi.ac.uk/metabolights/index
Advanced Mass Spectral Database (mzCloud)	Collection of high-resolution MS spectra improved with a third-generation spectra correlation algorithm	https://mzcloud.org/
NIST/EPA/NIH Mass Spectral Library	Commercial mass spectral library that includes sublibraries with MS/MS and retention index data	http://www.nist.gov/srd/nist1a.cfm
Disease & Physiology Databases		
Online Inheritance in Man (OMIM)	Database of human disease-related genes and genetic phenotypes	http://www.ncbi.nlm.nih.gov/omim/
Metagene	Database for inborn errors of metabolism	http://www.metagene.de/program/a.prg
Online Metabolic and Molecular Bases of Inherited Disease (OMMBID)	Encyclopedic database of human disorders comprising the genetic variation to the clinical descriptions and treatment	http://ommbid.mhmedical.com/
Comprehensive Metabolomic Databases		
Human Metabolome Database (HMDB)	Integrative metabolome database that contains detailed information about small human metabolites linked to chemical, clinical, and biochemical data	http://www.hmdb.ca/
BiGG Models	Database of genome-scale metabolic network reconstructions from several organism, including human	http://bigg.ucsd.edu/bigg/main.pl

CE-MS, Capillary electrophoresis–mass spectrometry; *EPA*, US Environmental Protection Agency; *FIE-MS*, Flow injection electrospray–mass spectrometry; *GC-MS*, gas chromatography–mass spectrometry; *HPLC-ESI-MS*, high-performance liquid chromatography–electrospray ionization–mass spectrometry; *LC-MS*, liquid chromatography–mass spectrometry; *MS*, mass spectrometry; *NIH*, National Institutes of Health; *NIST*, National Institute of Standards and Technology; *NMR*, nuclear magnetic resonance.

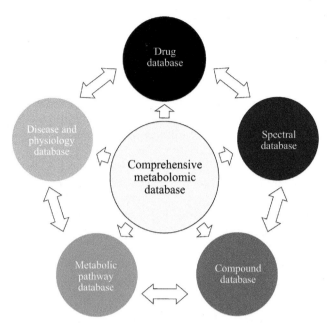

FIGURE 7.1 Oversimplified model of database relationship. All databases are connected with each other to some extent, represented by their proximity in the illustration. The content of comprehensive databases is hyperlinked with all mentioned databases.

Disease/physiology databases contain descriptions of the causes, clinical symptoms, diagnostic indicators, or genetic mutations associated to specific diseases. An example of these databases are the Online Inheritance in Man database (http://www.ncbi.nlm.nih.gov/omim), edited by the McKusick-Nathans Institute of Genetic Medicine, Johns Hopkins University School of Medicine, which contains information on all known mendelian disorders (see chapter Epigenetics and Epigenomics in Human Health and Disease) and over 12,000 genes involved in human genetic diseases [31]. Finally, comprehensive organism-specific metabolomic databases, such us the Human Metabolome database (http://www.hmdb.ca/), have been created in the attempt to combine the information from most of those databases into a single resource [32,33], turning such databases into a helpful resource in biomarker detection and metabolomics diagnosis (see Fig. 7.1).

BIOMARKER DETECTION

Biomarker discovery often starts with the study of animal models, helping researchers understand a disease's biochemistry. Biomarkers of several human diseases have been discovered using a metabolomics approach, notably certain cancers. For appropriate selection of biomarkers, the data bulk generated must be carefully mined in an efficient way. For this, it is necessary to adequately perform a number of processes: noise filtering, baseline correction, centering, normalization, peak picking, peak integration, and alignment [34]. The data generated must be represented in

a two-dimensional data matrix using open-source [35–37] or manufacturer software. However, technical difficulties may hinder the characterization of a tumor metabolome; for example, sample-to-sample variation, sensitivity, and the physiological status of the tumor [9]. For this reason, subsequent to biomarker selection, the process of result validation is an absolute requirement. Different sample sets may be used to guarantee reproducibility and specificity. Because metabolomics is an exploratory approach, final biochemical validations must be performed to confirm that concentrations of proposed biomarkers are different between the test and control samples [34]. In addition, metabolite profile variation among the different tumor types makes difficult to generalize findings across tumor groups [3]. Finally, the accuracy and sensitivity of NMR or MS could be improved by the use of stable isotopes (eg, ^{13}C-labeled metabolites), leading to the development of new diagnostic tools [38].

GROUP DISCRIMINATION (DIAGNOSIS)

Metabolomics (and metabonomics) are effective approaches to visualize an individual's metabolic hyperspace. Both analyses are useful as diagnostic parameters and metrics of therapeutic efficacy [23]. Hence statistical techniques (multivariate analysis) or machine learning tools are used in any metabolomics experiment. For a more comprehensive analysis, most discrimination models use principal component analysis (PCA), a statistical technique that converts a multivariate data set (possibly correlated) into fewer dimensions (two or three) defined by linearly uncorrelated variables called *principal components*. This allows data clustering in regard to these new variables, summarizing properties that allow distinct pattern recognition and group classification [39,40].

Machine learning algorithms are a more recent class of multivariate analysis technique [41]. These algorithms can be trained to recognize patterns from training data and then build a linear classifier for the test data. The lack of restrictive assumptions (eg, normally distribution) and the advantage of using it in nonlinear cases give support vector machines (SVMs) a better generalization performance when compared with other statistical multivariate methods. Often when the normality of a data set is analyzed, it becomes clear that a presumption is not valid. Lastly, SVMs generate a better predictive model than statistics procedures from the subset of features describing the phenotype using the lowest number of possible features, placing SVMs as a more efficient tool for diagnostic purposes than statistics procedures [39].

INBORN METABOLIC DISORDERS

Inborn metabolic disorders (IMDs) are caused by several genetic abnormalities that result in the abnormal accumulation of metabolites, such as the alteration in enzyme activities or the use of an alternative metabolic pathway.

This accumulation (or depletion) of a particular metabolite or a series of metabolites could produce a toxic or deleterious effect in the patient that may manifest clinically with different symptoms [13]. However, most infants with an IMD look apparently normal at birth or in the first few weeks of life, because most of the toxic metabolites are cleared through the placenta during gestation and evidence of an IMD requires excessive accumulation of metabolites to clinically significant levels [13]. Some of these metabolic by-products could be screened by either urinary or plasma assays or both. Some of the earlier newborn screening methods relied on this approach. However, before the introduction of MS or NMR, early detection and diagnosis was often missed or delayed. Therefore MS and NMR were proposed for newborn metabolic screening for early diagnosis with the objective to reduce morbidity and mortality through targeted therapeutic or early dietary intervention [13,42]. The main advantage of using this approach is that it allows the rapid screening of multiple metabolic disorders, instead of only one, within a short time [43].

Amino acid disorders, organic acid disorders, and fatty acid oxidation defects are some of the IMDs detected by MS/NMR newborn screening programs [43]. This approach takes advantage of the high sensitivity and specificity of metabolomics methods. However, because of the contribution of confounding factors such as prematurity, low birth weight, or maternal nutritional vitamin B12 deficiency, the positive predictive value is only around a 10% [43]. Furthermore, MS/NMR cannot distinguish isomeric compounds. For this reason, it is important to set up the cutoff values to minimize false-positive cases and avoid false-negatives.

CANCER METABOLOMICS

One of the most important applications of metabolomics is in the field of cancer cell and molecular biology [3]. Tumor cells have special metabolic requirements when compared with normal cells and often lose many regulatory functions [44]. Therefore in most cases the combination of metabolomics with other genomic and/or proteomic techniques is extremely useful for diagnosis of cancer; for example, the application of cytokine antibody microarrays as reliable biomarkers for hepatocellular carcinoma [45].

The use of ultraperformance liquid chromatography–electrospray ionization and time-of-flight MS (TOFMS), are used for the identification of several metabolites involved in tumor cell pathology. Patterson et al. [46] described an alteration of the lipid metabolism profile, with an increase of glycodeoxycholate, deoxycholate-3-sulfate, and bilirubin, and decreased levels of lysophosphocholine. Additionally, polar amino acids such as arginine, proline, alanine, lysine, and aspartate were found to be altered in many cases of liver cancer [47]. Brain tumors, gliomas, and neuroepithelial tumors also present with increased alanine

levels [48]. In breast cancer, the estrogen receptor and the human epidermal growth factor receptor 2 are usually used as biomarkers for prognostic or predictive purposes [9]. In addition, several compounds such as choline, phosphocholine, myoinositol, glycine, taurine, and glycerophosphocholine are altered in certain types of breast cancers. Moreover, other compounds such as myo-inositol are also found to be increased in prostate cancer, colon adenocarcinoma, and ovarian carcinoma [38]. Cheng and colleagues have described the use a high-resolution magic-angle spinning proton nuclear magnetic resonance (^{1}H NMR) spectroscopy to examine different types of brain tumors (glioblastomas, schwannomas, and meningiomas) based on the concentration of 11 metabolites and their associated relaxation properties [49,50].

Gastric cancer has a high morbidity and mortality. An early diagnosis is essential to ensure a greater chance of cure and/or long-term ·survival. Metabolomic analysis of gastric cancer tissue samples demonstrates a relationship of cancer cells to the Warburg effect. The Warburg effect is the phenomenon in which cancer cells produce additional energy through increased oxygen-dependent glycolysis followed by lactic acid fermentation with secretion of lactate. Alternatively, the conversion of glucose to pyruvate with consequent production of adenosine triphosphate does not require oxygen [51].

There is accumulating evidence that genetic mutations in cancer-driver genes, tumor suppressors, and amplified oncogenes are linked to specific alterations in glucose metabolism. Several studies demonstrate that lactate is the most consistently elevated carbohydrate pathway biomarker with consistent glucose depletion. In addition to glycolysis, the tricarboxylic acid (TCA) cycle is also affected, and mainly represented by an increased malate content [40]. Furthermore, using a capillary electrophoresis (CE)–connected electrospray ionization TOFMS and CE-MS/MS systems, Makinoshima et al. [52] demonstrated that the epidermal growth factor receptor–tyrosine kinase inhibitors decreased lactate production, increased glucose consumption, and the glucose-induced extracellular acidification rate in lung adenocarcinoma cells. This approach could help develop targeted therapy focused in a final metabolic regulation to treat patients with specific metabolic profiles in cancer cells.

BIOMARKERS FOR PSYCHIATRIC DISEASES

Currently, the diagnosis of psychiatric diseases primarily depends on the subjective judgment and interpretation of the heterogeneous clinical symptoms, often with high rates of misdiagnosis or delayed diagnosis [45]. In recent years, metabolomics has been successfully applied to describe the metabolite state of many psychiatric diseases, aiming to provide mechanistic insights into the pathophysiology of several neuropsychiatric diseases; for example, the early-onset Alzheimer dementia [53].

Brain tissues and cerebrospinal fluid theoretically seem to be the most appropriate biological samples for research; however, it is impractical to collect those samples from living patients owing to ethical and safety constraints. For this reason, efforts have been made to find the best biological samples for different psychiatric disorders. For example, several studies show the ease of collecting peripheral blood mononuclear cells for analyzing certain brain metabolites in making a reliable diagnosis of schizophrenia [45]. In addition, assay for urinary metabolite excretion may also assist in neuropsychiatric diagnosis [54].

In schizophrenic patients, a differential metabolite profile was observed when compared with healthy control subjects. In this profile, a variation of metabolites involved in energy metabolism, oxidative stress, and neurotransmitter metabolism was elucidated including increased levels of hydroxylamine, octanoic acid, glycerol, aspartic acid, 2-hydroxyethyl palmitate, benzoic acid, and homoserine, and decreased levels of methylphosphate, valine, fumaric acid, pyroglutamic acid, creatinine, sorbitol, inositol, dopamine, maltose, γ-tocopherol, and α-tocopherol [45].

Bipolar disorder (BD) is one of the common and debilitating mental disorders that currently has no reliable biomarkers for early diagnosis. A gas chromatography (GC)–MS based metabonomic method was used to characterize the urinary metabolic profile of BD subjects vs. healthy controls to identify and validate several urinary biomarkers. From a total of 37 differential urinary metabolites, 2,4-dihydroxypyrimidine was identified as a promising biomarker. This may facilitate the development of a urine-based diagnostic test for BD [55]. Furthermore, an NMR approach was used to characterize the urinary metabolic profile of BD showing four highly promising biomarkers including α-hydroxybutyrate, choline, isobutyrate, and N-methylnicotinamide [56].

Major depressive disorder (MDD) is a socially detrimental psychiatric disorder that contributes to increased healthcare expenditures and suicide rates, the latter being the major serious consequence of this disease. However, no empirical laboratory-based tests are available to support the diagnosis of MDD. A few studies used an NMR-based plasma metabonomic method for biomarker detection. In one of these studies, the levels of low-density lipoprotein (LDL), very–low-density lipoprotein (VLDL), cholesterol, unsaturated lipid, glucose, pyruvate, lactate, and amino acids (alanine, glycine, and taurine) were differentially altered [57]. In a subsequent study, the depressed suicide-attempter group was significantly distinguishable from the control group through several key metabolites: leucine, valine, LDL, VLDL, lactate, alanine, acetate, N-acetyl-cysteine, pyruvate, glutamine, creatine, creatinine, taurine, myo-inositol, glycine, glucose, and unsaturated lipid, suggesting an evident alteration of lipid metabolism [58]. Urine samples were also used to identify biomarkers in MDD patients who also underwent NMR-based metabonomics. Five urinary metabolites: malonate, formate, N-methyl-nicotinamide, m-hydroxyphenyl acetate, and alanine were successfully utilized to distinguish depressed subjects from healthy controls [59].

Finally, a few experimental approaches were used to investigate how chronic stressful events (key risk factors) affect MDD using animal models. Thus rats were exposed to a stressful event and then the group that developed anhedonia was contrasted with the control group (chronic mild stress–resilient and control group). The cerebellums of these rats were compared using a GC-MS approach. The four cerebellar metabolites (proline, lysine, glutamine, and dihydroxyacetone phosphate) were the key differential metabolites associated with stress resilience [60].

METABOLOMICS IN ENVIRONMENTAL AND PUBLIC HEALTH

Metabolites have commonly been classified simply as endogenous or xenobiotic, according to their origin. However, in truly *endogenous* processes, the primary determinants of the biochemical transformations are under the direct control of enzymes and transporters of the host genome and have been optimized by evolution. This covers the principal intermediary biosynthetic, energetic, and catabolic pathway. For this reason, several are used to clarify the origin and the destiny of those metabolites [23]. *Symendogenous* compounds are essential and can be metabolized, but not synthesized by the host (eg, vitamins and essential amino acids). Those involved in the cometabolism by two or more organisms that are commensal or symbiotic represent *symxenobiotic* metabolites (eg, bile acid metabolism). However, certain metabolites are not necessarily essential to the host but can influence endogenous or other xenobiotic metabolic processes. *Transxenobiotic* compounds present an extragenomic (or chemical) origin but are metabolically converted to endogenous species or metabolites that can be utilized directly in endogenous processes (eg, ethanol). In this classification, a *xenobiotic* compound is external to the endogenous processes and has no intrinsic biological function. However, they could have significant effects on host metabolic pathways control and can be extensively metabolized by the host enzyme systems with relatively low substrate specificity. Finally, *metabonate* is defined as a compound that is produced by a superficial chemical rearrangement within an organism and then can be excreted or further metabolized [23].

From various family studies and twin brothers [61,62], epigenetics [63], and the alteration of gene expression by the environment [64], it becomes evident that the human genome cannot explain by itself all the epidemiology related to a disease. Therefore *exposome* is defined as the characterization of all the chemical exposures from both endogenous and exogenous sources [65], with the objective to provide a holistic approach to understanding the primary cause of a chronic disease [65,66]. Until recently, external and internal exposures (including endogenous chemicals) have focused on a limited number of molecules and metals, leading to the isolated study of human metabolic response to a specific external agent; for example, aflatoxin [67] and *Helicobacter pylori* exposure [68]. Despite the technical difficulties of measuring exposures influenced by the spatial, temporal, and intraindividual and interindividual variation [69], metabolomics represents a promising tool for studying the effects of xenobiotics on human health. Therefore using the correct number of samples, appropriate experimental conditions, and statistical analysis, these limitations could be overcome and some generalization could be achieved. Bonvallot and collaborators [10] review several applications of metabolomics in the field of environmental health, mainly dividing those studies in two groups based on the experimental subjects (animal or human), the latter being less common.

HUMAN NUTRITION

The effects of diet and lifestyle were noted to have the greatest influence on the metabolomic profile—to such an extent as to obscure a great source of variability like gender [13]. In the postgenomic era, three associated terms were created in order to refer to the effects of foods on gene expression (*nutrigenetics* and *nutrigenomics*), and to refer to the effects of food on individual metabolic responses (*nutrimetabolomics*) [70]. The identification of nutritional compounds and the elucidation of the mechanism by which they affect the expression of target genes is extremely important in preventing diseases [71] and could provide the foundation for the discovery of bioprotective foods [22]. However, it is important to remark that the biomarkers associated with specific food intake must be properly validated because of the great amount of confounding variables, such as the occurrence of the same precursor in various foods. Scalbert and collaborators [72] reviewed an extensive list of novel dietary biomarkers identified through a metabolomic approach.

INTESTINAL METABOLOME

Gut microbiota is the home of trillions of microorganisms that in most cases develop a mutualistic relationship with the host. It must be distinguished from microbiome in which this last is specific to the collective genomes of the microorganisms that reside in a specific niche [73,74]. The gut microbiome, which may contain about 150-fold more genes than the human genome, provides us with other metabolic capabilities [75]. Bacteria colonize a human host from birth and potentially in utero [76]. Also, skin-to-skin contact and breastfeeding play a crucial role in microbiota

stabilization, which in terms of its effects has been compared with a previously unknown organ. However, the gut microbiota is rather flexible and it can be altered through different factors such as the environment or diet [75].

The paradigm of intestinal microflora study has suffered several changes throughout modern medical history. First, classical microbiology approaches focus on isolating and identifying individual bacterial species associated with the gut or a disease. More recently, metagenomic methods (see chapter Medical and Health Aspects of Genetics and Genomics) have been used to characterize the entire composition of microbial communities [77,78]. In the past, using biofluids (blood and urine), stools, and tissues (intact biopsies or extracts), metabolomic analyses have been used to provide novel insights into the molecular basis of these host–microbial relationships and their influence on health and disease risks [70]. This approach is based on the fact that specific metabolic activities of a single gut bacterial species can provide the host with new metabolites in sufficient quantity to be detected in blood. Using an untargeted MS metabolomics approach, Wikoff and collaborators have demonstrated the large effect of the intestinal microbiome on a mammalian plasma metabolic profile. Comparing germ-free and conventionally raised mice, they found that more than 10% of the total plasma metabolite concentration was altered by at least 50% [79].

Ulcerative colitis (UC) and Crohn disease (CD) are included into a group of inflammatory bowel diseases (IBDs), and these are characterized by a chronic disorder of the gastrointestinal track resulting from an autoimmune response where commensal microflora have been shown to play a key part in the disease process [80]. To date, the IBD diagnosis was only possible in an advanced disease state when symptoms are detectable by radiology or endoscopy with a biopsy of pathological lesions [81]. Thus metabolomics is focused on prognostics, diagnostics, and monitoring of the metabolic state of the intestine of IBD patients [70].

In human IBDs, the type of sample used in metabolomic techniques may have a major impact on the metabolic profile [82]. The urine metabolic profile has been used to identify hippurate as a primary biomarker and to generate a model that leads to the discrimination of IBD patients from healthy individuals, and between IBD patients [82,83]. Furthermore, hippurate levels also correlate with the presence of *Clostridia* in the gut of CD patients [84,85]. However, several studies were not successful in discriminating IBD from control patients using urine metabolomics because of the great contribution of confounding factors [82]. On the other hand, the use of a metabolic plasma profiles in mouse models with disease has provided satisfactory separation of UC and CD patients [86,87]. Nevertheless, several studies showed that metabolomics was successfully applied in the observation of differential plasma and urine metabolic profiles through time in interleukin (IL)-10 knockout mice in comparison with normal mice. IL-10 is an intestinal immunoregulator that downregulates inflammation [88], and the knockout mutants developed chronic colonic inflammation that mimics human IBD [89]. Despite chronic inflammation appearing from 8 weeks onward, differential profiles where observed at the age of 1 week. At this age, IL-10–deficient mice have shown higher levels of citrate, glutamine, fumarate, and lower levels of glucose and dimethylglycine than control mice. Finally, targeted metabolomics of amino acid profiles were able to discriminate IBD from healthy individuals, and UC from CD patients [90]. This suggests that the amino acid profiles and TCA cycle–related molecules could be useful in monitoring disease activity during progression of IBD [91].

Although biopsies from IBD lesions provide conflicting evidence, the results obtained from fecal samples allow the discrimination of gut microbiota disturbances [92] and lead to the discrimination of diseased and healthy individuals based on the short fatty acid composition [93].

Irritable bowel syndrome (IBS) is one of the major digestive disorders affecting around 20% of the industrialized adult population, for which underlying mechanisms must be further clarified [94]. IBS patients have an alteration of gut microbiota that correlates with an increase in amino acid levels (alanine and pyroglutamic acid) and phenolic compounds (hydroxyphenyl acetate and hydroxyphenyl propionate) [95].

Celiac disease is a multifactorial gastrointestinal disorder in which the presence of genetic factors *(DQ2/DQ8)* and an environmental trigger (gluten) are known and necessary but not sufficient for its development [96]. In individuals with a genetic predisposition, the accumulation of proline-/glutamine-rich peptides in the intestine triggers an abnormal innate and adaptive immune response mediated by gliadin-reactive CD4+ T cells [97]. It is known that gut microbiota is altered in individuals with celiac disease, and that this is not reverted with the implementation of gluten-free diet [98,99]. The use of metabolomics has clarified the dietary and physiological conditions of patients with celiac disease, with or without a strict gluten-free diet. The metabolic profile of celiac disease patients' plasma shows a decrease in the levels of pyruvate and lactate and an increased level of glucose, whereas the urine metabolic profile presents higher levels of indoxyl sulfate, *meta*-[hydroxyphenyl]propionic acid, phenylacetylglycine, and *p*-cresolsulfate [100]. Fecal and urine samples of treated celiac disease children showed elevated levels of free amino acids (proline, methionine, histidine, and tryptophan) and low levels of some short fatty acids (butyric, isocaproic, and propanoic acids) compared with healthy children [99], supporting the hypothesis that a gluten-free diet is not capable of modifying the gut microbiota of celiac disease patients. A PCA classification model presents a good accuracy (~84%) when discriminating healthy patients from those with celiac disease. However,

potential celiac disease patients [100] were classified with the overt celiac disease patients, whereas almost all patients adhering to a gluten-free diet were classified as healthy. This suggests that celiac disease–related abnormal metabolism precedes the intestinal damage and that the change in the plasma metabolite profile is caused mainly by an energetic misbalance that results from a malabsorption of nutritional elements [97].

Probiotic therapy is the attempt to alter the existing gut microbial environment through the ingestion of live consumable cultures of beneficial bacteria [79]. On the other hand, *prebiotics* refers to chemicals that induce the growth and/or activity of the human microbiota [101]. Furthermore, both approaches combined (called *synbiotics*) can be used to alter the microbiota composition [102]. Different combinations of probiotics and prebiotics promotes different metabolic profiles. For example, a germ-free mouse reinoculated with a model of human baby microbiota (HBM) supplemented with *Lactobacillus paracasei* and two different prebiotics (two galactosyl-oligosaccharides) shows a reduction of lactose levels in fecal samples. On the other hand, an HBM mouse supplemented with *Lactobacillus rhamnosus* and the same prebiotics was associated with a reduction in the content of lactose, glucose, glutamate, and octinoxate in fecal samples [103].

There is increased evidence of the interaction between the gut microbiota and the metabolism of xenobiotics [104]. For example, metabolites of microbial origin, such as *p*-cresol, may interfere with host metabolism of xenobiotics, reducing the capacity of the liver to metabolize efficiently the widely used analgesic acetaminophen. The gut microbiota can also have a more direct role in xenobiotic metabolism by catalyzing several reactions that influence the fate of these compounds [75].

However, there is an important number of metabolites for which their alteration resulting from gut flora modification has no direct contribution to the bacteria metabolism, suggesting a yet undefined host–microbe interaction. For example, serotonin plasma levels were altered in mice with gut flora modification, even when the production of serotonin by characterized enteric bacterial species has not been described [79].

CONCLUSIONS

Metabolomic studies require knowledge in different areas such as biochemistry, biology, physiology, and bioinformatics. In addition, metabolomics has different applications in areas such as human health, plant biotechnology, toxicology, and pharmacology, among others. The comprehension of metabolomic data requires the use of bioinformatics analyses, as well as the integration with genomic and proteomic data.

The study of an individual metabolome may be a powerful tool in public health. Using more sensitive technologies and new computational tools for statistical analysis of the results, metabolomics can help us to know in more detail the molecular mechanisms of a particular disease. Moreover, the identification and characterization of new biomarker molecules will allow us to improve the diagnosis and prevention of many diseases and also aid in the discovery and development of new drugs. Nevertheless, in most diseases it is essential to carry out the analysis of other biomarkers such as protein molecules, as well also the study of the physiological state of the patient for diagnosis and treatment [3].

ACKNOWLEDGMENT

This work was supported by grants from ANPCyT (PICT 2010-0543, 2011-0512, 2011-0982 and 2013-2188) and CONICET (PIP 134). MJG is a Doctoral fellow from CONICET. MVB and DGC are research members from CONICET.

REFERENCES

[1] Fiehn O. Metabolomics: the link between genotypes and phenotypes. Plant Mol Biol 2002;48(1/2):155–71.

[2] Stitt M, Fernie AR. From measurements of metabolites to metabolomics: an 'on the fly' perspective illustrated by recent studies of carbon–nitrogen interactions. Curr Opin Biotechnol 2003;14(2): 136–44.

[3] Gomez-Casati DF, Zanor MI, Busi MV. Metabolomics in plants and humans: applications in the prevention and diagnosis of diseases. Biomed Res Int 2013;2013:792527.

[4] Hall R, Beale M, Fiehn O, Hardy N, Sumner L, Bino R. Plant metabolomics: the missing link in functional genomics strategies. Plant Cell 2002;14(7):1437–40.

[5] Bino RJ, Hall RD, Fiehn O, Kopka J, Saito K, Draper J, et al. Potential of metabolomics as a functional genomics tool. Trends Plant Sci 2004;9(9):418–25.

[6] Weckwerth W. Metabolomics in systems biology. Annu Rev Plant Biol 2003;54(1):669–89.

[7] Lei Z, Huhman DV, Sumner LW. Mass spectrometry strategies in metabolomics. J Biol Chem 2011;286(29):25435–42.

[8] Griffin JL, Shockcor JP. Metabolic profiles of cancer cells. Nat Rev Cancer 2004;4(7):551–61.

[9] Spratlin JL, Serkova NJ, Eckhardt SG. Clinical applications of metabolomics in oncology: a review. Clin Cancer Res 2009;15(2):431–40.

[10] Bonvallot N, Tremblay-Franco M, Chevrier C, Canlet C, Debrauwer L, Cravedi JP, et al. Potential input from metabolomics for exploring and understanding the links between environment and health. J Toxicol Environ Health Part B Crit Rev 2014;17(1):21–44.

[11] Bothwell JH, Griffin JL. An introduction to biological nuclear magnetic resonance spectroscopy. Biol Rev Camb Philos Soc 2011;86(2):493–510.

[12] Cubbon S, Antonio C, Wilson J, Thomas-Oates J. Metabolomic applications of HILIC-LC-MS. Mass Spectrom Rev 2010;29(5):671–84.

[13] Ellis DI, Dunn WB, Griffin JL, Allwood JW, Goodacre R. Metabolic fingerprinting as a diagnostic tool. Pharmacogenomics 2007;8(9):1243–66.

[14] Beckonert O, Keun HC, Ebbels TM, Bundy J, Holmes E, Lindon JC, et al. Metabolic profiling, metabolomic and metabonomic procedures for NMR spectroscopy of urine, plasma, serum and tissue extracts. Nat Protoc 2007;2(11):2692–703.

[15] Liu M, Nicholson JK, Lindon JC. High-resolution diffusion and relaxation edited one- and two-dimensional 1H NMR spectroscopy of biological fluids. Anal Chem 1996;68(19):3370–6.

[16] Want EJ, Wilson ID, Gika H, Theodoridis G, Plumb RS, Shockcor J, et al. Global metabolic profiling procedures for urine using UPLC-MS. Nat Protoc 2010;5(6):1005–18.

[17] Zamboni N, Fendt SM, Ruhl M, Sauer U. (13)C-based metabolic flux analysis. Nat Protoc 2009;4(6):878–92.

[18] Ellis DI, Goodacre R. Metabolic fingerprinting in disease diagnosis: biomedical applications of infrared and Raman spectroscopy. Analyst 2006;131(8):875–85.

[19] Fiehn O. Combining genomics, metabolome analysis, and biochemical modelling to understand metabolic networks. Comp Funct Genomics 2001;2(3):155–68.

[20] Harrigan GG, Goodacre R. Metabolic profiling: its role in biomarker discovery and gene function analysis: its role in biomarker discovery and gene function analysis. 2003.

[21] Goodacre R, Vaidyanathan S, Dunn WB, Harrigan GG, Kell DB. Metabolomics by numbers: acquiring and understanding global metabolite data. Trends Biotechnol 2004;22(5):245–52.

[22] Goodacre R. Metabolomics of a superorganism. J Nutr 2007;137(1 Suppl.):259S–66S.

[23] Nicholson JK, Wilson ID. Opinion: understanding 'global' systems biology: metabonomics and the continuum of metabolism. Nat Rev Drug Discov 2003;2(8):668–76.

[24] Álvarez-Sánchez B, Priego-Capote F. Metabolomics analysis. I. Selection of biological samples and practical aspects preceding sample preparation. TrAC Trends 2010;29(2):111–9.

[25] Álvarez-Sánchez B, Priego-Capote F. Metabolomics analysis II. Preparation of biological samples prior to detection. TrAC Trends Anal 2010;29(2):120–7.

[26] Dumas ME, Maibaum EC, Teague C, Ueshima H, Zhou B, Lindon JC, et al. Assessment of analytical reproducibility of 1H NMR spectroscopy based metabonomics for large-scale epidemiological research: the INTERMAP Study. Anal Chem 2006;78(7):2199–208.

[27] Bictash M, Ebbels TM, Chan Q, Loo R, Yap IKS, Brown IJ, et al. Opening up the "Black Box": metabolic phenotyping and metabolome-wide association studies in epidemiology. J Clin Epidemiol 2010;63(9):970–9.

[28] Kanehisa M, Goto S, Hattori M, Aoki-Kinoshita KF, Itoh M, Kawashima S, et al. From genomics to chemical genomics: new developments in KEGG. Nucleic Acids Res 2006;34(Database issue):D354–7.

[29] Cui Q, Lewis IA, Hegeman AD, Anderson ME, Li J, Schulte CF, et al. Metabolite identification via the Madison Metabolomics Consortium Database. Nat Biotechnol 2008;26(2):162–4.

[30] Kopka J, Schauer N, Krueger S, Birkemeyer C, Usadel B, Bergmuller E, et al. GMD@CSB.DB: the Golm Metabolome Database. Bioinformatics 2005;21(8):1635–8.

[31] Hamosh A, Scott AF, Amberger JS, Bocchini CA, McKusick VA. Online Mendelian Inheritance in Man (OMIM), a knowledge base of human genes and genetic disorders. Nucleic Acids Res 2005;33(Database issue):D514–7.

[32] Wishart DS, Tzur D, Knox C, Eisner R, Guo AC, Young N, et al. HMDB: the Human Metabolome Database. Nucleic Acids Res 2007;35(Database issue):D521–6.

[33] Wishart DS, Knox C, Guo AC, Eisner R, Young N, Gautam B, et al. HMDB: a knowledgebase for the human metabolome. Nucleic Acids Res 2009;37(Database issue):D603–10.

[34] Gika HG, Theodoridis GA, Plumb RS, Wilson ID. Current practice of liquid chromatography-mass spectrometry in metabolomics and metabonomics. J Pharm Biomed Anal 2014;87:12–25.

[35] Katajamaa M, Miettinen J, Oresic M. MZmine: toolbox for processing and visualization of mass spectrometry based molecular profile data. Bioinformatics 2006;22(5):634–6.

[36] Smith CA, Want EJ, O'Maille G, Abagyan R, Siuzdak G. XCMS: processing mass spectrometry data for metabolite profiling using nonlinear peak alignment, matching, and identification. Anal Chem 2006;78(3):779–87.

[37] Lommen A. MetAlign: interface-driven, versatile metabolomics tool for hyphenated full-scan mass spectrometry data preprocessing. Anal Chem 2009;81(8):3079–86.

[38] Serkova NJ, Gamito EJ, Jones RH, O'Donnell C, Brown JL, Green S, et al. The metabolites citrate, myo-inositol, and spermine are potential age-independent markers of prostate cancer in human expressed prostatic secretions. Prostate 2008;68(6):620–8.

[39] Mahadevan S, Shah SL, Marrie TJ, Slupsky CM. Analysis of metabolomic data using support vector machines. Anal Chem 2008;80(19):7562–70.

[40] Chan AW, Gill RS, Schiller D, Sawyer MB. Potential role of metabolomics in diagnosis and surveillance of gastric cancer. World J Gastroenterol WJG 2014;20(36):12874–82.

[41] Cortes C, Vapnik V. Support-vector networks. Mach Learn 1995;20(3):273–97.

[42] Millington DS, Kodo N, Norwood DL, Roe CR. Tandem mass spectrometry: a new method for acylcarnitine profiling with potential for neonatal screening for inborn errors of metabolism. J Inherit Metab Dis 1990;13(3):321–4.

[43] Garg U, Dasouki M. Expanded newborn screening of inherited metabolic disorders by tandem mass spectrometry: clinical and laboratory aspects. Clin Biochem 2006;39(4):315–32.

[44] Hanahan D, Weinberg RA. The hallmarks of cancer. Cell 2000;100(1):57–70.

[45] Liu ML, Zheng P, Liu Z, Xu Y, Mu J, Guo J, et al. GC-MS based metabolomics identification of possible novel biomarkers for schizophrenia in peripheral blood mononuclear cells. Mol Biosyst 2014;10(9):2398–406.

[46] Patterson AD, Maurhofer O, Beyoglu D, Lanz C, Krausz KW, Pabst T, et al. Aberrant lipid metabolism in hepatocellular carcinoma revealed by plasma metabolomics and lipid profiling. Cancer Res 2011;71(21):6590–600.

[47] Chen J, Wang W, Lv S, Yin P, Zhao X, Lu X, et al. Metabonomics study of liver cancer based on ultra performance liquid chromatography coupled to mass spectrometry with HILIC and RPLC separations. Anal Chim Acta 2009;650(1):3–9.

[48] Masuo Y, Imai T, Shibato J, Hirano M, Jones OA, Maguire ML, et al. Omic analyses unravels global molecular changes in the brain and liver of a rat model for chronic Sake (Japanese alcoholic beverage) intake. Electrophoresis 2009;30(8):1259–75.

[49] Cheng LL, Chang IW, Louis DN, Gonzalez RG. Correlation of high-resolution magic angle spinning proton magnetic resonance spectroscopy with histopathology of intact human brain tumor specimens. Cancer Res 1998;58(9):1825–32.

[50] Cheng LL, Anthony DC, Comite AR, Black PM, Tzika AA, Gonzalez RG. Quantification of microheterogeneity in glioblastoma multiforme with ex vivo high-resolution magic-angle spinning (HRMAS) proton magnetic resonance spectroscopy. Neuro-oncology 2000;2(2):87–95.

[51] Archetti M. Evolutionary dynamics of the Warburg effect: glycolysis as a collective action problem among cancer cells. J Theor Biol 2014;341:1–8.

[52] Makinoshima H, Takita M, Matsumoto S, Yagishita A, Owada S, Esumi H, et al. Epidermal growth factor receptor (EGFR) signaling regulates global metabolic pathways in EGFR-mutated lung adenocarcinoma. J Biol Chem 2014;289(30):20813–23.

[53] Keller WR, Fischer BA, Carpenter Jr WT. Revisiting the diagnosis of schizophrenia: where have we been and where are we going? CNS Neurosci Ther 2011;17(2):83–8.

[54] Yap IK, Angley M, Veselkov KA, Holmes E, Lindon JC, Nicholson JK. Urinary metabolic phenotyping differentiates children with autism from their unaffected siblings and age-matched controls. J Proteome Res 2010;9(6):2996–3004.

[55] Xu XJ, Zheng P, Ren GP, Liu ML, Mu J, Guo J, et al. 2,4-Dihydroxypyrimidine is a potential urinary metabolite biomarker for diagnosing bipolar disorder. Mol Biosyst 2014;10(4):813–9.

[56] Zheng P, Wei Y-D, Yao G-E, Ren G-P, Guo J, Zhou C-J, et al. Novel urinary biomarkers for diagnosing bipolar disorder. Metabolomics 2013;9(4):800–8.

[57] Zheng P, Gao HC, Li Q, Shao WH, Zhang ML, Cheng K, et al. Plasma metabonomics as a novel diagnostic approach for major depressive disorder. J Proteome Res 2012;11(3):1741–8.

[58] Zheng P, Gao H-C, Qi Z-G, Jia J-M, Li FF, Chen JJ, et al. Peripheral metabolic abnormalities of lipids and amino acids implicated in increased risk of suicidal behavior in major depressive disorder. Metabolomics 2012;9(3):688–96.

[59] Zheng P, Wang Y, Chen L, Yang D, Meng H, Zhou D, et al. Identification and validation of urinary metabolite biomarkers for major depressive disorder. Mol Cell Proteomics 2013;12(1):207–14.

[60] Shao W-H, Fan S-H, Lei Y, Yao G-E, Chen J-J, Zhou J, et al. Metabolomic identification of molecular changes associated with stress resilience in the chronic mild stress rat model of depression. Metabolomics 2012;9(2):433–43.

[61] Lichtenstein P, Holm NV, Verkasalo PK, Iliadou A, Kaprio J, Koskenvuo M, et al. Environmental and heritable factors in the causation of cancer: analyses of cohorts of twins from Sweden, Denmark, and Finland. N Engl J Med 2000;343(2):78–85.

[62] Hemminki K, Lorenzo Bermejo J, Forsti A. The balance between heritable and environmental aetiology of human disease. Nat Rev Genet 2006;7(12):958–65.

[63] Gluckman PD, Hanson MA, Cooper C, Thornburg KL. Effect of in utero and early-life conditions on adult health and disease. N Engl J Med 2008;359(1):61–73.

[64] Preininger M, Arafat D, Kim J, Nath AP, Idaghdour Y, Brigham KL, et al. Blood-informative transcripts define nine common axes of peripheral blood gene expression. PLoS Genet 2013;9(3):e1003362.

[65] Wild CP. Complementing the genome with an "exposome": the outstanding challenge of environmental exposure measurement in molecular epidemiology. Cancer Epidemiol Biomarkers Prev 2005;14(8):1847–50. a publication of the American Association for Cancer Research, cosponsored by the American Society of Preventive Oncology.

[66] Rappaport SM, Smith MT. Epidemiology. Environment and disease risks. Science 2010;330(6003):460–1.

[67] Wild CP, Turner PC. The toxicology of aflatoxins as a basis for public health decisions. Mutagenesis 2002;17(6):471–81.

[68] Uemura N, Okamoto S, Yamamoto S, Matsumura N, Yamaguchi S, Yamakido M, et al. *Helicobacter pylori* infection and the development of gastric cancer. N Engl J Med 2001;345(11):784–9.

[69] Vineis P, van Veldhoven K, Chadeau-Hyam M, Athersuch TJ. Advancing the application of omics-based biomarkers in environmental epidemiology. Environ Mol Mutagen 2013;54(7):461–7.

[70] Martin FP, Collino S, Rezzi S, Kochhar S. Metabolomic applications to decipher gut microbial metabolic influence in health and disease. Front Physiol 2012;3:113.

[71] Riscuta G, Dumitrescu RG. Nutrigenomics: implications for breast and colon cancer prevention. Methods Mol Biol 2012;863:343–58.

[72] Rappaport SM, Barupal DK, Wishart D, Vineis P, Scalbert A. The blood exposome and its role in discovering causes of disease. Environ Health Perspect 2014;122(8):769–74.

[73] Backhed F, Ley RE, Sonnenburg JL, Peterson DA, Gordon JI. Host-bacterial mutualism in the human intestine. Science 2005;307(5717):1915–20.

[74] Qin J, Li R, Raes J, Arumugam M, Burgdorf KS, Manichanh C, et al. A human gut microbial gene catalogue established by metagenomic sequencing. Nature 2010;464(7285):59–65.

[75] Ursell LK, Haiser HJ, Van Treuren W, Garg N, Reddivari L, Vanamala J, et al. The intestinal metabolome: an intersection between microbiota and host. Gastroenterology 2014;146(6):1470–6.

[76] Ray K. Gut microbiota: married to our gut microbiota. Nat Rev Gastroenterol Hepatol 2012;9(10):555.

[77] Gill SR, Pop M, Deboy RT, Eckburg PB, Turnbaugh PJ, Samuel BS, et al. Metagenomic analysis of the human distal gut microbiome. Science 2006;312(5778):1355–9.

[78] Xie J, Song L, Li X, Yi X, Xu H, Li J, et al. Site-directed mutagenesis and thermostability of xylanase *XYNB* from *Aspergillus niger* 400264. Curr Microbiol 2011;62(1):242–8.

[79] Wikoff WR, Anfora AT, Liu J, Schultz PG, Lesley SA, Peters EC, et al. Metabolomics analysis reveals large effects of gut microflora on mammalian blood metabolites. Proc Natl Acad Sci USA 2009;106(10):3698–703.

[80] Swidsinski A, Ladhoff A, Pernthaler A, Swidsinski S, Loening-Baucke V, Ortner M, et al. Mucosal flora in inflammatory bowel disease. Gastroenterology 2001;122(1):44–54.

[81] Ahmadi AA, Polyak S. Endoscopy/surveillance in inflammatory bowel disease. Surg Clin North Am 2007;87(3):743–62.

[82] Schicho R, Shaykhutdinov R, Ngo J, Nazyrova A, Schneider C, Panaccione R, et al. Quantitative metabolomic profiling of serum, plasma, and urine by (1)H NMR spectroscopy discriminates between patients with inflammatory bowel disease and healthy individuals. J Proteome Res 2012;11(6):3344–57.

[83] Stephens NS, Siffledeen J, Su X, Murdoch TB, Fedorak RN, Slupsky CM. Urinary NMR metabolomic profiles discriminate inflammatory bowel disease from healthy. J Crohns Colitis 2013;7(2):e42–8.

[84] Li M, Wang B, Zhang M, Rantalainen M, Wang S, Zhou H, et al. Symbiotic gut microbes modulate human metabolic phenotypes. Proc Natl Acad Sci USA 2008;105(6):2117–22.

[85] Nagalingam NA, Lynch SV. Role of the microbiota in inflammatory bowel diseases. Inflamm Bowel Dis 2012;18(5):968–84.

[86] Williams HR, Willsmore JD, Cox IJ, Walker DG, Cobbold JF, Taylor-Robinson SD, et al. Serum metabolic profiling in inflammatory bowel disease. Dig Dis Sci 2012;57(8):2157–65.

[87] Schicho R, Nazyrova A, Shaykhutdinov R, Duggan G, Vogel HJ, Storr M. Quantitative metabolomic profiling of serum and urine in DSS-induced ulcerative colitis of mice by (1)H NMR spectroscopy. J Proteome Res 2010;9(12):6265–73.

[88] Kuhn R, Lohler J, Rennick D, Rajewsky K, Muller W. Interleukin-10-deficient mice develop chronic enterocolitis. Cell 1993;75(2):263–74.

[89] Martin FP, Rezzi S, Philippe D, Tornier L, Messlik A, Holzlwimmer G, et al. Metabolic assessment of gradual development of moderate experimental colitis in IL-10 deficient mice. J Proteome Res 2009;8(5):2376–87.

[90] Hisamatsu T, Okamoto S, Hashimoto M, Muramatsu T, Andou A, Uo M, et al. Novel, objective, multivariate biomarkers composed of plasma amino acid profiles for the diagnosis and assessment of inflammatory bowel disease. PLoS One 2012;7(1):e31131.

[91] Ooi M, Nishiumi S, Yoshie T, Shiomi Y, Kohashi M, Fukunaga K, et al. GC/MS-based profiling of amino acids and TCA cycle-related molecules in ulcerative colitis. Inflamm Res 2011;60(9):831–40.

[92] Le Gall G, Noor SO, Ridgway K, Scovell L, Jamieson C, Johnson IT, et al. Metabolomics of fecal extracts detects altered metabolic activity of gut microbiota in ulcerative colitis and irritable bowel syndrome. J Proteome Res 2011;10(9):4208–18.

[93] Marchesi JR, Holmes E, Khan F, Kochhar S, Scanlan P, Shanahan F, et al. Rapid and noninvasive metabonomic characterization of inflammatory bowel disease. J Proteome Res 2007;6(2):546–51.

[94] Loftus Jr EV. Clinical epidemiology of inflammatory bowel disease: incidence, prevalence, and environmental influences. Gastroenterology 2004;126(6):1504–17.

[95] Ponnusamy K, Choi JN, Kim J, Lee SY, Lee CH. Microbial community and metabolomic comparison of irritable bowel syndrome faeces. J Med Microbiol 2011;60(Pt 6):817–27.

[96] Sellitto M, Bai G, Serena G, Fricke WF, Sturgeon C, Gajer P, et al. Proof of concept of microbiome-metabolome analysis and delayed gluten exposure on celiac disease autoimmunity in genetically at-risk infants. PLoS One 2012;7(3):e33387.

[97] Calabro A, Gralka E, Luchinat C, Saccenti E, Tenori L. A metabolomic perspective on coeliac disease. Autoimmune Dis 2014;2014:756138.

[98] Nadal I, Donat E, Ribes-Koninckx C, Calabuig M, Sanz Y. Imbalance in the composition of the duodenal microbiota of children with coeliac disease. J Med Microbiol 2007;56(Pt 12):1669–74.

[99] Di Cagno R, De Angelis M, De Pasquale I, Ndagijimana M, Vernocchi P, Ricciuti P, et al. Duodenal and faecal microbiota of celiac children: molecular, phenotype and metabolome characterization. BMC Microbiol 2011;11:219.

[100] Bernini P, Bertini I, Calabro A, la Marca G, Lami G, Luchinat C, et al. Are patients with potential celiac disease really potential? The answer of metabonomics. J Proteome Res 2011;10(2):714–21.

[101] Schloss PD. Microbiology: an integrated view of the skin microbiome. Nature 2014;514(7520):44–5.

[102] Schrezenmeir J, de Vrese M. Probiotics, prebiotics, and synbiotics: approaching a definition. Am J Clin Nutr 2001;73(2 Suppl.):361S–4S.

[103] Martin FP, Wang Y, Sprenger N, Yap IK, Rezzi S, Ramadan Z, et al. Top-down systems biology integration of conditional prebiotic modulated transgenomic interactions in a humanized microbiome mouse model. Mol Syst Biol 2008;4:205.

[104] Clayton TA, Baker D, Lindon JC, Everett JR, Nicholson JK. Pharmacometabonomic identification of a significant host–microbiome metabolic interaction affecting human drug metabolism. Proc Natl Acad Sci USA 2009;106(34):14728–33.

Chapter 8

Microbial Genomics: Diagnosis, Prevention, and Treatment

A. Mutreja

MSD-Wellcome Trust Hilleman Laboratories, New Delhi, India; Wellcome Trust Sanger Institute, Cambridge, United Kingdom

Chapter Outline

Background	101	Treatment and Diagnostics	103
Control and Prevention	101	References	105

BACKGROUND

Our understanding of diseases and disorders is nowhere close to the finish line; in fact it has just begun. The threat of new infections and emergence of antibiotic-resistant microbes has kept the scientific community on its toes. However, from the early days of molecular biology to the current era of genomics, the global model of diagnosis, control, prevention, and treatment is changing rapidly. The availability of highly sensitive and specific diagnostic techniques such as high-resolution single nucleotide polymorphism (SNP)–level tracking of disease spread and transmission, accurate molecular epidemiology of pathogens; reverse genomics–based vaccines for prevention, and modern genomics-based therapeutics have only become possible because of introduction of translational "-omics" in today's commercial and academic research and development.

Whole genome sequencing (WGS) has potential to transform the way diagnostic and clinical microbiology is practiced, although it has not been fully realized yet. Partly it is because of the lack of available expertise in its technological operations; additionally, it is because of the lack of training in analysis of the output from next generation sequencing machines that in its raw form is almost unreadable by the user. Until recently, another factor that discouraged the use of WGS by the scientific and medical communities was its turnaround time and cost. However, the increase in speed of sequencing and drastic decrease in its cost has now meant that even less well-funded laboratories are starting to accommodate these technologies into their systems for increased efficiency. There are multiple tiers of information that this single technology provides; eg, data on antimicrobial resistance (AMR) of microbes before the phenotypic results are confirmed, epidemiological characterization, evolutionary patterns, and phylogenetic correlation with geography and time. All this information is vital for the local, national, and international public health agencies during outbreak situations when each ticking minute is of importance. Moreover, in endemic areas this information can guide the long-term strategy for keeping populations healthy and infections at bay.

AMR is listed by the World Health Organization as one of the top three threats to human health [1] and WGS provides insight into data on the resistance markers, which otherwise would be a time-consuming and laborious task. This information proves useful in making early public health preparations before the confirmatory phenotypic results become available.

CONTROL AND PREVENTION

The impact disease outbreaks have on society is devastating. All the systems like hospitals, health centers, and governments do their best to cut the chances of outbreaks happening in first place, clearly, more needs to be done. There are countries endemic to even those diseases for which protective vaccines are available. Moreover, sometimes epidemics are reported from countries that did not have that disease for almost a century [2], which indicates either a breach of a system being routinely followed to maintain a disease-free environment or introduction of disease by a non-native person or traveler, for which the system was unprepared.

Medical and Health Genomics. http://dx.doi.org/10.1016/B978-0-12-420196-5.00008-3

Once an outbreak (ie, a series of disease cases because of identical or closely linked isolates) is reported, major efforts are made in two directions: (1) understanding the chain of events and reach the source that could be targeted [3,4], and (2) taking lessons from the outbreak and making arrangements for future prevention of the same disease [3–5]. In both efforts, WGS can make a significant contribution. The best way to find the source of a disease is to do the epidemiological characterization of the strains isolated from the patients. The use of low-resolution typing techniques like multilocus sequence typing, multilocus variable number tandem repeat analysis, or, the current gold standard, pulse field gel electrophoresis does a reasonable job. However, there are circumstances when the outbreak investigation requires higher resolution [6,7]. At times, slowly evolving pathogens are wrongly marked as identical by the use of these traditional methodologies because their housekeeping genes do not undergo any natural evolution during the window of an outbreak. They can be differentiated by WGS, the highest resolution possible. The nucleotide level resolution from WGS could be used to plot the micro- or short-term phylogenetic evolution of the pathogen [6,8,9], which is exactly what is required during outbreak situations. In addition, WGS provides information on pathogen biology, antibiotic resistance markers [10,11], gene content and genomic evolution [12–14]. Other than guiding the public health authorities on management of the outbreak in the most rational way, this data has utility in drug design. Researchers compared methicillin-resistant *Staphylococcus aureus* (MRSA) collected during a hospital outbreak to nonoutbreak isolates and highlighted the genomic determinants behind the success of epidemic clones [15]. In another study, researchers identified the *sasX* gene as the major determinant that helped the outbreak isolates in better nasal colonization of MRSA in China [16].

Genomic data can also be used to draw models of evolution based on the nucleotide substitution rate [5,6,8,9,13,17–20]. Using SNP-based phylogenetic trees and linking bayesian statistics with the sequence and metadata, the realistic age of ancestral strains can be determined. Annual SNP acquisition rate by a bacterium that evolves in a clocklike manner can be estimated most accurately by the use of this technology [13,17,18].

Using a suitable collection, historical questions that have intrigued the scientific community for long time are also being answered by WGS. By enriching and genome sequencing a strain isolated from tissue samples archived in 1849 from a Philadelphia outbreak patient who died of cholera, one group checked the assumption that O1 serogroup classical biotype strains caused the second pandemic of cholera [21]. Using the previously estimated rates of evolution, they also predicted that the ancestors of classical lineage originated between 1843 and 1860 [21]. By sequencing a *Yersinia pestis* isolate from the time of Black Death (bubonic plague) between 1347 and 1351, Canadian researchers did not find any such genomic determinants that could differentiate the *Y. pestis* of medieval times from the modern strains [22]. Instead they propose that for the epidemiological investigations of *Y. pestis* outbreaks, environmental factors, host susceptibility, and vector dynamics should be given more importance [22]. Researchers in Germany compared medieval and modern strains of *Mycobacterium leprae* [23]. The skeletons of five medieval leprosy cases from the United Kingdom, Sweden, and Denmark were sampled for *M. leprae* isolation. They showed that the sudden decline in leprosy in the 16th century was most likely not caused by the loss of virulence from the outbreak strain, because they showed high similarity with the modern leprosy bacteria. Improved social conditions and decline in other diseases like plague and tuberculosis along with the development of host immunity instead could have been the factors behind the decline [23].

Based on the metadata available on the isolates sequenced, the genomic information could be brought to use in multiple ways. For instance, geographical and temporal information when combined with the genome sequence data can reveal the pattern of strain movement both during outbreaks and in endemic areas [8,9,18]. In a study on pandemic cholera, scientists sequenced 154 global *Vibrio cholerae* isolates spanning 100 years and reported that the seventh pandemic of cholera spread from the Bay of Bengal to the rest of the world in independent but temporally overlapping waves [13]. Strains in each wave had different genomic characteristics, but all radiated from one common seventh pandemic ancestor as clonal expansion. They showed that the acquisition of the SXT element, a multiple antibiotic resistance cassette integrative conjugative element of the R391 family has played a major role in the success of currently circulating isolates globally. Also their genomic data was able to highlight the possibilities that cholera in Latin America was seeded by West Africa and not by Asian ship ballasts carrying *V. cholerae* from south Asia. The study also answered historical questions on the evolution of classical and El Tor lineages of O1 *V. cholerae*, concluding that the two lineages were different by 20,000 SNPs and evolved independently. A molecular evolutionary clock rate of three SNP acquisitions/year was established using the Bayesian Evolutionary Analysis Sampling Trees (BEAST) platform [24] on whole genome data, and this rate is now used in all micro and global evolutionary studies on the seventh cholera pandemic. The same study also dwelled upon the source of a destructive cholera outbreak in Haiti that followed a major earthquake in the region. Their sample set indicated that the introduction of cholera in Haiti was from South Asia, but detailed information controversially confirming the role of Nepalese peacekeepers sent by the United Nations in seeding cholera in Haiti came from another study [25]. This study showed that isolates collected during the same time from Nepal were identical to the Haitian outbreak isolates.

In another study on *Shigella sonnei*, researchers sequenced 132 genomes and showed that the currently circulating population radiates from ancestors that existed less than 500 years ago [26]. *Shigella sonnei* has spread to other continents from Europe, where most of its diversification occurred in first place, and now it is established locally in several continents, where it has undergone localization by acquiring unique genomic determinants. This study proposes that although the Millennium Development Goal (MDG) of the United Nations, if achieved, will reduce the morbidity and mortality across the board, it will lead to increase in the *S. sonnei* incidence rate. A vaccine against *S. sonnei* would be ideal because all the strains are clonal and carry the same O antigen that has also been proven to be a successful vaccine target [26].

A sub-Saharan African study on invasive nontyphoidal *Salmonella* (iNTS), which causes a fever distinct from typhoid and gastroenteritis and indistinguishable from malaria, showed that iNTS has two major lineages globally. The first lineage was replaced by the second one most likely because of increased use of chloramphenicol for the treatment of iNTS infections. They also found that the transmission of this infection in Africa was temporally associated with the HIV pandemic [20].

Another area where genomics has shown great promise is tracking the transmission of diseases that can spread from humans to animals, and vice versa. In some professions such as farming, dairy, poultry, and meat industries, close contact with animals is unavoidable. Thus over time some pathogens that used to be species limited have adapted to the alternate options available to them. One study in the Netherlands sequenced and analyzed 65 *Clostridium difficile* 078 isolates collected from pigs ($n=19$), asymptomatic farmers ($n=15$), and hospitalized patients ($n=31$) between 2002 and 2011 and observed that sympatric farmers and pigs shared identical isolates with identical antibiotic resistance determinants, clearly indicating transmission between the two populations [27].

The two best public health interventions available to keep most diseases away are hygiene maintenance and vaccination. Because it is not so easy to implement hygiene policies in densely populated and cut-out rural areas where most of the disease burden is, vaccination is inarguably the most convenient way of lowering the risk of outbreaks. However, whole genome studies have showed that vaccination alone cannot be considered a "for good" measure, because pathogens evolve with time. Following the principle similar to the gaining of antibiotic resistance, as soon as a selection pressure is applied on a pathogen population, pathogens adapt to the change and thrive by mutation and recombination, leading to vaccine escape [17,28,29]. For example, makers of the vaccine against the pneumococcal disease target the immunogenic capsular region and vaccines include multiple capsule types to provide broader protection. However, under the selection pressure, rampant capsular switching by recombination and change in the molecular genotype of the bacteria leads to nonrecognition of mutated pathogens by vaccinated individuals, thus making the vaccine ineffective [17,29]. By sequencing a global collection of *Shigella flexneri*, researchers from Cardiff University proposed that owing to serotype switching, clean water instead of vaccines would be the most effective measure for controlling *S. flexneri* dysentery, highlighting the importance of achieving the clean water MDG of the United Nations [28].

TREATMENT AND DIAGNOSTICS

It has been known for years that many gastroenteric and infectious diseases can be best treated by fecal therapy, which involves a transplant of healthy microbiota in the gut of a patient suffering from a recurring and/or relapsing disorder. The first proof of concept was done by a scientist at the University of Colorado more than 50 years ago when he successfully treated a severe diarrhea patient infected with *Pseudomonas enterocolitis*. However, the concept of transplanting a healthy fecal graft did not gain wider acceptance from society and regulatory bodies until high-throughput metagenomics studies using next generation sequencing machines highlighted the promise of this therapy in a more scientific way. The best advocacy for the feces transplant came from a clinical trial that had to be stopped early because the difference between success of fecal therapy and vancomycin therapy in *C. difficile* relapse patients was huge [30]. Out of 43 subjects recruited 16 were delivered fecal infusions to their duodenum and 26 were given standard 2-week vancomycin treatment. The feces from which infusions were made came from 15 healthy donors who were comprehensively checked for transmissible diseases. Surprisingly, while 15 of the group containing 16 were cured, only seven of the group of 26 were cured. The stark difference between the two groups resulted in early termination of the trial followed by delivery of fecal infusions to all the patients recruited for the trial, curing the complete volunteer pool. The metagenomic analysis of all the volunteers showed that their gut microflora became more diverse after the infusion. This kind of difference in diversity is now being carefully examined by researchers to identify and isolate the bacterial species that truly made the difference in curing [31].

Genomic characterization of gut microbiota and finding that more than 90% of microbes in our gut cannot be cultured tells that the human body is a complex ecosystem of bacteria, fungi, and viruses that play an important role in the regulation of our physiological processes and keeping the body healthy. Our gut itself is home to more than 100 trillion bacteria, and researchers who are characterizing the gut microbiome using sequencing have named their efforts *microbiomics.* Microbiomics does not just characterize but also gives information on crucial genes of nonculturable bacteria, which in turn leads to better understanding of culture conditions that these microbes require. Once a nonculturable but viable bacterium is cultured and its individual and synergistic role with other microbes is established, disease treatment moves a step closer toward designing a more rational medical therapy. Along the lines of the Human Genome Project, in 2007 the National Institutes of Health fueled a Human Microbiome Project (HMP), in which a total of 3000 reference microbial genomes, isolated from different sites of human body, are being sequenced [32]. The information from HMP and more genomic data from the International Human Microbiome Consortium [33] studies will strengthen the concept of the fecal microbiota transplant and provide answers to classic unanswered questions such as whether the gut is the gateway to all the human diseases and whether diseases lead to change in the microbiota or change in microbiota leads to diseases.

Personalized medicine based on genetic markers from the human genome is progressing toward clinics at a lightening speed but it is out of the scope of this chapter focusing on microbial genomics.

Vaccine research and molecular epidemiology go hand in hand for design and redesign (as per change in the circulating genotype) of new vaccines and successful implementation of vaccination programs. Integration of genomics in various reference laboratories and in quick designing of difficult-to-make vaccines through -omics–based reverse vaccinology is must, because the pace of microbial evolution is much faster than the research activities. Thus design of a vaccine has to be of highest standards, with chances of final product failure being minimal. WGS data can cater precisely to that. Because of the reduction in number of steps required for screening of different antigens to find the best vaccine candidate gene or strain, comparative genomics has become an indispensable part vaccinology [34]. The traditional approach to vaccine antigen discovery involves cultivation of bacteria and mutating or knocking out of gene to screen for its function before choosing final antigens, which means that only culturable pathogens can be manipulated. This process takes time and is generally only performed on one strain type. Instead, reverse vaccinology starts with in silico analysis of DNA sequenced from multiple strains of the same pathogen followed by comparative genomics to find the genes or novel coding DNA sequence that could be of interest for a vaccine formulation [35–41].

The availability of benchtop sequencing machines has facilitated the use of technology in clinical and diagnostic laboratories. Other than the real-time outbreak investigations using WGS (as discussed), this data has utility in aiding the design of SNP-based diagnostic assays [42,43]. By using high-resolution melting assay, phylogenies can be redrawn as researchers did for *S. sonnei* based on just six SNPs [44]. These SNPs were selected from a robust global phylogeny based on genome-wide variation and, because *S. sonnei* does not recombine, the long-term stability of these SNPs can be relied upon for rapid detection of *S. sonnei* clones without the need of sequencing and analysis.

A group in Vietnam designed multiplex ligation-dependent probe amplification assay based on 11 highly informative phylogenetic markers from the *Salmonella typhi* genome [45]. The assay probes could detect multiple antibiotic-resistant H58 haplotype and mutations in AMR markers like *gyrA* and *parC* genes with more than 90% confidence, eliminating the need to sequence each and every isolate collected during epidemiological studies.

A WGS phylotyping assay was developed to distinguish between closely related MRSA isolates that are inseparable using traditional typing techniques [46]. The assay could easily differentiate between invasive (blood stream) and noninvasive (nasal) isolates, information that proves very important in clinical scenarios when a treatment approach has to be decided quickly based on the nature of the infection.

In genetically homogenous populations like that of *Mycoplasma pneumoniae*, no typing techniques other than the WGS differentiates their genetic types. Housekeeping genes that are used in most epidemiological characterization studies are uninformative in this case. Thus researchers selected eight canonical SNPs and used Applied Biosystems' SnaPshot technology [43] to differentiate between almost identical *M. pneumoniae* clones [47]. Although these assays are limiting in terms of new information that they provide, they definitely hasten the process of detection with highest confidence.

WGS has revolutionized the way we do molecular epidemiology, outbreak investigation, infection control strategies, and vaccine research. However, despite the big leaps that the next generation sequencing technology and assays based on WGS have made, one major gap that still remains is easy-to-use software for the downstream analysis of raw genomic data. More people need to be trained in bioinformatics and computational biology. Sequencing platforms that require different software for analysis somehow need to come to the same page, readable and in one font. Also, although enrichment technologies for the direct analysis of clinical specimens are coming into the picture [48,49], more needs to be done to ease the process for a technician who has to follow set hospital guidelines without compromising quality and time. Furthermore, databases containing streamlined and curated information on genomic markers should be made available, because a large amount of data in databases presently is either of poor quality or even wrong. The best use of genomic information on microbes can only be made if the role of host factors is also known. Genome-wide association studies and laboratory-based host–pathogen interaction experiments would be vital to better understand any pathogen or disease in a set context.

REFERENCES

[1] Infectious Diseases Society of A. The 10× '20 Initiative: pursuing a global commitment to develop 10 new antibacterial drugs by 2020. Clin Infect Dis 2010;50(8):1081–3.

[2] Seas C, Miranda J, Gil AI, Leon-Barua R, Patz J, Hug A, et al. New insights on the emergence of cholera in Latin America during 1991: the Peruvian experience. Am J Trop Med Hyg 2000;62(4):513–7.

[3] Quick J, Ashton P, Calus S, Chatt C, Gossain S, Hawker J, et al. Rapid draft sequencing and real-time nanopore sequencing in a hospital outbreak of *Salmonella*. Genome Biol 2015;16(1):114.

[4] Halachev MR, Chan JZ, Constantinidou CI, Cumley N, Bradley C, Smith-Banks M, et al. Genomic epidemiology of a protracted hospital outbreak caused by multidrug-resistant *Acinetobacter baumannii* in Birmingham, England. Genome Med 2014;6(11):70.

[5] Bryant JM, Harris SR, Parkhill J, Dawson R, Diacon AH, van Helden P, et al. Whole-genome sequencing to establish relapse or re-infection with *Mycobacterium tuberculosis*: a retrospective observational study. Lancet Respir Med 2013;1(10):786–92.

[6] Abd El Ghany M, Chander J, Mutreja A, Rashid M, Hill-Cawthorne GA, Ali S, et al. The population structure of *Vibrio cholerae* from the Chandigarh region of Northern India. PLoS Negl Trop Dis 2014;8(7):e2981.

[7] Walker TM, Ip CL, Harrell RH, Evans JT, Kapatai G, Dedicoat MJ, et al. Whole-genome sequencing to delineate *Mycobacterium tuberculosis* outbreaks: a retrospective observational study. Lancet Infect Dis 2013;13(2):137–46.

[8] Shah MA, Mutreja A, Thomson N, Baker S, Parkhill J, Dougan G, et al. Genomic epidemiology of *Vibrio cholerae* O1 associated with floods, Pakistan, 2010. Emerg Infect Dis 2014;20(1):13–20.

[9] Kiiru J, Mutreja A, Mohamed AA, Kimani RW, Mwituria J, Sanaya RO, et al. A study on the geophylogeny of clinical and environmental *Vibrio cholerae* in Kenya. PLoS One 2013;8(9):e74829.

[10] Holden MT, Hsu LY, Kurt K, Weinert LA, Mather AE, Harris SR, et al. A genomic portrait of the emergence, evolution, and global spread of a methicillin-resistant *Staphylococcus aureus* pandemic. Genome Res 2013;23(4):653–64.

[11] Reuter S, Ellington MJ, Cartwright EJ, Koser CU, Torok ME, Gouliouris T, et al. Rapid bacterial whole-genome sequencing to enhance diagnostic and public health microbiology. JAMA Intern Med 2013;173(15):1397–404.

[12] Harris SR, Feil EJ, Holden MT, Quail MA, Nickerson EK, Chantratita N, et al. Evolution of MRSA during hospital transmission and intercontinental spread. Science 2010;327(5964):469–74.

[13] Mutreja A, Kim DW, Thomson NR, Connor TR, Lee JH, Kariuki S, et al. Evidence for several waves of global transmission in the seventh cholera pandemic. Nature 2011;477(7365):462–5.

[14] Sabat AJ, Budimir A, Nashev D, Sa-Leao R, van Dijl J, Laurent F, et al. Overview of molecular typing methods for outbreak detection and epidemiological surveillance. Euro Surveill 2013;18(4):20380.

[15] Harris SR, Cartwright EJP, Torok ME, Holden MTG, Brown NM, Ogilvy-Stuart AL, et al. Whole-genome sequencing for analysis of an outbreak of methicillin-resistant *Staphylococcus aureus*: a descriptive study. Lancet Infect Dis 2013;13(2):130–6.

[16] Li M, Du X, Villaruz AE, Diep BA, Wang D, Song Y, et al. MRSA epidemic linked to a quickly spreading colonization and virulence determinant. Nat Med 2012;18(5):816–9.

[17] Croucher NJ, Harris SR, Fraser C, Quail MA, Burton J, van der Linden M, et al. Rapid pneumococcal evolution in response to clinical interventions. Science 2011;331(6016):430–4.

[18] He M, Miyajima F, Roberts P, Ellison L, Pickard DJ, Martin MJ, et al. Emergence and global spread of epidemic healthcare-associated *Clostridium difficile*. Nat Genet 2013;45(1):109–13.

[19] Holt KE, Baker S, Weill FX, Holmes EC, Kitchen A, Yu J, et al. *Shigella sonnei* genome sequencing and phylogenetic analysis indicate recent global dissemination from Europe. Nat Genet 2012;44(9):1056–9.

[20] Okoro CK, Kingsley RA, Connor TR, Harris SR, Parry CM, Al-Mashhadani MN, et al. Intracontinental spread of human invasive *Salmonella typhimurium* pathovariants in sub-Saharan Africa. Nat Genet 2012;44(11):1215–21.

[21] Devault AM, Golding GB, Waglechner N, Enk JM, Kuch M, Tien JH, et al. Second-pandemic strain of *Vibrio cholerae* from the Philadelphia cholera outbreak of 1849. N. Engl J Med 2014;370(4):334–40.

[22] Bos KI, Schuenemann VJ, Golding GB, Burbano HA, Waglechner N, Coombes BK, et al. A draft genome of *Yersinia pestis* from victims of the Black Death. Nature 2011;478(7370):506–10.

[23] Schuenemann VJ, Singh P, Mendum TA, Krause-Kyora B, Jager G, Bos KI, et al. Genome-wide comparison of medieval and modern *Mycobacterium leprae*. Science 2013;341(6142):179–83.

[24] Drummond AJ, Rambaut A. BEAST: bayesian evolutionary analysis by sampling trees. BMC Evol Biol 2007;7:214.

[25] Hendriksen RS, Price LB, Schupp JM, Gillece JD, Kaas RS, Engelthaler DM, et al. Population genetics of *Vibrio cholerae* from Nepal in 2010: evidence on the origin of the Haitian outbreak. mBio 2011;2(4):e00157–00111.

[26] Kaminski RW, Oaks EV. Inactivated and subunit vaccines to prevent shigellosis. Expert Rev Vaccin 2009;8(12):1693–704.

[27] Knetsch CW, Connor TR, Mutreja A, van Dorp SM, Sanders IM, Browne HP, et al. Whole genome sequencing reveals potential spread of *Clostridium difficile* between humans and farm animals in the Netherlands, 2002 to 2011. Euro Surveill 2014;19(45):20954.

[28] Connor TR, Barker CR, Baker KS, Weill FX, Talukder KA, Smith AM, et al. Species-wide whole genome sequencing reveals historical global spread and recent local persistence in *Shigella flexneri*. eLife 2015;4:e07335.

[29] Hu FZ, Eutsey R, Ahmed A, Frazao N, Powell E, Hiller NL, et al. In vivo capsular switch in *Streptococcus pneumoniae*: analysis by whole genome sequencing. PLoS One 2012;7(11):e47983.

[30] van Nood E, Vrieze A, Nieuwdorp M, Fuentes S, Zoetendal EG, de Vos WM, et al. Duodenal infusion of donor feces for recurrent *Clostridium difficile*. N. Engl J Med 2013;368(5):407–15.

[31] Lawley TD, Clare S, Walker AW, Stares MD, Connor TR, Raisen C, et al. Targeted restoration of the intestinal microbiota with a simple, defined bacteriotherapy resolves relapsing *Clostridium difficile* disease in mice. PLoS Pathog 2012;8(10):e1002995.

[32] Group NHW, Peterson J, Garges S, Giovanni M, McInnes P, Wang L, et al. The NIH human microbiome Project. Genome Res 2009;19(12):2317–23.

[33] Anonymous. A framework for human microbiome research. Nature 2012;486(7402):215–21.

[34] Rappuoli R. Reverse vaccinology, a genome-based approach to vaccine development. Vaccine 2001;19(17–19):2688–91.

[35] Madampage CA, Rawlyk N, Crockford G, Wang Y, White AP, Brownlie R, et al. Reverse vaccinology as an approach for developing *Histophilus somni* vaccine candidates. Biol J Int Assoc Biol Stand 2015;43.

[36] Davies DH, Duffy P, Bodmer JL, Felgner PL, Doolan DL. Large screen approaches to identify novel malaria vaccine candidates. Vaccine 2015;33.

[37] Guimaraes L, Soares S, Trost E, Blom J, Ramos R, Silva A, et al. Genome informatics and vaccine targets in *Corynebacterium urealyticum* using two whole genomes, comparative genomics, and reverse vaccinology. BMC genomics 2015;16(Suppl. 5):S7.

[38] Vernikos G, Medini D. Bexsero(R) chronicle. Pathog Glob Health 2014;108(7):305–16.

[39] Bager RJ, Kudirkiene E, da Piedade I, Seemann T, Nielsen TK, Pors SE, et al. In silico prediction of *Gallibacterium anatis* pan-immunogens. Veterinary Res 2014;45:80.

[40] Bohles N, Bohles N, Busch K, Busch K, Hensel M, Hensel M. Vaccines against human diarrheal pathogens: current status and perspectives. Hum Vaccin Immunother 2014;10(6):1522–35.

[41] Talukdar S, Zutshi S, Prashanth KS, Saikia KK, Kumar P. Identification of potential vaccine candidates against *Streptococcus pneumoniae* by reverse vaccinology approach. Appl Biochem Biotechnol 2014;172(6):3026–41.

[42] Ragoussis J. Genotyping technologies for genetic research. Annu Rev Genomics Hum Genet 2009;10:117–33.

[43] Sobrino B, Brion M, Carracedo A. SNPs in forensic genetics: a review on SNP typing methodologies. Forensic Sci Int 2005;154(2–3):181–94.

[44] Sangal V, Holt KE, Yuan J, Brown DJ, Filliol-Toutain I, Weill FX, et al. Global phylogeny of *Shigella sonnei* strains from limited single nucleotide polymorphisms (SNPs) and development of a rapid and cost-effective SNP-typing scheme for strain identification by high-resolution melting analysis. J Clin Microbiol 2013;51(1):303–5.

[45] Pham Thanh D, Tran Vu Thieu N, Tran Thuy C, Loden M, Tuin K, Campbell JI, et al. Identification of *Salmonella enterica* serovar Typhi genotypes by use of rapid multiplex ligation-dependent probe amplification. J Clin Microbiol 2013;51(9):2950–8.

[46] Hamed M, Nitsche-Schmitz DP, Ruffing U, Steglich M, Dordel J, Nguyen D, et al. Whole genome sequence typing and microarray profiling of nasal and blood stream methicillin-resistant *Staphylococcus aureus* isolates: clues to phylogeny and invasiveness. Infect Genet Evol 2015;36.

[47] Touati A, Blouin Y, Sirand-Pugnet P, Renaudin H, Oishi T, Vergnaud G, et al. Molecular epidemiology of *Mycoplasma pneumoniae*: genotyping using single nucleotide polymorphisms and SNaPshot technology. J Clin Microbiol 2015;53(10):3182–94.

[48] Miya F, Kato M, Shiohama T, Okamoto N, Saitoh S, Yamasaki M, et al. A combination of targeted enrichment methodologies for whole-exome sequencing reveals novel pathogenic mutations. Sci Rep 2015;5:9331.

[49] Hedges DJ, Guettouche T, Yang S, Bademci G, Diaz A, Andersen A, et al. Comparison of three targeted enrichment strategies on the SOLiD sequencing platform. PLoS One 2011;6(4):e18595.

Chapter 9

A Metagenomic Insight Into the Human Microbiome: Its Implications in Health and Disease

R. Saxena, V.K. Sharma

Indian Institute of Science Education and Research Bhopal, Madhya Pradesh, India

Chapter Outline

Outline of the Chapter	107	Type 2 Diabetes Mellitus	113
Introduction	107	Major Depressive Disorder	114
Human-Associated Microflora	109	Cardiovascular Diseases	114
Alterations in the Healthy Human Microbiome: Association With		Other Diseases	115
Diseases	110	Novel Therapeutic Strategies Based on the Human-Associated	
The Oral Microbiome	110	Microbiome	115
The Skin Microbiome	111	Conclusion	116
The Gut Microbiome	111	Glossary	117
Diseases Associated With Variation in the Gut Microbial		List of Acronyms and Abbreviations	117
Community	112	References	117
Inflammatory Bowel Disease	112		
Obesity	112		

OUTLINE OF THE CHAPTER

This chapter underlines the applications of metagenomics and next generation sequencing (NGS) in understanding the diversity and role of human microbiome in human health and disease.

INTRODUCTION

The successful completion of Human Genome Project in the year 2000 was a breakthrough in the field of biomedical genomics [1]. The revelation of the human genome sequence helped researchers to understand the biological processes at molecular and genomic levels for providing precise answers to many biological questions related to human health. Interestingly, the human body is known to harbor 10 times more bacterial cells than its own cells. These microbes are present at various anatomical sites in the human body, such as skin, the oral cavity, and the gut, and contribute about 100 times more genes than the human genome. These microbes help in several metabolic processes, activation of host immune response, and disease pathogenesis, thus playing an indispensable role in human health and well-being. Therefore a comprehensive analysis and understanding of our second genome, i.e., the human-associated microbiome, is much needed to understand the interplay between the human host and its microflora to decipher their respective roles in human physiology and treatment of diseases.

However, it is known that more than 98% of these microorganisms associated with the human body, and in other environments, are unculturable using known culture media and laboratory conditions, and hence cannot be sequenced using conventional methods. The need to sequence the environmental flora has led to the emergence of metagenomics [2]. The term *metagenomics* was first used in late 1990s and was defined as the genomic analysis of DNA isolated from microbial community in an environment through culture-independent techniques. Since 2005, metagenomics has transformed the sequencing of cloned DNA fragments using Sanger technology to direct sequencing of

Medical and Health Genomics. http://dx.doi.org/10.1016/B978-0-12-420196-5.00009-5

DNA without heterologous cloning [3]. With the technological developments in NGS technologies in recent years, the cost of DNA sequencing has dropped exponentially, making it more efficient and affordable for carrying out large-scale genomic and metagenomic studies. *NGS* refers to parallel high-throughput DNA sequencing by synthesis, which enables rapid sequencing of millions of DNA base pairs spanning the entire genome and producing millions of sequences (reads) in a single sequencing run, thus lowering the time and cost. The earliest metagenomic datasets were produced using Sanger sequencing technology, then replaced by pyrosequencing-based Roche 454 sequencing in the subsequent years, and have now mostly been replaced by Illumina sequencing technology. A number of efficient NGS platforms based on Illumina sequencing technology are now available, such as HiSeq X Ten, NextSeq 500, and MiSeq, each having their own distinct capabilities and applications. To assess the diversity of microbial species, 16S rRNA gene sequencing and analysis is the most commonly used method [4]. Similarly, for assessing the functional profile and composition of a metagenome, shotgun sequencing is used to get unbiased data by directly sequencing the whole metagenome as short reads, followed by assembly into contigs to identify the genes of interest and unknown genes inherent in the dataset (Fig. 9.1).

Several bioinformatics tools and pipelines are now available for metagenomic analysis (such as taxonomic identification [5–8], functional assignment, [9,10] and comparative metagenomics [11,12]) of the metagenomic

data obtained from different environments. One of the first metagenomic projects was the acid mine drainage project, which helped in the identification of microbial communities that can drive acidification in extremely acidic outflows from metal mines [13]. Among the largest metagenomic projects, the sequencing of the ocean samples (Sargasso Sea) by Craig Venter's group (JCVI, Rockville, MD.) revealed the complex microbial diversity present in deep oceans [14]. The Earth Microbiome Project was later established as a global initiative to analyze the microbial community for the benefit of the planet and humankind [15].

The study of the human microbiome has gained much attention recently. One of the pioneering studies was performed by Kurokawa et al. [15a], in which the researchers carried out the comparison of gut metagenomes from 13 healthy Japanese individuals of various ages including infants, which revealed the compositional differences between the gut flora of adults and infants. Many large-scale projects, such as the Human Microbiome Project and the Genome 10K Project are underway to reveal the human-associated microbiome [16–18]. From these findings, it can be inferred that metagenomic approaches appear to be greatly promising in understanding the role of human-associated microbiota in human health and in the etiology of human diseases. Furthermore, it would be helpful in the development of new probiotics, evaluation of their efficacy, and in the development of personalized medicines as novel therapeutic agents.

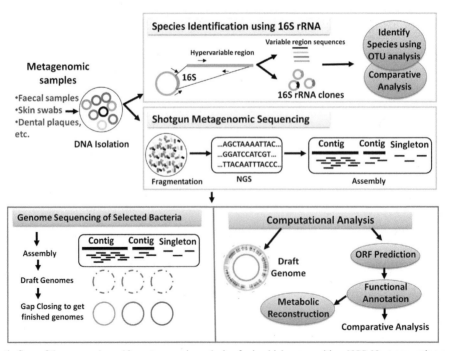

FIGURE 9.1   Schematic flow of the approach used in metagenomic analysis of microbial communities. *NGS*, Next generation sequencing; *ORF*, open reading frame; *OTU*, operational taxonomic unit.

HUMAN-ASSOCIATED MICROFLORA

Throughout the life-span, human beings live in constant association with the microbes distributed across different body sites, which constitute the human microbiome. The microbiota mainly consist of bacteria, however, viruses, fungi, and protozoans also contribute to the complexity of the human microflora. Numerous microbial cells reside on the epithelial surfaces, such as skin or in the cavities of our body, such as, human gut [18]. The microbiome of the human distal gut is among the most complex communities studied ever, with an estimated 1000 different microbial species found across different human populations [19]. These microbes encode millions of different genes, largely exceeding the total number of human genes [20]. The composition of the microbiota in a community can vary substantially as a result of different environmental sites, changes in diet, host genetics, early microbial exposure of the individual, and as a consequence of disease [21,22]. Because the microbiota is very complex and dynamic, it is known to have a profound influence on human physiology and nutrition. The need to understand the dynamic diversity and the effects of proportional differences of the human-associated microbial microbiome poses a great challenge in the field of medical and health genomics. The microbiome shows considerable changes throughout our lives caused by various factors such as growth, diet, nutritional status, lifestyle, travel, illness, medical treatments, and aging [23] which directly or indirectly affect human health. The nutritional value of food is affected in part by a person's microbiota and its component genes (microbiome), and vice versa [24]. One of the aims of human–microbiome association studies is to reveal how changes in the microbiome are directly or indirectly related to the nutritional status and diet of an individual.

To study the role of the microbiome in human metabolism, germ-free mice or gnotobiotic mice with no exposure to microbes are used, and are colonized with different microbial communities at specific life stages. This provides an excellent system for carrying out controlled studies using the same host genotype, microflora composition, diet, and environmental conditions. In a study by Turnbaugh et al. [24] a well-defined representative animal model of the human gut ecosystem was constructed in an attempt to understand the association between the host environmental factors and the microbiome. They transplanted fresh or frozen adult human fecal microbial communities into germ-free C57BL/6J mice. By metagenomic analysis of bacterial colonization, the gnotobiotic humanized mice were found to stably and heritably colonize and reproduce much of the bacterial diversity of the donor's microbiota. They also observed that switching to a high-sugar and high-fat "Western" diet from a low-fat, plant polysaccharide–rich diet during a single day altered the representation of metabolic pathways in the microbiome and its gene expression

in mice [24]. Humanized gnotobiotic mice have thus proven to be very useful for conducting clinical trials to confirm the effects of environmental and host-driven factors on the gut microbiota and host physiology.

Another study was carried out using fecal samples of healthy American human volunteers adhering to a non-vegetarian diet. This study was performed by using polymerase chain reaction amplicons of the V4 region of the *16S rRNA* gene amplified from metagenomic DNA and subsequent sequencing on the Illumina HiSeq platform. The results showed an overrepresentation of bile-tolerant microorganisms (*Alistipes*, *Bilophila*, and *Bacteroides*) and low abundance of Firmicutes, which metabolizes dietary plant polysaccharides (*Roseburia*, *Eubacterium rectale* and *Ruminococcus bromii*). This change in microbial composition mirrored differences between herbivorous and carnivorous mammals, suggesting trade-offs between protein and carbohydrate fermentation. Food-borne microbes, including bacteria, fungi, and also viruses, from both the diets were shown to transiently colonize the gut. The overrepresentation and activity of *Bilophila wadsworthia* in the animal-based diet suggested a link between bile acids, dietary fat, and the prominence of microorganisms that have the potential to induce inflammatory bowel disease (IBD). The results demonstrated that the gut microbiome rapidly responds to altered diet, thus, facilitating the diversity of human dietary lifestyles [25]. Therefore, it can be concluded that diet plays a prominent role in forming the community structure of the gut microbiome.

Variations in the gut microbial community in humans have also been observed as a consequence of age and demography [26]. Yatsunenko et al. carried out a *16S rRNA* gene–based analysis of bacterial species' diversity in fecal samples from 531 individuals, and also performed the functional analysis of 110 individuals (out of 531) using their gene content. The cohort consisted of healthy children and adults, including monozygotic and dizygotic twins from the rural Malawi, the Amazon region of Venezuela, and metropolitan areas in the United States. DNA was extracted from each fecal sample, and the V4 region of bacterial *16S rRNA* gene was amplified from the isolated community DNA. The amplicons were sequenced using the Illumina HiSeq 2000 instrument to identify the phylotypes present in the samples. The functions encoded in community DNA were characterized by performing multiplex shotgun 454 pyrosequencing of fecal DNA from a subset of 110 fecal samples, comprising 43 families with members matched as closely as possible for age. A closed resemblance in functional maturation of the gut microbiome was observed during the first 3 years of life in all three populations. These included age-associated changes in the genes involved in vitamin biosynthesis and metabolism. Notable differences in bacterial assemblages and functional gene repertoires were observed between US residents and those in the other two countries.

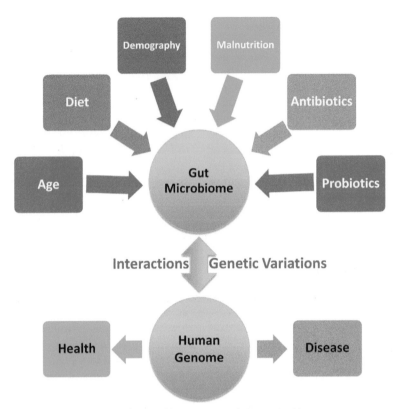

FIGURE 9.2 Complex interplay of the human gut microbiome and human genome in human health.

Urease gene representation was remarkably higher in Malawian and Amerindian (from Venezuela) baby microbiomes and found to decrease with age in these two populations, unlike in the United States, where it had shown to remain low from infancy to adulthood. The bacterial diversity was found to increase with age in every population. *Bifidobacterium* dominated the baby microbiota, whereas in adults, *Prevotella* was observed to be the dominating genus. But the ratio of these dominant species was variable in different populations. These findings underscore the need to understand the role of the microbiome, its variation, and its association with different life stages in humans to infer its effect on human health (Fig. 9.2).

ALTERATIONS IN THE HEALTHY HUMAN MICROBIOME: ASSOCIATION WITH DISEASES

The diversity and abundance of microbes within a body site have been associated with several human diseases. Some of the major anatomical sites in a human body that have been used for studying the human-associated microbiota include specimens from the *oral cavity and oropharynx*: saliva, buccal mucosa (cheek), keratinized gingiva (gums), palate, tonsils, supragingival and subgingival dental plaque (tooth biofilm above and below gums), throat and tongue soft

tissues; the *skin*: the two retroauricular creases (behind each ear), the two antecubital fosse (inner elbows), the anterior nares (nostrils), and the scalp; the *lower gastrointestinal tract*: stool specimen; and the *vaginal* specimen in females: specimens from the vaginal introitus, midpoint, and posterior fornix [27–29]. Studies of the human microbiome have revealed that even healthy individuals differ remarkably in microbes that occupy the major habitats, such as the gut, skin, oral cavity, and vagina [29]. Therefore, to determine the role of the human-associated microbiota in disease predisposition and pathogenesis, it is imperative to first understand its basic inherent composition.

THE ORAL MICROBIOME

The oral cavity in humans harbors approximately 700 prokaryote species, out of which about 49% are officially named, 34% are known only as uncultured phylotypes, and 17% are unnamed (but culturable) [30]. Studies of the oral cavity microbiome in healthy individuals have shown that Firmicutes is the most abundant phylum, followed by Proteobacteria, Bacteroidetes and Fusobacteria [31]. The signature clades at the genus level that have been observed by the Human Microbiome Project Consortium to be dominant in the oral cavity are *Streptococcus*, followed by *Haemophilus* in the buccal mucosa, and *Actinomyces* and *Prevotella* in the gingival plaque [29]. However, a metagenomic sequencing analysis of the microbiota associated with periodontal plaques has shown a variation from this "normal" distribution in diseased individuals. The dominance of bacterial communities has been observed to start majorly from Bacteroidetes to Firmicutes, Proteobacteria, and Actinobacteria in a periodontal plaque sample [32]. The gram-negative *Porphyromonas gingivalis*, *Treponema denticola*, and *Tannerella forsythia* (members of Bacteroidetes) have often been isolated from dental plaques in periodontal patients and are considered to be the specific pathogens of periodontal disease [33]. A recent metagenomic analysis carried out using Illumina HiSeq 2000 with 2×100 bp paired-end sequencing has revealed *Prevotella* to be the most abundant followed by *Capnocytophaga*, *Streptococcus*, *Corynebacterium*, and, lastly, *Actinomyces* in periodontal plaque samples [32]. These findings shed light on the contribution of microbes in the formation and progression of dental plaques and oral diseases. Still, more intensive analysis is required to explore the heterogeneity of periodontal microbial communities within individuals.

THE SKIN MICROBIOME

The skin is an ecosystem composed of $1.8\,m^2$ of diverse habitats with a large number of invaginations, folds, and specialized niches, thus, supporting the growth of a wide range of microorganisms including bacteria, viruses, and fungi, as well as mites. Symbiotic microorganisms occupying a wide range of skin niches protect against invasion of pathogenic organisms. They also have a role in the formation of memory cells of the immune system, so that the immune system responds similarly to their pathogenic counterparts [28]. The sebaceous glands connected to the hair follicle at the skin secrete a lipid-rich substance, sebum, which provides an antibacterial shield on the skin surface. Sebaceous glands are anoxic and support the growth of facultative anaerobes such as *Propionibacterium acnes*, a common skin commensal bacterium. The whole genome sequencing of *P. acnes* has revealed many genes that degrade the skin lipids of sebum [34]. *16S rRNA* gene sequencing performed on skin samples from different sites has shown that Betaproteobacteria, *Propionibacterium*, *Corynebacterium*, and *Staphylococcus* are some of the major bacterial taxa that colonize a healthy skin microbiome. Apart from these, Clostridiales, Bacteroidales, and Lactobacillales are scarcely found on the skin surface. Fungal species of *Malassezia* are found to be some of the major fungal inhabitants on the skin surface. A variation in the distribution of these bacterial and fungal taxa is very well evident in skin disorders such as atopic dermatitis, eczema, psoriasis and dandruff [28]. Dandruff has also been found to be associated with the disequilibrium in the proportion of the major bacterial and fungal populations colonizing the scalp. An overrepresentation of bacterial species such

as *Staphylococcus epidermidis*, and fungal species *Malassezia restricta* has been found to be associated with dandruff in hair. A dandruff scalp shows lowered level of *P. acnes* when compared with the healthy scalps [35]. Thus, host–microbe interactions at the skin surface also play a major role in disease predisposition and pathogenesis and these interactions should be taken into consideration when developing novel atopic drugs and ointments.

THE GUT MICROBIOME

The human gut harbors a diverse milieu of microbes that play a key role in the well-being of the host. These microorganisms are known to interact with one another and with the host immune system in various ways, influencing normal as well as diseased states in individuals. The gut microbiota metabolizes the substrate inputs from the host, such as diet and xenobiotics, into metabolites that can affect the host peripherally by entering the host's blood stream. Alternatively, the gut microbiota also helps in the metabolism of some complex nutritional compounds to simpler metabolites, hence being beneficial for the host [36].

The signature clade at the phylum level that was observed to be dominant in a healthy human gut is Bacteroidetes (shown by the Human Microbiome Project Consortium) [29]. This *16S rRNA* gene sequence–based study was performed in an American population with a cohort of 300 individuals. *Prevotella copri* was found to be overrepresented at the species level [29]. A similar study was performed by Arumugam et al. with 22 healthy individuals belonging to different groups of Europeans, such as Danish, French, Italian, and Spanish population [27]. The Sanger-sequenced metagenomes of these 22 individuals were combined with existing pyrosequencing (2 American) and Sanger (13 Japanese, 2 American) gut metagenomic data sets, forming a total of 39 gut metagenomes. The results represented concordance to the finding that Bacteroidetes is the most abundant phylum found in the gut microbiota.

The belief of the stomach being sterile was overturned after the discovery of *Helicobacter pylori* having adapted to survive in the acidic gastric environment. It has been found that individuals not having *H. pylori* as a member of their gastric microbiota have a high diversity of other microorganisms in their stomach [37]. Most of the prominent gastric phylotypes (*Streptococcus*, *Actinomyces*, *Prevotella*, and *Gemella*) are also abundant in the oropharynx of these individuals, indicating that close relatives of the oral microbiota colonize more distally in the human body. In contrast to this, *H. pylori*–positive individuals have *H. pylori* dominating among the gastric microbiota, thus, reducing the overall diversity in the stomach. Proteobacteria is found to be the most abundant phylum in the stomach of these individuals, leaving others to occupy a very small fraction of the same.

The *H. pylori* is a classical amphibiont; ie, it may have a pathogenic or a symbiotic association with other organisms, such as its host. The presence or absence of *H. pylori* has been found to be strongly associated with some diseases that show important age-related differences. Its presence has been known to increase the risks for developing gastric mucosa-associated lymphoid tissue tumors, peptic ulcer disease, and gastric adenocarcinomas. Contradictory to this, it is found to decrease the risk of reflux esophagitis and childhood-onset asthma, thus, demonstrating the complex biological interactions between hosts and microbiota [21].

The variation in gut flora is also known to occur largely as a consequence of different metabolic and immune-related disorders [21,38]. With the understanding of the healthy gut microbiome, the following text is focused on the association of human diseases with the gut microbiota, followed by a discussion of its predisposition and pathogenesis. Such association studies aim at adopting a holistic approach for studying the microbiota that goes beyond characterization of community composition and encompasses dynamic interrelations between all components of the microbiota and host tissue. Understanding these associations will be crucial over time for building predictive models for diagnosis and management of diseases allied to imbalances in our microbiota.

Diseases Associated With Variation in the Gut Microbial Community

Inflammatory Bowel Disease

IBD or irritable bowel syndrome is among the several diseases in which changes in microbial composition in the human gut is observed to be associated with inflammatory conditions in the colon and the small intestine [39]. IBD is a persistent gastrointestinal disorder with an unknown etiology and is characterized by an inflammatory reaction of the intestinal mucosa. The principal types of IBD are Crohn's disease and ulcerative colitis [38]. The development of IBD can partly be linked to the host immune system and recognition of microbes. It is known that the microbiome is also essential for the activation of host immune responses [21]. For example, helper T-17 cell differentiation in the mouse lamina propria is observed to occur in the presence of segmented filamentous bacteria and polysaccharide A produced by *Bacteroides fragilis* is known to contribute to the conversion of CD4+ T cells into regulatory T cells [21]. IBD has been considered to mirror interactions between microbes and the host for a long time. IBD receptiveness is associated with host polymorphisms in bacterial sensor genes, such as nucleotide-binding oligomerization domain-containing 2 [(*NOD2*) also known as caspase recruitment domain-containing protein 15 (*CARD15*)] and Toll-like receptor 4 (*TLR4*) [40,41]. The symptoms in patients with IBD sometimes improve with antibiotic treatment. Early childhood exposure to antibiotics is observed to be associated with a notably increased risk for Crohn's disease. This suggests that gut microbiome perturbations are imperative for disease risk [21]. The gut microbiome community in patients with ulcerative colitis or Crohn's disease is different from the normal distribution, and also differs prominently between the two diseases. Microbial diversity is found to be significantly diminished in Crohn's disease, suggesting a decreased gut microbiome flexibility that could affect immune interactions [38]. Specific bacteria of the Enterobacteriaceae family are known to act together with a perturbed microbiome to amplify the risk of ulcerative colitis. An influential study on twins at variance for ulcerative colitis described remarkably reduced bacterial diversity in the affected siblings but showed increased proportions of Proteobacteria and Actinobacteria [41]. Patients with Crohn disease show an overrepresentation of *Enterococcus faecium* and of several Proteobacteria when compared with controls [38]. The microbial patterns observed for the conditions described are yet in the preliminary stages and a deeper understanding about specific host–microbe association is required to develop a probiotic-based approach for the treatment of IBD.

Obesity

Obesity has become a major metabolic disorder in developed and developing countries. It has also been found to be associated with several types of cancer and has drawn the attention of many biologists working in the field of human health and diseases [42]. The relative abundance of the two dominant phyla in the gut flora, the Bacteroidetes and the Firmicutes, has been found to be correlated with obesity in genetically obese mice and their lean littermates, as well as that of lean and obese human individuals [43,44].

Microbes help in the metabolism of indigestible polysaccharides in our diet. For example, the microbial fermentation of carbohydrates in the gut produces short-chain fatty acids (SCFAs): acetate, propionate, butyrate, and lactate [45]. There is a drop in the pH driven by SCFAs, which has been proposed to be involved in the maintenance of microbial ecosystems. Because a single genus of bacteria cannot hydrolyze all metabolic substrates or cannot produce all four SCFAs upon carbohydrate fermentation, the profile of SCFAs represents the metabolic cooperation within the gut flora. SCFAs have different metabolic features. Butyrate is an energy substrate for colonocytes; propionate is a gluconeogenic substrate in the liver, and acetate is potentially used as a cholesterol or fatty acid precursor. More recently, SCFAs have been identified as the physiological ligands of G-protein coupled receptors, which get expressed in several cell types (immune cells, endocrine cells, and adipocytes) and in a wide variety of host tissues for cell–cell signaling mechanisms. Thus SCFAs, which are considered to be indirect metabolites produced by the gut microbiota, not only play a role in the regulation of immunity and energy metabolism, but also in adipose tissue expansion [45], thereby leading to obesity.

The first demonstration of a specific change in the gut microbial community between obese and lean phenotypes was made in genetically obese (*ob/ob*) mice. It was observed that, genetically obese (*ob/ob*) mice have decreased Bacteroidetes-to-Firmicutes ratios (ie, fewer Bacteroidetes and more Firmicutes, contradictory to the healthy gut flora) compared with their lean (*ob/+* and *+/+* wild-type) siblings [43]. Transplantation of gut microbiota from the obese (*ob/ob*) to germ-free mice lead to an obese phenotype, suggesting the transmissibility of metabolic phenotypes [21]. Altogether, such animal studies suggested that obesity and high-fat diets might be directly involved in the alteration of gut microbiota composition at both phylum as well as genus levels. Several other bacteria, namely Proteobacteria and *Bifidobacterium*, were also found to be inversely associated with obesity [45].

In humans, the relative proportions of members of the Bacteroidetes phylum have been observed to increase with loss in weight [21]. Soon after their first observation in experimental animals, Ley et al. confirmed that human obese subjects also demonstrate a larger proportion of Firmicutes and relatively fewer Bacteroidetes than their lean counterparts. In this study, the gut microbiota composition of 12 obese American individuals, who were randomly assigned to either a fat-restricted or a carbohydrate-restricted low calorie diet, were monitored over the course of 1 year by sequencing *16S rRNA* genes from fecal samples. Similar to earlier studies, members of the Bacteroidetes and Firmicutes divisions dominated the microbiota (92.6% of all *16S rRNA* sequences), with the larger fraction coming from Firmicutes as compared with Bacteroidetes [44]. Furthermore, another study was carried out focused on the gut microbiota of 154 monozygotic or dizygotic American twin-pair individuals and their mothers concordant for their lean or obese phenotype. A decrease in phylogenic microbial diversity in obese subjects was observed, along with a reduced representation of the Bacteroidetes and more representation of Actinobacteria [46].

Thus, it appears that obesity is associated with phylum-level changes in the microbiota, lowered bacterial diversity, and abnormal representation of bacterial genes and metabolic pathways. Therefore, it can be concluded that modern lifestyles with consumption of high-fat diet alter the selection pressures on microbiomes, hence changing the early life exposures of hosts to bacteria and thereby inducing obesity. With weight loss, there is an alteration in the gut microbial diversity, and the research in this area is booming to identify probiotics-based approaches or more contemporary approaches to incline the balance of the gut microbiota toward a less obesogenic composition [47].

Type 2 Diabetes Mellitus

Resistance to insulin action in the muscles and adipose tissues is mainly found to occur in type 2 diabetes mellitus (T2D). Several studies have shown the association between the gut flora and metabolic disorders like obesity and T2D [48]. Regulation of the flora depends on the proportion of nutrients like saturated and unsaturated fatty acids, which can alter the bacterial metabolic activities [49]. High-fat diet alters the flora and leads to enhanced intestinal permeability and susceptibility to microbial antigens. It also increases fatty acid oxidation in liver and adipose tissues and generates reactive oxygen species, which reduce mucus production [50]. Oxidation of polyunsaturated fatty acid produces malonaldehyde, which induces damage to the epithelial cell membrane and enhances the intestinal tight junction permeability [51,52]. These consequences correlate with the metabolic endotoxemia and insulin resistance [53].

A human metagenome-wide association study reveals a significant correlation of specific microbes present in the gut, certain bacterial genes, and their respective metabolic pathways with T2D [53a,b,c]. Several studies have suggested that the population of butyrate-producing bacteria, such as *Roseburia intestinalis* and *Faecalibacterium prausnitzii* were less abundant in the gut flora. Corresponding to this, sulfate reducing species (*Desulfovibrio*) and mucin-degrading species (*Akkermansia muciniphila*, *Lactobacillus* species including *Lactobacillus gasseri*, *Streptococcus mutans*, and Proteobacteria) were more often observed in T2D. Microbiota in individuals with T2D show some potential associated functions like enhancement in membrane transport of sugar, branched chain amino acid transport, sulfate reduction, and microbial gene expression (oxidative stress). However, there is a decrease in butyrate biosynthesis and genes involved in synthesis of vitamins like riboflavin [54]. Another study shows lowered count of *Bifidobacterium* and *F. prausnitzii* in diabetic individuals. Both of these bacteria are Gram-positive with antiinflammatory properties [55].

Perturbation in the gut flora of diabetic individuals has been observed previously, but it is still necessary to know whether the alteration in the gut microbiota, intestinal barrier, and metabolic endotoxemia are causes or consequences of the diabetes. Interestingly, in one study, intravenous administration of lipopolysaccharide (LPS) showed metabolic endotoxemia and systemic insulin resistance [56]. In several other studies, a relationship between LPS and the development of T2D have been observed [57].

Intake of a high-fat diet induces the death of Gram-negative bacteria, which leads to production of LPS in the lumen and its translocation into the intestinal capillaries, and finally activation of Toll-like receptors (TLRs) [58,59].

TLR4 is present in tissues targeted for insulin action. The inflammatory pathway induced by LPS-*TLR4* enhances the expression of inducible nitric oxide synthase. This nitric oxide can inhibit insulin transduction signal via phosphorylation of insulin receptor 1 substrate (*IRS1*) in serine, [60] causing the insulin production to be repressed, which leads to increase in blood glucose levels. In another pathway, TLRs directly induce inflammatory responses, resulting in activation of cytokines. These proinflammatory cytokines can also induce the phosphorylation of *IRS1* in serine, [61] leading to insulin repression as observed in the previous study.

Evidence suggests that certain drugs might affect the gut microbiota. Metformin is an oral antidiabetic drug in the biguanide class that is widely used in the therapy of T2D. When *Caenorhabditis elegans* is cocultured with *Escherichia coli*, metformin is able to slow down aging in *C. elegans* by changing the metabolism (disruption of folate cycle) in *E. coli* [62]. A decreased level of s-adenosylmethionine is observed when the folate cycle is disrupted in the bacteria, which affects aging in worms. Because of the widespread use of this drug in patients with T2D, similar effects can also be suggested in the metabolic pathways assisted by the human gut flora. These alterations can either help in maintaining homeostasis in diseased state or perturb the normal metabolic mechanisms taking place with the help of gut microbiota in humans.

Major Depressive Disorder

Depression, a word unheard of many years ago, now poses a serious socioeconomic threat to the modern societies worldwide. Major depressive disorder, or clinical depression, a mental disorder, is characterized by a pervasive and persistent low mood that is accompanied by low self-esteem and loss of interest in regular activities [63]. Studies in germ-free mice and mice exposed to pathogenic bacterial infections, probiotic bacteria, or antibiotic drugs highlight the role of gut microbiota in the regulation of anxiety, mood, cognition, and pain [64]. There are many potential direct and indirect pathways by which the gut microbiota can affect the central nervous system (CNS). They can have direct effects on the immune system, which shares a bidirectional communication with the CNS [64]. Tryptophan, an essential amino acid, is a precursor to many neurotransmitters like serotonin. There is some evidence suggesting that the probiotic *Bifidobacterium infantis*, when injected in mice, shows an increased level of tryptophan and kynurenine, representing antidepressant properties [65]. A recent report shows reduced stress-induced corticosterone and anxiety- and depression-related behavior in mice on the administration of *Lactobacillus rhamnosus*. This reduction was found to be a cause of the alteration in the expression of γ-aminobutyric acid receptor, which is involved in the major

physiological and psychological processes in the body [66]. Many similar findings using mice models suggest that the vagus nerve is a major neuromodulatory communication pathway between the gut bacteria and the brain [67]. Such findings highlight the importance of microorganisms in the bidirectional communication of the gut–brain axis.

Many probiotics have shown the capacity to generate neuroactive compounds. It has been seen that *Escherichia* spp., *Bacillus*, spp. and *Saccharomyces* spp. produce noradrenalin; *Candida*, spp., *Streptococcus spp.*, *Escherichia* spp., and *Enterococcus* spp. produce serotonin; *Bacillus* spp. produce dopamine and *Lactobacillus* spp. produce acetylcholine. These probiotics alter the levels of opioid and cannabinoid receptors in the gut epithelium. However, many pathways related to this still remain unclear [68]. Thus the concept of a microbiome-gut-brain axis is arising, which suggests that modulation of the gut microbiota may be a well brought-up strategy for developing novel therapeutics for multifaceted stress-related CNS disorders, which is much required to improve the present era of cognitive medicine.

Cardiovascular Diseases

Cardiovascular diseases (CVDs) are among the leading noncommunicable diseases. During past decades, periodontitis and infectious agents have also been associated with increased threat of cardiovascular diseases [69]. Clinical trials to reduce atherosclerosis through antibiotics have largely been unsuccessful. The underlying reason remains unclear, but it has been observed that these studies were generally aimed to eradicate only *Chlamydia pneumonia*. A study has revealed that bacterial diversity in human atheromas is very high and *C. pneumonia* constitutes about 50% of the plaque community [70]. Thus, antibiotic treatments that target specific bacteria may be ineffective. However, more suitable targets could be the metabolic pathways that are signaled by the bacteria. It has been reported that gut microbial (hippurate)– and dietary (alanine)–derived urinary metabolites are correlated with blood pressure in humans [71]. These findings suggest that perturbations in the gut microbiota may influence the progress of cardiovascular diseases.

Atherosclerosis is coupled with inflammation and lipid accumulation in the arterial walls, and bacteria have been suggested as a causal agent of this disease. Gut metagenome sequencing of Swedish patients with symptomatic atherosclerosis has shown enrichment of genus *Collinsella*. Whereas, healthy subjects show enriched *Roseburia* and *Eubacterium* in their gut flora. This study was performed with shotgun sequencing of the gut metagenome using the Illumina HiSeq 2000 platform. Further characterization of the functional repertoire of the patient gut metagenomes revealed upregulation of peptidoglycan synthesis encoding genes, whereas, there was downregulation of phytoene dehydrogenase encoding genes. Patients also showed reduced serum level of β-carotene. These findings suggest that the gut metagenome is associated with inflammatory status of patients with symptomatic atherosclerosis and these individuals harbor notable changes in their gut metagenome [72].

A metabolomics approach had been carried out to generate unbiased small-molecule metabolic profiles to predict the risk of CVD through plasma of symptomatic atherosclerosis patients in the United States, comparing them with healthy controls. Metabolites of the dietary lipid phosphatidylcholine (PC)–trimethylamine N-oxide (TMAO), choline, and betaine were found to demonstrate the risk for CVD. Intestinal microbial metabolism of dietary lecithin is known to produce TMAO, which is suggested to cause formation of atherosclerotic plaques through interactions with a variety of cells such as macrophages and plasma cells. This cohort study concluded that higher the levels of TMAO, the greater the risk for CVD. The researchers also recommended that TMAO ratio can predict heart risk better than other factors such as hyperlipidemia and high blood pressure. Strikingly, the metabolism of dietary PC was observed to be gut flora–dependent. The pathway identified was: dietary PC/choline → gut flora–formed TMA → hepatic flavin monooxygenase-formed TMAO → atherosclerosis → heart attack, stroke, and death [73]. Therefore, TMAO can prove to be an effective therapeutic target in the management of CVD. This vital finding also suggests restricted consumption of lecithin-containing food, especially for the individuals who are at increased risk for CVD. Probiotics may prove to be a solution to let TMAO-inhibiting bacteria grow or target the TMAO-producing bacteria.

Other Diseases

Apart from the diseases listed, perturbations in the gut microbiota have been observed to be associated with many other diseases recently. *16S rRNA* sequencing and shotgun sequencing of the gut metagenome of patients with rheumatoid arthritis revealed an overrepresentation of *P. copri*, which was strongly found to be correlated with new-onset untreated rheumatoid arthritis patients. This enrichment was accompanied with a decrease in the abundance of Bacteroidetes and other beneficial bacteria, such as *Clostridia*, *Blautia*, and *Lachnospiraceae*, which have been previously reported to be associated with an antiinflammatory state and regulatory T-cell production [74].

A recent study revealed the abundance of *Streptococcus* spp. and *Veillonella* spp. in the gut metagenome of patients with liver cirrhosis, suggesting that the two genera might play an important role in the development of the disease. Using a quantitative metagenomics approach, it was concluded that most (54%) of taxonomically assigned species enriched in the patient were of buccal origin, suggesting the translocation of oral microbiota to the gut [75].

The microbiome analysis of stool samples of African Americans with colon polyps was performed by Brim et al. [76]. These colon polyps are known to subsequently develop into colon cancer. The study suggested that at the preneoplastic stages, there is a trend showing microbiota changes between healthy and colon polyp patients at the subgenus level in the *Bacterodides* group [76].

The gut microbiome is also known to detoxify dietary compounds. Many plants produce oxalate, which chelates with calcium ions (Ca^{2+}) and other free metal cations, leading to the development of kidney stones, renal failure, and hyperoxaluria. The enzyme required to detoxify this oxalate is not found in mammals, and thus they rely on the microbial biotransformation. *Oxalobacter formigenes* is one of the key bacteria that participates in the catabolism of oxalate in the human gut. Deficiency of *O. formigenes* has been found to be associated with an increased risk of hyperoxaluria and kidney stones and administering *O. formigenes* is known to reduce urinary oxalate concentrations [77,78] (Table 9.1).

NOVEL THERAPEUTIC STRATEGIES BASED ON THE HUMAN-ASSOCIATED MICROBIOME

All the studies discussed ultimately reveal that the human-associated microbiome regulates many functions, ranging from toxin neutralization and immunity against pathogens to metabolic regulation and cognitive processes. Consequently, alterations in the composition of the gut microbiota have shown to contribute to the development of various chronic diseases. Whereas some of the indigenous flora possesses the potential to cause disease, many of them exhibit health-promoting properties as well. For example, *Bifidobacteria* and *Lactobacilli* are associated with good health, whereas, *Clostridia* is considered to be detrimental to health. The identification of healthy human microflora with the understanding of the relative proportion of constituent microbes is helpful in the development of probiotics aimed at maintaining optimal health and treating chronic inflammatory and immune-related diseases [79]. Probiotics are not only helpful in the treatment of diseases, but also in restoring the intestinal homeostasis that has been altered as a consequence of the disease. A study has shown that consumption of probiotics (*Bifidobacterium bifidum* and *Lactobacillus plantarum 8PA3*) in human alcohol-induced liver injury restore intestinal flora and improve the efficiency of liver enzymes [80].

The World Health Organization defined probiotics as "A live organism which provides benefit to the host when consumed in adequate quantities." The *E. coli* strain Nissle

TABLE 9.1 Summary of Major Diseases Correlated With Alterations in Human-Associated Microbiota

S. No.	Disease Name	Altered Microbiota	Anatomical Site	References
1	Periodontal plaque	Bacteroidetes ↑ *Porphyromonas gingivalis* ↑ Actinobacteria ↓ *Treponema denticola* ↓	Oral cavity	[32,33]
2	Scalp-associated dandruff	*Staphylococcus epidermidis* ↑ *Malassezia restricta* ↑ *Propeonibacterium acnes* ↓	Skin	[35]
3	Inflammatory bowel disease	Proteobacteria ↑ Actinobacteria ↑ *Enterococcus faecum* ↑	Gut	[41]
4	Obesity	Bacteroidetes ↓ Firmicutes ↑	Gut	[44]
5	Type2 diabetes mellitus	Proteobacteria ↑ *Lactobacillus* spp. ↑ Bifidobacterium ↓	Gut	[55]
6	Cardiovascular disease	*Collinsella* ↑	Gut	[72]
7	Rheumatoid arthritis	*Prevotella copri* ↑ Bacteroidetes ↓	Gut	[74]
8	Liver cirrhosis	*Vellonella* ↑	Gut	[75]
9	Colon cancer	Bacteroidetes ↑	Gut	[76]
10	Kidney stones	*Oxalobacter formigenes* ↓	Gut	[77]

1917, and a number of *Bifidobacterium* and *Lactobacillus* strains represent the efficacy of probiotic agents in the management of various diseases such as IBD, diarrhea, atopic dermatitis, and allergic diseases [81]. The beneficial effects of probiotics have been related to their capacity to produce vitamins, antioxidants, and defensins against microbial pathogens. Probiotics are also known to produce SCFAs and are characterized by the absence in toxins [45]. Many probiotic bacteria inhibit the growth of pathogens by various mechanisms. For example, Gram-negative *Bifidobacteria* and *Lactobacilli*, being devoid of LPS, reduce the risk of infection by competing with pathogens for dietary nutrients and receptors on the gut wall. *Bifidobacterium longum* and *Bifidobacterium breve* have been known for the treatment and prevention of acute diarrhea in infants [81]. A very recent finding suggests that *Lactobacillus helveticus R0052* and *B. longum R0175*, when taken in combination, display anxiolytic-like activity in mice and beneficial physiological effects in human volunteers, thus posing to be a promising psychotropic-like probiotic agent in the coming time [82]. Other than probiotics, the concept of fecal transplantation also represents a potential therapy against many diseases including anorexia nervosa, autoimmunity infections, insulin resistance, IBD, obesity, and multiple sclerosis. Among the bacterial community, *Clostridium difficile* has shown many significant findings as a source of fecal transplantation [81].

CONCLUSION

Combinations of metagenomic approaches supported by NGS technologies appear very promising in understanding the role of human microbiome in human health and implications in diseases. Because a majority of human-associated bacteria are unculturable, the metagenomic approach helps in accessing the genetic information inherent in the microflora and to identify the key microbial players associated with different microbiome-related diseases. More efforts are needed to validate the role of microbes in the implicated diseases, such as investigation in larger cohorts or experiments in animal models, to establish a strong link between human-associated microbiome and diseases. This knowledge will be helpful primarily for the development of biomarkers for disease diagnosis and application of prebiotics, probiotics, and personalized medicine for chronic diseases. In summary, metagenomics has an unprecedented potential to revolutionize the field of health and medicine by providing novel insights into our second genome, "the human microbiome."

GLOSSARY

Amplicon Polymerase chain reaction–amplified product from genomic DNA.

Atheromas Degenerative accumulation of lipid-containing plaques on the innermost layer of an artery wall.

Atherosclerosis A condition of narrowed and hardened arteries due to the development atheromas.

Clade A taxonomic group of organisms that includes a common ancestor and all the descendants of that ancestor.

Gnotobiotic mice Mice free of germs or associated with few specified germs.

Human Microbiome Project A US National Institutes of Health initiative that aims to characterize microbial communities found at multiple human body sites and to look for correlations between changes in the microbiome and human health.

Hyperoxaluria State in which there is excessive urinary excretion of oxalate.

Metagenome Collection of genomes and genes from the members of a microbiota.

Metagenomics Study of genomes of a microbial community taken directly from an environment.

Microbiome Group of microbial genomes present in an ecosystem.

Microbiota/microflora Microorganisms inhabiting in an environment.

Next generation sequencing High-throughput, non-Sanger–based sequencing technology.

Periodontitis A gum infection damaging the tissues and bones that supports a teeth.

Phylotype Observed similarity that classifies a group of organisms by their phenetic relationship.

Plaque Biofilm of bacteria growing on tooth surface.

Prebiotics Nondigestible fiber compounds that act as a substrate for advantageous bacteria colonizing the gut and stimulate them.

Probiotics Live microorganisms that provide health benefits on consumption.

LIST OF ACRONYMS AND ABBREVIATIONS

CVD Cardiovascular disease
HGP Human Genome Project
IBD Inflammatory Bowel Disease
MDD Major Depressive Disorder
NGS Next Generation Sequencing
SCFAs Short Chain Fatty Acids

REFERENCES

[1] Finishing the euchromatic sequence of the human genome. Nature 2004;431(7011):931–45.

[2] Gilbert JA, et al. Microbial metagenomics: beyond the genome. Annu Rev Mar Sci 2011;3:347–71.

[3] Desai N, et al. From genomics to metagenomics. Curr Opin Biotechnol 2012;23(1):72–6.

[4] Tringe SG, et al. Metagenomics: DNA sequencing of environmental samples. Nat Rev Genet 2005;6(11):805–14.

[5] Chaudhary N, et al. 16S classifier: a tool for fast and accurate taxonomic classification of *16S rRNA* hypervariable regions in metagenomic datasets. PLoS One 2015;10(2):e0116106.

[6] Sharma VK, et al. Fast and accurate taxonomic assignments of metagenomic sequences using MetaBin. PLoS One 2012;7(4): e34030.

[7] Quast C, et al. The SILVA ribosomal RNA gene database project: improved data processing and web-based tools. Nucleic Acids Res 2013;41(Database issue):D590–6.

[8] Overbeek R, et al. The subsystems approach to genome annotation and its use in the project to annotate 1000 genomes. Nucleic Acids Res 2005;33(17):5691–702.

[9] Sharma VK, et al. MetaBioME: a database to explore commercially useful enzymes in metagenomic datasets. Nucleic Acids Res 2010;38(Database issue):D468–72.

[10] Gupta A, et al. MP3: a software tool for the prediction of pathogenic proteins in genomic and metagenomic data. PLoS One 2014;9(4):e93907.

[11] Caporaso JG, et al. QIIME allows analysis of high-throughput community sequencing data. Nat Methods 2010;7(5):335–6.

[12] Meyer F, et al. The metagenomics RAST server: a public resource for the automatic phylogenetic and functional analysis of metagenomes. BMC Bioinforma 2008;9:386.

[13] Tyson GW, et al. Community structure and metabolism through reconstruction of microbial genomes from the environment. Nature 2004;428(6978):37–43.

[14] Venter JC, et al. Environmental genome shotgun sequencing of the Sargasso Sea. Science 2004;304(5667):66–74.

[15] Gilbert JA, et al. The Earth Microbiome project: successes and aspirations. BMC Biol 2014;12:69.

[15a] Kurokawa K, Itoh T, Kuwahara T, Oshima K, Toh H, Toyoda A, et al. Comparative metagenomics revealed commonly enriched gene sets in human gut microbiomes. DNA Res 2007;14(4):169–81.

[16] Peterson J, et al. The NIH Human Microbiome project. Genome Res 2009;19(12):2317–23.

[17] Koepfli KP, et al. The Genome 10K project: a way forward. Annu Rev Animal Biosci 2015;3:57–111.

[18] Tlaskalova-Hogenova H, et al. The role of gut microbiota (commensal bacteria) and the mucosal barrier in the pathogenesis of inflammatory and autoimmune diseases and cancer: contribution of germ-free and gnotobiotic animal models of human diseases. Cell Mol Immunol 2011;8(2):110–20.

[19] Fodor AA, et al. The "most wanted" taxa from the human microbiome for whole genome sequencing. PLoS One 2012;7(7):e41294.

[20] Qin J, et al. A human gut microbial gene catalogue established by metagenomic sequencing. Nature 2010;464(7285):59–65.

[21] Cho I, et al. The human microbiome: at the interface of health and disease. Nat Rev Genet 2012;13(4):260–70.

[22] Penders J, et al. New insights into the hygiene hypothesis in allergic diseases: mediation of sibling and birth mode effects by the gut microbiota. Gut Microbes 2014;5(2):239–44.

[23] Gerber GK. The dynamic microbiome. FEBS Lett 2014;588(22): 4131–9.

[24] Turnbaugh PJ, et al. The effect of diet on the human gut microbiome: a metagenomic analysis in humanized gnotobiotic mice. Sci Transl Med 2009;1(6). 6ra14.

[25] David LA, et al. Diet rapidly and reproducibly alters the human gut microbiome. Nature 2014;505(7484):559–63.

[26] Yatsunenko T, et al. Human gut microbiome viewed across age and geography. Nature 2012;486(7402):222–7.

[27] Arumugam M, et al. Enterotypes of the human gut microbiome. Nature 2011;473(7346):174–80.

[28] Grice EA, et al. The skin microbiome. Nat Rev Microbiol 2011;9(4):244–53.

[29] Structure, function and diversity of the healthy human microbiome. Nature 2012;486(7402):207–14.

[30] Chen T, et al. The Human Oral Microbiome Database: a web accessible resource for investigating oral microbe taxonomic and genomic information. Database 2010;2010:baq013. J Biol Database Cur.

[31] Costello EK, et al. Bacterial community variation in human body habitats across space and time. Science 2009;326(5960):1694–7.

[32] Wang J, et al. Metagenomic sequencing reveals microbiota and its functional potential associated with periodontal disease. Sci Rep 2013;3:1843.

[33] Zambon JJ, et al. Black-pigmented *Bacteroides* spp. in the human oral cavity. Infect Immun 1981;32(1):198–203.

[34] Bruggemann H, et al. The complete genome sequence of *Propionibacterium acnes,* a commensal of human skin. Science 2004;305(5684):671–3.

[35] Clavaud C, et al. Dandruff is associated with disequilibrium in the proportion of the major bacterial and fungal populations colonizing the scalp. PLoS One 2013;8(3):e58203.

[36] Ursell LK, et al. The intestinal metabolome: an intersection between microbiota and host. Gastroenterology 2014;146(6):1470–6.

[37] Andersson AF, et al. Comparative analysis of human gut microbiota by barcoded pyrosequencing. PLoS One 2008;3(7):e2836.

[38] Clemente JC, et al. The impact of the gut microbiota on human health: an integrative view. Cell 2012;148(6):1258–70.

[39] Li J, Feng Q. Analysis of gut microbiome and diet modification in patients with Crohn's disease. SOJ Microbiol Infect Dis 2014;2(3):1–4.

[40] Franchimont D, et al. Deficient host–bacteria interactions in inflammatory bowel disease? The Toll-like receptor *(TLR)-4* Asp299gly polymorphism is associated with Crohn's disease and ulcerative colitis. Gut 2004;53(7):987–92.

[41] Lepage P, et al. Twin study indicates loss of interaction between microbiota and mucosa of patients with ulcerative colitis. Gastroenterology 2011;141(1):227–36.

[42] Yoshimoto S, et al. Obesity-induced gut microbial metabolite promotes liver cancer through senescence secretome. Nature 2013;499(7456):97–101.

[43] Ley RE, et al. Obesity alters gut microbial ecology. Proc Natl Acad Sci USA 2005;102(31):11070–5.

[44] Ley RE, et al. Microbial ecology: human gut microbes associated with obesity. Nature 2006;444(7122):1022–3.

[45] Delzenne NM, et al. Interaction between obesity and the gut microbiota: relevance in nutrition. Annu Rev Nutr 2011;31:15–31.

[46] Turnbaugh PJ, et al. A core gut microbiome in obese and lean twins. Nature 2009;457(7228):480–4.

[47] Santacruz A, et al. Interplay between weight loss and gut microbiota composition in overweight adolescents. Obesity 2009;17(10):1906–15.

[48] Erejuwa OO, et al. Modulation of gut microbiota in the management of metabolic disorders: the prospects and challenges. Int J Mol Sci 2014;15(3):4158–88.

[49] Serino M, et al. Intestinal microflora and metabolic diseases. Diabetes Metab 2009;35(4):262–72.

[50] Brownlee IA, et al. Action of reactive oxygen species on colonic mucus secretions. Free Radic Biol Med 2007;43(5):800–8.

[51] Hall DM, et al. Mechanisms of circulatory and intestinal barrier dysfunction during whole body hyperthermia. Am J Physiol Heart Circ Physiol 2001;280(2):H509–21.

[52] Muccioli GG, et al. The endocannabinoid system links gut microbiota to adipogenesis. Mol Syst Biol 2010;6(1).

[53] Cani PD, et al. Metabolic endotoxemia initiates obesity and insulin resistance. Diabetes 2007;56(7):1761–72.

[53a] Remely M, Dworzak S, Hippe B, Zwielehner J, Aumüller E, Brath H, et al. Abundance and diversity of microbiota in type 2 diabetes and obesity. J Diabetes Metab 2013;4(253):2.

[53b] Larsen N, Vogensen FK, van den Berg FW, Nielsen DS, Andreasen AS, Pedersen BK, et al. Gut microbiota in human adults with type 2 diabetes differs from non-diabetic adults. PLoS One 2010;5(2):e9085.

[53c] Gomes AC, Bueno AA, de Souza R, Mota JF. Gut microbiota, probiotics and diabetes. Nutr J 2014;13:60.

[54] Tilg H, et al. Microbiota and diabetes: an evolving relationship. Gut 2014;63(9):1513–21. http://dx.doi.org/10.1136/gutjnl-2014-306928.

[55] Furet J-P, et al. Differential adaptation of human gut microbiota to bariatric surgery–induced weight loss links with metabolic and low-grade inflammation markers. Diabetes 2010;59(12):3049–57.

[56] Mehta NN, et al. Experimental endotoxemia induces adipose inflammation and insulin resistance in humans. Diabetes 2010;59(1):172–81.

[57] Creely SJ, et al. Lipopolysaccharide activates an innate immune system response in human adipose tissue in obesity and type 2 diabetes. Am J Physiol Endocrinol Metab 2007;292(3):E740–7.

[58] Neal MD, et al. Enterocyte *TLR4* mediates phagocytosis and translocation of bacteria across the intestinal barrier. J Immunol 2006;176(5):3070–9.

[59] Spruss A, et al. Toll-like receptor 4 is involved in the development of fructose-induced hepatic steatosis in mice. Hepatology 2009;50(4):1094–104.

[60] Shinozaki S, et al. Liver-specific inducible nitric-oxide synthase expression is sufficient to cause hepatic insulin resistance and mild hyperglycemia in mice. J Biol Chem 2011;286(40):34959–75.

[61] Kim JJ, et al. *TLR4* and insulin resistance. Gastroenterol Res Pract 2010;2010.

[62] Cabreiro F, et al. Metformin retards aging in *C. elegans* by altering microbial folate and methionine metabolism. Cell 2013;153(1):228–39.

[63] Labermaier C, et al. Biomarkers predicting antidepressant treatment response: how can we advance the field? Dis Markers 2013;35(1):23–31.

[64] Cryan JF, et al. Mind-altering microorganisms: the impact of the gut microbiota on brain and behaviour. Nat Rev Neurosci 2012;13(10):701–12.

[65] Desbonnet L, et al. The probiotic *Bifidobacteria infantis*: an assessment of potential antidepressant properties in the rat. J Psychiatr Res 2008;43(2):164–74.

[66] Bravo JA, et al. Ingestion of *Lactobacillus* strain regulates emotional behavior and central GABA receptor expression in a mouse via the vagus nerve. Proc Natl Acad Sci USA 2011;108(38):16050–5.

[67] Cryan JF, et al. The microbiome-gut-brain axis: from bowel to behavior. Neurogastroenterology and motility, 2011;23(3):187–92.

[68] Lyte M. Probiotics function mechanistically as delivery vehicles for neuroactive compounds: microbial endocrinology in the design and use of probiotics. BioEssays: News and Reviews in Molecular, Cellular and Developmental Biology, 2011;33(8):574–81.

[69] Mattila KJ, et al. Dental infections and cardiovascular diseases: a review. J Periodontol 2005;76(Suppl. 11):2085–8.

[70] Ott SJ, et al. Detection of diverse bacterial signatures in atherosclerotic lesions of patients with coronary heart disease. Circulation 2006;113(7):929–37.

[71] Backhed F. 99th Dahlem conference on infection, inflammation and chronic inflammatory disorders: the normal gut microbiota in health and disease. Clin Exp Immunol 2010;160(1):80–4.

[72] Karlsson FH, et al. Symptomatic atherosclerosis is associated with an altered gut metagenome. Nat Commun 2012;3:1245.

[73] Wang Z, et al. Gut flora metabolism of phosphatidylcholine promotes cardiovascular disease. Nature 2011;472(7341):57–63.

[74] Scher JU, et al. Expansion of intestinal *Prevotella copri* correlates with enhanced susceptibility to arthritis. eLife 2013;2:e01202.

[75] Qin N, et al. Alterations of the human gut microbiome in liver cirrhosis. Nature 2014;513(7516):59–64.

[76] Brim H, et al. Microbiome analysis of stool samples from African Americans with colon polyps. PloS One 2013;8(12):e81352.

[77] Carmody RN, et al. Host–microbial interactions in the metabolism of therapeutic and diet-derived xenobiotics. J Clin Invest 2014;124(10):4173–81.

[78] Miller AW, et al. The metabolic and ecological interactions of oxalate-degrading bacteria in the mammalian gut. Pathogens 2013;2(4):636–52.

[79] Gill HS, et al. Probiotics and human health: a clinical perspective. Postgrad Med J 2004;80(947):516–26.

[80] Kirpich IA, et al. Probiotics restore bowel flora and improve liver enzymes in human alcohol-induced liver injury: a pilot study. Alcohol 2008;42(8):675–82.

[81] Lin CS, et al. Impact of the gut microbiota, prebiotics, and probiotics on human health and disease. Biomed J 2014;37(5):259–68.

[82] Messaoudi M, et al. Assessment of psychotropic-like properties of a probiotic formulation (*Lactobacillus helveticus R0052* and *Bifidobacterium longum R0175*) in rats and human subjects. Br J Nutr 2011;105(5):755–64.

Chapter 10

Pharmacogenetics and Pharmacogenomics

A. Alfirevic, M. Pirmohamed

University of Liverpool, Liverpool, United Kingdom

Chapter Outline

Introduction	**121**	Oncology	128
Pharmacogenomic Information in Drug Labeling	**122**	Infection and Inflammation	130
Pharmacogenetic Guidelines	124	Psychiatry	133
Pharmacogenetic Study Design	**124**	**Conclusions**	**133**
Clinical Areas	126	**References**	**134**
Cardiology	126		

INTRODUCTION

Pharmacogenetics is the study of the genetic factors determining variability in response to drugs in terms of both efficacy and toxicity. More recently, the term *pharmacogenomics* has been coined, which refers to the wider genome, but has also been used in the field of new drug development to identify new drug targets in the human genome. However, the two terms, *pharmacogenetics* and *pharmacogenomics*, are often used interchangeably (Fig. 10.1). Vogel first used the term *pharmacogenetics* in 1959 [1], but the concept of "chemical individuality" was introduced at the start of the 20th century by Sir Archibald Garrod [2].

The aim of pharmacogenomic research is to identify robust genetic predictors of drug response that can be used in clinical practice to identify patients at risk of developing adverse drug reactions (ADRs), those who may not benefit from medicines, and those who may require alternative drug treatment. The ultimate goal is to tailor medicines to individuals or groups of patients who will derive maximum benefit from the drug, and will have a reduced risk of drug toxicity, thereby maximizing the benefit-risk ratio of drugs. This concept has evolved into *personalized medicine*, an emerging practice of medicine that uses an individual's genetic profile (or other nongenetic predictors) to guide decisions made with regard to the prevention, diagnosis and treatment of disease (https://ghr.nlm.nih.gov/primer/genomicresearch/pharmacogenomics; see chapter: Cancer Genetics and Genomics). Given that tailoring drugs to individual patients is often not practical or prohibitively

expensive, two similar terms have been used to describe similar clinical approaches: *stratified medicine*, an innovative treatment concept based on the use of genetic or other molecular information to select the best therapeutic strategy in order to improve health outcomes, such as effectiveness and safety, for a targeted group of patients sharing similar biological characteristics (ABPI, 2014; http://www.abpi.org.uk/our-work/library/medical-disease/Documents/strat_med.pdf) and *precision medicine*, an approach to medical treatment that combines advanced research and technologies with detailed information about the individual patient's condition in order to improve outcomes (US National Research Council, 2011; https://www.aamc.org/advocacy/washhigh/highlights2011/266622/nationalresearchcouncilmakesrecommendationsforprecisionmedicine.html). In essence, with all these terms the idea is to progress from our current paradigm of prescribing drugs on the basis of population level data and hoping for a favorable response, to one where drug choice is dependent on a diagnostic test that predicts the response with a more favorable outcome for the patient.

There have been many studies of genetic factors determining drug response, but most have either yielded negative results or positive results that could not be replicated in subsequent studies. There are, however, several important findings of genetic factors in diverse clinical areas, which have improved our knowledge of the mechanisms of drug action, including toxicity, for which pretreatment testing is recommended. In this chapter, we focus on areas that have already changed, or may in the future change, clinical practice.

Medical and Health Genomics. http://dx.doi.org/10.1016/B978-0-12-420196-5.00010-1

FIGURE 10.1 **Number of publications over the last 25 years that have used different terms for improving drug response in patients.** A sharp increase in the number of publications using the terms *pharmacogenetics* and *pharmacogenomics* occurred around the time of the completion of the human genome sequencing project in 2001 [90], with a decrease seen in 2014. Recently, the greatest increase in the number of publications has been seen with the term *personalized medicine*. The diagram was constructed using the number of hits in PubMed for each term over from 1990 to 2014.

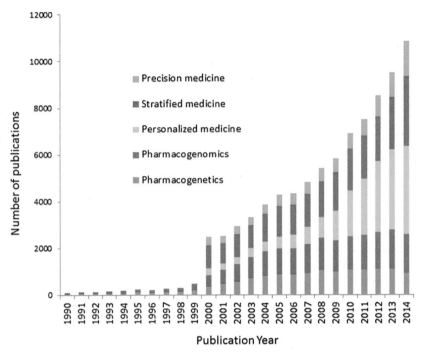

PHARMACOGENOMIC INFORMATION IN DRUG LABELING

In the United States, the US Food and Drug Administration (FDA) website (http://www.fda.gov/drugs/scienceresearch/researchareas/pharmacogenetics/ucm083378.htm) regularly updates information on genetic/genomic biomarkers associated with variability in clinical response to drugs that have been included in labels, and issues warnings about the increased risk of adverse events and variable dose requirements in individuals with specific genotypes. In Europe and Japan, drug labeling is coordinated by the European Medicines Agency (EMA) and the Pharmaceuticals and Medical Devices Agency, respectively.

Annotation of drug labels containing pharmacogenetic information approved by these drug regulatory agencies is undertaken by the Pharmacogenomics Knowledge Implementation Group (PharmGKB) (http://www.pharmgkb.org/view/drug-labels.do). In PharmGKB, there are four levels of pharmacogenomic drug labeling include:

- Drug labels for which genetic testing is *required* (labels say testing "should be performed").
- Drug labels for which genetic testing is *recommended* (labels say testing "should be considered"). The label states or implies that some sort of gene, protein, or chromosomal testing, including genetic testing, functional protein assays, or cytogenetic studies, should be conducted before using the drug. This requirement may only be for a particular subset of patients.
- *Actionable* pharmacogenetic information relates to the label that does not discuss genetic or other testing for gene/protein/chromosomal variants, but does contain information about changes in efficacy, dosage, or toxicity because of such variants. The label may mention contraindication to a drug in a particular subset of patients but does not require or recommend gene, protein, or chromosomal testing.
- Similarly, the *informative* pharmacogenetic label mentions a gene or protein that is involved in the metabolism or pharmacodynamics of the drug, but there is no information to suggest that variation in these genes/proteins leads to different responses.

In Table 10.1, we have included a list of drugs for which pharmacogenetic testing is required or recommended before the drug can be administered.

There are substantial differences in the pharmacogenomic information included in package inserts from the three regulatory agencies. The differences may be attributed to the type and purpose of the biomarkers and the strength of the evidence that supports the use of the biomarkers [3]. The evidence supporting pharmacogenetic biomarker testing has been investigated [4]. Wang and colleagues evaluated 119 FDA drug labels that described the use of a biomarker and they found that 43 (36.1%) provided clinical validity evidence (ie, the ability to predict phenotype), whereas only 18 (15.1%) provided convincing evidence of clinical utility (ie, the ability to improve clinical outcomes). Although recommendations were made on how to base clinical decisions on the results of a biomarker test in 61 (51.3%) drug labels, only 36 (30.3%) of those contained convincing clinical utility data. Biomarkers for cancer drugs were most likely to demonstrate convincing clinical utility evidence (37.8%), whereas neuropsychiatry biomarkers were the least likely (0%) [4]. They concluded that testing recommendations should not be based on clinical validity alone, because they do not add clarity to the clinical decision-making process.

TABLE 10.1 List of Drugs for Which Genetic Testing is Required and/or Recommended by at Least One of the Three Regulatory Agencies, the US Food and Drug Administration, the European Medicines Agency, and the Pharmaceuticals and Medical Devices Agency

Drug	Therapeutic Area	Test	FDA	EMA	PMDA
Required					
Abacavir	Infectious diseases (AIDS)	*HLA-B*57:01* carrier	Rec	Req	Inf
Afatinib	Oncology	*EGFR* positive	Req	Req	
Anastrozole	Oncology	*ESR1, PGR* positive	Req		
Arsenic trioxide	Oncology	*PML/RARA* positive	Req	Req	Req
Bosutinib	Oncology	*BCR/ABL1* positive	Req	Req	
Brentuximab vedotin	Oncology	*TNFRSF8* (CD30 positive)	Inf	Req	
Carbamazepine	Neurology	*HLA-B*15:02* carriers, *HLA-A*31:01* carriers	Req		Inf
Carglumic acid	Metabolic disorders	*NAGS* deficient	Req	Req	
Cetuximab	Oncology	*EGFR* positive, *KRAS* negative	Req	Req	Req
Crizotinib	Oncology	*ALK* positive	Req	Req	Req
Dabrafenib	Oncology	*BRAF, G6PD* positive	Act	Req	
Dasatinib	Oncology	*BRC/ABL1*	Req	Req	Req
Denileukin diftitox	Oncology	*IL2RA* (CD25 positive)	Req		
Eliglustat	Genetics, hematology	*CYP2D6*	Req		
Erlotinib	Oncology	*EGFR* positive	Req	Req	
Ethinyl estradiol	Neurology	*CYP2C19* poor metabolizers	Inf	Req	
Everolimus	Oncology	*ERBB2* (Her2 negative), *ESR1* positive	Req	Act	
Exemestane	Oncology	*ESR1* positive	Req		
Fulvestrant	Oncology	*ESR1* positive	Req	Req	Req
Gefitinib	Oncology	*EGFR* positive	Inf	Req	Req
Ibritumomab	Oncology	*MS4A1* (CD20 positive)	Inf	Req	
Imatinib	Oncology	*KIT* negative, *BCR/ABL1* positive, *PDGFRB* positive, *FIP1L/PDGFRA* positive	Req	Req	Req
Ivacaftor	Pulmonary	*CFTR* mutation carriers	Req	Req	
Lapatinib	Oncology	*ERBB2* (Her2 positive)	Req	Req	Req
Letrozole	Oncology	*ESR1, PGR* positive	Req		
Lomitapide	Endocrinology	*LDLR*	Inf	Req	
Maraviroc	Infectious diseases (AIDS)	*CCR5* positive	Req	Req	
Nilotinib	Oncology	*BCR/ABL1* positive, *UGT1A1*28* homozygous	Req	Req	
Norelgestromin	Gynecology	Factor V Leiden (rs6025)		Req	
Panitumumab	Oncology	*EGFR* positive, *KRAS* negative	Req	Req	Req
Pegloticase	Rheumatology	*G6PD* deficient	Act	Req	
Pertuzumab	Oncology	*ERBB2* (Her2 positive)	Req	Req	

Continued

TABLE 10.1 List of Drugs for Which Genetic Testing is Required and/or Recommended by at Least One of the Three Regulatory Agencies, the US Food and Drug Administration, the European Medicines Agency, and the Pharmaceuticals and Medical Devices Agency—cont'd

Drug	Therapeutic Area	Test	FDA	EMA	PMDA
Pimozide	Psychiatry	CYP2D6 poor metabolizers	Req		
Ponatinib	Oncology	BCR-ABL T315I	Act	Req	
Rasburicase	Oncology	G6PD deficient	Req	Act	Req
Rituximab	Oncology	MS4A1 (CD20 positive)	Inf	Req	
Tetrabenazine	Neurology	CYP2D6 poor metabolizers	Req		
Trametinib	Oncology	BRAF positive	Req		
Trastuzumab	Oncology	ERBB2 (Her2 positive)	Req	Req	Req
Tretinoin	Oncology	PML/RARA positive	Req		Inf
Valproic acid	Neurology	POLG positive, NAGS, CPS1, ASS1, OTC, ASL, ABL2 deficient	Req		Act
Velaglucerase alfa	Genetics, Haematology	GBA	Req	Req	
Vemurafenib	Oncology	BRAF positive	Req	Req	
Recommended					
Atazanavir	Infectious diseases (AIDS)	CYP2C19 if voriconazole treatment required		Rec	
Azathioprine	Rheumatology	TPMT poor metabolizers	Rec		Act
Clopidogrel	Cardiology	CYP2C19 poor metabolizers	Rec	Act	Act
Dextromethorphan	Neurology	CYP2D6 poor metabolizers	Rec	Act	
Mercaptopurine	Oncology	TPMT poor metabolizers	Rec	Act	
Quinidine	Cardiology	CYP2D6 poor metabolizers	Rec	Act	
Vandetanib	Oncology	RET		Rec	

Act, Actionable pharmacogenetics; *EMA*, European Medicines Agency; *FDA*, US Food and Drug Administration; *Inf*, informative pharmacogenetics; *PMDA*, Pharmaceuticals and Medical Devices Agency; *Rec*, genetic testing recommended; *Req*, genetic testing required.

Pharmacogenetic Guidelines

A large number of published studies have highlighted the potential of genetic testing to improve the use of medications. But despite evidence of clinical validity in some cases, the slow adoption of pharmacogenetics into clinical practice has been obstructing implementation even though the cost of tests is decreasing and the availability has been increasing. In order to address the slow adoption, various organizations have produced guidelines. For instance, the Dutch Pharmacogenetics Working Group established in 2005 by the Royal Dutch Pharmacist's Association is multidisciplinary and includes clinical pharmacists, physicians, clinical pharmacologists, clinical chemists, epidemiologists, and toxicologists [5]. In addition, the Clinical Pharmacogenetics Implementation Consortium was established in 2009 and consists of members from the National Institutes of Health's Pharmacogenomics Research Network (http://www.pgrn.org) and PharmGKB (http://www.pharmgkb.org), and experts in specific areas of pharmacogenetics/genomics and laboratory medicine [6]. The consortium has already published guidance for several classes of drugs that may be helpful for clinicians and laboratories (Table 10.2). The guidelines contain a systematic analysis of how genetic variants are associated with drug response phenotypes, their severity, and the availability of alternative treatments and laboratory tests.

PHARMACOGENETIC STUDY DESIGN

In the past, most studies conducted in pharmacogenetics were case-control association studies, which often

TABLE 10.2 Clinical Pharmacogenetics Implementation Consortium Guidelines Were Published for Several Classes of Drugs, for Which Strong Evidence for Clinical Utility of Pharmacogenetic Testing Exists

Drug/Drug Class	Therapeutic Area	Genetic Test	References
Warfarin	Cardiology	CYP2C9 and VKORC1	Johnson et al. [22]
Codeine	Neurology	CYP2D6	Crews et al. [91,92]
Simvastatin	Cardiology	SLCO1B1	Wilke et al. [93,94]
Thiopurine	Oncology, inflammation	TPMT	Relling et al. [95,96]
Carbamazepine	Neurology	HLA-B*15:02	Leckband et al. [97]
Peginterferon alpha-2b	Infection	IFNL3 (IL28B)	Muir et al. [98]
Abacavir	Infection	HLA-B*57:01	McLaren et al. [99], Martin et al. [100]
Ivacaftor	Pulmonary	CFTR	Clancy et al. [101]
Phenytoin	Neurology	CYP2C9 and HLA-B*15:02	Caudle et al. [102]
Rasburicase	Oncology	G6PD	Relling et al. [55]
Fluoropyrimidine	Oncology	DPDH	Caudle et al. [103]
Clopidogrel	Cardiology	CYP2C19	Scott et al. [104,105]
Tricyclic antidepressants	Psychiatry	CYP2D6 and CYP2C19	Hicks et al. [106]
Allopurinol	Rheumatology	HLA-B*58:01	Hershfield et al. [107], Saito et al. [108]

investigated several variants in single genes. These studies were hypothesis driven and required a priori knowledge of a gene/variant function. They largely focused on drug metabolizing enzymes and drug transporters and occasionally on the drug targets. This has led to many important discoveries that have been replicated; however, the candidate gene approach suffers from the fact that a priori knowledge of how a drug works or how it causes an adverse reaction is needed, but this is lacking in most cases. This has therefore also led to many false-positive and probably false-negative associations.

With the rapid progress in genotyping technology, hypothesis-free approaches have become possible, and single nucleotide polymorphisms (SNPs) and copy number variants can now be genotyped across the whole genome in genome-wide association studies (GWASs). A catalog of all published GWASs exists (http://www.genome.gov/page.cfm?pageID=26525384&start=2451&clearquery=1#result_table), but since 2005 when the first GWAS study on macular degeneration was published [7], only approximately 10% of all studies (2100 in January 2015) in the catalog relate to pharmacogenomics. Using GWAS approaches has been particularly successful in detecting novel associations for serious idiosyncratic ADRs including drug induced skin, muscle, and liver injury (DILI) [8–12].

GWAS depends on the availability of comprehensive data on variability of the human genome and mapping of human haplotypes in the HapMap project (http://hapmap.

ncbi.nlm.nih.gov/), where common variants with minor allele (incidence of >5%) are well characterized. Given that most genome-wide studies simultaneously interrogate more than 1 million genetic variants, for complex diseases it has become the norm to include a large number of cases and controls to ensure that the study has statistical power to achieve genome-wide significance. In addition, any hits found on GWAS need to be validated in an independent replication cohort. However, large numbers and replication cohorts are not always achievable in pharmacogenomics studies, particularly where the phenotype is rare. For example, ADRs such as drug hypersensitivity or liver injury occur in approximately 1 in 10,000 to 1 in 100,000 individuals exposed to the culprit drug. Fortunately, for some of these phenotypes, especially the immune-mediated ADRs, the genetic effect size is much greater than that for complex disease, which has meant that genome-wide significance has sometimes been achieved with as few as 23 patients. The scientific community in this area has also started forming large international consortia, which can then collect DNA from well-phenotyped patients throughout the world. Examples of such consortia include EUDRAGENE (European collaboration to establish a case-control DNA collection for studying the genetic basis of adverse drug reactions), International Drug-Induced Liver Injury Consortium, and ITCH (International Hypersensitivity/Skin Injury Network). An alternative, although complementary, approach in the absence of a replication cohort is to undertake functional studies to provide a biological underpinning to the genetic association identified.

In the future, electronic health records may also provide a solution. One such recent effort, eMERGE-PGx, is very promising as it involves not only genomic and pharmacogenomic research networks but also a partnership with electronic medical records [13]. The aims of the eMERGE-PGx project are to: (1) assess sequence variation in a set of 84 proposed pharmacogenes in approximately 9000 patients, (2) integrate well-established and clinically validated pharmacogenomic variants into the electronic health records to facilitate implementation into clinical practice, and (3) develop a repository of pharmacogenomic variants with currently unknown significance that can be linked with an electronic health record–phenotype repository.

Clinical Areas

Cardiology

Cardiovascular disease is a leading cause of death worldwide. Although many pharmacological interventions are very successful in cardiology, there is significant interindividual variability in drug response. Pharmacogenetics is beginning to help in optimizing dosing and selection of cardiovascular drugs to maximize their benefit-risk ratio. Indeed, cardiovascular drugs have been used as paradigms for pharmacogenetic study design and application and, in particular, warfarin and clopidogrel have been extensively studied.

Oral Anticoagulants

Warfarin, a coumarin derivative, has been a standard of care for a wide variety of indications, including deep vein thrombosis, atrial fibrillation, and pulmonary embolism. It is administered as a racemic mixture of two enantiomers S-warfarin and R-warfarin, of which the S-enantiomer is approximately 3 times more potent. These two enantiomers are metabolized by different cytochrome P450 enzymes, *CYP2C9* being the most important gene responsible for the metabolism of S-warfarin [14]. Two variants, *CYP2C9*2* and *CYP2C9*3* have reduced enzyme activity compared with wild type *CYP2C9*1* allele, [15] and individuals who carry variant alleles require significantly lower dosages of warfarin. Warfarin acts by inhibiting vitamin K epoxide reductase subunit 1 in the liver, which is essential for the posttranslational γ-carboxylation of vitamin K–dependent clotting factors II (prothrombin), VII, IX, and X. The *VKORC1* gene is also polymorphic and approximately 25% of the variance in dose requirements is attributable to common haplotypes in *VKORC1* [16]. The prevalences in Japanese and African populations differ from those of white European populations, but the tendency that carriers of variant alleles are at increased risk of bleeding is the same

across different populations [17,18]. Warfarin dosaging is difficult because it has a narrow therapeutic index, and monitoring is usually performed to adjust the dosage based on the measurements of the International Normalized Ratio (INR), a standardized ratio of patient's prothrombin time vs. a normal control sample. Irrespective of drug monitoring, warfarin is one of the most common causes of emergency hospitalization for ADRs in the elderly [19]. In addition to genetic factors, dietary and environmental factors can lead to variability in warfarin response.

Recently, two multicenter randomized controlled clinical trials were conducted to assess the clinical utility of genotype-guided warfarin dosaging compared with the current standard practice. The two trials, named Clarification of Optimal Coagulation through Genetics (COAG) and European Pharmacogenetics of Anticoagulant Therapy (EU-PACT) had contradictory findings [20,21]. Whereas COAG collaborators concluded that genotype-guided dosaging of warfarin did not improve anticoagulation control during the first 4 weeks of therapy, the EU-PACT trial demonstrated that individuals who were prescribed warfarin based on genotypes had improved time in the therapeutic INR range than those with standard dosaging over the 3-month follow-up period. The different findings may be for various reasons, including differences in subject characteristics, algorithmic strategy, indications for warfarin use, ethnicity (in COAG, 27% of participants were of African origin, whereas in EU-PACT 98% were white), the use of a loading dose, genotype availability before the first dose, and duration of follow-up monitoring. COAG used a clinical algorithm and EU-PACT used usual standard dosaging, but this is unlikely to be the full reason for the different results, given that in the multitude of studies that have been conducted on warfarin pharmacogenetics, the contribution of clinical factors was significantly less than that of genetic factors. The conflicting data regarding benefits associated with warfarin pharmacogenetic testing have led to confusion. Further implementation of warfarin pharmacogenetics will probably need to be country-specific, taking into account the pertaining clinical practice in that country. It is likely that in the United States, at least, ordering new warfarin pharmacogenomics test before therapy initiation is not going to be the norm. However, if genetic information is already available, and this will become more common as exome and whole genome sequencing become affordable, it will be difficult to ignore the genetic data [22].

Another randomized clinical trial in the EU-PACT project investigated genotype-guided prescribing of *acenocoumarol* and *phenprocoumon*, which are preferred to warfarin in many European countries. The study design was similar to COAG and the results showed significantly higher time in the INR therapeutic range than the clinical algorithm at 4 weeks, but not at 12 weeks [23]. The data from the acenocoumarol and phenprocoumon arms were combined, as each of the arms did not have adequate power; given the major differences in the pharmacokinetics of the two drugs, whether this is scientifically valid is debatable.

The pharmacogenetics of newer anticoagulants such as the direct thrombin inhibitor dabigatran is only starting to emerge now. In a GWAS of nearly 3000 participants in the Randomized Evaluation of Long-term Anticoagulation Therapy study, polymorphisms in the *CES1* and *ABCB1* genes were associated with a lower risk of dabigatran-induced bleeding [24]. Dabigatran etexilate is an oral pro-drug that is rapidly converted by esterases *(CES1)* to its active form dabigatran. In addition, dabigatran etexilate, but not dabigatran is a substrate for the P-glycoprotein *(ABCB1)* efflux transporter.

Antiplatelet Agents

The thienopyridine *clopidogrel* is a prodrug that requires metabolism by *CYP2C19* to convert it into its active form, which inhibits adenosine diphosphate ADP–mediated platelet aggregation by binding to the platelet ADP recep-tor *P2Y12*. There is significant interindividual variability in response to clopidogrel; more than 20% of patients do not respond to clopidogrel, which increases their risk of cardiovascular events after procedures such as percutane-ous coronary intervention (PCI) [25]. Loss-of-function genetic variants in the *CYP2C19* gene (*CYP2C19*2, *3, *5*) have been associated with adverse clinical outcomes including myocardial infarction and stroke [26], whereas for *CYP2C19*17*, a gain-of-function allele has been associ-ated with an increased risk of bleeding [27]. These findings have been confirmed in a GWAS study where the effect of *CYP2C19* genotype was found to be associated with both the antiplatelet effect, measured using platelet activity assays, and clinical outcomes [28]. A recent prospective randomized proof-of-concept trial investigated point-of-care genetic testing for the *CYP2C19*2* variant in patients undergoing PCI [29]. Patients testing positive for this low-function allele were prescribed alternative antiplatelet ther-apy with prasugrel instead of clopidogrel. The trial showed that genotyping was feasible, could be done at the point-of-care, and had a short turnaround time. A recent analysis has suggested that genotype-guided personalization may improve the cost-effectiveness of therapy with the newer antiplatelet agents after PCI, with the most favorable result seen with ticagrelor [30].

Aspirin (acetylsalicylic acid) is an antiplatelet drug that reduces platelet aggregation by irreversibly inhibiting cyclooxygenase *(COX-1)*. Several studies have been con-ducted to determine the genetic contribution to variability in platelet aggregation and nonresponsiveness to aspirin treat-ment, which is termed *aspirin resistance*. Greater inhibition of platelet aggregation has been found in the carriers of the *COX-1* A842 allele [31], and the P1A1/A2 variant in the platelet glycoprotein *GPIIb/IIIa*, responsible for activation

of fibrinogen binding site and cross-linking of platelets and von Willebrand factor, has been associated with aspirin resistance [32]. Reports claim that between 2% and 50% of patients, depending on the platelet function test used, may be resistant to antiplatelet action of aspirin. Clinical aspirin resistance is defined as a rate of on-treatment myocardial infarction or other cardiovascular events. Although non-compliance with therapy, malabsorption, or concomitant medication can contribute to resistance, a group of genes (n = 60) included in the "aspirin response signature" appear to be highly reflective of platelet function [33]. Whether these can be utilized in a panel of genes/SNPs to assess responsiveness of individuals to aspirin treatment remains to be investigated in terms of both clinical utility and cost-effectiveness.

Lipid-Lowering Drugs

Statins are widely prescribed drugs effective for lowering low-density lipoprotein cholesterol (LDLC) and risk of cardiovascular disease. They inhibit 3-hydroxy-3-methyl-glutaryl coenzyme A reductase, the enzyme that catalyzes the rate-limiting step in cholesterol biosynthesis. However, the magnitude of statin-induced LDLC reduction and the individual risk of muscle toxicity are variable. Variability in response to statins has been attributed to clinical factors such as age, ancestry, concomitant medications, and gender. Genetic studies, including several GWASs, have identified that the extent of LDLC reduction is associated with poly-morphisms in the *LPA* and *APOE/TOMM40* loci [34]. In addition, in a GWAS published in 2008, [10] statin-induced myotoxicity was associated with loss-of-function genetic variants in the *SLCO1B1* gene, which encodes the solute carrier organic anion transporter family member 1B1. This finding has been replicated, and at least for simvastatin, this may lead to patient risk stratification according to genotype. Statin-induced myopathy and rhabdomyolysis are rare (3.4–11 per 100,000 patient-years), but given the large numbers of individuals taking statins worldwide, it is an important issue. Because pharmacokinetic pathways differ greatly for different statin types, more research is needed to elucidate all genetic factors that may contribute to efficacy and toxic-ity of different statins.

Antihypertensive Agents

Angiotensin-converting enzyme (ACE) inhibitors, which reduce the activity of the renin-angiotensin-aldosterone system, are first-line therapy for hypertension in younger patients (under the age of 55). Over 40 million people are treated with ACE inhibitors worldwide. These drugs are safe and effective in the majority of patients, but in some they can cause adverse reactions including dry cough (9% of patients) or, more rarely (0.1–0.7%), potentially life-threatening angioedema affecting the head and neck region. In the past, several studies have reported conflicting findings on the correlation between ACE plasma concentrations and genetic variants in the *ACE* and angiotensinogen (*AGT*) genes [35,36]. More recently, a pharmacogenetic score that combined three SNPs in the bradykinin type I (*BK1*) and angiotensin II type I receptor (*AT1*) genes accurately predicted patients with stable coronary artery disease who are most likely to benefit from the ACE inhibitor perindopril [37]. This approach is promising, but requires replication and careful consideration about generalizability. In terms of safety, a recent systematic review reported that one gene region *(XPNPEP2)* has been associated with ACE-inhibitor angioedema, but there were no associations with ACE-inhibitor cough [38]. In a functional assessment of the association, it was suggested that the angioedema was related to the lower plasma aminopeptide P activity in individuals who carry functional polymorphisms in the regulatory region of the *XPNPEP2* gene and a common *ATG* haplotype.

β-Blockers

β-Adrenoreceptor antagonists, which competitively bind to β1 adrenoreceptors, are indicated in the treatment of heart failure, hypertension, and angina pectoris. Pharmacogenetic variants in adrenergic system genes, in particular *ADRB1*, *ADRB2, ADRA2C*, and G protein-coupled receptor kinase 5 *(GRK5)* and 4 *(GRK4)*, have been extensively studied in relation to clinical outcomes and response to β-blockers. A lot of conflicting data exists on the role of these genes in β-blocker pharmacogenetics and the evidence is not yet sufficient to support clinical utility. Perhaps the most compelling evidence of treatment benefit with β-blockers has been seen in African American patients with heart failure who carry the *GRK5* Leu41 genotype. The presence of *GRK5* 41Leu was associated with decreased mortality in patients with heart failure and cardiac ischemia, and, in vitro, in transfected cells and transgenic mice, *GRK5* Leu41 uncoupled isoproterenol stimulated responses more effectively than *GRK5*-Gln41 [39]. Furthermore, the effect of atenolol in lowering systolic and diastolic blood pressure is reduced in patients with the *GRK4* 3SNP haplotype and is associated with an increased risk of cardiovascular events including death, myocardial infarction, and stroke [40].

Antiarrhythmics

Antiarrhythmic medications have been associated with severe ADRs, specifically with a prolongation in QT interval and drug-induced torsade de pointes (DITdP), as a consequence of perturbed cardiac repolarization through blockade of ion channels. To facilitate research into pharmacogenomic risk markers, the Phenotype Standardization Project sponsored by the International Serious Adverse Event Consortium, has recently established the minimum set of criteria for a DITdP case. It has been estimated that approximately 10% of DITdP

is predisposed to the same rare mutations (for example, in the *KCNE2* gene) that have been implicated in congenital long QT syndrome. A recent next generation sequencing study discovered rare variants in 22 congenital arrhythmia genes linked with DITdP [41]. No recommendations have been issued on the clinical application of these variants.

Oncology

In oncology, testing for pharmacogenomic markers is required for many drugs, and particularly for molecularly targeted anticancer therapy agents (see Table 10.1). Tests are based on both inherited (germline) and acquired (somatic) mutations to guide decision making. Germline DNA is obtained by peripheral blood samples and can help predict pharmacokinetic and pharmacodynamics responses. On the other hand, somatic DNA is obtained by tumor biopsy, is dependent on tumor tissue sampling site selection, and is useful in predicting pharmacodynamics and tumor response to treatment. A large effort to categorize somatic mutations, copy number variations, expression, and epigenetic modification within various tumor types has been undertaken by the Cancer Genome Atlas Network [42]. The latest comprehensive genomic characterization of 279 head and neck squamous cell carcinomas has identified therapeutic candidate alterations in most of the tumor types, which may accelerate progress in prevention and therapy across all tumors [43]. Herceptin, the poster child of efficacy pharmacogenetics, is a good example of how a prospective research strategy based on robust tumor genomic data linked to the target disease phenotype and codevelopment of drug-test companion diagnostics, can lead to fast implementation into clinical practice. The number of pharmacogenetic assays with robust evidence of drug–gene associations is continuously expanding and panels of genes are available. This is particularly important because many anticancer drugs have a narrow therapeutic index, low or unpredictable efficacy, and often severe systemic toxicity.

Germline DNA Mutations

Thiopurines One of the best characterized pharmacogenomic markers is the thiopurine-S-methyltransferase (*TPMT*) gene and treatment response to *6-mercaptopurine (6-MP)*. 6-MP is an important component of treatment for children with acute lymphoblastic leukemia, and its prodrug azathioprine is also used for nonmalignant indications. Several variants in the *TPMT* gene cause reduced function of the enzyme, leading to high levels of cytotoxic thioguanine nucleotides, which increase the risk of severe myelosuppression [44]. Depending on the ethnicity, one in 180 to one in 3700 individuals are homozygous for nonfunctional *TPMT* variants, and 3–14% of the world population is heterozygous. The diagnosis of *TPMT* deficiency is associated with a high risk of developing mercaptopurine toxicity when the drug is administered conventionally. A rational reduction in dosage has been shown to reduce toxicity, whereas other concurrent cytotoxic agents used in combination therapy remain at their usual unadjusted dosages to maintain efficacy. In addition to *TPMT*, other genetic factors can influence the effects of mercaptopurine, but their clinical relevance is not well understood.

Topoisomerase I Inhibitors *Irinotecan*, used in colorectal cancer, is bioactivated into SN-38 by carboxyesterases 1 and 2 and then bioinactivated by glucuronidation via uridine diphosphate glucuronosyltransferase isoform 1A1 (*UGT1A1*). The variable number of tandem repeat variant TA7, termed *UGT1A1*28*, diminishes glucuronidation and patients carrying that variant have increased SN-38, which is responsible for both efficacy and toxicity of irinotecan. Approximately 35% of patients receiving irinotecan experience ADRs such as severe diarrhea and neutropenia. The US FDA endorsed a cautionary notice in the drug label in 2005. The appropriate action in patients with the low-activity alleles is to modify the dosage of the drug. Several clinical trials have investigated genotype-guided dose escalation of irinotecan in combination with 5-fluorouracil (5-FU) and leucovorin (FOLFIRI treatment regimen), showing that a higher-dosage regimen including irinotecan (260 mg/m^2) was safe in individuals who carried the wild type *UGT1A1*1* allele [45]. Most studies of irinotecan pharmacogenetics have focused on the effect of *UGT1A1*28* on irinotecan toxicity. To investigate clinical utility of the *UGT1A1*28* testing, a recent collaborative meta-analysis focused on patient survival after irinotecan therapy. There was no statistically significant difference in the survival between patients of different *UGT1A1*8* genotypes [46]. Therefore *UGT1A1*28* is not prognostic of overall survival for individuals treated with irinotecan, which is in strong contrast to the association between this allele and irinotecan-induced toxicity.

Pyrimidine Analogues *5-FU*, an analogue of uracil, is converted to its cytotoxic nucleotides that inhibit thymidylate synthase. 5-FU is metabolized by dihydropyrimidine dehydrogenase (encoded by the *DPYD* gene) into its inactive form. More than 40 *DPYD* polymorphisms have been identified, of which 17 have been reported in patients with 5-FU–induced toxicity, which includes mucositis, neutropenia, and neurological symptoms [47]. Decreased *DPYD* activity can lead to the accumulation of 5-FU and toxicity; however, up to two-thirds of patients with toxicity do not have polymorphisms in *DPYD*. Several other genetic variants have been linked with tumor response, disease progression, and drug toxicity, including the variants in thymidylate synthase (*TYMS*) and 5,10-methylenetetrahydrofolate reductase (*MTHFR*) [48].

Estrogen Receptor Antagonists *Tamoxifen* is used in the treatment and chemoprevention of estrogen receptor–positive breast cancer. Tamoxifen is metabolized by several CYP450 enzymes, *CYP2D6* being the most important for the formation of endoxifen, an active metabolite mostly responsible for the therapeutic effect of tamoxifen. At least 88 allelic variants in *CYP2D6* have been described, many of which have reduced or nonfunctional catalytic activity. Initial studies suggested that women with decreased *CYP2D6* activity had poor clinical outcomes when treated with tamoxifen [49]. However, two retrospective analyses of breast cancer trials failed to replicate these findings [50,51]. These two studies have led to intense discussions in the field and the International Tamoxifen Pharmacogenetics Consortium was established to address the controversy [52]. Meta-analysis of data on nearly 5000 women treated with tamoxifen was performed and although *CYP2D6* was a strong predictor of invasive disease-free survival, the Consortium concluded that a prospective study was required to assess the role of *CYP2D6* in tamoxifen response.

Poly(ADP-Ribose) Polymerase Inhibitor Selective preventative strategies, including bilateral mastectomy, have been available to women who carry germline mutations in the *BRCA1* and *BRCA2* genes, which result in a 45–65% chance of developing breast cancer, but also increase the risk of ovarian, colon, and prostate cancer [53]. In December 2014, the FDA approved *olaparib*, a poly(ADP-ribose) polymerase (PARP) inhibitor, for patients with germline *BRCA* ovarian cancer who have been treated with three chemotherapy regimens. Olaparib is one of the first drugs to inhibit PARP, an enzyme involved in DNA repair. Selective toxicity of olaparib is achieved by preventing base-excision DNA repair, which produces tumor cell death in *BRCA*-deficient cells but not in normal cells with functional homologous recombination pathways. Cells with *BRCA* mutations have a nonfunctioning homologous recombination DNA repair mechanism [54].

Urate Oxidase Enzymes *Rasburicase*, a recombinant urate oxidase enzyme that breaks down uric acid to hydrophilic allantoin and hydrogen peroxide, is approved by the various regulatory authorities for the prophylaxis and treatment of hyperuricemia during chemotherapy in adults and children with lymphoma, leukemia, and solid tumors. A pegylated form of urate oxidase, *pegloticase* is approved for the treatment of refractory gout. Both drugs are contraindicated for use in patients with known glucose-6-phosphate dehydrogenase (G6PD) deficiency. G6PD is the enzyme that converts glucose-6-phosphate into 6-phosphogluconolactone, the first step of the pentose phosphate pathway, and at the same time, G6PD produces nicotinamide adenine dinucleotide phosphate-oxidase (NADPH) from nicotinamide adenine dinucleotide phosphate. G6PD is particularly important in erythrocytes, because it is the only available source of NADPH, which is required to protect erythrocytes from oxidative stress. Administration of rasburicase to G6PD-deficient patients can result in hemolytic anemia, methemoglobinemia and death [55]. Patients should therefore be tested for *G6PD* variants before the use of these drugs. Of course, evidence has also accumulated over many decades that G6PD deficiency can predispose a patient to hemolysis with a number of other drugs including primaquine, dapsone, and sulfonamides, and also with certain foods such as fava beans.

Somatic DNA Mutations

Monoclonal Antibodies Monoclonal antibodies such as trastuzumab, rituximab, and cetuximab, comprise an increasingly important class of anticancer therapy. They bind to tumor cells and cause antibody-dependent cell-mediated toxicity. *Trastuzumab* is used for the treatment of *HER2* overexpressing breast cancer and metastatic breast and gastric cancer. Test results demonstrating *HER2* protein overexpression are required before initiating therapy. *HER2* overexpression can be determined using two techniques: immunohistochemistry, which measures protein levels, or through fluorescence in situ hybridization, which detects gene amplification. *Cetuximab* is used alone or in combination therapy to treat advanced squamous cell carcinoma of the head and neck and to treat *KRAS*-negative, epidermal growth factor receptor *(EGFR)*–expressing metastatic colon cancer. *KRAS* mutation and *EGFR* expression status have to be determined before treatment. *Rituximab* is used to treat patients with malignant diseases such as non-Hodgkin lymphoma, and chronic lymphocytic leukemia, but also in nononcological conditions such as rheumatoid arthritis (RA). Rituximab is a monoclonal antibody directed against CD20 (coding gene *MS4A1*), expressed on B lymphocytes and on B-cell non-Hodgkin lymphoma. There is discrepancy in recommendations on pretreatment testing with only the EMA recommending treatment only in patients with CD20-positive tumors, whereas the FDA provides pharmacogenetic information and no recommendations.

Receptor Tyrosine Kinase Inhibitors The decision on whether to treat patients with *imatinib* is based on the presence of genetic biomarkers, which include the Philadelphia chromosome *(BCR-ABL)*, KIT *(CD117)*, and *PDGFR* gene rearrangements. Imatinib mesylate is a small-molecule inhibitor of protein tyrosine kinases used to treat mostly Philadelphia chromosome–positive chronic myeloid leukemia (CML). Imatinib binds to and inhibits the activity of *BCR-ABL* tyrosine kinase, which is the critical pathogenetic event in CML. Imatinib also inhibits other protein kinases such as *KIT* and *PDGFR*. Variants in P-glycoprotein *(ABCB1)*, which effluxes imatinib, may be associated with its efficacy and resistance [56]. In addition, imatinib is transported into CML cells by *OCT1*, an influx transporter. Both expression of *OCT1* mRNA and polymorphisms in the *OCT1 (SLC22A1)* gene have been associated with response to imatinib [57,58].

Erlotinib and *gefitinib* are synthetic anilinoquinazolines that bind to the adenosine triphosphate (ATP) binding site of *EGFR* and also inhibit ATP-binding cassette (ABC) transporter–mediated drug efflux. *Crizotinib* is an inhibitor of the anaplastic lymphoma kinase *(ALK)* and *EML4-ALK* fusion protein. All these compounds are targeted agents approved for treatment of non–small-cell lung cancer. Although effective, owing to their high cost, these drugs have a significant financial effect on patients and society [59].

B-raf/MEK/ERK Pathway Inhibitors *Vemurafenib* was developed for the treatment of late-stage melanoma. Vemurafenib causes apoptosis in melanoma cells by interrupting the B-raf/*MEK/ERK* pathway only in the presence of the common *BRAF* V600E mutation. A *BRAF* V600E FDA-approved test exists and in addition to vemurafenib, the test is also required before prescribing *dabrafenib* and *trametinib*; both drugs are used in patients with unresectable or metastatic melanoma.

The long-term efficacy of the new targeted cancer therapies is limited by resistance to therapy, which may be primary or acquired, with failure to respond occurring after initial response in the latter situation. For example, in patients with *BRAF* V600E melanoma treated with vemurafenib, resistance and tumor recurrence develop within 5–7 months. This is because of tumor cell heterogeneity and diminished treatment efficacy in subpopulations of cells, which gain additional mutations in *BRAF* or in downstream effectors such as *ERK* and *MEK* [60]. To address the issue of resistance, novel strategies using combination therapy to target disease-modifying molecular mutations is required.

Infection and Inflammation

Rheumatology

Disease-Modifying Antirheumatic Drugs The most important predictor of long-term disease outcome in RA is patient responsiveness to the first-line treatment used [61]. Only a small number of strong associations have been revealed in genetic studies and only those associated with tumor necrosis factor (TNF)-inhibitors have reached genome-wide significance. They will be discussed here together with several weaker associations that have been reported.

Disease-modifying antirheumatic drugs are important for preventing severe complications of RA, of which *methotrexate (MTX)* is the most commonly used. MTX inhibits several enzymes in the folate pathway, including dihydrofolate reductase *(DHFR)*, *TYMS*, and 5-aminoimidazole-4-carboxamide ribonuclease transformylase [62]. Variability in response to MTX has been associated with polymorphisms in genes coding for *DHFR*, *TYMS*, *MTHFR*, reduced folate carrier 1 *(RFC1)*, and ABC transporters. Polymorphisms in *DHFR* have been associated with resistance to MTX, whereas with MTX sensitivity and toxicity has been associated with *MTHFR* polymorphisms. The *MTHFR* C677T polymorphism, which leads to decreased enzyme activity, has been investigated in relation to MTX efficacy, but with some contradictory findings. Discontinuation of MTX treatment occurs in 10–30% of RA patients [63]; a recent meta-analysis showed that polymorphisms in the *MTHFR* gene were not reliable predictors of toxicity [64].

Response to TNF antagonists *etanercept, infliximab*, and *adalimumab* has been associated with haplotypes in the major histocompatibility complex (MHC) on chromosome 6 spanning the region from *HLA-DRB1* to *TNF* [65]. In addition, receptor-type tyrosine protein phosphatase C (*PTPRC* or *CD45*), was associated with response to anti-TNF drugs in large studies; however, consistent replication of associations between genetic variants and outcome of therapy is lacking [61,66]. Two associations that have reached the genome-wide significant level involve variants in the leukocyte differentiation antigen *CD84* gene and variants in the *PDE3A* and *SLCO1C1* gene locus [67,68]; however, their predictive value in clinical setting needs to be evaluated.

Azathioprine, a prodrug converted into 6-MP after oral administration, is inactivated by TPMT. Polymorphisms in the *TPMT* gene (described previously in this chapter) can lead to azathioprine-induced toxicity [69]. One of the newer drugs used in the treatment of RA is *leflunomide*, which inhibits de novo synthesis of pyrimidine ribonucleotides, leading to a decreased lymphocyte proliferation. Leflunomide is also a prodrug that is activated by cytochrome P450 *CYP1A2, 2C19*, and *3A4*. Variants associated with decreased or increased activity of these enzymes have been associated with reduced efficacy or toxicity, respectively [70], but these data need to be replicated.

Sulfasalazine is also used to treat RA, but its use is limited because of severe adverse effects such as agranulocytosis. After ingestion, sulfasalazine is cleaved by intestinal bacteria into 5-aminosalicylic acid and sulfapyridine, which is acetylated in the liver by N-acetyl transferase 2 (*NAT2*). Plasma levels of sulfapyridine are altered by functional polymorphisms in *NAT2* and also the *ABCG2* transporter gene. Individuals with the *NAT2* variants *NAT2*5A, *5B, *5C*, and *6* that lead to slow acetylator status are at increased risk of developing toxicity [71].

Inflammatory Bowel Disease

Almost all classes of drugs that have been used in the treatment of inflammatory bowel disease (IBD) have been investigated using pharmacogenomic approaches, but very few findings have been replicated. Research into IBD drug efficacy and safety mainly has focused on polymorphisms in the genes encoding enzymes involved in metabolic pathways. Recently however, a GWAS in patients with IBD who had developed thiopurine-induced pancreatitis within 3 months of starting therapy identified strong association with two

HLA Class II alleles: *HLA-DRB1*07:01* and *DQA1*02:01* in linkage disequilibrium, tagged by rs2647087 [72]. These results are likely to be generalizable to many other patient populations in which thiopurines are used. Several other classes of drugs other than thiopurines, azathioprine, and 6-MP have been extensively investigated, including MTX, aminosalicylates, and TNF-α inhibitors. All of these have been mentioned previously in this chapter. Despite the intensive research, only polymorphisms in the *TPMT* gene are used clinically to guide thiopurine therapy.

Immune-Mediated Adverse Drug Reactions

Severe drug-induced hypersensitivity and drug-induced liver injury are a heterogeneous group of type B or off-target adverse reactions induced by more than 200 different chemically unrelated compounds. These reactions are rare, but can be potentially lethal and have considerable economic implications. Clinical manifestations of these reactions are also variable and may range from mild skin rashes to severe systemic symptoms with fever, hepatotoxicity, and blistering skin rashes such as Stevens-Johnson syndrome (SJS) and toxic epidermal necrolysis (TEN). Many other organ systems can also be affected, especially in hypersensitivity syndrome, also called *drug reaction with eosinophilia and systemic symptoms syndrome*. The pathogenesis is not fully understood, but immune factors have been implicated. The genetic association with immune ADRs can be drug specific, phenotype specific, or ethnicity specific. The strongest associations have been with *HLA* alleles, which reside in the polymorphic MHC region on chromosome 6 in the human genome (Table 10.3). Clinical application has been achieved with some of these drug associations, in that pretreatment genotyping is recommended.

Abacavir, a nucleoside reverse transcriptase inhibitor, causes hypersensitivity reactions in 5–7% of patients with HIV. The association between the *HLA B*57:01* and abacavir hypersensitivity was reported in 2002 and replicated in numerous studies afterward [73,74]. Pretreatment genetic testing for *B*57:01* reduces the incidence of hypersensitivity; it is cost-effective, and therefore guidance for its use has been issued by the FDA and several other regulatory agencies [75]. Furthermore, the binding of abacavir with the antigen-binding cleft of *HLA-B*57:01* has been demonstrated using radiographic crystallography [76]. Abacavir binding resulted in up to a 25% change in the peptide repertoire presented by *B*57:01*. This finding confirmed the causative role of *B*57:01* in the mechanism of abacavir hypersensitivity and together with the epidemiological and clinical data represents a paradigm for translation of pharmacogenomics into clinical practice.

The aromatic antiepileptic drug *carbamazepine (CBZ)* is widely used in the treatment of trigeminal neuralgia and bipolar disorder. Although well tolerated by the majority of patients, it can cause severe hypersensitivity reactions

TABLE 10.3 HLA Alleles Associated With Immune-Mediated Adverse Drug Reactions

Drug	HLA-A	HLA-B	HLA-C	HLA-DRB1	HLA-DQA1	HLA-DQB1	HLA-DPB1
Abacavir		57:01		07:01			
Allopurinol	33:03	58:01	03:02	03:01			
Amoxicillin-clavulanic acid	30:02			15:01		06:02	
Anti-TB drugs						02:01	
Aspirin				13:02		03:01	05:01
						06:09	13:01
Azathioprine				07:01	02:01		
Bucillamine				08:02		04:02	
Carbamazepine	31:01	15:02	01:01	12:02			
		15:11	08:01	07:01			
Clozapine						05:02	04:01
Flucloxacillin		57:01					
Lamotrigine	68:01	15:02	07:18	13:01		06:09	
	33:03	13:02					
		58:01					
Lapatinib				07:01	02:01	02:02	
Lumiracoxib				15:01	01:02	06:02	
Methazolamide, acetazolamide		59:01					
Nevirapine	68:01	07:05	04:01	01:02			
		35:05	03:02				
Phenytoin		15:02	08:01	16:02			
Ticlopidine	33:03	44:03	14:03	13:02		06:04	
Trichlorethylene		13:01					
Ximelagatran				07:01	02:01		

in approximately one in 10,000 individuals. CBZ-induced SJS has been strongly associated with *HLA-B*15:02* in several Asian populations [8], but not in Caucasians and Japanese [77]. Subsequent to these findings, the drug label for CBZ now includes a recommendation for pretreatment genetic testing to prevent SJS in Southeast Asians (see Tables 10.1 and 10.2) [78]. Cross-reactivity to structurally related aromatic antiepileptic drugs has been demonstrated and, interestingly, *B*15:02* is important risk factor for SJS with *phenytoin* and *lamotrigine* in Asians [79]. In Northern Europeans and Japanese, *HLA-A*31:01* has been associated with CBZ-induced hypersensitivity reactions, both mild and severe [11,12]. *A*31:01* is present in many worldwide populations, whereas *B*15:02* has a high prevalence in Southeast Asia, but a very low incidence in the rest of the world. *HLA-A*31:01* has been included in the drug labels in the United States, European Union, and Japan, but is included for information only. Whether *HLA-B*15:02* and *HLA-A*31:01* are both causative or whether they are in linkage disequilibrium with a causative variant in the MHC remains to be seen.

Allopurinol, a drug commonly used for the treatment of gout, can cause hypersensitivity reactions that have been strongly associated with the *HLA-B*58:01* allele [80]. The association has subsequently been replicated in diverse worldwide populations. Recently, a prospective study was conducted to evaluate the usefulness of *B*58:01* testing in preventing allopurinol-induced hypersensitivity [81]. The study design was similar to the study from Taiwan, which evaluated the clinical utility of *B*15:02* testing [78]. Allopurinol naive patients were tested for *B*58:01* before the initiation of therapy. Patients who tested negative for *B*58:01* were started on 50–100 mg of allopurinol according to their creatinine clearance. Patients positive for *B*58:01* were given a small dose (50 mg) of allopurinol that was slowly escalated using the tolerance induction protocol. Patients who did not consent to this protocol were given alternative medications such as benzbromarone or febuxostat. All patients were observed for 90 days to detect the occurrence of hypersensitivity reactions. A historical retrospective control cohort of patients was used to assess the incidence of allopurinol hypersensitivity. No severe hypersensitivity was observed in any of the prospective patients. Compared with historical controls, a significant reduction in severe hypersensitivity (18%) was observed in *HLA-B*58:01*-positive patients [81].

The human leukocyte antigen (HLA) system has also been shown to be an important predisposing genetic factor for DILI. A number of diverse classes of drugs, including the β-lactam antibiotics flucloxacillin and co-amoxiclav, the anticoagulant ximelagatran, the anticancer drug lapatinib, and the *COX2*-inhibitor lumiracoxib, can cause DILI, which has been linked to a number of *HLA* alleles [9,82–85]. No genetic testing is currently recommended to prevent DILI, largely because the predictive values are not strong enough. For instance, with flucloxacillin, only one in 500 to 1000 individuals who carry the risk allele would develop DILI, and thus genotyping for *HLA-B*57:01* before flucloxacillin treatment is unlikely to be clinically useful or cost-effective.

Psychiatry

Several recent studies have demonstrated that pharmacogenetic testing for a number of genes associated with treatment response can reduce treatment costs in patients with mental health illness [86,87]. Cost-effectiveness of genetic testing has been investigated in patients prescribed antidepressants using the Genecept Assay [88]. The assay examines 10 genes associated with treatment response of many psychiatric medications, including the cytochrome P450s, *CYP2D6*, *CYP2C19*, and *CYP3A4*, and the pharmacodynamic genes encoding *SLC6A4*, *5HT2C*, *CACNA1C*, *DRD2*, *COMT*, *ANK3*, and *MTHFR*. This and similar studies have shown that genotyping not only leads to cost savings, but also increases drug adherence and drug tolerability and decreases medication discontinuation resulting from ADRs [89]. Importantly, both psychiatrists and patients in this study consider genetic testing that can predict an increased risk of ADRs or treatment efficacy very useful and these tests had no negative impact on the patients' quality of life. However, despite these results, genotyping before the use of drugs in psychiatry is not used routinely, with testing occasionally undertaken in refractory patients.

CONCLUSIONS

In this chapter, we summarized some promising applications of pharmacogenetic markers in clinical medicine. Although many aspects of clinical implementation of genetic testing into clinical practice are challenging, it is hoped that novel powerful technologies together with the high-quality clinical information available through electronic medical records in large cohorts of patients will enable the identification of genetic markers associated with drug response, and subsequently their faster adoption. Through the development of panels of genetic markers that all contribute a small percentage to the overall drug response and through education on drug selection, dose selection, and patient selection, it may be possible to introduce genetic testing before drug prescription in many areas but, of course, further research is still needed in this area.

It is also important to emphasize that the approach to drug individualization should not only be based on genomics, but other "-omics" strategies including proteomics, transcriptomics, and metagenomics should also be employed. Current challenges in integrating diverse -omics data with the national electronic health information infrastructure are

FIGURE 10.2   Genetic marker implementation into clinical practice.
There are many barriers to successful application of genomics in the clinic;
some of those mentioned here are currently being addressed by researchers
and those interested in public health issues, including the regulatory bodies, patients, and general public.

surmountable, and it is possible that in the future our medical records will contain information on our whole genomes,
which will influence health providers' decision making
(Fig. 10.2). Global collaboration and concerted efforts from
all stakeholders is needed to adopt and effectively implement pharmacogenetics into the clinic.

REFERENCES

[1] Vogel F. Moderne Probleme der Humangenetik. Ergeb Inn Med
Kinderheilkd 1959;12:52–125.

[2] Meyer UA. Pharmacogenetics: five decades of therapeutic lessons
from genetic diversity. Nat Rev Genet 2004;5(9):669–76.

[3] Shimazawa R, Ikeda M. Differences in pharmacogenomic biomarker information in package inserts from the United States, the
United Kingdom and Japan. J Clin Pharm Ther 2013;38(6):468–75.

[4] Wang B, Canestaro WJ, Choudhry NK. Clinical evidence supporting pharmacogenomic biomarker testing provided in US Food and
Drug Administration drug labels. JAMA Intern Med 2014;174(12):
1938–44.

[5] Swen JJ, Wilting I, de Goede AL, et al. Pharmacogenetics: from
bench to byte. Clin Pharmacol Ther 2008;83(5):781–7.

[6] Relling MV, Klein TE. CPIC: Clinical Pharmacogenetics Implementation Consortium of the pharmacogenomics research network.
Clin Pharmacol Ther 2011;89(3):464–7.

[7] Klein RJ, Zeiss C, Chew EY, et al. Complement factor H
polymorphism in age-related macular degeneration. Science
2005;308(5720):385–9.

[8] Chung WH, Hung SI, Hong HS, et al. Medical genetics: a marker
for Stevens-Johnson syndrome. Nature 2004;428(6982):486.

[9] Daly AK, Donaldson PT, Bhatnagar P, et al. HLA-B*5701 genotype
is a major determinant of drug-induced liver injury due to flucloxacillin. Nat Genet 2009;41(7):816–9.

[10] Group SC, Link E, Parish S, et al. SLCO1B1 variants and statin-
induced myopathy: a genomewide study. N Engl J Med 2008;359(8):
789–99.

[11] McCormack M, Alfirevic A, Bourgeois S, et al. HLA-A*3101 and
carbamazepine-induced hypersensitivity reactions in Europeans. N
Engl J Med 2011;364(12):1134–43.

[12] Ozeki T, Mushiroda T, Yowang A, et al. Genome-wide association
study identifies HLA-A*3101 allele as a genetic risk factor for carbamazepine-induced cutaneous adverse drug reactions in Japanese
population. Hum Mol Genet 2011;20(5):1034–41.

[13] Rasmussen-Torvik LJ, Stallings SC, Gordon AS, et al. Design and
anticipated outcomes of the eMERGE-PGx project: a multicenter
pilot for preemptive pharmacogenomics in electronic health record
systems. Clin Pharmacol Ther 2014;96(4):482–9.

[14] Kaminsky LS, Zhang ZY. Human P450 metabolism of warfarin.
Pharmacol Ther 1997;73(1):67–74.

[15] Rettie AE, Wienkers LC, Gonzalez FJ, Trager WF, Korzekwa KR.
Impaired (S)-warfarin metabolism catalysed by the R144C allelic
variant of CYP2C9. Pharmacogenetics 1994;4(1):39–42.

[16] Rieder MJ, Reiner AP, Gage BF, et al. Effect of VKORC1 haplotypes on transcriptional regulation and warfarin dose. N Engl J Med
2005;352(22):2285–93.

[17] Momary KM, Shapiro NL, Viana MA, Nutescu EA, Helgason CM,
Cavallari LH. Factors influencing warfarin dose requirements in
African-Americans. Pharmacogenomics 2007;8(11):1535–44.

[18] Obayashi K, Nakamura K, Kawana J, et al. VKORC1 gene variations are the major contributors of variation in warfarin dose in Japanese patients. Clin Pharmacol Ther 2006;80(2):169–78.

[19] Budnitz DS, Lovegrove MC, Shehab N, Richards CL. Emergency
hospitalizations for adverse drug events in older Americans. N Engl
J Med 2011;365(21):2002–12.

[20] Kimmel SE, French B, Kasner SE, et al. A pharmacogenetic versus a clinical algorithm for warfarin dosing. N Engl J Med 2013;
369(24):2283–93.

[21] Pirmohamed M, Burnside G, Eriksson N, et al. A randomized trial of genotype-guided dosing of warfarin. N Engl J Med
2013;369(24):2294–303.

[22] Johnson JA, Gong L, Whirl-Carrillo M, et al. Clinical pharmacogenetics implementation consortium guidelines for CYP2C9 and
VKORC1 genotypes and warfarin dosing. Clin Pharmacol Ther
2011;90(4):625–9.

[23] Verhoef TI, Ragia G, de Boer A, et al. A randomized trial of genotype-guided dosing of acenocoumarol and phenprocoumon. N Engl
J Med 2013;369(24):2304–12.

[24] Pare G, Eriksson N, Lehr T, et al. Genetic determinants of dabigatran plasma levels and their relation to bleeding. Circulation
2013;127(13):1404–12.

[25] Snoep JD, Hovens MM, Eikenboom JC, van der Bom JG, Jukema
JW, Huisman MV. Clopidogrel nonresponsiveness in patients
undergoing percutaneous coronary intervention with stenting: a
systematic review and meta-analysis. Am Heart J 2007;154(2):
221–31.

[26] Simon T, Verstuyft C, Mary-Krause M, et al. Genetic determinants
of response to clopidogrel and cardiovascular events. N Engl J Med
2009;360(4):363–75.

[27] Wallentin L, James S, Storey RF, et al. Effect of CYP2C19 and
ABCB1 single nucleotide polymorphisms on outcomes of treatment
with ticagrelor versus clopidogrel for acute coronary syndromes:
a genetic substudy of the PLATO trial. Lancet 2010;376(9749):
1320–8.

[28] Shuldiner AR, O'Connell JR, Bliden KP, et al. Association of cytochrome *P450* 2C19 genotype with the antiplatelet effect and clinical efficacy of clopidogrel therapy. JAMA 2009;302(8):849–57.

[29] Roberts JD, Wells GA, Le May MR, et al. Point-of-care genetic testing for personalisation of antiplatelet treatment (RAPID GENE): a prospective, randomised, proof-of-concept trial. Lancet 2012; 379(9827):1705–11.

[30] Kazi DS, Garber AM, Shah RU, et al. Cost-effectiveness of genotype-guided and dual antiplatelet therapies in acute coronary syndrome. Ann Intern Med 2014;160(4):221–32.

[31] Halushka MK, Walker LP, Halushka PV. Genetic variation in cyclooxygenase 1: effects on response to aspirin. Clin Pharmacol Ther 2003;73(1):122–30.

[32] Goodman T, Ferro A, Sharma P. Pharmacogenetics of aspirin resistance: a comprehensive systematic review. Br J Clin Pharmacol 2008;66(2):222–32.

[33] Voora D, Cyr D, Lucas J, et al. Aspirin exposure reveals novel genes associated with platelet function and cardiovascular events. J Am Coll Cardiol 2013;62(14):1267–76.

[34] Deshmukh HA, Colhoun HM, Johnson T, et al. Genome-wide association study of genetic determinants of LDL-c response to atorvastatin therapy: importance of Lp(a). J Lipid Res 2012;53(5):1000–11.

[35] Harrap SB, Tzourio C, Cambien F, et al. The *ACE* gene I/D polymorphism is not associated with the blood pressure and cardiovascular benefits of ACE inhibition. Hypertension 2003;42(3):297–303.

[36] Rigat B, Hubert C, Alhenc-Gelas F, Cambien F, Corvol P, Soubrier F. An insertion/deletion polymorphism in the angiotensin I-converting enzyme gene accounting for half the variance of serum enzyme levels. J Clin Invest 1990;86(4):1343–6.

[37] Brugts JJ, Isaacs A, Boersma E, et al. Genetic determinants of treatment benefit of the angiotensin-converting enzyme-inhibitor perindopril in patients with stable coronary artery disease. Eur Heart J 2010;31(15):1854–64.

[38] Mahmoudpour SH, Leusink M, van der Putten L, et al. Pharmacogenetics of ACE inhibitor-induced angioedema and cough: a systematic review and meta-analysis. Pharmacogenomics 2013;14(3):249–60.

[39] Liggett SB, Cresci S, Kelly RJ, et al. A *GRK5* polymorphism that inhibits beta-adrenergic receptor signaling is protective in heart failure. Nat Med 2008;14(5):510–7.

[40] Vandell AG, Lobmeyer MT, Gawronski BE, et al. G protein receptor kinase 4 polymorphisms: beta-blocker pharmacogenetics and treatment-related outcomes in hypertension. Hypertension 2012;60(4): 957–64.

[41] Ramirez AH, Shaffer CM, Delaney JT, et al. Novel rare variants in congenital cardiac arrhythmia genes are frequent in drug-induced torsades de pointes. Pharmacogenomics J 2013;13(4):325–9.

[42] Collins FS, Barker AD. Mapping the cancer genome: pinpointing the genes involved in cancer will help chart a new course across the complex landscape of human malignancies. Sci Am 2007;296(3):50–7.

[43] Cancer Genome Atlas N. Comprehensive genomic characterization of head and neck squamous cell carcinomas. Nature 2015;517(7536): 576–82.

[44] Paugh SW, Stocco G, McCorkle JR, Diouf B, Crews KR, Evans WE. Cancer pharmacogenomics. Clin Pharmacol Ther 2011;90(3): 461–6.

[45] Freyer G, Duret A, Milano G, et al. Pharmacogenetic tailoring of irinotecan-based first-line chemotherapy in metastatic colorectal cancer: results of a pilot study. Anticancer Res 2011;31(1): 359–66.

[46] Dias MM, Pignon JP, Karapetis CS, et al. The effect of the *UGT1A1*28* allele on survival after irinotecan-based chemotherapy: a collaborative meta-analysis. Pharmacogenomics J 2014;14(5): 424–31.

[47] van Kuilenburg AB, Meinsma R, van Gennip AH. Pyrimidine degradation defects and severe 5-fluorouracil toxicity. Nucleosides, Nucleotides Nucleic Acids 2004;23(8–9):1371–5.

[48] Schwab M, Zanger UM, Marx C, et al. Role of genetic and nongenetic factors for fluorouracil treatment-related severe toxicity: a prospective clinical trial by the German 5-FU Toxicity Study Group. J Clin Oncol 2008;26(13):2131–8.

[49] Schroth W, Goetz MP, Hamann U, et al. Association between *CYP2D6* polymorphisms and outcomes among women with early stage breast cancer treated with tamoxifen. JAMA 2009;302(13): 1429–36.

[50] Rae JM, Drury S, Hayes DF, et al. *CYP2D6* and *UGT2B7* genotype and risk of recurrence in tamoxifen-treated breast cancer patients. J Natl Cancer Inst 2012;104(6):452–60.

[51] Regan MM, Leyland-Jones B, Bouzyk M, et al. *CYP2D6* genotype and tamoxifen response in postmenopausal women with endocrine-responsive breast cancer: the breast international group 1-98 trial. J Natl Cancer Inst 2012;104(6):441–51.

[52] Province MA, Goetz MP, Brauch H, et al. *CYP2D6* genotype and adjuvant tamoxifen: meta-analysis of heterogeneous study populations. Clin Pharmacol Ther 2014;95(2):216–27.

[53] Chen S, Parmigiani G. Meta-analysis of *BRCA1* and *BRCA2* penetrance. J Clin Oncol 2007;25(11):1329–33.

[54] Fong PC, Boss DS, Yap TA, et al. Inhibition of poly(ADP-ribose) polymerase in tumors from *BRCA* mutation carriers. N Engl J Med 2009;361(2):123–34.

[55] Relling MV, McDonagh EM, Chang T, et al. Clinical Pharmacogenetics Implementation Consortium (CPIC) guidelines for rasburicase therapy in the context of *G6PD* deficiency genotype. Clin Pharmacol Ther 2014;96(2):169–74.

[56] Illmer T, Schaich M, Platzbecker U, et al. P-glycoprotein-mediated drug efflux is a resistance mechanism of chronic myelogenous leukemia cells to treatment with imatinib mesylate. Leukemia 2004;18(3):401–8.

[57] Alves R, Fonseca AR, Goncalves AC, et al. Drug transporters play a key role in the complex process of imatinib resistance in vitro. Leuk Res 2015;39(3):355–60.

[58] Koren-Michowitz M, Buzaglo Z, Ribakovsky E, et al. *OCT1* genetic variants are associated with long term outcomes in imatinib treated chronic myeloid leukemia patients. Eur J Haematol 2014;92(4):283–8.

[59] Chouaid C, Crequit P, Borget I, Vergnenegre A. Economic evaluation of first-line and maintenance treatments for advanced non-small cell lung cancer: a systematic review. Clin Outcomes Res 2015;7:9–15.

[60] Sullivan RJ, Flaherty KT. Resistance to *BRAF*-targeted therapy in melanoma. Eur J cancer 2013;49(6):1297–304.

[61] Plant D, Wilson AG, Barton A. Genetic and epigenetic predictors of responsiveness to treatment in RA. Nat Rev Rheumatol 2014;10(6):329–37.

[62] Zhang LL, Yang S, Wei W, Zhang XJ. Genetic polymorphisms affect efficacy and adverse drug reactions of DMARDs in rheumatoid arthritis. Pharmacogenet Genomics 2014;24(11):531–8.

[63] Taniguchi A, Urano W, Tanaka E, et al. Validation of the associations between single nucleotide polymorphisms or haplotypes and responses to disease-modifying antirheumatic drugs in patients with rheumatoid arthritis: a proposal for prospective pharmacogenomic study in clinical practice. Pharmacogenet Genomics 2007;17(6):383–90.

[64] Owen SA, Lunt M, Bowes J, et al. *MTHFR* gene polymorphisms and outcome of methotrexate treatment in patients with rheumatoid arthritis: analysis of key polymorphisms and meta-analysis of *C677T* and *A1298C* polymorphisms. Pharmacogenomics J 2013;13(2):137–47.

[65] Plenge RM, Criswell LA. Genetic variants that predict response to anti-tumor necrosis factor therapy in rheumatoid arthritis: current challenges and future directions. Curr Opin Rheumatol 2008;20(2): 145–52.

[66] Pappas DA, Oh C, Plenge RM, Kremer JM, Greenberg JD. Association of rheumatoid arthritis risk alleles with response to anti-TNF biologics: results from the CORRONA registry and meta-analysis. Inflammation 2013;36(2):279–84.

[67] Acosta-Colman I, Palau N, Tornero J, et al. GWAS replication study confirms the association of *PDE3A-SLCO1C1* with anti-TNF therapy response in rheumatoid arthritis. Pharmacogenomics 2013;14(7):727–34.

[68] Cui J, Stahl EA, Saevarsdottir S, et al. Genome-wide association study and gene expression analysis identifies *CD84* as a predictor of response to etanercept therapy in rheumatoid arthritis. PLoS Genet 2013;9(3):e1003394.

[69] Colleoni L, Kapetis D, Maggi L, et al. A new thiopurine s-methyltransferase haplotype associated with intolerance to azathioprine. J Clin Pharmacol 2013;53(1):67–74.

[70] Wiese MD, Schnabl M, O'Doherty C, et al. Polymorphisms in cytochrome *P450* 2C19 enzyme and cessation of leflunomide in patients with rheumatoid arthritis. Arthritis Res Ther 2012;14(4): R163.

[71] Wiese MD, Alotaibi N, O'Doherty C, et al. Pharmacogenomics of *NAT2* and *ABCG2* influence the toxicity and efficacy of sulphasalazine containing DMARD regimens in early rheumatoid arthritis. Pharmacogenomics J 2014;14(4):350–5.

[72] Heap GA, Weedon MN, Bewshea CM, et al. *HLA-DQA1-HLA-DRB1* variants confer susceptibility to pancreatitis induced by thiopurine immunosuppressants. Nat Genet 2014;46(10):1131–4.

[73] Mallal S, Nolan D, Witt C, et al. Association between presence of *HLA-B*5701, *HLA-DR7*, and *HLA-DQ3* and hypersensitivity to HIV-1 reverse-transcriptase inhibitor abacavir. Lancet 2002; 359(9308):727–32.

[74] Mallal S, Phillips E, Carosi G, et al. *HLA-B*5701 screening for hypersensitivity to abacavir. N Engl J Med 2008;358(6):568–79.

[75] Hughes DA, Vilar FJ, Ward CC, Alfirevic A, Park BK, Pirmohamed M. Cost-effectiveness analysis of *HLA B*5701 genotyping in preventing abacavir hypersensitivity. Pharmacogenetics 2004;14(6): 335–42.

[76] Illing PT, Vivian JP, Dudek NL, et al. Immune self-reactivity triggered by drug-modified HLA-peptide repertoire. Nature 2012;486(7404): 554–8.

[77] Alfirevic A, Jorgensen AL, Williamson PR, Chadwick DW, Park BK, Pirmohamed M. *HLA-B* locus in Caucasian patients with carbamazepine hypersensitivity. Pharmacogenomics 2006;7(6):813–8.

[78] Chen P, Lin JJ, Lu CS, et al. Carbamazepine-induced toxic effects and *HLA-B*1502 screening in Taiwan. N Engl J Med 2011;364(12):1126–33.

[79] Bloch KM, Sills GJ, Pirmohamed M, Alfirevic A. Pharmacogenetics of antiepileptic drug-induced hypersensitivity. Pharmacogenomics 2014;15(6):857–68.

[80] Hung SI, Chung WH, Liou LB, et al. *HLA-B*5801 allele as a genetic marker for severe cutaneous adverse reactions caused by allopurinol. Proc Natl Acad Sci USA 2005;102(11):4134–9.

[81] Jung JW, Kim DK, Park HW, et al. An effective strategy to prevent allopurinol-induced hypersensitivity by HLA typing. Genet Med 2015;17(10).

[82] Kindmark A, Jawaid A, Harbron CG, et al. Genome-wide pharmacogenetic investigation of a hepatic adverse event without clinical signs of immunopathology suggests an underlying immune pathogenesis. Pharmacogenomics J 2007;8(3):186–95.

[83] Lucena MI, Molokhia M, Shen Y, et al. Susceptibility to amoxicillin-clavulanate–induced liver injury is influenced by multiple *HLA* class I and II alleles. Gastroenterology 2011;141(1):338–47.

[84] Singer JB, Lewitzky S, Leroy E, et al. A genome-wide study identifies *HLA* alleles associated with lumiracoxib-related liver injury. Nat Genet 2010;42(8):711–4.

[85] Spraggs CF, Budde LR, Briley LP, et al. *HLA-DQA1*02:01 is a major risk factor for lapatinib-induced hepatotoxicity in women with advanced breast cancer. J Clin Oncol 2011;29(6):667–73.

[86] Herbild L, Andersen SE, Werge T, Rasmussen HB, Jurgens G. Does pharmacogenetic testing for *CYP450 2D6* and *2C19* among patients with diagnoses within the schizophrenic spectrum reduce treatment costs? Basic Clin Pharmacol Toxicol 2013;113(4):266–72.

[87] Ruano G, Szarek BL, Villagra D, et al. Length of psychiatric hospitalization is correlated with *CYP2D6* functional status in inpatients with major depressive disorder. Biomarkers Med 2013;7(3):429–39.

[88] Winner J, Allen JD, Altar CA, Spahic-Mihajlovic A. Psychiatric pharmacogenomics predicts health resource utilization of outpatients with anxiety and depression. Transl Psychiatry 2013;3:e242.

[89] Fagerness J, Fonseca E, Hess GP, et al. Pharmacogenetic-guided psychiatric intervention associated with increased adherence and cost savings. Am J Manag Care 2014;20(5):e146–56.

[90] Lander ES, Linton LM, Birren B, et al. Initial sequencing and analysis of the human genome. Nature 2001;409(6822):860–921.

[91] Crews KR, Gaedigk A, Dunnenberger HM, Klein TE, Shen DD, Callaghan JT, et al. Clinical pharmacogenetics implementation consortium (CPIC) guidelines for codeine therapy in the context of cytochrome P450 2D6 (CYP2D6) genotype. Clin Pharmacol Ther February 2012;91(2):321–6. Epub 2011 Dec 28. PMID: 22205192; http://dx.doi.org/10.1038/clpt.2011.287.

[92] Crews KR, Gaedigk A, Dunnenberger HM, Leeder JS, Klein TE, Caudle KE, et al. Clinical pharmacogenetics implementation consortium guidelines for cytochrome P450 2D6 genotype and codeine therapy: 2014 update. Clin Pharmacol Ther April 2014;95(4): 376–82. Epub 2014 Jan 23. Review. PMID: 24458010; http://dx.doi. org/10.1038/clpt.2013.254.

[93] Wilke RA, Ramsey LB, Johnson SG, Maxwell WD, McLeod HL, Voora D, et al. The clinical pharmacogenomics implementation consortium: CPIC guideline for SLCO1B1 and simvastatin-induced myopathy. Clin Pharmacol Ther July 2012;92(1):112–7. Epub 2012 May 23. PMID: 22617227; http://dx.doi.org/10.1038/clpt.2012.57.

[94] Ramsey LB, Johnson SG, Caudle KE, Haidar CE, Voora D, Wilke RA, et al. The clinical pharmacogenetics implementation consortium guideline for SLCO1B1 and simvastatin-induced myopathy: 2014 update. Clin Pharmacol Ther October 2014;96(4):423–8. Epub 2014 Jun 11. PMID: 24918167; http://dx.doi.org/10.1038/ clpt.2014.125.

[95] Relling MV, Gardner EE, Sandborn WJ, Schmiegelow K, Pui CH, Yee SW, et al. Clinical pharmacogenetics implementation consortium guidelines for thiopurine methyltransferase genotype and thiopurine dosing. Clin Pharmacol Ther March 2011;89(3):387–91. Epub 2011 Jan 26. PMID: 21270794. http://dx.doi.org/10.1038/clpt.2010.320.

[96] Relling MV, Gardner EE, Sandborn WJ, Schmiegelow K, Pui CH, Yee SW, et al. Clinical pharmacogenetics implementation consortium guidelines for thiopurine methyltransferase genotype and thiopurine dosing: 2013 update. Clin Pharmacol Ther March 2013;93(4):324–5. Epub 2013 Jan 17. PMID: 23422873; http://dx.doi.org/10.1038/clpt.2013.4.

[97] Leckband SG, Kelsoe JR, Dunnenberger HM, George Jr AL, Tran E, Berger R, et al. Clinical pharmacogenetics implementation consortium guidelines for HLA-B genotype and carbamazepine dosing. Clin Pharmacol Ther September 2013;94(3):324–8. Epub 2013 May 21. PMID: 23695185; http://dx.doi.org/10.1038/clpt.2013.103.

[98] Muir AJ, Gong L, Johnson SG, Lee MT, Williams MS, Klein TE, et al. Clinical phrmacogenetics implementation consortium (CPIC) guidelines for IFNL3 (IL28B) genotype and PEG interferon-a-based regimens. Clin Pharmacol Ther February 2014;95(2):141–6. Epub 2013 Oct 4. PMID: 24096968; http://dx.doi.org/10.1038/clpt.2013.203.

[99] McLaren PJ, Ripke S, Pelak K, Weintrob AC, Patsopoulos NA, Jia X, et al. Fine-mapping classical HLA variation associated with durable host control of HIV-1 infection in African Americans. Hum Mol Genet October 1, 2012;21(19):4334–47. Epub 2012 Jun 19. PMID: 22718199; http://dx.doi.org/10.1093/hmg/dds226.

[100] Martin MA, Hoffman JM, Freimuth RR, Klein TE, Dong BJ, Pirmohamed M, et al. Clinical pharmacogenetics implementation consortium guidelines for HLA-B genotype and abacavir dosing: 2014 update. Clin Pharmacol Ther May 2014;95(5):499–500. Epub 2014 Feb 21. PMID: 24561393; http://dx.doi.org/10.1038/clpt.2014.38.

[101] Clancy JP, Johnson SG, Yee SW, McDonagh EM, Caudle KE, Klein TE, et al. Clinical pharmacogenetics Implementation Consortium (CPIC) guidelines for ivacaftor therapy in the context of CFTR genotype. Clin Pharmacol Ther June 2014;95(6):592–7. Epub 2014 Mar 5. PMID: 24598717; http://dx.doi.org/10.1038/clpt.2014.54.

[102] Caudle KE, Rettie AE, Whirl-Carrillo M, Smith LH, Mintzer S, Lee MT, et al. Clinical pharmacogenetics implementation consortium guidelines for CYP2C9 and HLA-B genotypes and phenytoin dosing. Clin Pharmacol Ther November 2014;96(5):542–8. Epub 2014 Aug 6. PMID: 25099164; http://dx.doi.org/10.1038/clpt.2014.159.

[103] Caudle KE, Thorn CF, Klein TE, Swen JJ, McLeod HL, Diasio RB, et al. Clinical pharmacogenetics implementation consortium guidelines for dihydropyrimidine dehydrogenase genotype and fluoropyrimidine dosing. Clin Pharmacol Ther December 2013;94(6):640–5. Epub 2013 Aug 29. PMID: 23988873; http://dx.doi.org/10.1038/clpt.2013.172.

[104] Scott SA, Sangkuhl K, Gardner EE, Stein CM, Hulot JS, Johnson JA, et al. Clinical pharmacogenetics implementation consortium guidelines for cytochrome P450-2C19 (CYP2C19) genotype and clopidogrel therapy. Clin Pharmacol Ther August 2011;90(2): 328–32. Epub 2011 Jun 29. Review. PMID: 21716271; http://dx.doi.org/10.1038/clpt.2011.132.

[105] Scott SA, Sangkuhl K, Stein CM, Hulot JS, Mega JL, Roden DM, et al. Clinical pharmacogenetics implementation consortium guidelines for CYP2C19 genotype and clopidogrel therapy: 2013 update. Clin Pharmacol Ther Sep 2013;94(3):317–23. Epub 2013 May 22. PMID: 23698643; http://dx.doi.org/10.1038/clpt.2013.105.

[106] Hicks JK, Bishop JR, Sangkuhl K, Müller DJ, Ji Y, Leckband SG, et al. Clinical pharmacogenetics implementation consortium (CPIC) guideline for CYP2D6 and CYP2C19 genotypes and dosing of selective serotonin reuptake inhibitors. Clin Pharmacol Ther August 2015;98(2):127–34. Epub 2015 Jun 29. Review. PMID: 25974703; http://dx.doi.org/10.1002/cpt.147.

[107] Hershfield MS, Callaghan JT, Tassaneeyakul W, Mushiroda T, Thorn CF, Klein TE, et al. Clinical pharmacogenetics implementation consortium guidelines for human leukocyte antigen-B genotype and allopurinol dosing. Clin Pharmacol Ther February 2013;93(2):153–8. Epub 2012 Oct 17. PMID: 23232549; http://dx.doi.org/10.1038/clpt.2012.209.

[108] Saito Y, Stamp LK, Caudle KE, Hershfield MS, McDonagh EM, Callaghan JT, et al. Clinical pharmacogenetics implementation consortium (CPIC) guidelines for human leukocyte antigen B (HLA-B) genotype and allopurinol dosing: 2015 update. Clin Pharmacol Ther January 2016;99(1):36–7. Epub 2015 Jul 16. PMID: 26094938; http://dx.doi.org/10.1002/cpt.161.

Chapter 11

Medical and Health Aspects of Genetics and Genomics

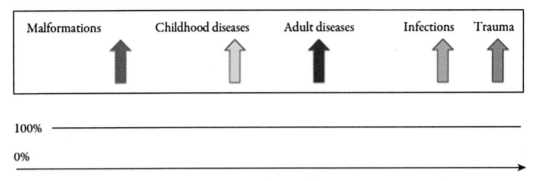

FIGURE 11.1   Genetic factors in human disease.

that affect both expression and modification of gene products in mediating the adult form of the disease [5]. Similarly, many genetically determined molecules and pathways have been characterized that are crucial in the pathogenesis of bronchial asthma [6]. It is now widely believed that a clearer understanding of the mechanisms and pathways of a disease will assist us in delineating distinct disease subtypes and may resolve many questions related to variable disease symptoms, progression, and response to therapy. This might help in revising the current diagnostic criteria. Eventually, genetics may contribute a new taxonomy of human disease in clinical practice.

Although genetics is acknowledged to be an important aspect of understanding the pathogenesis of disease, the genetic classification of human disease has not yet received full recognition. There is ample evidence in support of the argument that genetic factors are probably associated with all human diseases except for trauma (Fig. 11.1). However, underlying genetic and genomic factors such as genetically determined connective-tissue disorders, host response to infection, and tissue damage or inflammation could influence the outcome of trauma. Various categories of genetic disorders are considered to be rare, with a tendency to be included under the broad title of "organ-system diseases." Often these are listed as simply etiological factors rather than as a distinct disease category. This concept and approach is now rapidly being outdated, however, and replaced with new classes of diseases; for example disorders of genome architecture. This progress is seriously hampered by the lack of formal education at all levels and integration of appropriate technologies into the modern medical diagnostic and therapeutic infrastructure.

Traditionally, genetic diseases are classified as chromosomal (numerical or structural), mendelian or single-gene disorders, multifactorial/polygenic complex diseases or congenital anomalies, and diseases associated with specific mitochondrial gene mutations (Table 11.1). Apart from chromosomal disorders, essentially all genetic disorders result from some form of alteration or mutation

TABLE 11.1 The Classification of Genetic Disorders

Chromosomal	Numerical: aneuploidy
	Structural: deletion, duplication, inversion, isochromosome and ring chromosome
	Rearrangement-reciprocal or robertsonian translocation
Mendelian	Autosomal recessive
	Autosomal dominant
	X-linked recessive
	X-linked dominant
Epigenetic	Imprinting/parent of origin effect; indirect influence on gene function
Oligogenic	Distinct phenotype caused by two or more genes
Polygenic	Environmental interaction with several hundreds of low-risk alleles, genetic polymorphisms, and genomic copy number variations
Mitochondrial	Deletion, point mutations, polymorphic variants in mtDNA
Genomic variation	Copy number variation, single-nucleotide polymorphisms

occurring in a specific gene (single-gene diseases) or involving multiple loci spread across the human genome (polygenic disorders). The major impact of chromosomal disorders occurs before birth and inflicts a serious health burden throughout childhood and during the early years of life (Fig. 11.2). On the other hand, single-gene diseases can pose a real medical and health burden from the perinatal period to adult age, with a peak around mid-childhood. In contrast, the polygenic/multifactorial diseases tend to present late, except for developmental anomalies that will require active multidisciplinary care during a child's early life. A brief description of the major types of genetic diseases is included in this chapter. Any leading medical genetics textbook will contain a detailed description of all these group of genetic disorders.

CHROMOSOMAL DISORDERS

The entire human genome is spread around 23 pairs of chromosomes, including one pair specifically assigned to male (XY) or female (XX) gender, designated the "sex-chromosome pair." The chromosomal constitution of man is complex and comprises variable amounts of euchromatin and heterochromatin that exhibit a characteristic "banding-pattern" and are essential for the physical and distinctive appearance of a particular chromosome. Typically,

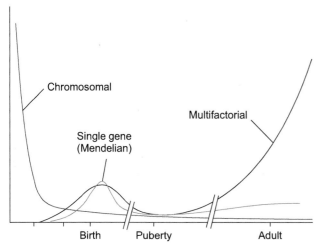

FIGURE 11.2 Distribution of different genetic disorders in various age groups. *Adapted with permission from Gelehrter TD, Collins FS, Ginsburg D. Principles of Medical Genetics. Lippincott Williams & Wilkins; 1998.*

a chromosome pair includes two homologues, each comprising a short arm (p) and a long arm (q) separated by the central heterochromatin-G-C–rich region designated the "centromere." A detailed account of the chromosome structure and fundamental changes that occur during meiosis and mitosis can be found in any leading textbook on basic genetics.

Chromosomal disorders are essentially disorders of the genome resulting from either loss or addition of a whole chromosome (aneuploidy) or parts of chromosomes (structural). A chromosome abnormality results in major disturbance in the genomic arrangement because each chromosome or part thereof consists of thousands of genes and several noncoding or methylation-sensitive polymorphic DNA sequences. The physical manifestations of chromosome disorders are often quite striking, characterized by growth retardation, developmental delay, and a variety of somatic abnormalities. Many chromosomal syndromes are now described and more are added with new genome diagnostic methods. The diagnosis and genetic management of these disorders fall within the scope of the subspecialty of clinical cytogenetics or, more specifically, molecular cytogenomics.

The management of chromosomal disorders requires a coordinated and dedicated team approach involving a wide range of clinicians and health professionals. A typical example is Down syndrome, resulting from either three copies of chromosome 21 (trisomy) (Fig. 11.3) or an addition to the long arm of chromosome 21, usually owing to an unbalanced meiotic rearrangement of a parental chromosomal translocation between chromosome 21 and one of the other acrocentric (centromere located at the end) chromosomes (robertsonian translocation). Down syndrome occurs in about one in 800 live births and increases in incidence with advancing maternal age. It is characterized by growth

FIGURE 11.3 Karyotype of a female (XX) with Down syndrome. Note trisomy 21 indicated by the *arrow*.

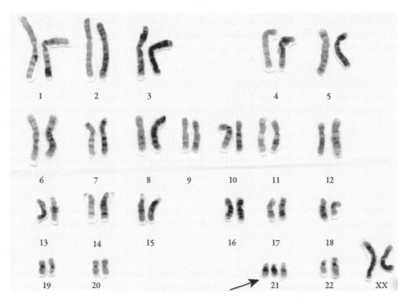

and developmental delay, moderate to severe mental retardation, and the characteristic facial appearance recognized by upward-slanting eyes. A major cause of death in these individuals is associated congenital heart defects that can complicate the clinical management in a significant proportion of Down syndrome cases. Prenatal diagnosis and prenatal assessment of the maternal risk for Down syndrome employing a variety of imaging and biochemical parameters is now established clinical and public health practice in most countries.

Clinically significant chromosome abnormalities occur in nearly 1% of live births and account for about 1% of pediatric hospital admissions and 2.5% of childhood mortality [7]. The loss or gain of whole chromosomes is often incompatible with survival, and such abnormalities are a major cause of spontaneous abortions or miscarriages. Almost half of the spontaneous miscarriages are associated with a major chromosomal abnormality. It is estimated that about a quarter of all conceptions may suffer from major chromosome problems, because approximately 50% of all conceptions may not be recognized as established pregnancies, and 15% of these end in a miscarriage. Essentially, the major impact of chromosomal disorders occurs before birth or during early life (see Fig. 11.2).

The delineation of rare and uncommon chromosomal disorders has been crucial in the gene-mapping of several mendelian (single-gene) disorders such as X-linked Duchenne muscular dystrophy and type 1 neurofibromatosis. The chromosomal regions involved in deletion, duplication, inversion, and breakpoints are part of a complex chromosomal rearrangement that provides an important clue and assists the keen researcher in focusing on genes located within the chromosomal segment.

MENDELIAN (SINGLE-GENE) DISORDERS

About 4000 human diseases are caused by mutations in single genes, and these constitute a major health burden. Single-gene disorders account for approximately 5–10% of pediatric hospital admissions and childhood mortality. The major impact of these disorders occurs in the newborn period and early childhood. However, these also constitute a significant proportion of adulthood diseases, notably late-onset neurodegenerative diseases and various forms of familial cancer. Although the majority of single-gene diseases are rare, some are relatively common and pose a major health problem; for example, familial hypercholesterolemia, a major predisposing factor in premature coronary artery disease, occurs in one in 500 people. Other good examples would be familial breast and colorectal cancers, which affect approximately one in 300 individuals. Some single-gene disorders are specific for certain populations, like Tay-Sachs disease among Ashkenazi Jews, cystic fibrosis in Caucasians, thalassemias among people from Southeast Asia and the Mediterranean countries, and sickle cell disease in people of Western African origin. Techniques in molecular biology have enabled the characterization of a number of mutated genes. Sickle cell disease was the first single-gene disorder to be defined at the molecular level. This has revolutionized the diagnosis and management of these disorders. The single-gene disorders are inherited in a simple mendelian manner, and hence justifiably are called *Mendelian disorders*. The genetic transmission of altered genes or traits follows principles set out by the Austrian monk Gregor Mendel in 1865 and based on his seminal work on garden pea plants [8]. Mendel inferred that "those characteristics that are transmitted entire, or almost unchanged by hybridization, and therefore constitute the characters of the hybrid, are termed dominant, and those that become latent in the process, recessive."

The nomenclature of these disorders reflects their gender-specific transmission and is supported by localization of an altered gene on either an autosome (1–22) or the X chromosome. Mendelian disorders are described as autosomal dominant, autosomal recessive, and X-linked recessive (Fig. 11.4) or X-linked dominant (Fig. 11.5). The latter pattern differs from X-linked recessive by having an excess of affected females in a family because the heterozygous mutation on the X chromosome can be transmitted to the daughter from an affected mother as well as the affected father. Sporadic X-linked dominant diseases are predominantly encountered in a female rather than a male because they are often lethal in the latter. A detailed family history and careful interpretation of the pedigree are essential prerequisites in the diagnosis of a mendelian disease. Accurate risk estimates, for use in genetic counseling, are impossible without a reliable and comprehensive pedigree. The major features of the individual inheritance pattern are described in leading genetic texts [1]. All human disorders and traits that follow the mendelian principles are listed in a major resource, Online Mendelian Inheritance in Man (OMIM), which is regularly revised online (www.OMIM.org).

POLYGENIC OR MULTIFACTORIAL DISORDERS

This group of disorders includes the most common and probably least understood human diseases. These diseases result from the interaction of known or unknown environmental factors (acquired or nongenetic factors) with multiple genes or pathogenic gene-specific or genomic polymorphisms (genetic factors), some of which may have a major effect, but the majority of which carry only a relatively minor causative or modulating effect. The minor additive effect of these multiple loci lowers the threshold of an organ or body system's ability to withstand environmental pressures, resulting in either a developmental anomaly or an abnormal disease state, commonly referred to as the *threshold effect*.

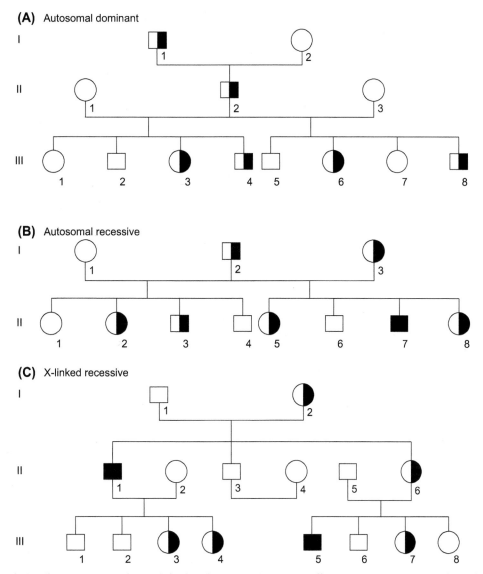

FIGURE 11.4 Typical pedigree appearances in mendelian inheritance. *Blank square*, unaffected male; *open circle*, unaffected female; *black-filled*, affected (homozygous); *half black-filled*, carrier (heterozygous).

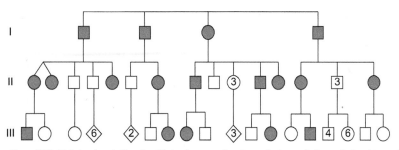

FIGURE 11.5 A pedigree with an X-linked dominant disorder. Note absence of male–male transmission; all daughters of an affected male would be heterozygous and thus could be symptomatic. *Adapted with permission from Gelehrter TD, Collins FS, Ginsburg D. Principles of Medical Genetics. Lippincott Williams & Wilkins; 1998.*

Examples include common congenital anomalies such as cleft lip, cleft palate, neural tube defects, and most congenital heart diseases. The common chronic medical diseases fall within this category of genetic disorders, including diabetes mellitus, coronary heart disease, hypertension, arthritis, bipolar affective disorder, schizophrenia, and most nonfamilial dementia. Understanding the genetic basis of common diseases remains the major challenge facing modern genetics and genomics.

The clinical impact of multifactorial diseases is significant both in the neonatal period as well as in adult life. It is estimated that about 25–50% of pediatric hospital admissions are related to these groups of disorders and are associated with 25–35% of childhood mortality. There is an even greater medical and health burden from these disorders during adult life because of the sufferers' chronic natural history of resulting medical diseases. For instance, diabetes mellitus and obesity account for about 40% of the adult medical problems in the developed and developing world.

Identification of any such disorder or condition is important in assessing risks to close relatives. A comparison of general population and multiple cases in a family would indicate a shift of the bell-shaped gaussian curve to the right, reflecting a lowered threshold with an increased incidence (Fig. 11.6). The precise additional risk would depend on the degree of relationship with the index case in the family. In addition, the gender of the index case is also important in assessing the liability. The genetic liability is estimated to be greater if the index case is of the gender with lowest incidence. For example, in the case of pyloric stenosis, greater risk would be applicable if the index case were a female, which carries the lowest birth prevalence. Finally, recurrence risks for a given population group are estimated to equal the square root of the birth incidence. For instance,

birth incidence of ventricular septal defect is approximately three per 1000; the recurrence risk to a first-degree relative, such as the next child, would be the square root of 0.003, or 3%. These figures are useful in genetic counseling to the couple after the birth of a child with a congenital anomaly.

This group of diseases poses the challenge of working out the mechanisms that determine the additive or interactive effects of many genes creating predisposition to diseases, which in turn manifest only in the presence of certain environmental factors. It is hoped that a combination of molecular genetic approaches, gene mapping, and functional genomics will enable a clearer definition of these genetic diseases. Several sections in this book will address this issue at length and focus on specific disease groups and systems.

MITOCHONDRIAL GENETIC DISORDERS

Apart from nuclear DNA (nDNA), a small proportion of DNA is also found in mitochondria in the cytoplasm of cells [mitochondrial DNA (mtDNA)]. Each cell contains 2–100 mitochondria, each of which contains 5–10 circular chromosomes. The 16.5-kb mtDNA molecule is free from any noncoding intronic regions and encodes two ribosomal RNA genes, 22 transfer RNAs, and 13 polypeptides that are parts of multisubunit enzymes involved in oxidative phosphorylation (see chapter: A Metagenomic Insight into the Human Microbiome: Its Implications in Health and Disease; Fig. 11.7). In comparison to the nDNA, the mtDNA is 20 times more prone to recurrent mutations, resulting in generation of mutagenic oxygen radicals in the mitochondria. The inheritance of mtDNA is exclusively maternal because of its cytoplasmic location. The mature sperm head contains very little mtDNA, because it is almost completely lost during the fertilization process, apparently with the loss of the tail that carried the bulk mtDNA in the cytoplasm. As a result of the wholly maternal cytoplasmic location, only females can transmit mitochondrial diseases to their offspring of either gender (see Fig. 11.7).

Because mtDNA replicates separately from the nDNA, and mitochondria segregate in daughter cells independently of the nuclear chromosomes (replicative segregation), the proportion of mitochondria carrying the mtDNA mutation can differ among somatic cells. This mitochondrial heterogeneity is also called *heteroplasmy* and plays an important part in the variable and tissue-specific phenotype of mitochondrial disease. Because different tissues have varying degrees of dependence on oxidative phosphorylation, with heart, muscle, and the central nervous system being the most dependent, the common manifestations of mitochondrial disease include cardiomyopathy, myopathy, and encephalopathy. Furthermore, oxidative phosphorylation declines with age, probably related to the accumulation of successive mtDNA mutations. Thus the clinical phenotype

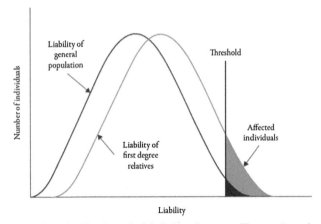

FIGURE 11.6 The "gaussian" bell-shaped curve to illustrate "genetic threshold," indicated by liability in the general population (black). A shift to the right (gray) indicates increased liability in first-degree relatives with an increased risk of recurrence. *With permission from Weatherall DJ. The new genetics and clinical practice, vol. 12. Oxford, UK: Oxford University Press; 1991.*

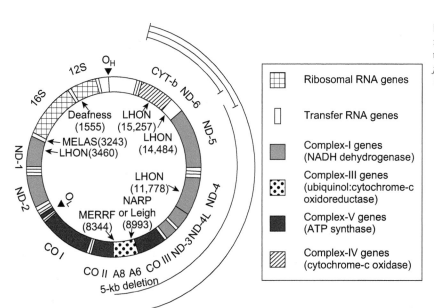

FIGURE 11.7 The human mitochondrial DNA molecule with examples of point mutations with their associated clinical phenotypes. *Adapted from Pulst SM. Neurogenetics: XA-GB. Oxford University Press; 2000.*

in a mitochondrial disease is not simply or directly related to mtDNA genotype but reflects several factors, including the overall capacity for oxidative phosphorylation determined by mtDNA and nDNA genes, the accumulation of somatic mtDNA mutations and degree of heteroplasmy, tissue-specific requirements of oxidative phosphorylation, and age.

Several mitochondrial diseases have now been characterized (Table 11.2). One of the best characterized is Leber hereditary optic neuropathy, which exclusively affects males. There is loss of central vision secondary to optic nerve degeneration. The vision loss usually occurs in the patient's 20s and can progress rapidly in some men. Eleven different missense mtDNA mutations in three different mitochondrial genes encoding respiratory chain enzyme subunits have been described. The phenotype in other mitochondrial diseases tends to include a combination of heart, muscle, and central nervous system manifestations, with considerable intrafamilial/interfamilial variability for the same mtDNA mutation. In addition, mitochondrial dysfunction can be part of the phenotype in some mendelian diseases where the mutant gene product presumably has a pathogenic influence on the mitochondrially mediated metabolic pathway. Examples of this are the autosomal recessive respiratory enzyme disorders. Genetic counseling and decision for prenatal diagnosis can be difficult in mitochondrial disorders because of the difficulty in predicting the phenotype in the affected pregnancy.

Finally, a high degree of sequence variation (polymorphism) is known to occur in the noncoding region of the mitochondrial chromosome (the D-loop). This polymorphism has been used in anthropological and evolutionary studies to trace the origins and links of human populations. In addition, this information has been applied in forensic analysis as well, to match maternal grandparents' mtDNA with

an orphaned child whose parents have "disappeared" during war, a natural disaster, or in mysterious circumstances.

GENOMIC DISORDERS

Recent advances in molecular genetics have enabled us to identify groups of disorders that result from characteristic mechanisms involving specific areas of the human genome. Often, these do not conform to the standard basic principles of genetics. A broad term, *genomic disorders*, has been coined to describe these conditions (Table 11.3) [11]. Essentially, in most of these clinical phenotypes there is underlying loss or gain of the genomic material belonging to a specific chromosomal region. The set of disrupted sequences might contain *haploinsufficient alleles* or polymorphisms (intergenic or intragenic) capable of altering or modulating gene function, often part of a *gene–molecule* family.

A number of hereditary disorders present with complex genetic pathology that do not follow the conventional principles of inheritance as outlined in the previous sections. There is now overwhelming evidence within these disorders that indicates unusual mechanisms, suggesting nontraditional inheritance. The mechanisms involve certain genomic regions that directly or indirectly influence regulation and expression of one or more genes manifesting in complex phenotypes. Currently, some of these disorders are listed either as chromosomal or as single-gene disorders. A detailed description of these disorders is beyond the scope of this chapter.

Disorders of Genomic Imprinting: Epigenetic Diseases

The term *epigenetics* refers to heritable factors that affect gene expression without any change in the gene coding

TABLE 11.2 Genetic Classification of Mitochondrial Disorders

Disorder	Major Clinical Features	Type of Gene	Mitochondrial DNA Mutation
Chronic progressive external ophthalmoplegia (CPEO)	External ophthalmoplegia, bilateral ptosis, mild proximal myopathy	tRNA	• A3243G, T8356C • Rearrangement (deletion/duplication)
Kearns-Sayre syndrome (KSS)	Progressive external ophthalmoplegia, onset <20 years, pigmentary retinopathy, cerebellar ataxia, heart block, cerebrospinal fluid protein >1 g/L		Rearrangement (deletion/duplication)
Pearson syndrome	Sideroblastic anemia of childhood, pancytopenia, renal tubular defects, exocrine pancreatic deficiency		Rearrangement (deletion/duplication)
Diabetes and deafness	Diabetes mellitus, sensorineural hearing loss	tRNA	• A3243G, C12258A • Rearrangement (deletion/duplication)
Leber hereditary optic neuropathy (LHON)	Subacute painless bilateral visual loss, age of onset 24 years, males>females (~4:1), dystonia, cardiac preexcitation syndromes	Protein encoding	G11778A, T14484C, G3460A
Neurogenic ataxia with retinitis pigmentosa (NARP)	Late-childhood or adult-onset peripheral neuropathy, ataxia, pigmentary retinopathy	Protein encoding	T8993 G/C
Leigh syndrome (LS)	Subacute relapsing encephalopathy, cerebellar and brainstem signs, infantile onset	Protein encoding	T8993 G/C
Exercise intolerance and myoglobulinuria	Exercise-induced myoglobulinuria	Protein encoding	Cytochrome B mutations
Mitochondrial encephalomyopathy with lactic acidosis and strokelike episodes (MELAS)	Strokelike episodes before 40 years, seizures and/or dementia, ragged-red fibers and/or lactic acidosis, diabetes mellitus, hypertrophic or dilated cardiomyopathy, deafness, cerebellar ataxia	tRNA	A32343G, T3271C, A3251G
Myoclonic epilepsy with ragged-red fibers (MERRF)	Myoclonus, seizures, cerebellar ataxia, myopathy, dementia, optic atrophy, bilateral deafness, peripheral neuropathy, spasticity, multiple lipomata	tRNA	A8344G, T8356C
Cardiomyopathy	Hypertrophic cardiomyopathy progressing to dilated cardiomyopathy	tRNA	A3243G, A4269G
Infantile myopathy/encephalopathy	Early-onset progressive muscle weakness with developmental delay	tRNA	T14709C, A12320G, G1606A, T10010C
Nonsyndromic sensorineural deafness	Early-onset, progressive, bilateral, moderate to severe sensorineural hearing loss	rRNA	A7445G
Aminoglycoside-induced nonsyndromic deafness	Early-onset, nonprogressive sensorineural deafness secondary to aminoglycoside administration	rRNA	A1555G

rRNA, Ribosomal RNA; *tRNA*, transfer RNA.

TABLE 11.3 Classification of Genomic Disorders

Disorders of genomic imprinting (epigenetic diseases)

Disorders of genome architecture (loss or gain of variable genomic segments)

Trinucleotide repeat disorders (variable number of nucleotide repeats with effect on gene function/expression)

Genomic variation (copy number variation; single-nucleotide polymorphisms)

sequence. These factors could be operational either during meiosis or mitosis and are often selective and preferential on the basis of their parent of origin. The term *imprinting* is commonly used to describe this important biological mechanism that is recognized to influence wide-ranging physical and molecular phenotypes. Numerous human diseases have now been confirmed to result from epigenetic changes in various parts of the genome. The term *epigenetic diseases* (or *genomic imprinting disorders*) refer to this group of diseases. Basic mechanisms related to the phenomenon

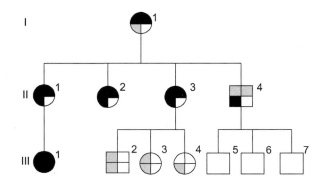

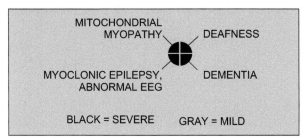

FIGURE 11.8 Pedigree of a family with mitochondrial encephalopathy with ragged-red muscle fibers (MERRF). Note segregation of different features with variable severity in the affected family members. *EEG*, Electroencephalograph.

of epigenetics or epigenomics are reviewed separately (see chapter: The Significance of Metabolomics in Human Health).

Epigenetic initiation and silencing is regulated by the complex interaction of three systems, including DNA methylation, RNA-associated silencing, and histone modification [12]. The relationship between these three components is vital for the expression or silencing of genes (Fig. 11.8). Disruption of one or another of these interacting systems can cause inappropriate expression or silencing of genes, leading to epigenetic diseases. Methylation of the C^5 position of cytosine residues in DNA has long been recognized as an epigenetic silencing mechanism of fundamental importance [13]. The methylation of CpG sites within the human genome is maintained by a number of DNA methyltransferases (DNMTs) and plays multifaceted roles in the silencing of transportable elements for defense against viral sequences and for transcriptional repression of certain genes. A strong suppression of the CpG methyl-acceptor site in human DNA results from mutagenic changes in 5-methylcytosine, causing C:G to T:A transitions. Normally, CpG islands, which are GC-rich, evolutionarily conserved regions of more than 500 base pairs, are kept free of methylation. These stretches of DNA are located within the promoter region of about 40% of mammalian genes and, when methylated, cause stable, heritable transcriptional silencing. Aberrant de novo methylation of CpG islands is a hallmark of human cancers and is found early during carcinogenesis [14].

In addition to DNA methylation, histone modifications have also been found to have epigenetic effects. Acetylation and methylation of conserved lysine residues of the amino-terminal tail domains are the key elements in histone modification. Generally, the acetylation of histones marks active, transcriptionally competent regions, whereas hypoacetylation histones are found in transcriptionally inactive euchromatic and heterochromatic regions. On the other hand, histone methylation can be a marker for both active and inactive regions of chromatin. Methylation of lysine residue 9 on the N terminus of histone 3 (H3K9) is a hallmark of silent DNA and is evenly distributed throughout the heterochromatic regions such as centromeres and telomeres, including the inactive X chromosome. In contrast, methylation of lysine 4 of histone 3 (H3K4) denotes activity and is predominantly found at promoter regions of active genes [15]. This constitutes a "histone code" that can be read and interpreted by different cellular factors. There is evidence that DNA methylation depends on methylation of H3K9 and can also be a trigger for its methylation. Recently, evidence has accumulated on the role of RNA in posttranscriptional silencing. In addition, RNA in the form of antisense transcripts (*XIST* or RNA interference) can also lead to mitotically heritable transcriptional silencing by the formation of heterochromatin. For example, transcription of antisense RNA leads to gene silencing and to the methylation of the structurally normal α-globin gene in patients with α-thalassemia. This could be one of the many human diseases resulting from epigenetic silencing caused by antisense RNA transcripts [16].

Mutations in genes that affect genomic epigenetic profiles can give rise to human diseases that can be inherited or somatically acquired (Table 11.4). These epigenetic mutations can be caused either by hypermethylation (silencing) of a regulating gene or by loss of methylation (activation) of another gene that has a positively modifying effect on the phenotype. The parental imprinting effect can be inferred by demonstrating the parental origin of the mutant allele. Similarly, either a loss or a gain of a chromosomal segment can result in the same situation. Confirmation of a specific chromosomal deletion or duplication is usually possible by using the fluorescence in situ hybridization (FISH) method. The paternal imprinting in this situation is commonly demonstrated by genotyping a set of polymorphic markers located within the chromosomal segment. Inheritance of the whole chromosomal homologue from one parent effectively confirms imprinting phenomenon, because the regulatory gene sequences for the pathogenic gene would be missing from the other parent. This characteristic abnormality is commonly referred to as *uniparental disomy (UPD)*. This could either be isodisomy (similar parental homologues) or heterodisomy (parental and grandparental homologues) (Fig. 11.9). The origin of UPD is believed to result from the loss of the additional chromosomal homologue, failing

TABLE 11.4 Recognizable Epigenetic Dysmorphic Syndromes [12]

Disease	Main Features	Epigenetic Mechanism
α-Thalassemia X-linked mental retardation (ATR-X) syndrome	α-Thalassemia, facial dysmorphic features, neurodevelopmental disabilities	Mutations in *ATRX* gene, hypomethylation of repeat and satellite sequences
Fragile X syndrome	Chromosome instability, physical and learning/behavioral difficulties	Expansion and methylation of CGG repeat in *FMR1* 5′ untranslated region, promoter methylation
Immunodeficiency, chromosome instability, and facial anomalies (ICF) syndrome	Chromosome instability, immunodeficiency	*DNMT3* mutations, DNA hypomethylation
Angelman syndrome	Seizures and intellectual disabilities	Deregulation of one or more imprinted genes at 15q11-13 (maternal)
Prader-Willi syndrome	Obesity, intellectual disabilities	Deregulation of one or more imprinted genes at 15q11-13 (paternal)
Beckwith-Wiedemann syndrome (BWS)	Organ overgrowth, childhood tumors	Deregulation of one or more syndrome imprinted genes at 11p15.5 (*IGF2, CDKN1C, KvDMR1*, etc.)
Russel-Silver syndrome	Growth delay, body asymmetry	Deregulation of one or more imprinted genes at 7p (maternal)
Rett syndrome	Seizures, intellectual disabilities	*MeCP2* mutations
Rubinstein-Taybi syndrome	Facial dysmorphism, intellectual disabilities	Mutation in cAMP-response-element-binding protein (CREB)–binding protein (histone acetylation)
Coffin-Lowry syndrome	Facial dysmorphism, developmental delay	Mutation in *RSk-2* (histone phosphorylation)

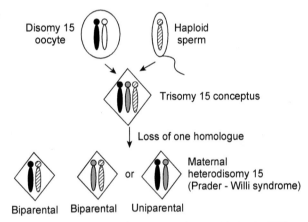

FIGURE 11.9 The origin of uniparental disomy 15 in Prader-Willi syndrome through trisomic rescue during early embryogenesis. Note different homologues (maternal heterodisomy).

which the conceptus would be trisomic. This mechanism is also called *trisomic rescue*.

For a maternally imprinted disorder, paternal UPD would be confirmatory, and maternal UPD diagnostic for the paternally imprinted condition. For example, maternal UPD is diagnostic for Prader-Willi syndrome, and paternal UPD for Angelman syndrome, both conditions being associated with a microdeletion of the 15q11 region. The parental origin of the 15q microdeletion follows the expected epigenetic pattern and is in keeping with the clinical diagnosis.

Recurrence risk estimates vary, depending on the specific epigenetic pattern. This information is crucial to obtain in order to offer accurate genetic counseling in any genomic imprinting disorder.

Many epigenetic diseases are associated with chromosomal alterations and manifest with physical and learning difficulties. For example, mutations in X-linked mental retardation with the alpha thalassemia phenotype *(ATRX)* result in consistent changes in the methylation pattern of ribosomal DNA, Y-specific repeats, and subtelomeric repeats. Another X-linked recessive mental retardation syndrome associated with a visible "fragile site" on the terminal part of the long arm of the X chromosome [fragile X syndrome (FRAXA)], results from de novo silencing of the pathogenic gene *FMR1*. The syndrome is characteristically associated with an abnormal expansion of CGG triplet repeats in the *FMR1* 5′ untranslated terminal region. Methylation of the expansion leads to silencing of the *FMR1* gene and under certain cultural conditions creates the visible fragile site on the X chromosome.

Epigenetic silencing is probably also significant in other neurodevelopmental disorders. For example, in Rett syndrome, a common cause of intellectual disability in young girls, mutations of the *MeCP2* gene are seen in about 80% of cases. The MeCP protein binds to methylcytosine residues and causes derepression of genes normally suppressed by DNA methylation. Despite the lack of firm evidence, it is thought likely that *MeCP2* might have a key role in the control of neuronal gene activity resulting in the pathology of Rett syndrome [17]. Interaction with another pathogenic gene (*CTKL5* or *STK9*) in Rett syndrome is likely to be important in the pathogenesis of this neurodevelopmental disorder [18]. On a wider genomic level, mutations in the *DNMT3b* gene, causing the immunodeficiency, centromeric region instability, and facial anomalies syndrome, result in deregulation of DNA methylation patterns. A notable example is that of Beckwith-Wiedemann syndrome (BWS), an overgrowth syndrome predisposing to Wilms tumor and other childhood tumors, which is associated with duplications and rearrangements of a small chromosomal region on the short arm of the chromosome (11p15.5). This region contains a cluster of genes that is susceptible to a number of epigenetic alterations, manifesting with the BWS phenotype and tumorigenesis, particularly Wilms tumor and other childhood embryonal tumors (Fig. 11.10). Loss of methylation in imprinting control regions (such as

KvDMR1) can cause deregulation of imprinting and either biallelic expression (*IGF2* and *H19*) or silencing (such as *CDKN1C*) of imprinted genes, which is seen in most sporadic BWS cases [19].

The epigenetic phenomenon is probably significant for the phenotypical manifestations in some other hereditary tumors. For example, transmission of autosomal dominant familial chemodectomas (nonchromaffin paragangliomas or glomus tumors) is exclusively via the paternal line (Fig. 11.11) [20]. The maternally derived gene is inactivated during oogenesis and can be reactivated only during spermatogenesis. This genetically heterogeneous cancer family syndrome is associated with germline mutations in succinate dehydrogenase subunits B (*SDHB*) and D (*SDHD*) [21].

Thus epigenetic changes are probably significant in a number of other complex phenotypes, particularly those associated with cancer and a number of degenerative diseases.

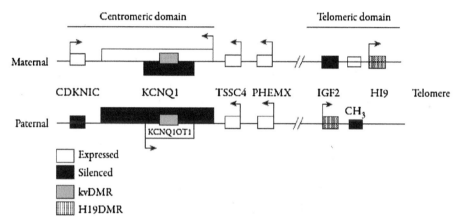

FIGURE 11.10 The cluster of genes on 11p15.5 associated with the phenotype of Beckwith-Wiedemann syndrome. The methylated region *KvDMR1* is indicated by the *gray box* within the gene *KCNQ1OT1* and marked *CH₃* on the maternal homologue. The methylated region between the *IGF2* and *H19* genes is indicated by the *hatched box* and marked *CH₃* on the paternal homologue. *With permission from Weksberg R, Nishikawa J, Caluseriu O, Fei YL, Shuman C, Wei C, et al. Tumor development in the Beckwith-Wiedemann syndrome is associated with a variety of constitutional molecular 11p15 alterations including imprinting defects of* KCNQ1OT1. *Hum Mol Genet 2001;10(26):2989–3000.*

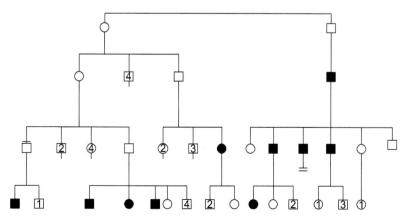

FIGURE 11.11 Pedigree showing paternal transmission of paraganglioma in a family. Note no maternal transmission among at-risk family members [20].

Disorders of the Genome Architecture

Recent completion of the Human Genome Project and sequencing of the total genomes of yeast and other bacterial species have enabled investigators to view genetic information in the context of the entire genome. As a result, it is now possible to recognize mechanisms of some genetic diseases at the genomic level. Among several biological processes, duplication of genes, gene segments, and repetitive gene clusters have helped in the evolution of mammalian genomes [22]. This aspect of genome architecture provides recombination hot spots between nonsyntenic regions of chromosomes that are distributed across the whole genome. These genomic regions become susceptible to further DNA rearrangements that may be associated with an abnormal phenotype. Such disorders are collectively grouped under the broad category of "genome architecture disorders" [11].

The term *genome architecture disorder* refers to a disease that is caused by an alteration of the genome that results in complete loss, gain, or disruption of the structural integrity of one or more dosage-sensitive gene(s) (Fig. 11.12). Notable examples include a number of chromosome deletion/duplication syndromes (Table 11.5). In these conditions, there is a critical rearranged genomic segment flanked by large (usually >10 kb), highly homologous, low-copy repeat (LCR) structures that can act as recombination substrates. Meiotic recombination between nonallelic LCR copies, also known as *nonallelic homologous recombination (NAHR)*, can result in deletion or duplication of the intervening segment.

Similarly, other chromosomal rearrangements, including reciprocal, robertsonian, and jumping translocations; inversions; isochromosomes; and small marker chromosomes, may also involve susceptibility to rearrangement related to genome structure or architecture. In several cases, LCRs, A-T–rich palindromes, and pericentromeric repeats are located at such rearrangement breakpoints. This susceptibility to genomic rearrangements is implicated not only in disease etiology, but also in primate genome evolution [25].

An increasing number of mendelian diseases (Table 11.6) are recognized to result from recurrent interchromosomal and intrachromosomal rearrangements involving unstable genomic regions facilitated by LCRs [26]. These genomic regions are predisposed to NAHR between paralogous genomic segments. LCRs usually span approximately 10–400 kb of genomic DNA, share 97% or greater sequence identity, and provide the substrates for NAHR, thus predisposing to rearrangements. LCRs have been shown to facilitate meiotic DNA rearrangements associated with several multiple malformation syndromes and some disease traits (Table 11.7). Seminal examples include microdeletion syndromes [Williams-Beuren syndrome (7q11del), DiGeorge syndrome (22q11del)]; autosomal dominant Charcot-Marie-Tooth disease type 1A (CTMD; *PMP22* gene duplication); hereditary neuropathy with liability to pressure palsy (HNPP; *PMP22* gene deletion) mapped to 17p11.2;

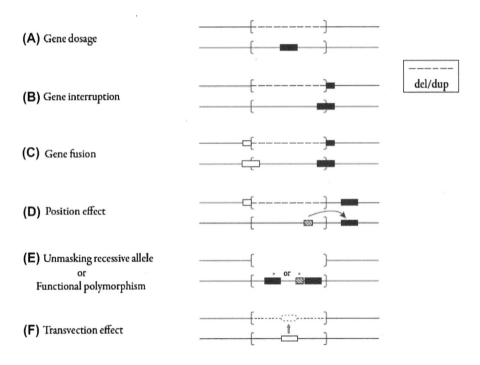

FIGURE 11.12   Molecular mechanisms for genomic disorders. *Asterisks*, point mutations; *brackets*, rearranged genomic interval; *dashed lines*, deleted or duplicated region; *filled horizontal rectangle*, gene; *horizontal hash-marked rectangle*, regulatory gene [24].

TABLE 11.5 Contiguous Gene Syndromes as Genomic Disorders [11]

Disorder (OMIM)	Inheritance Pattern Orientation	Locus	Gene	Rearrangement		Recombination Substrates		
				Type	Size (kb)	Repeat	%	Identity
William-Beuren syndrome (194050)	AD	7q11.23	ELN	del; inv	1600	>320	98	C
Prader-Willi syndrome (176270)	AD	5q11.2q13	?	del	3500	>500		C
Angelman syndrome (105830)	AD	15q11.2q13	UBE3A	del	3500	>500		C
Dup(15)(q11.2q13)		15q11.2q13	?	dup	3500	>500		C
Triplication 15q11.2q13		15q11.2q13	?	trip		>500		C
Smith-Magenis syndrome (18290)		17p11.2	RA13	del	4000	~250	98	C
Dup(17)(p11.2p11.2)	AD	17p11.2	PMP22	dup	4000	~250	98	C
DiGeorge/velocardiofacial syndrome (VCFS) (188400)	AD 192430	22q11.2	TBX1	del	3000/1500	~225–400	97–98	C
Male infertility (415000)	YL	Yq11.2	DBY	del	800	~10		D
AZFa microdeletion			USP9Y					
AZFc microdeletion 400024	YL	Yq11.2	RBMY	del	3500	~220	99.9	C
			DAZ?					

AD, Autosomal dominant; C, complex; D, direct; del, deletion; dup, duplication; inv, inversion; OMIM, Online Mendelian Inheritance in Man; trip, triplication; YL, Y-linked.

TABLE 11.6 Mendelian Genomic Disorders [11]

Disorders	Inheritance	Chromosome Location	Gene(s)	Rearrangement		Recombination Substrates		
				Type	Size (kb)	Repeat	%	Identity Orientation
Barter syndrome type III	AD	1p36	CLCNKA/8	del	11		91	D
Gaucher disease	AR	1q21	GBA	del	16	14		D
Familial juvenile nephronophthisis	AR	2q13	NPHP1	del	290	45	>97	D
Facioscapulohumeral muscular dystrophy	AD	4q35	FRG1?	del	25–222	3.3		D
Spinal muscular dystrophy	AR	5q13.2	SMN	inv/dup	500			I
Congenital adrenal hyperplasia caused by 21 hydroxylase deficiency	AR	6p21.3	CYP21	del	30		96–98	D
Glucocorticoid remediable aldosteronism	AD	8q21	CYP11B1/2	dup	45	10	95	D
β-Thalassemia	AR	11p15.5	HBB	del	4,(7)?			D
α-Thalassemia	AR	16p13.3	HBA	del	3,7,4.2?	4		D
Polycystic kidney disease type 1	AD	16p13.3	PKD1			50	95	
Charcot-Marie-Tooth syndrome type 1 (CMT1)	AD	17p12	PMP22	dup	1400	24	98.7	D
Hereditary neuropathy with liability to pressure palsy (HNPP)	AD	17p12	PMP22	del	1400	24	98.7	D
Neurofibromatosis type 1 (NF1)	AD	17q11.2	NF1	del	1500	85		D
Pituitary dwarfism	AR	17q23.3	GH1	del	6.7	2.24	99	D
CYP2D6 pharmacogenetic trait	AR	22q13.1	CYP2D6	del/dup	9.3	2.8		
Ichthyosis	XL	Xq28	STS	del	1900	20		D
Red-green color blindness	XL	Xq28	RCP/GCP	del	0	39	98	D
Incontinentia pigmenti	XL	Xq28	NEMO	del	10	0.870		D
Hemophilia A	XL	Xq28	F8	inv	300–500	9.5	99.9	I
Emery-Dreifuss muscular dystrophy (EMD)	XL	Xq28	EMD/FLN1	del/dup/inv	48	11.3	99.2	I
Hunter syndrome	XL	Xq28	IDS	inv/del	20	3	>88	

AD, autosomal dominant; AR, autosomal recessive; C, complex; D, direct; del, deletion; dup, duplication; I, inverted; inv, inversion.

TABLE 11.7 Disorders With Trinucleotide (Triplet) Repeats Expansion [11]

Disorder	Triplet	Location	Normal#	Mutation#
Fragile X syndrome	CGG	5′ Untranslated region	10–50	200–2000
Friedreich ataxia	GAA	Intronic	17–22	200–900
Kennedy syndrome/spinobulbar muscular atrophy (SBMA)	CAG	Coding	17–24	40–55
Spinocerebellar ataxia type 1 (SCA1)	CAG	Coding	19–36	43–81
Huntington disease	CAG	Coding	9–35	37–100
Dentatorubral-pallidoluysianatrophy (DRPLA)	CAG	Coding	7–23	49–>75
Machado-Joseph disease/ spinocerebellar ataxia type 3 (SCA3)	CAG	Coding	12–36	67–>79
Spinocerebellar ataxia type 2 (SCA2)	CAG	Coding	15–24	35–39
Spinocerebellar ataxia type 6 (SCA6)	CAG	Coding	4–16	21–27
Spinocerebellar ataxia type 7 (SCA7)	CAG	Coding	7–35	37–200
Spinocerebellar ataxia type 8 (SCA8)	TGG	Untranslated region	16–37	100–>500
Myotonic dystrophy	CTG	3′ Untranslated region	5–35	50–4000
Fragile site E (FRAXE)	CCG	Promoter	6–25	>200
Fragile site F (FRAXF)	GCC	?	6–29	>500
Fragile site 16 A (FRA16A)	CCG	?	16–49	1000–2000

and Smith-Magenis syndrome (SMS), a contiguous gene syndrome with del (17)(p11.2p11.2). Dominantly inherited male infertility related to *AZF* gene deletion follows a similar mechanism. In addition, this LCR-based complex genome architecture appears to play a major role in primate karyotype evolution, the pathogenesis of complex traits, and human carcinogenesis.

A notable example includes genetically heterogeneous CMTD. The disorder is also known as *hereditary motor and sensory neuropathy* by virtue of being a peripheral neuropathy caused by involvement of either the axonal or myelinated segments of the peripheral nerve. Genetically autosomal dominant, autosomal recessive, and X-linked dominant types are recognized. The disorder is not uncommon, affecting approximately one in 2500 of the adult population. This could be an underestimate, because medically the condition is benign, often not requiring any medical or surgical intervention. However, some affected individuals experience increasingly progressive neuromuscular weakness of distal muscles of lower legs, feet, distal forearms, and hands, with onset in the early teens, and causing severe locomotor restrictions.

An affected person usually seeks treatment late with relative hypertrophy of the upper calf muscles, described as an "inverted Champagne bottle" appearance (Fig. 11.13) and associated with *pes cavus* related to wasting of the small muscles of the feet. Similarly, wasting of the small muscles of hand leads to "claw hands." Neurophysiological studies remain an essential method of differentiating the two major types of CMTD. A reduced motor-nerve-conduction velocity of less than 35 m/s helps in differentiating type 1 CMTD from type 2 CMTD, in which the motor-nerve-conduction velocity is usually normal but the sensory-nerve conduction is often slow. Although this distinction is undoubtedly helpful in determining clinical management, application for genetic counseling is limited because both types are genetically heterogeneous. For instance, molecular characterization and gene mapping have confirmed the existence of at least four types of type 1 CMTD: autosomal dominant types 1a, 1b, and 1c, and the X-linked type. Similarly, there are distinct genetic types within the type 2 CMTD group.

Approximately two-thirds of cases of type 1 CMTD (CMT1) have a detectable 1.5 Mb duplication within a proximal chromosomal segment of the short arm of chromosome 17 (17p12) [27]. This duplicated chromosomal segment contains a gene for peripheral myelin protein called *PMP22*. This duplication results in the disruption of the gene, leading to abnormal myelination of the peripheral nerves, an essential molecular pathological step resulting in the CMT1 phenotype designated as *CMT1A*. The *CMT1A* duplication was visualized by multiple molecular methods, including FISH, pulsed-field gel electrophoresis (PFGE),

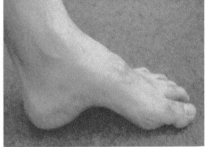

FIGURE 11.13 Lower legs and feet in Charcot-Marie-Tooth disease. Note characteristic lower-leg appearance and *pes cavus. Adopted with permission from Chapter 20, Genomic Medicine-Principles and Practice, Oxford University Press, New York, 2014.*

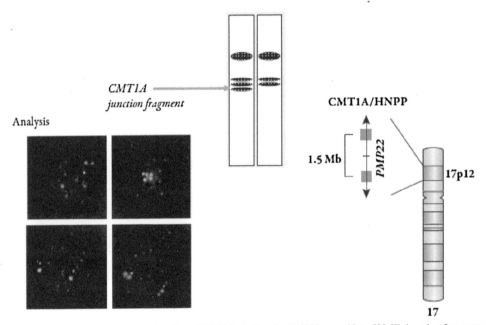

FIGURE 11.14 The 1.5-Mb duplicated chromosomal region of 17p12 including the *PMP22* gene. Note 500-Kb junction fragment allele flanking the *CMT1A* gene detected by pulsed-field gel electrophoresis (PFGE) and Southern analysis. Note additional 17p segment (dark gray) by metaphase (top two pictures) and interphase (lower two pictures) fluorescence in situ hybridization (FISH) [11].

and dosage differences of heterozygous alleles by restriction fragment length polymorphisms (Fig. 11.14). This finding led to further molecular studies on the origin of the 1.5 Mb duplicated 17p12 segment [28].

Studies by several investigators have revealed a significant variation in the size of marker alleles flanking the duplicated 17p12 region. It soon became apparent that a 500 kb allele cosegregated with 17p duplication in all affected individuals. This suggested a stable mutation and followed a precise recombination mechanism. However, in de novo duplication, the presence of repeated flanking marker alleles indicated the mechanism of unequal crossing-over, leading to duplication. Indeed, this was confirmed when a highly homologous >20 kb–size repeat sequence was confirmed flanking the 17 p duplication. It was appropriately named *CMT1A-REP*. As predicted by the unequal crossing-over model, *CMT1A-REP* was found to be present

in three copies on the *CMT1A* duplication-bearing chromosome. Interestingly, the presence of only one copy was soon demonstrated in another peripheral nervous system disorder known as *HNPP* [29]. Most clinically affected individuals with HNPP have mild to moderate episodic weakness of the lower limbs and occasionally of upper limbs when subjected to prolonged pressure such as sitting or sleeping. The disorder is dominantly inherited in an autosomal dominant manner. This is generally a clinically mild and benign hereditary neuropathy. The presence of only one copy results from a reciprocal deletion after unequal crossing-over involving the *CMT1A-REP* repeat (Fig. 11.15).

Similar observations were also made in relation to SMS, a contiguous gene syndrome associated with a microdeletion of the 17p11.2 segment [28]. Typically a child has with facial dysmorphic features, severe speech delay, an unusual sleep pattern, and behavioral problems with signs of self-harm. A specific junction fragment was detected by PFGE *(SMS-REP)*, involved in recurrent rearrangements resulting in either *SMS* or reciprocal 17p11.2 duplication. Pathogenic mutations in *RAI1* gene, mapped to the 17p11.2 chromosomal region, are now shown to be etiologically linked with SMS [18]. It is also possible to have both duplication and deletion at the same time, resulting from DNA rearrangements on both homologues of chromosome 17. This was demonstrated in a patient with mild delay and a family history of autosomal dominant carpel tunnel syndrome [30]. The occurrence of both the 17p11.2 duplication and the *HNPP* deletion in this patient reflects the relatively high rate at which these abnormalities arise and the underlying molecular characteristics of the genome in this region.

It is perfectly reasonable to accept the argument that similar molecular mechanisms apply in causing other genomic disorders (Table 11.7). The human genome has evolved an architecture (see chapter: The Human Genome) that may make us as a species more susceptible to rearrangements causing genomic disorders [28].

Disorders With Trinucleotide (Triplet) Repeats

Several disorders are recognized to have a phenomenon of earlier age-at-onset of disease in successive generations. This is known as *anticipation*. This observation failed to secure a valid biological explanation and had been put aside simply on the basis of biased ascertainment of probands or random variations in the age of onset. With the identification of unstable DNA repeats distributed across the genome, a molecular basis has been found for the phenomenon of anticipation. These unstable DNA repeats tend to increase in size during meiosis over successive generations. The abnormal expansion is correlated with reduced age of onset and increased severity with further expansion of DNA repeats. The characteristic pattern of the DNA repeat involving a set of three nucleotides is commonly referred to as *tri-nucleotide* or *triplet* repeats [31]. This soon became established as a novel class of mutation, and it offered a plausible explanation for the phenomenon of anticipation and variable clinical severity in a number of neurodegenerative diseases (see Table 11.7).

The X-linked recessive spinal bulbar atrophy (Kennedy syndrome; OMIM 313200) was one of the first hereditary neurological disorders recognized to be associated with CAG triplet repeats. The expanded region can occur anywhere in the gene and thus can disrupt the expression of the gene. In the case of FRAXA, the CGG repeats are found in the 5′ untranslated region of the first exon of *FMR1*, the pathogenic gene for FRAXA (Fig. 11.16). However, in the case of Friedreich ataxia, an autosomal recessive form of spinocerebellar ataxia (SCA), the expanded triplet repeat allele (GAA) occurs in the first intron of *X25*, the gene encoding frataxin. In Huntington disease (HD) and other inherited neurodegenerative disorders,

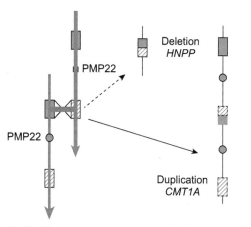

FIGURE 11.15 The unequal meiotic recombination (crossing-over), resulting in duplication *(CMT1A)* and deletion *(HNPP)* [11].

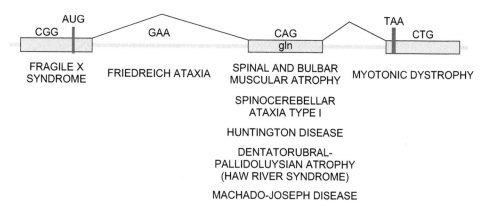

FIGURE 11.16 Location of four classes of triplet repeats in human diseases. Exons are shown in *light gray* with intervening introns as a *solid line*. The translation site AUG and termination signal TAA are indicated by *dark gray vertical bars*. *Adapted with permission from Gelehrter TD, Collins FS, Ginsburg D. Principles of Medical Genetics. Lippincott Williams & Wilkins; 1998.*

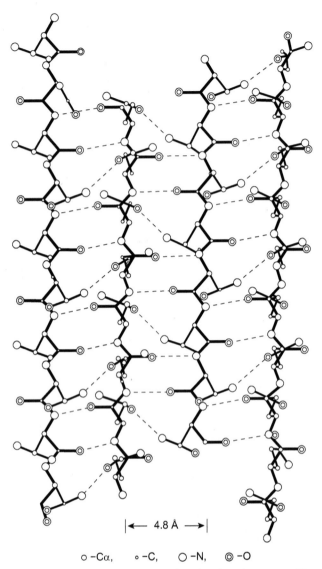

|← 4.8 Å →|

o −Cα,   o −C,   O −N,   ◎ −O

FIGURE 11.17 Schematic diagram of the polyglutamine tract resulting from abnormal expansion of CAG trinucleotide repeats [33].

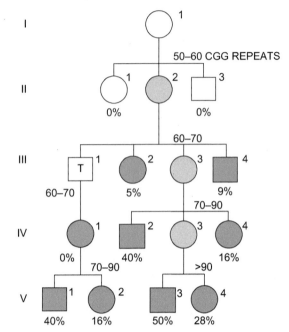

FIGURE 11.18 The Sherman paradox: a hypothetical pedigree showing affected members (dark gray) and carrier females (light gray); individual III.1 is a normal transmitting male; the percentage of risk for mental retardation is given for respective size of the triplet (CGG) repeats. *Adapted with permission from Gelehrter TD, Collins FS, Ginsburg D. Principles of Medical Genetics. Lippincott Williams & Wilkins; 1998 and Kumar D. Clinical medicine in the genome era: an introduction. Genomics Clin Med 2008;53:145.*

the CAG triplet repeats occur within exons and encode an elongated polyglutamine tract (Fig. 11.17). However, the expanded CTG triplet repeats of myotonic dystrophy (DM) are found in the 3′ untranslated region of the last exon of the DM protein kinase (myotonin) gene *(DM)*.

Each class of trinucleotide repeats exists in normal individuals. A pathogenic expansion is the one that is seen in clinically symptomatic individuals. Carriers for an X-linked disease also have an expanded allele (premutation) that does not usually result in an abnormal phenotype. However, it is likely that some carrier females might exhibit some manifestations, such as in FRAXA. An expanded allele in the premutation range in a male would not be associated with any clinical manifestations [normal transmitting male (NTM)], but this could further expand, resulting in all his daughters being carriers. However, recent studies have

provided data on the existence of late-onset gait ataxia in NTMs [32]. On the other hand, a normal-size CGG repeat in a normal male could undergo further expansion during meiosis, leading to a carrier daughter. This usually comes to light when a symptomatic grandson is confirmed to have pathogenic FRAXA expansion. Before availability of the molecular testing in FRAXA, this kind of unusual pedigree pattern in FRAXA was called the *Sherman paradox* (Fig. 11.18). Detailed molecular studies in the family are often necessary to offer accurate genetic counseling to at-risk carrier females. Carrier females are at an additional risk for developing premature ovarian failure, usually diagnosed when investigated for secondary infertility (see chapter: Genetics and Genomics of Reproductive Medicine and Health).

Genetic counseling in other neurodegenerative disorders with triplet repeats is often complicated. In particular, the clinical prediction in "borderline" expanded triplet repeats (intermediate allele) in HD is extremely difficult because of lack of reliable data. However, recent studies have produced some data that are likely to be helpful in genetic counseling.

Complex Genomic Diseases

All inherited disorders have a genetic abnormality present in the DNA of all cells in the body, including germ cells (sperm and egg), and can be transmitted to subsequent generations. In contrast, a genetic abnormality present only in specific somatic cells could not be transmitted. The genetic abnormality in a somatic cell can occur any time from the postconception stage to late adult life. The paradigm of somatic cell genetic disorders is cancer, where the development of malignancy is often the consequence of mutations in genes that control cellular growth. There are several such genes, and these are designated *oncogenes*. It is now accepted that all human cancer results from mutations in the nDNA of a specific somatic cell, making it the most common genetic disease. The various genetic mechanisms that can result in cancer are discussed in the chapter on cancer genomics (see chapter: Cancer Genetics and Genomics).

The clinical course and outcome of treatment in a number of acute and chronic medical conditions depend upon various factors. For instance, there is overwhelming evidence that highly polymorphic cytokine, interferon, and interleukin families of complex proteins influence the host's response to acute infection and physical injury or inflammation. Several genes encode these inflammatory pathway proteins. Similarly, association of human leukocyte antigens in the pathogenesis of a number of acute and chronic medical disorders is well known. In addition, interaction of mutations within these genes and with several other genomic polymorphisms, such as single nucleotide polymorphisms (SNPs), is probably important in several acute medical conditions, including trauma. This will have a major impact in critical care and acute medicine. The role of SNPs in modulating complex medical disorders such as diabetes mellitus, coronary heart disease, hypertension, and various forms of cancer is unclear. However, the complexity of interaction of SNPs with other genetic traits and loci is probably important in the prognosis of these disorders, in particular the outcome of therapeutic interventions. This argument probably justifies separating some of these disorders into the category of "complex genomic diseases."

Various cancers and degenerative diseases occur with increasing incidence in old age. However, these may also present at a younger age, such as childhood leukemias. The molecular mechanisms in these diseases are not entirely clear, but probably include defects in the DNA repair mechanisms, accelerated apoptosis, deregulation of imprinted genomic regions, and de novo chromosome rearrangements involving specific genomic regions. Although these disorders can be arguably included under the broad category of "multifactorial/polygenic diseases," the pattern of distribution and recurrence does not follow the agreed principles of multifactorial/polygenic inheritance as discussed elsewhere in this chapter.

As described in the previous section on epigenetics, epigenetic changes play a major role in the development of human cancer [12]. A high percentage of patients with sporadic colorectal cancer possess microsatellite instability and show methylation and silencing of the gene encoding *MLH1*. It is thus likely that epigenetic changes also predispose to genetic instability. In some cases, promoter-associated methylation of *MLH1* is found not only in the tumor but also in normal somatic tissues, including spermatozoa. These germline "epimutations" predispose individuals carrying abnormal methylation patterns to multiple cancers. Indeed, disruption of pathways that lead to cancer is often caused by the de novo methylation of the relevant gene's promoters [14]. Epigenetic silencing has been recognized as a third pathway satisfying Knudson's "two-hit" hypothesis for the silencing of tumor-suppressor genes [34].

Chromosomal rearrangements have long been associated with human leukemias. These result in the formation of fusion proteins, including histone acetyltransferases and histone methyltransferases, that influence upregulation of target genes. In acute promyelocytic leukemia, the oncogenic fusion protein promyelocytic leukemia–retinoic acid receptor-α (PML-RARα) causes the repression of genes that are essential for the differentiation of hematopoietic cells. Similarly, in acute myeloid leukemia, *AML-ETO* fusions recruit the repressive *N-COR-Sin3-HDAC1* complex and inhibit myeloid development [35]. There are further examples of complex genomic arrangements that result in other cancers and that can modify the therapeutic response. For example, mutations in genes for the adenosine triphosphatase complex are associated with poorer prognosis in patients with non–small cell lung cancer [36].

DISEASE SPECTRUM, BIOLOGICAL PATHWAYS, AND GENOTYPES

In modern medicine, diagnosis of any disease or morbid state relies on establishing the phenotype along the lines of agreed parameters (Table 11.8). The next logical step is to find evidence for likely pathophysiological changes that could be logically linked with one or more phenotypes. This could be demanding and challenging, because it might involve in-depth analysis and understanding of the complex biological (eg, metabolic or molecular) pathways implicated in the disease process (Table 11.9). Once this is achieved, then a correlation could be looked for with specific protein or enzyme systems recognized to be essential components of the core and successive biological pathways. Finding structural or functional abnormalities of any given protein or enzyme system would require undertaking investigations targeted at specific genes or genomic regions harboring particular genes. Establishing the precise genotype would then be the final piece in the complex jigsaw puzzle that is collectively viewed as a disease or syndrome. The individual genotype could be in any form (see Tables 11.8 and III), including gross chromosomal changes, specific genes or gene clusters, and extremely small segments of the genome. Thus the whole

TABLE 11.8 Taxonomy of Human Disease: Correlation of Clinical Phenotypes, Molecular Pathways, and Genotypes

I: Phenotypes

Clinical: symptoms and physical signs

Correlation and interpretation of the symptoms and signs with one or more of the following parameters with the aim of arriving at a diagnosis or most likely underlying mechanism of the disease:

Biochemical (eg, urea/electrolytes, blood gases)

Metabolic (eg, sugar/lipid/endocrine profiles)

Radiological: radiographs, ultrasound, CT/MRI scans, magnetic resonance spectroscopy, radioisotope, etc.

Pathological: histopathology, histochemistry, immune-histology, fluorescence microscopy, and electron microscopy

Hematological: hemoglobin, hematocrit, coagulation profile, etc.

Immunological: immunoglobulins (IgG, IgM, IgA, etc.); antibody profiles, eg, lupus; specific immunological investigation

Microbial/pathogens: battery of tests for bacterial, viral, protozoal, parasitic, and fungal infections, including specific pathogen profiles

Toxicology: poisons, alcohol, therapeutic and recreational drugs

Environmental (ecological): temperature extremes, high altitude, supersonic flying, and space travel

II: Correlation of the listed phenotypes with one or more of the biological/molecular pathways

Growth factors/growth factor receptors (eg, EGF/EGFRs, FGF/FGFRs, TGF/TGFRs, and VGF/VGFRs)

Dynamic cell/tissue factors (eg, protein kinase families such as P13, RAS/MAPK; tumor suppressor systems, etc.)

Respiratory chain/oxidative pathways (eg, mitochondrial and cyclooxygenase systems)

Cell/tissue response systems (eg, cytokines, interleukins, tissue necrosis factors, complement factors, etc.)

Apoptosis/senescence systems (eg, apoptotic pathways)

Scavenger/housekeeping systems (eg, lysosomal enzymes, α1 antitrypsin, DNA repair genes, etc.)

Metabolic regulatory systems (eg, insulin/glycemic regulation, lipids/hepatobiliary systems, Krebs cycle, etc.)

Energy regulation (eg, temperature regulation, energy conservation, nutritional state, etc.)

Hormonal regulation (eg, endocrine pathways, autocrine and paracrine pathways)

Vascular pathways (eg, angiogenetic systems, clotting/coagulation pathways, and platelet factors)

III: Correlation and/or interpretation with ALL of the listed phenotypes and biological/molecular pathways with one (or more) individual's genetic/genomic pathology

Chromosomal aberration: aneuploidy (eg, trisomy 21, 18, 13; triploidy), structural changes (microdeletion/duplication, inversion, ring chromosome, etc.)

OR

Specific gene mutations in a mendelian disorder (eg, βl thalassemia, cystic fibrosis, Duchenne muscular dystrophy)

OR

Mutations in two or more genes (oligogenic or multigenic) belonging to a gene/molecular family (eg, sarcomere genes in hypertrophic cardiomyopathy)

OR

Interaction of several hundreds and thousands of low risk alleles/genes with one or more environmental factors, including lifestyle (polygenic/multifactorial)

OR

Mitochondrial gene mutations and/or polymorphisms: several multisystem disorders that follow matrilineal inheritance pattern

TABLE 11.9 Genetic and Genomic Pathology in Human Disease

The following genetic/genomic pathology might be associated with one or more clinical phenotypes. Interpretation and precise diagnosis would depend on the natural history, family history, and sensitivity/specificity of genetic/genomic analyses:

- Epigenetic/epigenomic changes: mutations/deletions/duplication/inversion of genes or genomic segments adjacent to the promoter region of certain genes; genetic imprinting abnormality involving specific genes demonstrating "parent of origin effect," including complete, partial, or mosaic uniparental disomy
- Genome-wide abnormalities: pathogenic or disease-modification effect of structural variation across the genome; for example, single-nucleotide polymorphisms, copy number variations, deletions/duplications, nucleotide repeats (eg, trinucleotide repeats)
- Gene function/expression: specific "gain of function" or "loss of function" gene mutations (eg, increased risk of cancer/tumor resulting from a mutation in a tumor-suppressor gene); mutations in transcription factors associated with a range of developmental anomalies; abnormalities in RNA interference system associated with exaggerated or blunted therapeutic response; posttranslational modification/changes in the gene product associated with one or more phenotypes consistent with a disease diagnosis

landscape of the disease or diagnosis involves a closely linked network of three domains in the order of genotype–pathway–phenotype. In other words, a diagnosis of any disease or morbid state (including the mortal state) would be a cumulative process that should take into account all three of these domains: disease = genotype + molecular pathway + genotype.

SUMMARY

Developments in genetics and subsequently the sequencing of the human genome have provided us with an opportunity to review the taxonomy of human disease. Conventionally, the causation of human disease includes malformations, trauma, infection, immune dysfunction, metabolic abnormality, malignancy, and degenerative conditions associated with aging. Genetic factors have long been recognized in all of these disease groups. The traditional genetic categories of diseases include chromosomal disorders, single-gene or mendelian diseases, and several forms of multifactorial/polygenic conditions. In addition, somatic genetic changes and mutations of the mitochondrial genome probably account for a small, albeit important, number of diseases. These groups of disorders are well recognized and have an established place in the classification of human disease.

Recent developments in genome research have provided a wealth of data indicating different genomic mechanisms to explain complex pathogenesis in some disorders. The spectrum of these disorders is wide and includes both acute and chronic medical and surgical diseases. Perhaps it is reasonable to identify these disorders on the basis of their underlying molecular pathology, including genomic imprinting, genomic rearrangements, and gene–environment interactions involving multiple genes and genomic polymorphisms. This chapter has reviewed the genetic and genomic approaches in the classification of human disease. The stepwise approach is based on correlations of the clinical phenotype, supporting investigative phenotypes, and specific evidence from targeted genetic and genomic analyses. This approach would enable a modern clinician

to finally arrive at the final determining factor in the causation of human disease. The new taxonomy of human disease is likely to have a major impact on the practice of clinical medicine in the future.

REFERENCES

[1] Gelehrter TD, Collins FS, Ginsburg D. Principles of medical genetics. Lippincott Williams & Wilkins; 1998.

[2] Kumar D. Clinical medicine in the genome era: an introduction. Genomics Clin Med 2008;53:145.

[3] Bell JI. The double helix in clinical practice. Nature 2003;421(6921):414–6.

[4] Pearce JMS. Sir Thomas Lewis MD, FRS (1881–1945). J Neurol 2006;253(9):1246–7.

[5] Ioannidis JPA, Kavvoura FK. Concordance of functional in vitro data and epidemiological associations in complex disease genetics. Genet Med 2006;8(9):583–93.

[6] Daniels SE, Bhattacharrya S, James A, Leaves NI, Young A, Hill MR, et al. A genome-wide search for quantitative trait loci underlying asthma. Nature 1996;383(6597):247–50.

[7] Hall JG, Powers EK, McIlvaine RT, Ean VH. The frequency and financial burden of genetic disease in a pediatric hospital. Am J Med Genet 1978;1(4):417–36.

[8] Henig RM. The monk in the garden: the lost and found genius of Gregor Mendel, the father of genetics. Houghton Mifflin Harcourt; 2000.

[9] Weatherall DJ. The new genetics and clinical practice, vol. 12. Oxford, UK: Oxford University Press; 1991.

[10] Pulst SM. Neurogenetics: XA-GB. Oxford University Press; 2000.

[11] Kumar D. Disorders of the genome architecture: a review. Genomic Med 2008;2(3–4):69–76.

[12] Egger G, Liang G, Aparicio A, Jones PA. Epigenetics in human disease and prospects for epigenetic therapy. Nature 2004;429(6990):457–63.

[13] Jaenisch R, Bird A. Epigenetic regulation of gene expression: how the genome integrates intrinsic and environmental signals. Nat Genet 2003;33:245–54.

[14] Jones PA, Baylin SB. The fundamental role of epigenetic events in cancer. Nat Rev Genet 2002;3(6):415–28.

[15] Lachner M, O'Sullivan RJ, Jenuwein T. An epigenetic road map for histone lysine methylation. J Cell Sci 2003;116(11):2117–24.

[16] Lapidot M, Pilpel Y. Genome-wide natural antisense transcription: coupling its regulation to its different regulatory mechanisms. EMBO Rep 2006;7(12):1216–22.

[17] Martinowich K, Manji H, Lu B. New insights into *BDNF* function in depression and anxiety. Nat Neurosci 2007;10(9):1089–93.

[18] Carmona-Mora P, Molina J, Encina CA, Walz K. Mouse models of genomic syndromes as tools for understanding the basis of complex traits: an example with the Smith-Magenis and the Potocki-Lupski syndromes. Curr Genomics 2009;10(4):259.

[19] Murrell A, Heeson S, Cooper WN, Douglas E, Apostolidou S, Moore GE, et al. An association between variants in the *IGF2* gene and Beckwith-Wiedemann syndrome: interaction between genotype and epigenotype. Hum Mol Genet 2004;13(2):247–55.

[20] Petropoulos AE, Luetje CM, Camarata PJ, Whittaker CK, Lee G, Baysal BE. Genetic analysis in the diagnosis of familial paragangliomas. Laryngoscope 2000;110(7):1225–9.

[21] Dannenberg H, Dinjens WN, Abbou M, Van Urk H, Pauw BK, Mouwen D, et al. Frequent germ-line succinate dehydrogenase subunit D gene mutations in patients with apparently sporadic parasympathetic paraganglioma. Clin Cancer Res 2002;8(7):2061–6.

[22] Ovcharenko I, Loots GG, Nobrega MA, Hardison RC, Miller W, Stubbs L. Evolution and functional classification of vertebrate gene deserts. Genome Res 2005;15(1):137–45.

[23] Weksberg R, Nishikawa J, Caluseriu O, Fei YL, Shuman C, Wei C, et al. Tumor development in the Beckwith-Wiedemann syndrome is associated with a variety of constitutional molecular 11p15 alterations including imprinting defects of *KCNQ1OT1*. Hum Mol Genet 2001;10(26):2989–3000.

[24] Lupski JR, Stankiewicz P. Genomic disorders: molecular mechanisms for rearrangements and conveyed phenotypes. PLoS Genet 2005;1(6):e49.

[25] Gu W, Zhang F, Lupski JR. Mechanisms for human genomic rearrangements. Pathogenetics 2008;1(1):4.

[26] Zhang F, Carvalho C, Lupski JR. Complex human chromosomal and genomic rearrangements. Trends Genet 2009;25(7):298–307.

[27] Garcia CA, Malamut RE, England JD, Parry GS, Liu P, Lupski JR. Clinical variability in two pairs of identical twins with the Charcot-Marie-Tooth disease type 1A duplication. Neurology 1995;45(11):2090–3.

[28] Lupski JR. Genomic disorders: recombination-based disease resulting from genome architecture. Am J Hum Genet 2003;72(2):246.

[29] Murakami T, Garcia CA, Reiter LT, Lupski JR. Charcot-Marie-Tooth disease and related inherited neuropathies. Medicine 1996;75(5):233–50.

[30] Potocki L, Chen KS, Koeuth T, Killian J, Iannaccone ST, Shapira SK, et al. DNA rearrangements on both homologues of chromosome 17 in a mildly delayed individual with a family history of autosomal dominant carpal tunnel syndrome. Am J Hum Genet 1999;64(2):471–8.

[31] Margolis RL, McInnis MG, Rosenblatt A, Ross CA. Trinucleotide repeat expansion and neuropsychiatric disease. Archives General Psychiatry 1999;56(11):1019.

[32] Greco CM, Berman RF, Martin RM, Tassone F, Schwartz PH, Chang A, et al. Neuropathology of fragile X–associated tremor/ataxia syndrome (FXTAS). Brain 2006;129(1):243–55.

[33] Chen YW. Local protein unfolding and pathogenesis of polyglutamine-expansion diseases. Proteins Struct Funct Bioinforma 2003;51(1):68–73.

[34] Jones PA, Laird PW. Cancer-epigenetics comes of age. Nat Genet 1999;21(2):163–7.

[35] Jones LK, Saha V. Chromatin modification, leukaemia and implications for therapy. Br J Haematol 2002;118(3):714–27.

[36] Reisman D, Glaros S, Thompson EA. The SWI/SNF complex and cancer. Oncogene 2009;28(14):1653–68.

Chapter 12

Content and Variation of the Human Genome

S.E. Antonarakis[1,2,3,4]

[1]University of Geneva Medical School, Geneva, Switzerland; [2]University Hospitals of Geneva, Geneva, Switzerland; [3]iGE3 Institute of Genetics and Genomics of Geneva, Geneva, Switzerland; [4]HUGO (Human Genome Organization), Geneva, Switzerland

Chapter Outline

Introduction: "The Genome Anatomy" 161
  Protein-Coding Genes 165
    Size of Protein-Coding Genes 166
  Noncoding, RNA-Only Genes 166
    Ribosomal RNA (rRNA) Genes ~650–900 166
    Transfer RNA ~500 (49 Types) 167
    Small Nuclear RNA ~1912 168
    Small Nucleolar RNA ~978 168
    Micro RNAs 3828 168
    Long Noncoding RNAs 14,470 168
    Other Noncoding RNAs 168
  Regions of Transcription Regulation 168
  Conserved Elements Not Included in the Previous Categories 168
  Repetitive Elements 168
  Segmental Duplications 169

Special Genomic Structures Containing Selected Repeats 170
  Human Centromeres 170
  Human Telomeres 171
  Short Arms of Human Acrocentric Chromosomes 171
  Mitochondrial Genome 172
  Genomic Variability 172
    Single Nucleotide Polymorphisms 172
    Short Sequence Repeats 174
    Insertion/Deletion Polymorphisms 174
    Copy Number Variants 174
    Inversions 175
  Genome Variation as a Laboratory Tool to Understand the Genome 176
References 176

INTRODUCTION: "THE GENOME ANATOMY"

In order to be able to understand the biological importance of the genetic information in health and disease (assign a particular phenotype to a genome variant), we first needed to know the entire nucleotide sequence of the human genome. Thus an international collaborative project has been undertaken, named the *Human Genome Project*, to determine the nucleotide sequence of the human genome. The project was initiated on October 1, 1990, and was essentially completed in 2004. The potential medical benefits from the knowledge of the human genome sequence were the major rationale behind the funding of this international project. In addition, the involvement and contributions of the biotechnology company Celera (Alameda, Calif.) may have provided the necessary competition for the timely completion of the project. We now live in the "postgenome" era of medicine!

The goals of the different phases of the Human Genome Project were to:

1. determine the linkage map of the human genome [1,2]
2. construct a physical map of the genome by means of cloning all fragments and arrange them in the correct order [3,4]
3. determine the nucleotide sequence of the genome [5–8]
4. provide an initial exploration of the variation among human genomes.

As of October 2004, about 93% of the human genome (that corresponds to 99% of the euchromatic portion of the genome) has been sequenced to an accuracy of better than one error in 100,000 nucleotides [5–7]. The DNA that was used for sequencing from the public effort came from a number of anonymous donors [7], and that from the industrial effort came from five subjects, of which one is eponymous Dr. JC Venter [8]. The methodology used was also different between the two competing efforts; the public effort sequenced cloned DNA fragments that had been previously mapped, whereas that of Celera sequenced both ends of unmapped cloned fragments and subsequently

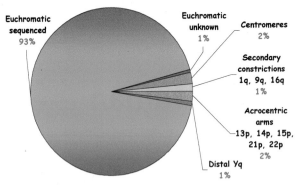

FIGURE 12.1   Pie chart of the fractions of the genome sequenced [*blue regions* (dark gray in print versions)] and not sequenced [*nonblue regions* (gray in print versions)].

assembled them in continuous genomic sequences. Fig. 12.1 shows the parts of the genome (mainly the heterochromatic fraction) that are not yet fully known: the pericentromeric regions; the secondary constrictions of 1q, 9q, and 16q; the short arms of acrocentric chromosomes (13p, 14p, 15p, 21p, 22p); and the distal Yq chromosome.

The total number of nucleotides of the finished sequence is 2,937,639,396, and the total estimated length that includes the current gaps is ~3,088,269,832 nucleotides (Table 12.1) from the last hg38 assembly of the human genome. The length of the human chromosomes ranges from ~46,7 (chromosome 21) to ~248,9 Mb (chromosome 1). The average guanine-cytosine (GC) content of the human genome is 41%. This varies considerably among the different

TABLE 12.1   Number of Nucleotides per Chromosome in the Reference Genome

Chromosome	Assembled Size	Sequenced Size	Total Gap Size
1	248956422	230481121	18475301
2	242193529	240548241	1645288
3	198295559	198100144	195415
4	190214555	189752667	461888
5	181538259	181265378	272881
6	170805979	170078524	727455
7	159345973	158970135	375838
8	145138636	144768136	370500
9	138394717	121790553	16604164
10	133797422	133263006	534416
11	135086622	134533742	552880
12	133275309	133137821	137488
13	114364328	97983128	16381200
14	107043718	90568149	16475569
15	101991189	84641328	17349861
16	90338345	81805944	8532401
17	83257441	82920216	337225
18	80373285	80089650	283635
19	58617616	58440758	176858
20	64444167	63944268	499899
21	46709983	40088623	6621360
22	50818468	39159782	11658686
X	156040895	154893034	1147861
Y	57227415	26415048	30812367
Total nuclear genome	3088269832	2937639396	150630436
M[a]	16569	16569	0

[a]*Chromosome M is the DNA of the mitochondrial genome.*
From ftp://ftp.ncbi.nlm.nih.gov/genomes/ASSEMBLY_REPORTS/All/GCF_000001405.26.stats.txt.

chromosomes and within the different bands of each chromosome. Chromosomal bands positive for Giemsa staining have lower average GC content of 37%, whereas in Giemsa-negative bands the average GC content is 42%. Interestingly, Giemsa negative bands are gene-rich regions of DNA.

Fig. 12.2A a schematic representation of the human chromosomes, depicts the ideogram of a banded karyotype. The DNA content of the blue and white bands has been determined, whereas that of the green and pink regions (heterochromatin) is still unknown.

Fig. 12.2B shows the current status of the "completion" of the human genome sequence [5]. Red bars above the chromosomes represent the sequence gaps. The DNA content of the red blocks (heterochromatin) is still unknown/unassembled. Heterochromatic regions of chromosomes are those that remain highly condensed throughout the cell cycle (see Section 3.2.1); it is thought that transcription is limited in those regions that contain a considerable number of repetitive elements, which renders the assembly of their sequence almost impossible.

The sequence of the human genome is freely and publicly available on the following genome browsers, which also contain many additional annotations:

http://genome.ucsc.edu/
http://www.ensembl.org/
http://www.ncbi.nlm.nih.gov/genome/guide/human/

Representative pages of two of these browsers are shown in Fig. 12.3.

There is now a considerable effort internationally to identify all the functional elements of the human genome. A collaborative project called ENcyclopedia Of DNA Elements (ENCODE) is currently in progress with the ambitious objective to identify all functional elements of the human genome [9–11]. The student is encouraged to look through the different options of the genome browsers for the different functional elements under study.

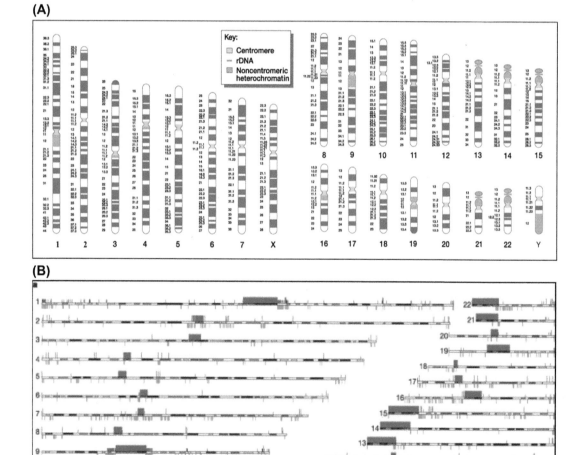

FIGURE 12.2 (A) Schematic representation of the human chromosomes. (B) Schematic representation of the completion of the human genome per chromosome. *Red regions* (gray in print versions) represent areas not sequenced; *blue regions* (dark gray in print versions) below the chromosomal line represent gaps in the sequences. The major blocks of unknown sequence include the short arms of acrocentric chromosomes, the pericentromeric sequences, and the large heterochromatic regions.

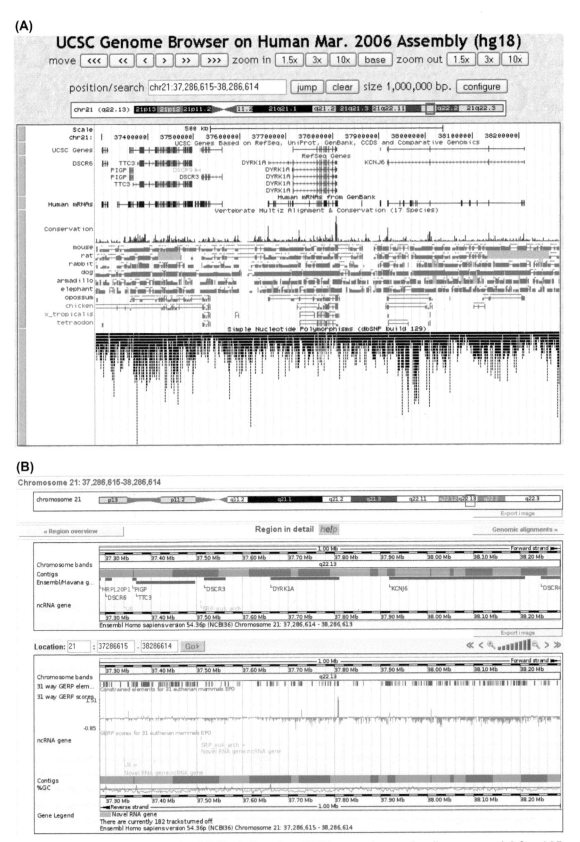

FIGURE 12.3 (A) A screenshot of the University of California Santa Cruz (UCSC) genome browser (http://genome.ucsc.edu/) for a 1-Mb region of chromosome 21 (21: 37,286,615–38,286,614). Among the many features that could be displayed, the figure shows genes, sequence conservation in 17 species, and single nucleotide polymorphisms (SNPs) that map in this 1-Mb region. The tracks shown from top to bottom include: a scale for the genomic region, the exact location in nucleotides, schematic representation of genes included in the UCSC database, the messenger RNA (mRNAs) from GenBank, the conservation in the species shown, and the location of SNPs. The color of some SNPs corresponds to synonymous and nonsynonymous substitutions. (B) A screenshot of the Ensembl genome browser (http://www.ensembl.org) for a 1-Mb region of chromosome 21 (21: 37,286,615–38,286,614). Among the many features that could be displayed, the figure shows genes (Ensembl/Havana gene track), noncoding RNAs (ncRNA gene track), sequence conservation in 31 species (31-way Genomic Evolutionary Rate Profiling track), and guanine-cytosine (GC) content in this 1-Mb region. The different browsers have similarities and differences, and some features could only be displayed in one browser.

The genome of today's humans, as a result of the evolutionary process, has similarities with the genomes of other species. The order of genomic elements has been conserved in patches in different species such that we could recognize today regions of synteny in different species; ie, regions that contain orthologous genes and other conserved functional elements. Fig. 12.4 shows a synteny map of conserved genomic segments in the human and the mouse.

The functional elements of the genome could be classified as:

1. protein-coding genes
2. noncoding RNA-only genes
3. regions of transcription regulation
4. conserved elements not included in the other categories.

Protein-Coding Genes

The total number of protein-coding genes is a moving target, because this number depends on the functional annotation of the genome, the comparative analysis with the genomes of other species, and the experimental validation. The different genome databases list slightly different numbers of human protein-coding genes; a consensus database called *Gencode* (http://www.gencodegenes.org/) contains the most updated and carefully curated gene numbers, and its last release was of March 2015. (http://www.gencodegenes.org/stats/current.html). Table 12.2 contains the summary statistics of gene numbers from the last release.

The human genes are not equally distributed in the chromosomes. Giemsa pale bands of chromosomes are gene rich, and this result in unequal numbers of genes per size unit for the different chromosomes. Fig. 12.5 displays the

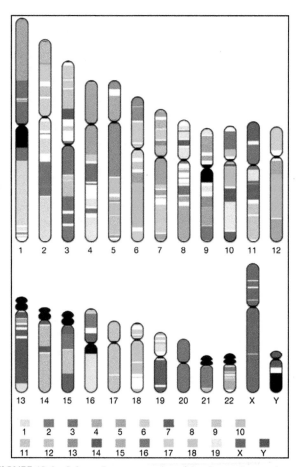

FIGURE 12.4 Schematic representation of the observed genomic segments between the human and mouse genomes. The color code of the human chromosomes corresponds to the different mouse chromosomes shown on the bottom. For example human chromosome 20 is all homologous to mouse chromosome 2; human chromosome 21 is homologous to mouse chromosomes 16, 17, and 10. Centromeric and heterochromatic regions, and acrocentric *p*-arms are shown in black. *From Lander ES, Linton LM, Birren B, Nusbaum C, Zody MC, Baldwin, J, et al. Initial sequencing and analysis of the human genome. Nature 2001; 409: 860–921.*

TABLE 12.2 Number of Protein-Coding and Other Genes in Humans Taken From the Last Release (March 2015) of Gencode

Genomic Element	Genes	Transcripts	Transcripts/Gen
Protein coding	19797	79795	4.0
Long noncoding RNAs	15931	27817	1.7
Long inter-genic noncoding RNAs	7678	13301	1.7
Pseudogenes	14477		
Small (<200 nt) noncoding RNAs	9882		
Micro RNAs	4093		

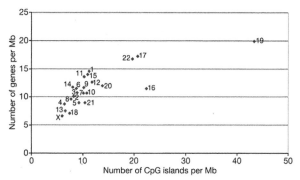

FIGURE 12.5 Gene density per chromosome, and correlation with CpG-rich islands of the genome. Chromosome 19, for example, has the highest gene content and the highest CpG island content. *From Lander ES, Linton LM, Birren B, Nusbaum C, Zody MC, Baldwin, J, et al. Initial sequencing and analysis of the human genome. Nature 2001; 409: 860–921.*

gene density per Mb for each chromosome and the correlation with CpG-rich islands.

Chromosomes 22, 17, and 19 are unusually gene-rich, whereas chromosomes 13, 18, and X are relatively gene-poor. (Interestingly, trisomies for chromosomes 13 and 18 are among the few human trisomies at birth.) Over the whole genome, the median size of an exon is 122 bp (mean = 145 bp), the median number of exons is 7 (mean = 8.8), and the median coding sequence encodes 367 amino acids (mean = 447 amino acids [7]). Thus the total number of annotated exons ranges from 183,000 to 268,000 (depending on the database) and the total exonic genome size is up to 78 Mb.

The mapping position of the genes can be seen in the genome browsers, and their names could be found in the gene nomenclature website (http://www.genenames.org/).

A single gene may have different isoforms resulting from alternative splicing of exons, alternative utilization of the first exon, and alternative 5′ and 3′ untranslated regions. There are on average four transcripts per protein-coding gene and this is likely to be an underestimate. The size of genes and number of exons varies enormously. The average genomic size of genes (according to the current annotation) is 27 kb. There are, however, small genes that occupy less than 1 kb, and large genes that extend to more than 2400 kb of genomic space. There are intronless genes [eg, histone *H1a* with a messenger RNA (mRNA) of 781 nt] and others with more than 360 exons (e.g., titin *TTN* with an mRNA of 80.7 kb).

Size of Protein-Coding Genes

The size of protein-coding genes within the human genome shows enormous variability (see Table 12.2). For example, the gene for histone H1a (*HIST1HIA*) is relatively small and simple, lacking introns and encoding mRNA sequences of 781 nt and a 215 amino acid protein (648 nt open reading frame). Dystrophin (*DMD*) is the largest protein-coding gene in the human reference genome, spanning a total of 2.2 MB, whereas Titin (*TTN*) has the longest coding sequence (80,780 bp), the largest number of exons (364), and the longest single exon (17,106 bp). Over the whole genome, the median size of an exon is 122 bp (mean = 145 bp); the median number of exons is 7 (mean = 8.8); and the median coding sequence encodes 367 amino acids (mean = 447 amino acids; see Table 21 in Ref. [7]).

Protein-coding genes can be grouped in families according to their similarity with other genes. These families of genes are the result of the evolutionary processes that shaped the genomes of the human and other species. The members of the gene families could be organized in a single cluster, or multiple clusters, or could be dispersed in the genome. Examples of gene families include the globin, immunoglobulin, histone, and olfactory receptor gene families. Furthermore, genes encode proteins with diverse but recognizable

domains. The database Pfam (http://pfam.sanger.ac.uk/, http://www.uniprot.org/) is a comprehensive collection of protein domains and families [12]; the current release of Pfam (27.0, March 2013) contains 14,831 protein families. For example, the WD40 domain family (PF00400) includes 277 human genes. The identification of domains helps in the prediction of the function and structure of a protein.

Pseudogenes are "dead," nonfunctional genes. These sequences that could be transcribed and spliced contain mutations that render them inactive.

Pseudogenes could be generated by several mechanisms that include:

1. Gene duplication events in which one of the duplicated copies accumulates inactivating mutations; alternatively the duplicated genes may be truncated. These pseudogenes are also called *nonprocessed pseudogenes*.
2. Transposition events in which a copy of complementary DNA is reinserted into the genome. These pseudogenes, called also *processed*, are not functional usually because they lack regulatory elements that promote transcription. In addition, inactivating mutations also occur in processed pseudogenes.

The current estimated number of human pseudogenes (according to the Gencode database) is 14,363 (10,736 are processed and 3202 unprocessed pseudogenes). These pseudogenes belong to 1422 families; eg, the immunoglobulin gene family has 226 pseudogenes (http://pseudofam. pseudogene.org/pages/psfam/overview.jsf).

The total number of human genes is not dramatically different from that of other "less" complex organisms. Fig. 12.6A depicts the current estimate of protein-coding gene number for selected species.

Noncoding, RNA-Only Genes

Besides the protein-coding genes, there is a growing number of additional genes (transcripts) that produce an RNA that is not translated to protein. The understanding of the functional consequences of these genes in health and disease is limited, and considerable research is devoted to this objective [13,14]. Table 12.2 contains the current number of these genes. The different classes of RNA-only genes are briefly discussed in the following sections.

Ribosomal RNA (rRNA) Genes   ~650–900

These are genes organized in tandemly arranged clusters in the short arms of the five acrocentric chromosomes (13, 14, 15, 21, and 22). The transcripts for 28S, 5.8S, and 18S ribosomal RNA (rRNAs) are included in one transcription unit, repeated 30–50 times per chromosome. These tandemly arranged genes are continuously subjected to concerted evolution, which results in homogeneous sequences caused by

(A)

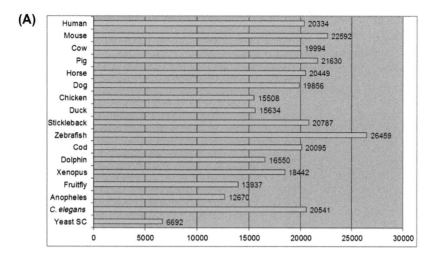

(B)

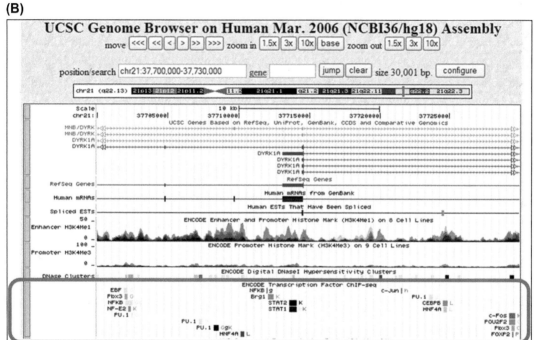

FIGURE 12.6   (A) Histogram of the September 2014 estimate of the number of protein-coding genes in selected species (ENSEMBL browser, http://www.ensembl.org/index.html). These numbers are subject to change. (B) depicts a 30-kb region on chromosome 21; *red bottom panel* (dark gray in print versions) schematically shows the location of the transcription factor binding sites likely to be involved in gene expression regulation. The names of the transcription factors studied are shown. *C elegans, Caenorhabditis elegans*; *SC, Saccharomyces cerevisiae*.

unequal homologous exchanges. The transcripts for the 5S rRNAs are also tandemly arranged, and the majority map to chromosome 1qter. There exist also several pseudogenes for all classes. The total number of these genes is polymorphic in different individuals. The best estimates of the number of rRNA genes are:

28S (components of the large cytoplasmic ribosomal subunit)	~150–200
5.8S (components of the large cytoplasmic ribosomal subunit)	~150–200

5S (components of the large cytoplasmic ribosomal subunit)	~200–300
18S (components of the small cytoplasmic ribosomal subunit)	~150–200

Transfer RNA   ~500 (49 Types)

At the last count there are 497 transfer RNA (tRNA) genes (usually 74–95 nucleotides long) encoded by the nucleus and transcribed by RNA polymerase III (additional tRNAs are encoded by the mitochondria genome). There are also

324 tRNA pseudogenes. The tRNA nuclear genes form 49 groups for the 61 different sense codons. Although the tRNA genes are dispersed throughout the genome, more than 50% of these map to either chromosomes 1 or 6.

Small Nuclear RNA   ~1912

Small nuclear RNAs (snRNAs) are heterogeneous small RNAs. A notable fraction of these are the spliceosome RNA genes, many of which are uridine-rich; the U1 group contains16 genes, whereas U2 contains four; U4, four, U6, 44, and the other subclasses are represented by one member.

Small Nucleolar RNA   ~978

This is a large class of RNA genes that process and modify the tRNAs and snRNAs. There are two main families: C/D box small nucleolar RNAs (snoRNAs) that are involved in specific methylations of other RNAs, and H/ACA snoRNAs that are mostly involved in site specific pseudouridylations. The majority of snoRNAs map in introns of protein-coding genes and can be transcribed by RNA polymerase II or III.

Micro RNAs   3828

These are single-stranded RNA molecules of about 21–23 nucleotides in length that regulate the expression of other genes. Micro RNAs (miRNAs) are encoded by RNA genes that are transcribed from DNA but not translated into protein; instead they are processed from primary transcripts known as *primary miRNA* to short stem-loop structures called *precursor miRNA*, and finally to functional miRNA. Mature miRNA molecules are complementary to regions in one or more mRNA molecules, which they target for degradation. A database of the known and putative miRNAs, and their potential targets, can be found in http://microrna. sanger.ac.uk/. Micro RNAs have been shown to be involved in human disorders.

Long Noncoding RNAs   14,470

This class contains RNA-only genes in which the transcript is more than 200 nucleotides.

The human genome encodes many thousands of long intergenic noncoding RNAs (lincRNAs), which have been annotated via the transcript evidence of active genes without protein-coding potential (7408 lincRNA genes; Gencode version 20). The functional information on lincRNAs remains limited. Based on a small number of well-studied cases, lincRNAs are involved in X chromosome inactivation, genomic imprinting, cell-cycle regulation, apoptosis, and establishment of cell identity. Despite the fact that the number of functionally annotated lincRNAs is rapidly growing, the question of whether the majority of the lincRNAs per se has a biological role is still unanswered.

Other Noncoding RNAs

The field of noncoding RNA series is constantly expanding. It is unknown how many and which types of additional RNA-only transcripts will be identified.

Regions of Transcription Regulation

The genome certainly contains information for the regulation of transcription. The current list of these regulatory elements includes promoters, enhancers, silencers, and locus control regions [15]. These elements are usually found in *cis* to the transcriptional unit, but there is growing evidence that there is also *trans* regulation of transcription. The objective of the ENCODE project [9,11] (http://www.genome.gov/ encode/) is to identify all the DNA regions involved in the regulation of transcription.

Fig. 12.6B shows a University of California Santa Cruz screenshot with the location of transcription factor binding sites in a small region of chromosome 21.

Conserved Elements Not Included in the Previous Categories

Since it is assumed that functional DNA elements are conserved, whereas nonfunctional DNA diverges rapidly, it is expected that all other conserved elements are of interest and should be studied for potential pathogenic variability. How much of the human genome is evolutionary conserved? The answer to this question depends on the species compared and the time of their common ancestor. Comparative genome analysis between the human and the mouse, for example, is particularly instructive, because the time of the common ancestor between these two species is estimated to be approximately 75 million years ago, and thus the conserved elements are likely to be functional. Approximately 5% of the human genome is conserved compared with mouse genome [16] (and to several other mammalian genomes). Of this, approximately 1–2% are the coding regions of protein-coding genes and approximately 3% are conserved noncoding DNA sequences (CNCs) [17,18] (Fig. 12.7). The function of the majority of CNCs is unknown. Please note that this 5% conserved fraction between human and mouse is an underestimate of the functional fraction of the human genome, which is likely to be bigger and to contain additional sequences not conserved with the mouse.

Repetitive Elements

The function of the majority of the human genome is unknown. Remarkably, approximately 45% of the genome is composed of repetitive elements, and roughly another 43% is not conserved and does not belong to the functional categories mentioned previously. The different interspersed repeats of the human genome are shown in Fig. 12.8.

- *Long interspersed nuclear elements (LINEs)* are autonomous transposable elements, mostly truncated nonfunctional insertions (average size of 900 bp). More than 20% of the human genome is polluted by LINEs. Transposable elements are mobile DNA sequences that can migrate to different regions of the genome. Autonomous elements are those that are capable of transposing by themselves. A small fraction of LINEs (~100) are still capable of transposing. The full LINE element is 6.1 kb long, has an internal PolII promoter, and encodes two open reading frames, an endonuclease, and a reverse transcriptase. Upon insertion, a target site duplication of 7–20 bp is formed. There are a few subclasses of LINEs, categorized according to their consensus sequence. The subfamily LINE1 is the only LINE capable of autonomous retrotransposition (copying itself and pasting copies back into the genome in multiple places). These LINEs enable transposition of short interspersed nuclear elements (SINEs), processed pseudogenes, and retrogenes. LINE retrotransposition has been implicated in human

disorders. LINEs are more abundant in G-dark bands of human chromosomes.
- *SINEs* mainly include the Alu repeats, which are the most abundant repeats in the human genome, occurring on average in every 3 kb. Thus 13% of the genome is polluted by Alu sequences and other SINEs. They are inactive elements originated from copies of tRNA or from single recognition particle (7SL) RNA. The full-length element is about 280 nt long and consists of two tandem repeats, each approximately 120 nt followed by polyA. Alu sequences are transcriptionally inactive and are GC-rich. SINEs can retrotranspose in a nonautologous way, because they use the LINE machinery for transposition. Because of their abundance, they could mediate deletion events in the genome that result in human disorders. SINEs are more abundant in G-light bands of human chromosomes.
- *Retrovirus-like (long terminal repeat [LTR]) transposons* are elements flanked by LTRs. Those that contain all the essential genes are theoretically capable of transposition, but that has not happened in the last several million years. Collectively they account for 8% of the genome. Most are known as human endogenous retrovirus sequences (HERVs) and are transposition defective. Transcription from the *HERV* genes may modulate the transcriptional activity of nearby protein-coding genes.
- *DNA transposon fossils* have terminal inverted repeats and are no longer active. They include two main families, *MER1* and *MER2*, and comprise 3% of the genome.

Updated information about repeats can be found in http://www.girinst.org/server/RepBase/.

Segmental Duplications

Approximately 5.2% of the human genome consists of segmental duplications or "duplicons"; ie, regions of more than 1 kb with >90% identified that are present more than once in the genome. Segmental duplications are either

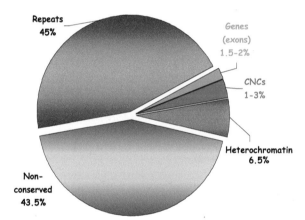

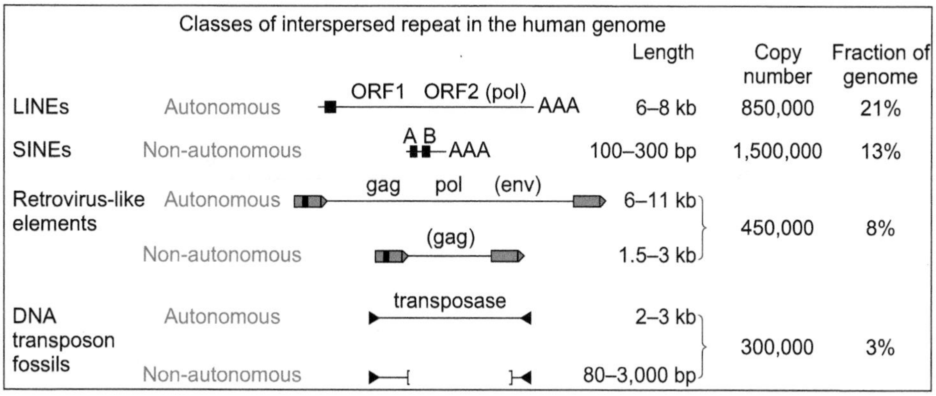

FIGURE 12.7   The pie chart depicts the different fractions of the genome. *CNC*, Conserved noncoding sequence.

Classes of interspersed repeat in the human genome			Length	Copy number	Fraction of genome
LINEs	Autonomous	ORF1 ORF2 (pol) ■━━━ AAA	6–8 kb	850,000	21%
SINEs	Non-autonomous	A B ■■━ AAA	100–300 bp	1,500,000	13%
Retrovirus-like elements	Autonomous	gag pol (env) ▭▭━━━━━▭	6–11 kb	450,000	8%
	Non-autonomous	(gag) ▭▭━━▭	1.5–3 kb		
DNA transposon fossils	Autonomous	transposase ▶━━━◀	2–3 kb	300,000	3%
	Non-autonomous	▶━┤ ├◀	80–3,000 bp		

FIGURE 12.8   Some basic characteristics of the classes of interspersed repeats in the human genome. *From Lander ES, Linton LM, Birren B, Nusbaum C, Zody MC, Baldwin, J, et al. Initial sequencing and analysis of the human genome. Nature 2001; 409: 860–921.*

intrachromosomal (on the same chromosome, 3.9%), or inter-chromosomal (on different chromosomes, 2.3%). Most of the duplicons are in the pericentromeric regions (Fig. 12.9).

Fig. 12.10 shows the distribution of intrachromosomal duplicons in the human genome [19,20]. These duplications are important in evolution and as risk factors for genomic rearrangements that cause human disorders because of unequal crossing-over in meiosis (pathogenic microdeletions and microduplications). Some examples of these include cases of α-thalassemia on chromosome 16p, Charcot-Marie-Tooth syndrome on chromosome 17p, velocardiofacial syndrome on chromosome 22q, Williams-Beuren syndrome on chromosome 7q, and Smith-Magenis syndrome on chromosome 17p.

These duplications are important in evolution and as risk factors for genomic rearrangements that cause human disorders because of unequal crossing-over in meiosis.

Fig. 12.11 provides a schematic representation of the products of a meiotic unequal crossing-over resulting in either duplication or deletion of the sequences bordered by "duplicons."

SPECIAL GENOMIC STRUCTURES CONTAINING SELECTED REPEATS

Human Centromeres

Human centromeres consist of hundreds of kilobases of repetitive DNA, some chromosome specific and some non-specific. Actually most of the remaining sequence gaps in the human genome map near and around centromeres. The structure of human centromeres is unknown, but the major repeat

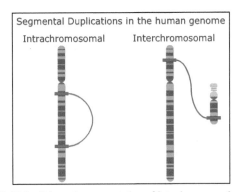

FIGURE 12.9 Schematic representation of intrachromosomal and inter-chromosomal segmental duplications. The repeat element is shown in *red region* (dark gray in print versions), and there is a *connecting line* indicating the highly homologous sequences.

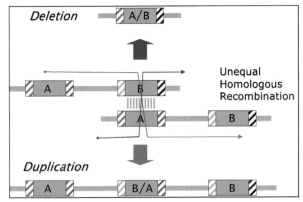

FIGURE 12.11 Unequal homologous recombination in meiosis results in two genomic alterations, duplication, or deletions. Regions A and B are identical in DNA sequence and they could misalign in meiosis. The crossing-over event will produce gametes with duplication (bottom alternative) and deletion (top alternative).

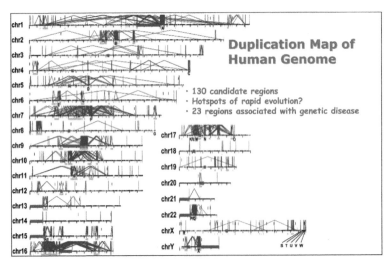

FIGURE 12.10 Schematic representation of the intrachromosomal segmental duplications. In each chromosome a *blue line* (gray in print versions) links a duplication pair. For example, on chromosome 21 there is only one duplicon shown; in contrast, on chromosome 22 there is a considerable number of duplications. *Richly blue areas* (black in print versions) are considered susceptible to microduplication/microdeletion syndromes. *From Bailey JA, Gu Z, Clark RA, Reinert K, Samonte RV, Schwartz S, et al. Recent segmental duplications in the human genome. Science 2002;297: 1003–07; Samonte RV, Eichler EE. Segmental duplications and the evolution of the primate genome. Nat Rev Genet 2002;3: 65–72.*

component of human centromeric DNA is α-satellite or alphoid sequence (a tandem repeat unit of 171 bp that contains binding sites for *CENP-B*, a centromeric binding protein). Fig. 12.12 shows an example of the structure of four human centromeres.

Human Telomeres

Human telomeres [21] consist of tandem repeats of a sequence (TTAGGG)n that spans about 3–20 kb, beyond which at the centromeric side there are about 100–300 kb

of subtelomeric-associated repeats [5] before any unique sequence is present. Fig. 12.13 schematically shows the sequence organization of six human subtelomeric regions.

Short Arms of Human Acrocentric Chromosomes

The finished sequence of the human genome does not include the short arms of acrocentric chromosomes (13p, 14p, 15p, 21p, and 22p). Cytogenetic data show that the

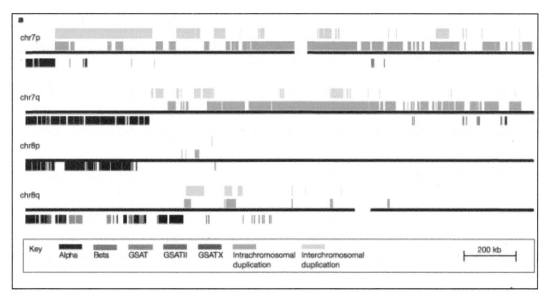

FIGURE 12.12   Examples of sequence organization of two human centromeres (chromosomes 7 and 8). Alphoid repeats are the major component of this special chromosomal structure; in addition, several other repetitive elements border the alphoid sequences. The length of these regions is also polymorphic in different individuals. *From Finishing the euchromatic sequence of the human genome. Nature 2004;431: 931–45.*

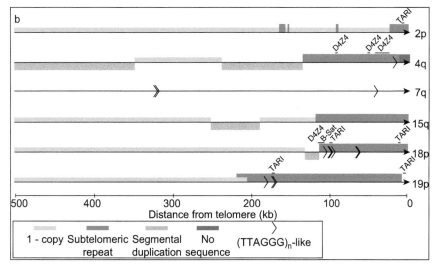

FIGURE 12.13   Examples of the sequence organization of six human telomeres (chromosomes 2pter, 4qter, 7qter, 15qter, 18pter, and 19pter). The *arrows* represent the TTAGGG repeat, and the *blue regions* (dark gray in print versions) depict the subtelomeric repeats that mainly consist of telomere associated repeat 1 family *(TAR1)*, D4Z4 (a 3.3-kb tandem repeat, each copy of which contains two homeoboxes and two repetitive sequences, LSau and hhspm3), and β-satellite sequences (a tandem repeat unit of 68 bp of the *Sau3A* family). *From Finishing the euchromatic sequence of the human genome. Nature 2004;431: 931–45.*

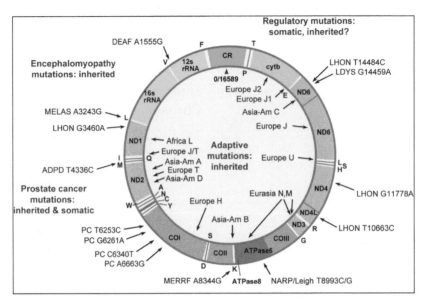

FIGURE 12.14   Schematic representation of circular mtDNA, its genes and clinical relevant mutations, and certain polymorphic markers. Letters within the ring depict the genes encoded. Letters on the outside indicate amino acids of the transfer RNA genes. The *CR* is the control of replication region that contains promoters for the heavy and light strands. *Arrows* outside show the location of pathogenic mutations.

p arms contain large heterochromatic regions of polymorphic length. Molecular analysis revealed that they are composed mainly of satellite and other repeat families, including satellites I (AT-rich repeat of a monomer of 25–48 bp), II (monomer repeat, 5 bp), III (monomer repeat, 5 bp), β-satellite (a tandem repeat unit of 68 bp of the *Sau3A* family), and other repeats. These repeats have a complex pattern and are often organized in subfamilies shared between different acrocentric chromosomes. The *p* arms encode the ribosomal genes.

Mitochondrial Genome

In human cells there is also the mitochondrial genome (Fig. 12.14) that is 16,568 nucleotides long and encodes for 13 protein-coding genes, 22 tRNAs, one 23S rRNA, and one 16S rRNA [22–24] (http://www.mitomap.org). The mitochondria genome–encoded genes are all essential for oxidative phosphorylation and energy generation in the cell. Each cell has hundreds of mitochondria and thousands (10^3–10^4) of mitochondrial DNA (mtDNA) copies. Human mtDNA has approximately a 20 times higher mutation rate than nuclear DNA. The inheritance of mtDNA is exclusively maternal. (The oocyte contains 10^5 mtDNA copies.) Several human phenotypes are caused by pathogenic mutations in the mitochondrial genome.

Genomic Variability

The human genome is polymorphic; ie, there are many DNA sequence variants among different individuals. These variants are the molecular basis of the genetic individuality of each member of our species. In addition, this genetic variability is the molecular substrate of the evolutionary process. Finally, this variability causes disease phenotypes or predisposes to common complex multifactorial phenotypes and traits.

Single Nucleotide Polymorphisms

The majority of the DNA variants are single nucleotide substitutions commonly known as *SNPs*. The first SNPs were identified in 1978 in the laboratory of YW Kan at 3′ to the β-globin gene [25]. These polymorphic sites have two alternative alleles. In Fig. 12.15, the depicted SNP has two alleles in the population: the blue C allele and the red T allele. The frequency of each allele could vary in different populations.

There is on average one SNP in approximately 1000 nucleotides between two randomly chosen chromosomes in the population. Many of these SNPs are common in the population (Fig. 12.16). A common SNP is that in which the minor allele frequency (MAF) is more than 5%. On average, two haploid genomes differ in about 3,000,000 SNPs. In addition, there is a large number of rare (MAF <1%) or near-rare (MAF between 1 and 5%) SNP variants, that could be identified by the genome sequencing of various individuals. The database of SNPs contains 62 million common and rare SNPs (http://www.ncbi.nlm.nih.gov/SNP/snp_summary.cgi, version 141, May 2014). Of those, approximately 443,000 are in the protein-coding regions of genes, and about 212,000 result in amino acid substitutions (nonsynonymous substitutions). International projects, such as the HapMap (http://www.hapmap.org/) [26–28], and the 1000 Genomes Project ("a deep catalog of human genetic variation"; http://www.1000genomes.org) have extensively studied the common and rare variability of genomes from different geoethnic groups and populations [29].

The informativeness of SNPs (and polymorphic variation in general) is usually measured by the number of heterozygotes in the population (*homozygotes* are individuals that contain the same allele in both chromosomes; *heterozygotes* are individuals that contain two different alleles in

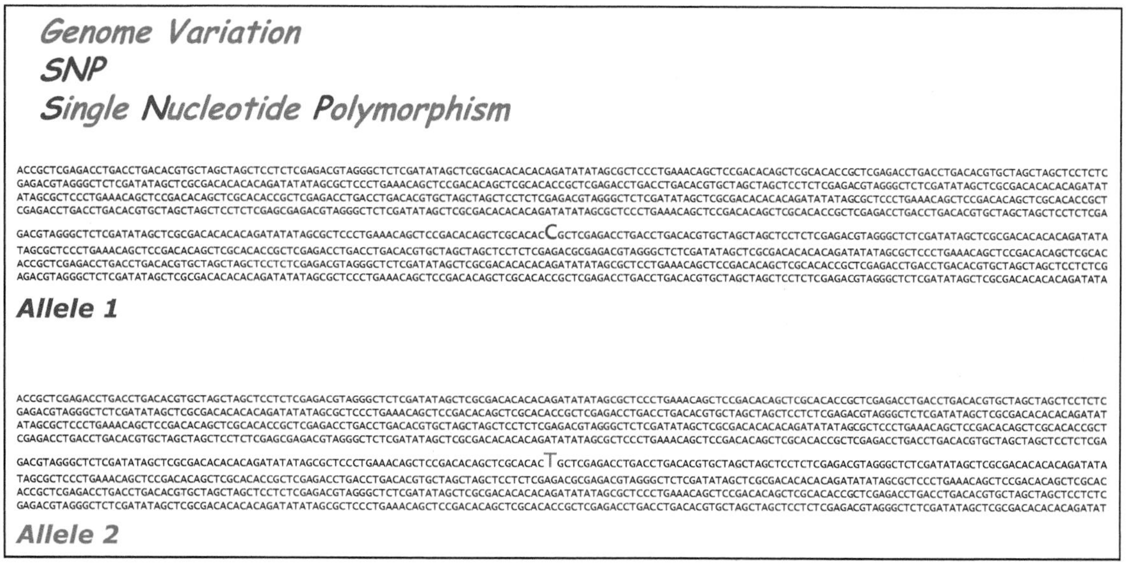

Genome Variation
SNP
Single Nucleotide Polymorphism

ACCGCTCGAGACCTGACCTGACACGTGCTAGCTAGCTCCTCTCGAGACGTAGGGCTCTCGATATAGCTCGCGACACACAGATATATAGCGCTCCCTGAAACAGCTCCGACACAGCTCGCACACCGCTCGAGACCTGACCTGACACGTGCTAGCTAGCTCCTCTC
GAGACGTAGGGCTCTCGATATAGCTCGCGACACACAGATATATAGCGCTCCCTGAAACAGCTCCGACACAGCTCGCACACCGCTCGAGACCTGACCTGACACGTGCTAGCTAGCTCCTCTCGAGACGTAGGGCTCTCGATATAGCTCGCGACACACAGATAT
ATAGCGCTCCCTGAAACAGCTCCGACACAGCTCGCACACCGCTCGAGACCTGACCTGACACGTGCTAGCTAGCTCCTCTCGAGACGTAGGGCTCTCGATATAGCTCGCGACACACAGATATATAGCGCTCCCTGAAACAGCTCCGACACAGCTCGCACACCGCT
CGAGACCTGACCTGACACGTGCTAGCTAGCTCCTCTCGAGCGAGACGTAGGGCTCTCGATATAGCTCGCGACACACAGATATATAGCGCTCCCTGAAACAGCTCCGACACAGCTCGCACACCGCTCGAGACCTGACCTGACACGTGCTAGCTAGCTCCTCGA

GACGTAGGGCTCTCGATATAGCTCGCGACACACAGATATATAGCGCTCCCTGAAACAGCTCCGACACAGCTCGCACACGCTCGAGACCTGACCTGACACGTGCTAGCTAGCTCCTCTCGAGACGTAGGGCTCTCGATATAGCTCGCGACACACAGATATA
TAGCGCTCCCTGAAACAGCTCCGACACAGCTCGCACCGCTCGAGACCTGACCTGACACGTGCTAGCTAGCTCCTCTCGAGACGCGAGACGTAGGGCTCTCGATATAGCTCGCGACACACAGATATATAGCGCTCCCTGAAACAGCTCCGACACAGCTCGCAC
ACCGCTCGAGACCTGACCTGACACGTGCTAGCTAGCTCCTCTCGAGACGTAGGGCTCTCGATATAGCTCGCGACACACAGATATATAGCGCTCCTGAAACAGCTCCGACACAGCTCGCACACCGCTCGAGACCTGACCTGACACGTGCTAGCTAGCTCCTCTCG
AGACGTAGGGCTCTCGATATAGCTCGCGACACACAGATATATAGCGCTCCCTGAAACAGCTCCGACACAGCTCGCACACCGCTCGAGACCTGACCTGACACGTGCTAGCTAGCTCCTCTCGAGACGTAGGGCTCTCGATATAGCTCGCGACACACAGATATA

Allele 1

ACCGCTCGAGACCTGACCTGACACGTGCTAGCTAGCTCCTCTCGAGACGTAGGGCTCTCGATATAGCTCGCGACACACAGATATATAGCGCTCCCTGAAACAGCTCCGACACAGCTCGCACACCGCTCGAGACCTGACCTGACACGTGCTAGCTAGCTCCTCTC
GAGACGTAGGGCTCTCGATATAGCTCGCGACACACAGATATATAGCGCTCCCTGAAACAGCTCCGACACAGCTCGCACACCGCTCGAGACCTGACCTGACACGTGCTAGCTAGCTCCTCTCGAGACGTAGGGCTCTCGATATAGCTCGCGACACACAGATAT
ATAGCGCTCCCTGAAACAGCTCCGACACAGCTCGCACACCGCTCGAGACCTGACCTGACACGTGCTAGCTAGCTCCTCTCGAGACGTAGGGCTCTCGATATAGCTCGCGACACACAGATATATAGCGCTCCCTGAAACAGCTCCGACACAGCTCGCACACCGCT
CGAGACCTGACCTGACACGTGCTAGCTAGCTCCTCTCGAGCGAGACGTAGGGCTCTCGATATAGCTCGCGACACACAGATATATAGCGCTCCCTGAAACAGCTCCGACACAGCTCGCACACCGCTCGAGACCTGACCTGACACGTGCTAGCTAGCTCCTCTCGA

GACGTAGGGCTCTCGATATAGCTCGCGACACACAGATATATAGCGCTCCCTGAAACAGCTCCGACACAGCTCGCACACTGCTCGAGACCTGACCTGACACGTGCTAGCTAGCTCCTCTCGAGACGTAGGGCTCTCGATATAGCTCGCGACACACAGATATA
TAGCGCTCCCTGAAACAGCTCCGACACAGCTCGCACCGCTCGAGACCTGACCTGACACGTGCTAGCTAGCTCCTCTCGAGACGCGAGACGTAGGGCTCTCGATATAGCTCGCGACACACAGATATATAGCGCTCCCTGAAACAGCTCCGACACAGCTCGCAC
ACCGCTCGAGACCTGACCTGACACGTGCTAGCTAGCTCCTCTCGAGACGTAGGGCTCTCGATATAGCTCGCGACACACAGATATATAGCGCTCCTGAAACAGCTCCGACACAGCTCGCACACCGCTCGAGACCTGACCTGACACGTGCTAGCTAGCTCCTCTCG
GAGACGTAGGGCTCTCGATATAGCTCGCGACACACAGATATATAGCGCTCCCTGAAACAGCTCCGACACAGCTCGCACACCGCTCGAGACCTGACCTGACACGTGCTAGCTAGCTCCTCTCGAGACGTAGGGCTCTCGATATAGCTCGCGACACACAGATAT

Allele 2

FIGURE 12.15 Schematic representation of a single nucleotide polymorphism (SNP). Allele 1 has a C in the sequence, whereas allele 2 contains a T in the same position.

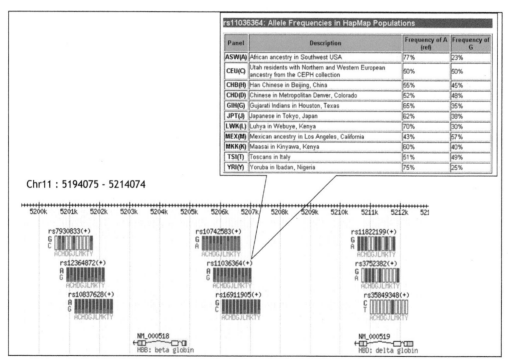

FIGURE 12.16   The genomic region of Chr11: 5194075–5214074 is shown. For each of the nine SNPs shown in the bottom, the frequency of the two alternative alleles is shown in different populations. For example, for SNP rs11036364 that maps between the *HBB* and *HBD* globin genes, the allele frequencies are shown in the callout. The four original populations of the HapMap project were *EUR, YRI, JPT,* and *CHB,* and the other populations were added in a later stage. *Adapted from http://www.hapmap.org/.*

their chromosomes). If p is the frequency of one allele and q the frequency of the other allele (obviously $p+q=1$), it follows that the frequency of homozygotes for one allele is p^2, and the frequency of the homozygotes for the other allele is q^2, whereas the frequency of the heterozygotes is $2pq$. In a large population with random mating (and some other assumptions), there is *Hardy-Weinberg equilibrium:*

$p^2+2pq+q^2=1$. This equation is useful for the calculation of the number of heterozygotes based on the allele frequencies; it also allows the estimation of the mutant gene frequency and carrier numbers for monogenic genetic disorders.

The pattern of DNA polymorphisms in a single chromosome is called *haplotype* (a contraction of *haploid*

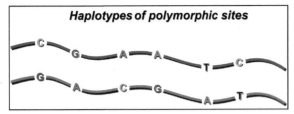

FIGURE 12.17 Schematic representation of haplotype of polymorphic variants in a segment of the genome. The parental origin is shown as the *blue* (dark gray in print versions) (paternally-inherited) and *red* (gray in print versions) (maternally-inherited) lines. SNPs are shown as letters interrupting the lines. The haplotype is defined as the combination of SNP alleles per haploid genome.

genotype; allelic composition of an individual chromosome). In the example of Fig. 12.17, the haplotype of polymorphic sites for the paternal (blue) chromosome is CGAATC, whereas for the maternally inherited red chromosome it is GACGAT.

When two (or more) polymorphic sites are in random association, they are in a state of linkage equilibrium; however, if two (or more) polymorphic sites are not in random association they are in a state of *linkage disequilibrium (LD)*. Figure shows two SNPs: the first C/G and the second A/C. There are four theoretical haplotypes with expected frequencies derived from the frequencies of the alleles of the two SNPs. If the allele frequencies of the first SNP are 0.6/0.4 and that of the second SNP are 0.3/0.7, respectively, then the expected frequencies of the four haplotypes are 0.18, 0.42, 0.12, and 0.28. However, the expected frequencies are different from the observed because there exists an LD between the two SNPs. The reasons for LD are numerous and include recombination rate, genetic distance between the two markers, age of the polymorphic site, genetic drift, and selection. Please note that allele A of the second SNP predicts the allelic state of the first SNP (which is always G because haplotype 1 in this example has not been observed in the population). If a new pathogenic mutation that causes a monogenic disorder occurs in haplotype 2, for example, it will also be in LD with the two SNPs.

Short Sequence Repeats

This is a polymorphic variation caused by different number of short sequence repeat (SSR) units, first described by Wyman and White [30]. Most common are the dinucleotide repeats, but SSRs could be tri-, tetra-, penta-,… repeats (often also called *microsatellites*, where the repeat unit $n=1–15$ nucleotides). SSRs with longer repeat units ($n=15–500$ nucleotides) are often termed *minisatellites*. These sequences comprise approximately 3% of the genome, and there is approximately one SSR per 5 kb. The most common dinucleotide SSR is the (GT)n, with an occurrence in the genome of about 28 times per Mb. The most common trinucleotide SSR is the (TAA)n that occurs about 4 times per Mb. The major advantage of SSRs (or microsatellites) is that there are more than two alleles per polymorphic site and a large fraction of human population is heterozygous for each SSR. Therefore SSRs are extremely useful in linkage mapping and subsequent positional cloning for monogenic disorders and other marking studies of the genome, including the development of genomic linkage maps. In addition, SSRs are extensively used in forensic studies. Fig. 12.18 shows an example of an SSR with three alleles in the population.

Insertion/Deletion Polymorphisms

This variation is caused by the presence or absence of certain sequences. These sequences could be a few nucleotides, but they could also be transposons or interspersed repeats such as LINE or SINE elements; alternatively they could be pseudogenes or other elements. Note that this category of variants is not completely separate from the next one; the arbitrary distinction is just the size of the variation in terms of base pairs. There are usually biallelic polymorphisms, not as common as SNPs, but useful for evolutionary studies and for the understanding of the dynamic structure of the human genome. In Fig. 12.19, the blue sequence was inserted in the DNA and created a variant with two alleles: the blue allele 1 with insertion, and the black allele 2 without insertion.

Copy Number Variants

This refers to large-scale structural variation of our genome, in which there are large tandems of 50 Kb to 5 Mb in length that are present in variable number of copies. This type of polymorphic variants includes large-scale duplications and deletions. These have been recognized since the studies of the α-globin genes in humans [31]. In Fig. 12.20, allele 1 contains three copies and allele 2 has five copies of a large repeat. A copy number variant (CNV) map of the human genome in 270 individuals has revealed that approximately 6% of the genome contains CNVs. A list of these variants could be found in http://projects.tcag.ca/variation/.

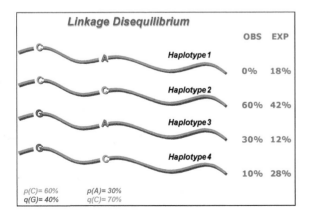

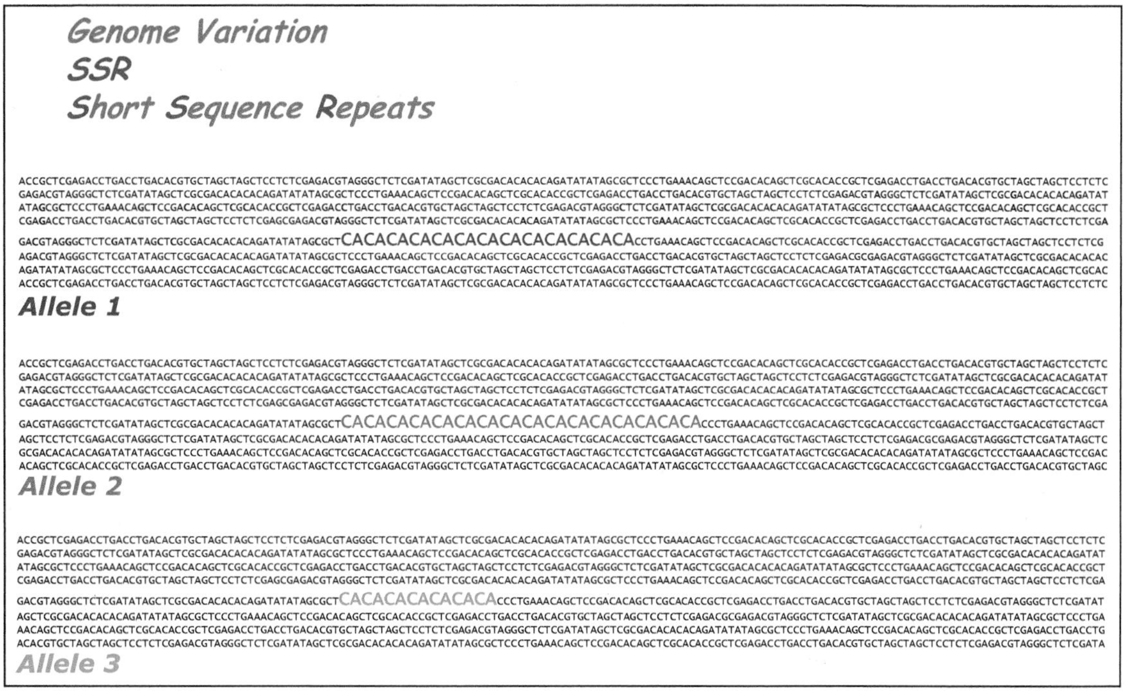

Genome Variation
SSR
Short Sequence Repeats

```
ACCGCTCGAGACCTGACCTGACACGTGCTAGCTAGCTCCTCTCGAGACGTAGGGCTCTCGATATAGCTCGCGACACACAGATATATAGCGCTCCCTGAAACAGCTCCGACACAGCTCGCACACCGCTCGAGACCTGACCTGACACGTGCTAGCTAGCTCCTCTC
GAGACGTAGGGCTCTCGATATAGCTCGCGACACACAGATATATAGCGCTCCCTGAAACAGCTCCGACACAGCTCGCACACCGCTCGAGACCTGACCTGACACGTGCTAGCTAGCTCCTCTCGAGACGTAGGGCTCTCGATATAGCTCGCGACACACAGATAT
ATAGCGCTCCCTGAAACAGCTCCGACACAGCTCGCACACCGCTCGAGACCTGACCTGACACGTGCTAGCTAGCTCCTCTCGAGACGTAGGGCTCTCGATATAGCTCGCGACACACAGATATATAGCGCTCCCTGAAACAGCTCCGACACAGCTCGCACACCGCT
CGAGACCTGACCTGACACGTGCTAGCTAGCTCCTCTCGAGCGGAGACGTAGGGCTCTCGATATAGCTCGCGACACACAGATATATAGCGCTCCCTGAAACAGCTCCGACACAGCTCGCACACCGCTCGAGACCTGACCTGACACGTGCTAGCTAGCTCCTCTCGA

GACGTAGGGCTCTCGATATAGCTCGCGACACACAGATATATAGCGCT**CACACACACACACACACACACACACA**CCTGAAACAGCTCCGACACAGCTCGCACACCGCTCGAGACCTGACCTGACACGTGCTAGCTAGCTCCTCTCG
AGACGTAGGGCTCTCGATATAGCTCGCGACACACAGATATATAGCGCTCCCTGAAACAGCTCCGACACCGCTCGAGACCTGACCTGACACGTGCTAGCTCCTCTCGAGACGCGAGACGTAGGGCTCTCGATATAGCTCGCGACACACAC
AGATATATAGCGCTCCCTGAAACAGCTCCGACACAGCTCGCACACCGCTCGAGACCTGACCTGACACGTGCTAGCTAGCTCCTCTCGAGACGTAGGGCTCTCGATATAGCTCGCGACACACACAGATATATAGCGCTCCCTGAAACAGCTCCGACACAGCTCGCAC
ACCGCTCGAGACCTGACCTGACACGTGCTAGCTAGCTCCTCTCGAGACGTAGGGCTCTCGATATAGCTCGCGACACACAGATATATAGCGCTCCCTGAAACAGCTCCGACACAGCTCGCACACCGCTCGAGACCTGACCTGACACGTGCTAGCTAGCTCCTCTC
```
Allele 1

```
ACCGCTCGAGACCTGACCTGACACGTGCTAGCTAGCTCCTCTCGAGACGTAGGGCTCTCGATATAGCTCGCGACACACAGATATATAGCGCTCCCTGAAACAGCTCCGACACAGCTCGCACACCGCTCGAGACCTGACCTGACACGTGCTAGCTAGCTCCTCTC
GAGACGTAGGGCTCTCGATATAGCTCGCGACACACAGATATATAGCGCTCCCTGAAACAGCTCCGACACAGCTCGCACACCGCTCGAGACCTGACCTGACACGTGCTAGCTAGCTCCTCTCGAGACGTAGGGCTCTCGATATAGCTCGCGACACACAGATAT
ATAGCGCTCCCTGAAACAGCTCCGACACAGCTCGCACACCGCTCGAGACCTGACCTGACACGTGCTAGCTAGCTCCTCTCGAGACGTAGGGCTCTCGATATAGCTCGCGACACACAGATATATAGCGCTCCCTGAAACAGCTCCGACACAGCTCGCACACCGCT
CGAGACCTGACCTGACACGTGCTAGCTAGCTCCTCTCGAGCGGAGACGTAGGGCTCTCGATATAGCTCGCGACACACAGATATATAGCGCTCCCTGAAACAGCTCCGACACAGCTCGCACACCGCTCGAGACCTGACCTGACACGTGCTAGCTAGCTCCTCTCGA

GACGTAGGGCTCTCGATATAGCTCGCGACACACAGATATATAGCGCT**CACACACACACACACACACACACACACACACACA**CCCTGAAACAGCTCCGACACAGCTCGCACACCGCTCGAGACCTGACCTGACACGTGCTAGCTAGCT
AGCTCCTCTCGAGACGTAGGGCTCTCGATATAGCTCGCGACACACAGATATATAGCGCTCCCTGAAACAGCTCCGACACAGCTCGCACACCGCTCGAGACCTGACCTGACACGTGCTAGCTAGCTCCTCTCGAGACGTAGGGCTCTCGATATAGCTC
GCGACACACAGATATATAGCGCTCCCTGAAACAGCTCCGACACAGCTCGCACACCGCTCGAGACCTGACCTGACACGTGCTAGCTAGCTCCTCTCGAGACGTAGGGCTCTCGATATAGCTCGCGACACACAGATATATAGCGCTCCCTGAAACAGCTCCGAC
ACAGCTCGCACACCGCTCGAGACCTGACCTGACACGTGCTAGCTAGCTCCTCTCGAGACGTAGGGCTCTCGATATAGCTCGCGACACACAGATATATAGCGCTCCCTGAAACAGCTCCGACACAGCTCGCACACCGCTCGAGACCTGACCTGACACGTGCTAGC
```
Allele 2

```
ACCGCTCGAGACCTGACCTGACACGTGCTAGCTAGCTCCTCTCGAGACGTAGGGCTCTCGATATAGCTCGCGACACACAGATATATAGCGCTCCCTGAAACAGCTCCGACACAGCTCGCACACCGCTCGAGACCTGACCTGACACGTGCTAGCTAGCTCCTCTC
GAGACGTAGGGCTCTCGATATAGCTCGCGACACACAGATATATAGCGCTCCCTGAAACAGCTCCGACACAGCTCGCACACCGCTCGAGACCTGACCTGACACGTGCTAGCTAGCTCCTCTCGAGACGTAGGGCTCTCGATATAGCTCGCGACACACAGATAT
ATAGCGCTCCCTGAAACAGCTCCGACACAGCTCGCACACCGCTCGAGACCTGACCTGACACGTGCTAGCTAGCTCCTCTCGAGACGTAGGGCTCTCGATATAGCTCGCGACACACAGATATATAGCGCTCCCTGAAACAGCTCCGACACAGCTCGCACACCGCT
CGAGACCTGACCTGACACGTGCTAGCTAGCTCCTCTCGAGCGGAGACGTAGGGCTCTCGATATAGCTCGCGACACACAGATATATAGCGCTCCCTGAAACAGCTCCGACACAGCTCGCACACCGCTCGAGACCTGACCTGACACGTGCTAGCTAGCTCCTCTCGA

GACGTAGGGCTCTCGATATAGCTCGCGACACACAGATATATAGCGCT**CACACACACACACA**CCCTGAAACAGCTCCGACACAGCTCGCACACCGCTCGAGACCTGACCTGACACGTGCTAGCTAGCTCCTCTCGAGACGTAGGGCTCTCGATAT
AGCTCGCGACACACAGATATATAGCGCTCCCTGAAACAGCTCCGACACAGCTCGCACACCGCTCGAGACCTGACCTGACACGTGCTAGCTAGCTCCTCTCGAGACGTAGGGCTCTCGATATAGCTCGCGACACACAGATATATAGCGCTCCCTGAAACAGCTCCGAC
AACAGCTCCGACACAGCTCGCACACCGCTCGAGACCTGACCTGACACGTGCTAGCTAGCTCCTCTCGAGACGTAGGGCTCTCGATATAGCTCGCGACACACAGATATATAGCGCTCCCTGAAACAGCTCCGACACAGCTCGCACACCGCTCGAGACCTGACCTG
ACACGTGCTAGCTAGCTCCTCTCGAGACGTAGGGCTCTCGATATAGCTCGCGACACACAGATATATAGCGCTCCCTGAAACAGCTCCGACACAGCTCGCACACCGCTCGAGACCTGACCTGACACGTGCTAGCTAGCTCCTCTCGAGACGTAGGGCTCTCGATA
```
Allele 3

FIGURE 12.18   An example of a dinucleotide SSR with three alleles in the population: the *blue* (dark gray in print versions) allele with $(CA)_{13}$ repeats, the *red* (light gray in print versions) allele with $(CA)_{16}$ repeats, and the *green* (gray in print versions) allele with $(CA)_7$ repeats.

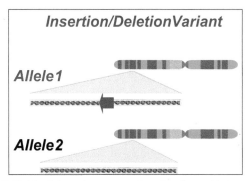

FIGURE 12.19   Schematic representation of a polymorphic locus caused by insertion/deletion of a genomic element, shown as a *blue arrow* (dark gray in print versions).

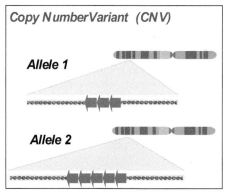

FIGURE 12.20   Schematic representation of copy number variation (CNV) in the human genome. Allele 1 in the population contains three copies of a sequence (three *red arrowheads*) and allele 2 contains five copies.

The extent of CNVs in the human genome is certainly an underestimate, because there are numerous additional CNVs of less than 50 kb. The current methodology for the detection of CNVs is comparative genomic hybridization on DNA microarrays [32]. Further improvement of the method will allow us to detect small CNVs. The most detailed CNV map of the human genome was recently established by the Genome Structural Variation Consortium. This consortium conducted a project to identify common CNVs greater than 500 bp in 20 females of European ancestry and 20 of African ancestry; the project identified 8599 copy number variant events. Data can be viewed at http://www.sanger.ac.uk/humgen/cnv/42mio/.

Inversions

Large DNA segments could have different orientation in the genomes of different individuals. These inversion polymorphisms predispose for additional genomic alterations (Fig. 12.21). An example of a common inversion polymorphism involves a 900 kb segment of chromosome 17q21.31, which is present in 20% of European alleles but it is almost absent or very rare in other populations [33].

Mixed Polymorphisms

There are combinations of SSRs and SNPs. Fig. 12.22 depicts such an example; the repeat units of an SSR contain

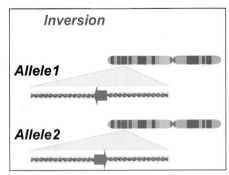

FIGURE 12.21 Schematic representation of a polymorphic inversion shown as a *red arrowhead* (gray in print versions).

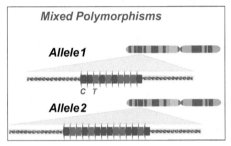

FIGURE 12.22 Schematic representation of a highly polymorphic region of the genome with mixed polymorphism that include single nucleotide polymorphisms (SNPs) in the copies of copy number variants (CNVs) or short sequence repeats (SSRs). The copies of the repeat are shown as *arrowheads*; the *blue* (dark gray in print versions)/*red* (gray in print versions) color of the repeats designate the SNP in them [blue (dark gray in print versions) for C and red (gray in print versions) for T].

an SNP and thus even alleles with the same repeat number could be distinguished based on their exact DNA sequence [34]. These highly polymorphic systems could be served as "recognition barcodes" in humans.

Genome Variation as a Laboratory Tool to Understand the Genome

DNA variants, besides their functional importance in health and disease, are very useful in human genetics research because they serve as genomic markers for a variety of studies. Some of the uses of DNA variants are to:

1. Create linkage (genetic) maps of human chromosomes. This has allowed the initial mapping of the human genome and it was a prerequisite for the sequence assembly.
2. Map the genomic location of monogenic phenotypes to human chromosomes by linkage analysis. A large number of such phenotypes have been mapped to small genomic intervals because of the genotyping of members of affected families. Positional cloning of pathogenic mutations was subsequently possible.

3. Map the genomic location of polygenic phenotypes to human chromosomes by genome-wide linkage and association studies.
4. Fetal diagnosis and carrier testing by linkage analysis of the cosegregation of a polymorphic marker and the phenotype of interest.
5. Perform paternity and forensic studies. A whole field was developed mainly with the use microsatellite SSR variants.
6. Study genome evolution and origin of pathogenic mutations.
7. Study the recombination rate and properties of the human genome.
8. Study the instability of the genome in tumor tissues.
9. Identify loss-of-heterozygosity in human tumors.
10. Study uniparental disomy and thus understanding genomic imprinting.
11. Study the parental and meiotic origin and deciphering the mechanisms of nondisjunction.
12. Study population history and substructure.

REFERENCES

[1] Gyapay G, Morissette J, Vignal A, Dib C, Fizames C, Millasseau P, et al. The 1993–94 Genethon human genetic linkage map. Nat Genet 1994;7:246–339.

[2] A comprehensive genetic linkage map of the human genome. NIH/CEPH Collaborative Mapping Group. Science 1992;258:67–86.

[3] Hudson TJ, Stein LD, Gerety SS, Ma J, Castle AB, Silva J, et al. An STS-based map of the human genome. Science 1995;270:1945–54.

[4] Cohen D, Chumakov I, Weissenbach J. A first-generation physical map of the human genome. Nature 1993;366:698–701.

[5] Finishing the euchromatic sequence of the human genome. Nature 2004;431:931–45.

[6] Venter JC, Adams MD, Myers EW, Li PW, Mural RJ, Sutton GG, et al. The sequence of the human genome. Science 2001;291:1304–51.

[7] Lander ES, Linton LM, Birren B, Nusbaum C, Zody MC, Baldwin J, et al. Initial sequencing and analysis of the human genome. Nature 2001;409:860–921.

[8] Levy S, Sutton G, Ng PC, Feuk L, Halpern AL, Walenz BP, et al. The diploid genome sequence of an individual human. PLoS Biol 2007;5:e254.

[9] Birney E, Stamatoyannopoulos JA, Dutta A, Guigo R, Gingeras TR, Marguiles EH, et al. Identification and analysis of functional elements in 1% of the human genome by the ENCODE pilot project. Nature 2007;447:799–816.

[10] The ENCODE (ENCyclopedia of DNA elements) project. Science 2004;306:636–40.

[11] ENCODE Project Consortium. An integrated encyclopedia of DNA elements in the human genome. Nature 2012;489:57–74. http://dx.doi.org/10.1038/nature11247.

[12] Finn RD, Bateman A, Clements J, Coggill P, Eberhardt RY, Eddy SR, et al. The Pfam protein families database. Nucleic Acids Res 2008;36:D281–8.

[13] Esteller M. Non-coding RNAs in human disease. Nat Rev Genet 2011;12:861–74. http://dx.doi.org/10.1038/nrg3074. pii: nrg3074.

[14] Fatica A, Bozzoni I. Long non-coding RNAs: new players in cell differentiation and development. Nat Rev Genet 2014;15:7–21. http://dx.doi.org/10.1038/nrg3606.

[15] Maston GA, Evans SK, Green MR. Transcriptional regulatory elements in the human genome. Annu Rev Genomics Hum Genet 2006;7:29–59.

[16] Waterston RH, Lindblad-Toh K, Birney E, Rogers J, Abril JF, Agarwal P, et al. Initial sequencing and comparative analysis of the mouse genome. Nature 2002;420:520–62.

[17] Dermitzakis ET, Reymond A, Antonarakis SE. Conserved non-genic sequences: an unexpected feature of mammalian genomes. Nat Rev Genet 2005;6:151–7.

[18] Dermitzakis ET, Reymonnd A, Lyle R, Scamuffa N, Ucla C, Deutsch S, et al. Numerous potentially functional but non-genic conserved sequences on human chromosome 21. Nature 2002;420:578–82.

[19] Bailey JA, Gu Z, Clark RA, Reinert K, Samonte RV, Schwartz S, et al. Recent segmental duplications in the human genome. Science 2002;297:1003–7.

[20] Samonte RV, Eichler EE. Segmental duplications and the evolution of the primate genome. Nat Rev Genet 2002;3:65–72.

[21] Riethman H. Human telomere structure and biology. Annu Rev Genomics Hum Genet 2008;9:1–19.

[22] Wallace DC. Mitochondrial diseases in man and mouse. Science 1999;283:1482–8.

[23] Wallace DC. A mitochondrial paradigm of metabolic and degenerative diseases, aging, and cancer: a dawn for evolutionary medicine. Annu Rev Genet 2005;39:359–407.

[24] Wallace DC. Why do we still have a maternally inherited mitochondrial DNA? Insights from evolutionary medicine. Annu Rev Biochem 2007;76:781–821.

[25] Kan YW, Dozy AM. Polymorphism of DNA sequence adjacent to human β-globin structural gene: relationship to sickle mutation. Proc Natl Acad Sci USA 1978;75:5631–5.

[26] International HapMap Consortium. The international HapMap project. Nature 2003;426:789–96.

[27] Altshuler D, et al. A haplotype map of the human genome. Nature 2005;437:1299–320.

[28] Frazer KA, Ballinger DG, Cox DR, Hinds DA, Stuve LL, Gibbs RA, et al. A second generation human haplotype map of over 3.1 million SNPs. Nature 2007;449:851–61.

[29] A map of human genome variation from population-scale sequencing. Nature 2010;467:1061–73. http://dx.doi.org/10.1038/nature09534. pii: nature09534.

[30] Wyman AR, White R. A highly polymorphic locus in human DNA. Proc Natl Acad Sci USA 1980;77:6754–8.

[31] Goossens M, Dozy AM, Embury SH, Zachariades Z, Hadjiminas MG, Stamatoyannopoulos G, et al. Triplicated alpha-globin loci in humans. Proc Natl Acad Sci USA 1980;77:518–21.

[32] Carter NP. Methods and strategies for analyzing copy number variation using DNA microarrays. Nat Genet 2007;39:S16–21.

[33] Stefansson H, Helgason A, Thorleifsson G, Steinthorsdottir V, Masson G, Barnard J, et al. A common inversion under selection in Europeans. Nat Genet 2005;37:129–37.

[34] Jeffreys AJ, MacLeod A, Tamaki K, Neil DL, Monckton DG. Minisatellite repeat coding as a digital approach to DNA typing. Nature 1991;354:204–9.

Chapter 13

Spectrum of Genetic Diseases and Management

B. Kerr

Manchester Academic Health Sciences Centre (MAHSC), Manchester, United Kingdom

Chapter Outline

Introduction	179
Molecular Approach to Genetic Disease Nosology	179
From Marfan Syndrome to Fibrillinopathies	180
Diagnosis of Marfan Syndrome	182
Differential Diagnosis	182
Management of Marfan Syndrome	183
Treatment	183
The Emergence of RASopathies	184
Malignancy Risk in RASopathies	185
Making a Diagnosis of a Genetic Disorder in the Era of Molecular Pathways	186
Conclusion	187
References	187

INTRODUCTION

Genomic technologies have altered the clinical definitions of many genetic disorders, as precise molecular testing has provided accurate diagnosis and illuminated the underlying cellular pathology of many disorders. This has implications for potential treatments, but also for clinical management. The relationship between disorders previously considered distinct has been clarified. Variable expressivity has been confirmed as a confounder of clinical criteria as the basis of disease definition. Many previously defined distinct disorders are now known to have perturbation of the same underlying cellular mechanisms as their basis, with clinical features on a spectrum. The consequences of this will be illustrated through two groups of disorders, fibrillinopathies and RASopathies.

MOLECULAR APPROACH TO GENETIC DISEASE NOSOLOGY

Until recently, the diagnosis of most genetic disease relied on clinical characteristics, with clinical diagnostic criteria available for many conditions [1,2]. Literature discussions over precise diagnostic boundaries were common [3] and clinicians could be broadly characterized as "lumpers versus splitters" [4].

It was in the areas of skeletal dysplasias that disease classification first encompassed the underlying molecular pathology [5]. Correct diagnosis for many disorders is increasingly reliant on molecular testing, and clinical management might encompass consideration of a range of symptomatology, previously described as accompanying separate, clinically defined, conditions.

In the most recent skeletal dysplasia classification [6], 456 different conditions are included, 316 of these associated with mutations in one or more of 226 genes. Of the 40 groupings, eight are based on the underlying gene or pathway perturbation. Two of these categories, *TRVP4* and aggrecan, are based on the recognition of a new gene or pathway.

In the case of *TRVP4*, subsequent to the finding of this gene as mutated in metatrophic dysplasia and spondylometaphyseal dysplasia, Kozlowski type [7], a continuous phenotypic spectrum was recognized [8], in addition to a number of rare variant skeletal phenotypes with less obviously overlapping features. Different mutations in *TRVP4* have also been associated with dominant neurological disorders, and a in small number of patients with both skeletal dysplasias and peripheral neuropathy [9].

Although the significance of genetic information in classifying disease can be debated [10], consideration of the possible underlying molecular pathogenesis enhances diagnosis by promoting consideration of disorders that have overlapping clinical characteristics. This approach also supports better clinical care. Recognition of disorders on a phenotypic spectrum improves management by either surveillance for severe complications or by demonstration that for particular conditions within a grouping, some aspects of the natural history are different.

Medical and Health Genomics. http://dx.doi.org/10.1016/B978-0-12-420196-5.00013-7

TABLE 13.1 Examples of Molecular Pathology–Based Disease Classifications

Classification	Disease/Syndrome(s)	Gene(s)
Ciliopathy [13]	Meckel-Gruber syndrome	15 genes
	Joubert syndrome	22 genes, unknown
	BardetBiedl syndrome	17 genes
	Orofaciodigital syndrome	*OFD1, TCTN3, C5orf42*, unknown
	Hydrolethalus syndrome	*HYLS1,KIF7,*
	Acrocallosal syndrome	*KIF7, GLI3*
	Pallister-Hall syndrome	*GLI3*
	Grieg cephalopolysyndactyly syndrome	*GLI3*
Fibrillinopathy	Marfan syndrome	*FBN1*
	Familial ectopia lentis	*FBN1*
	Familial thoracic aneurysm	*FBN1*
	Congenital contractural arachnodactyly (Beals syndrome)	*FBN2*
	Weill-Marchesani syndrome (autosomal dominant form)	*FBN1*
	Geleophysic dysplasia (autosomal dominant form)	*FBN1*
	Acromicric dysplasia	*FBN1*
Cohesinopathies [90,91]	Cornelia de Lange syndrome	*NIPBL, HDAC8, SMC1A, SMC3, RAD21*
	Roberts syndrome	*ESCO2*
	Warsaw breakage syndrome	*DDX11*
	X-linked α-thalassaemia mental retardation syndrome	*ATRX*

The pathway-based approach to genetic disease nosology has also illuminated the role of cellular organelles and pathways in human development. The recognition of a group of diverse multisystem diseases as "ciliopathies," for example, has increased understanding of the role of cilia as vital sensory organs in development and homeostasis [11]. These developmental disorders can be differentiated from primary ciliary dyskinesia, which remains the sole genetic disorder of motile cilia dysfunction [12].

The ciliopathies are also noteworthy because of the number of involved genes, the significance of multiallelic inheritance to the phenotype, and the variability of the phenotype when defined clinically, even among affected siblings [13].

There are now many other conditions that although individually described and defined, are now best considered to be a disturbance in a critical cellular pathway. Table 13.1 provides a representative, but not exhaustive, list. Two of these categories will be considered in detail: the fibrillinopathies and RASopathies.

FROM MARFAN SYNDROME TO FIBRILLINOPATHIES

Marfan syndrome (MFS) was first described in 1896 by Antoine Marfan, a French pediatrician, [14] in a 5-year-old girl. Although it was first recognized as a skeletal disorder, there may be manifestations in the heart (the major cause of premature mortality), eyes, skin, dura, and lung. It is an autosomal dominant disorder, with an incidence of 1–2/10,000, and a new mutation rate of around 25% [1].

Mutations in the fibrillin gene *(FBN)* were established as causative of MFS in 1991 [15].

FBN is a large glycoprotein that has an important role in the structural integrity of the extracellular matrix and in regulating the bioavailability of transforming growth factor β (TGF-β). There are three mammalian *FBN*s, but there are no human diseases currently associated with *FBN3*. Mutations in *FBN1* result in increased activity of TGF-β and structural failure of the extracellular matrix [16].

TABLE 13.1 Examples of Molecular Pathology–Based Disease Classifications—cont'd

Classification	Disease/Syndrome(s)	Gene(s)
Desmosome disorders [92]	Arrhythmogenic cardiomyopathy	*JUP, DSP, PKP2, DSG2, DSC2*
	Cardiomyopathy, dilated, with woolly hair and keratoderma	*DSP*
	Epidermolysis bullosa, lethal acantholytic	*DSP*
	Keratosis palmoplantaris striate II	*DSP*
	Skin fragility, woolly hair syndrome	*DSP*
Nuclear envelopathies [93]	Disorders of striated muscle, adipose tissue, peripheral nerve, and adipose tissue (progeroid)	*LMNA*
	Cardiomyopathy with muscular dystrophy	*EMD*
	Pelger-Huët anomaly/Greenberg skeletal dysplasia	*LBR*
	Cerebellar ataxia	*SYNE1*
	DYT1 dystonia	*TOR1A*
	Adult-onset leukodystrophy	*LMNB1*
	Aquired partial leukodystrophy	*LMNB2*
	Restrictive dermopathy and progeroid diseases	*ZNPSTE24*
RASopathies	Neurofibromatosis type 1	Neurofibromin
	Legius syndrome	*SPRED1*
	Noonan syndrome	*PTPN11,SOS1, KRAS,NRAS, RAF, RIT1*
	Noonan syndrome with multiple lentigines	*PTPN11, RAF1*
	Noonan syndrome with loose anagen hair	*SHOC2*
	Noonan syndrome–like disorder with or without juvenile myelomonocytic leukemia	*CBL*
	Cardiofaciocutaneous syndrome	*BRAF, MAP2K1, MAP2K2, KRAS*
	Costello syndrome	*HRAS*
Other *RAS/MAPK* disorders	Capillary malformation–arteriovenous malformation syndrome	*RASA1*
	Multiple hereditary gingivomatosis	*SOS1*
	Schimmelpenning syndrome, nevus sebaceous	Mosaic *HRAS* or *KRAS*
	Epidermal nevus	Mosaic *HRAS*

Formal diagnostic criteria for MFS were published in 1988 [17] and called the *Berlin criteria* after an international consensus meeting. These criteria were exclusively clinical and were based on the number of involved organ systems according to major and minor criteria. Family history was included. The criteria applied to adults only and at-risk children needed evaluation once they had completed their growth.

For clinicians using the criteria, the awareness of both variable expressivity and penetrance within families made exclusion of the diagnosis clinically difficult.

The diagnostic criteria were revised in 1996 and again in 2010, in the latter iteration to include mutation status in the scoring system [18]. In these criteria, more significance was given to aortic dilation and ectopia lentis (EL), with the presence of both of these features alone being sufficient for a diagnosis. A scoring system was described for systemic manifestations, and bonafide *FBN* mutations were given greater weight. The crucial importance of excluding other conditions in the presence of unusual clinical findings was also emphasized. Such features that might suggest a different diagnosis include craniosynostosis, arterial tortuosity, diffuse arterial aneurysms, bifid uvula, developmental disability, crumpled ears, dystrophic scars, easy bruising, and thin and dystrophic skin.

MFS had always been differentiated clinically from apparently isolated EL (IEL), with these patients having inherited lens dislocation without any of the systemic features of MFS. The revised Ghent scoring system of 2010, however, reclassified patients with IEL as MFS if their *FBN* mutation had previously been reported to be associated with aortic dilation and rupture.

In a recent review of all published cases of inherited EL with an *FBN* mutation [19], almost half of 123 probands were reclassified as MFS, and 38.5% of mutations associated with isolated EL were found in patients with aortic dilation/dissection.

These findings suggest that apparently isolated EL caused by mutations in *FBN* are actually part of the spectrum of fibrillinopathies, and patients with IEL with *FBN* mutations and no systemic features of MFS nevertheless should be monitored for aortic dilation.

If *FBN* mutation testing is normal, mutations in *LTBP2* or *ADAMTSL4* can be sought as the basis for EL syndrome, which is not associated with a risk of aortic dilation [18].

The Ghent criteria recognized that assessment for the clinical features of MFS is necessary in adult life, because the clinical manifestations increase with age. In a large pediatric series, only 56% of children could be clinically diagnosed, with molecular testing increasing this proportion to 85% [20]. The increase in clinical manifestations with age supports the need for clinical follow-up observation into adulthood, even if molecular testing is negative, in those at risk on the basis of a family history.

A longitudinal study of cardiovascular manifestations in a cohort of 1013 probands with pathogenic *FBN* mutations demonstrated that aortic events were rare before age 20 years [21], but that dilation increased with age, reaching 74% by age 60. Men were at greater risk of aortic dilation and aortic events. Mitral valve disease also increased with age.

Diagnosis of Marfan Syndrome

The multisystem nature of MFS dictates a multidisciplinary approach to management. However, a multidisciplinary approach is required even in achieving a diagnosis of a fibrillinopathy. A diagnostic tool [18] to support accurate diagnosis is available on www.marfan.org.

Requirements for a correct diagnosis are:

- awareness of the variable manifestations of a fibrillinopathy
- knowledge of conditions that may have overlapping clinical features

- knowledge of the Ghent criteria and how to use them, particularly with regard assessment of skeletal signs [18]
- expertise in assessing aortic diameter at the sinuses of Valsalva and calculation of a Z score
- ability to diagnose EL and myopia
- a molecular genetics service familiar with the spectrum of mutations seen in the *FBN* gene and the principles underlying attribution of causality [18].

The required expertise therefore encompasses clinical and molecular genetics, pediatrics, cardiology, ophthalmology, and radiology for determination of dural ectasia and protrusio acetabula when indicated by other clinical findings.

Differential Diagnosis

In pediatric practice, the most common differential diagnosis of MFS is normal tall stature. Should there be any doubt, reexamination after growth is completed is essential.

As mentioned, patients with apparent IEL should be evaluated for signs of MFS, including cardiac evaluation, and be referred for *FBN* mutation testing.

Patients referred for consideration of MFS on the basis of joint hypermobility may be found to have signs consistent with type III Ehlers-Danlos syndrome.

Clinical examination should specifically look for the features previously outlined that may suggest a syndromal diagnosis (craniosynostosis, arterial tortuosity, diffuse arterial aneurysms, bifid uvula, developmental disability, crumpled ears, dystrophic scars, easy bruising, thin and dystrophic skin). Syndromes to be considered include Loeys-Dietz syndrome, Shprintzen-Goldberg syndrome, homocystinuria, arterial tortuosity syndrome, and vascular (type IV) and kyphoscoliotic (type VI) Ehlers-Danlos syndrome.

Other fibrillopathies also form part of the differential diagnosis of MFS (see Table 13.1). Familial thoracic aneurysm may be associated with livido reticularis and iris flocculi [18], with no systemic signs of MFS.

Mutations in *FBN2* cause Beals syndrome or congenital contractural arachnodactyly, characterized by crumpled ears, early onset scoliosis, and contractures that tend to improve with age. Although the lack of aortic dilation was originally described as a discriminator from MFS, it has now been established that a risk of aortic dilation exists [22]. Differentiating between Beals syndrome and neonatal MFS is not always clear cut, but is very important for both prognosis and treatment [23].

Paradoxically, several of the acromelic dysplasia group of disorders have been found to be associated with *FBN* mutations, although they are not part of the differential diagnosis (see Table 13.1). Patients with these conditions share short stature, short hands and feet, and stiff joints [24].

Recent evidence suggests that aortic dilation may be part of these conditions, particularly if mutations in exons 41 or 42 are causative [25].

This supports the concept that despite ostensibly very different key diagnostic features (tall stature versus marked short stature in these fibrillinopathies) for diseases with a common molecular pathology, common disease manifestations should be considered in clinical management.

Management of Marfan Syndrome

The most critical aspect of management, given that aortic root dilation is progressive and may develop after diagnosis, is regular monitoring of aortic root diameter. Current recommendations are for annual echocardiography in those less than age 20 years and in those recently diagnosed; adults with repeatedly normal echocardiography can have the interval between examinations extended to 2–3 years [18,26]. More frequent examinations may be required if the aortic diameter is approaching surgical threshold (≥4.5 cm in adults), is rapidly increasing, or if there is concern about cardiac or valvular function.

Given that around 10% of dissection in MFS is of the thoracic aorta, regular imaging of the whole aorta is desirable in adulthood.

Management of acute dissection is a surgical emergency. Referral for elective surgery is considered when the diameter at the sinus of Valsalva is around 5.0 cm. The decision is also influenced by a family history of dissection, the rate of aortic growth, the severity of associated aortic or mitral valve regurgitation, ventricular dysfunction, pregnancy planning in women, and the desire for a valve-sparing operation [18].

Pregnancy in women is associated with an increased risk of vascular instability, especially in the third trimester and the immediate postpartum period. Management and monitoring of the pregnancy should occur in a center with fetal and cardiac expertise. The risk is greatest in women with an aortic diameter ≥4 cm [1] and, in these women, extends to the whole pregnancy.

The management of exercise in MFS is often difficult, particularly for young men whose tall stature may contribute to sporting expertise. General advice is to avoid contact sports, exercise to exhaustion, and isometric activities involving a Valsalva maneuver. Up-to-date guidance is available from the website of The Marfan Foundation, www.marfan.org. A very helpful guide to MFS for the school nurse is available to down load and includes practical advice about the approach to physical activity from early childhood.

Annual monitoring by an ophthalmologist for EL, strabismus, myopia, cataract, retinal detachment, and glaucoma is important. To prevent amblyopia, this is of particular importance in early life. Patients should be informed of the symptoms of retinal detachment and the referral pathway for immediate expert assessment.

Patients should be aware of the risk of pneumothorax, its symptoms, and the need to seek urgent medical advice in the presence of acute chest pain.

Management of scoliosis, hernia, joint hypermobility, and pectus excavatum are as in the general population.

Treatment

The value of β blockade in slowing the rate of aortic dilation in MFS was established as long ago as 1994 [27]. It is recommended for all patients with MFS unless specifically contraindicated [26].

More recently, mouse models have suggested that aortic root dilation in MFS is caused by excessive TGF-β signaling secondary to *FBN* deficiency, and that this can be mitigated by TGF-β antagonists, including angiotensin II type 1 receptor blockers (ARBs). In the mouse model, the ARB losartan had a beneficial effect on preventing or decreasing aortic dilation and on pathological changes within the aortic wall. Propranolol, in contrast, had no effect on intramural pathology and a moderate effect on slowing dilation.

Different effects of treatment with a β blocker versus an ARB on the biophysical properties of the aorta in MFS and Loeys-Dietz syndrome have been confirmed in humans [28].

A beneficial effect of losartan has been demonstrated in a small number of pediatric patients with preexisting aortic dilation and poor response to other therapies [29].

More recently, a randomized, controlled trial comparing losartan with a β blocker in 608 children and young adults with MFS found no significant difference in the rate of aortic dilation over 3 years [30].

However, a randomization study of losartan using a classification based on mutation type demonstrated a differential response based on the class of mutation. In this patient cohort, around two-thirds of mutations were dominant negative, with the remainder being causative of haploinsufficiency. Dominant negative mutations are those that result in incorporation of a mutant and abnormal protein, in this case *FBN* into the intracellular matrix. Haploinsufficiency mutations cause a decreased amount of normal protein. In this trial, over 3 years losartan had a clear benefit on aortic root dilation in patients with haploinsufficiency mutations, but no effect in patients with dominant negative mutations [31].

A number of other trials are ongoing. Other possible treatments are being considered experimentally, based on an understanding of genotype and molecular pathogenesis. It is certainly possible, if not likely, that illuminating molecular pathology in genetic diseases affecting the aorta will improve treatment for nongenetic aortic disease [32].

THE EMERGENCE OF RASOPATHIES

RAS proteins are essential components of complex and inter-connecting signaling pathways emanating from cell surface receptors. The discovery of *RAS* oncogenes in the early 1980s [33], as present as somatic mutations in around 30% of cancers, fundamentally altered the understanding of cancer biology to a model of cancer origin as an accumulation of genetic alterations all contributing to the loss of normal cellular controls and functions [34].

In humans, three *RAS* genes encode four homologous proteins: *HRAS, NRAS,* and *KRAS4A,* and *KRAS4B,* the latter being alternative splice variations of the KRAS gene. *Ras* proteins form the link between cell surface receptors and intracellular effector pathways. They cycle between on and off configurations, "on" when guanosine triphosphate (GTP) bound and "off" when guanosine diphosphate (GDP) bound. The transitions are mediated by guanine nucleotide exchange factors, which promote activation through stimulating GDP for GTP exchange and GTPase-activating proteins (*GAPs*) that accelerate GTP hydrolysis [34]. Prevention of inactivation by GAPs is the predominant mutation consequence.

The commonest disorder of the *RAS/MAPK* pathway is neurofibromatosis type 1 (NF1) [35], caused by mutations in the neurofibromin, a *RAS-GAP* protein gene. Clinically, a subgroup of patients were recognized who had similarities to patients with Noonan syndrome (NS), particularly facially, but also in the presence of pulmonary stenosis. Mutations or deletions in the neurofibromin gene have been found in many of these patients [36].

Legius syndrome is an NF1-like condition caused my mutations in *SPRED1* [37]. It is characterized by multiple café-au-lait patches and often macrocephaly, but without any of the other complications of NF1.

NS was first described by Jacqueline Noonan [38], who recognized a group of children with pulmonary stenosis, short stature, similar facial features, and, in boys, cryptorchidism. It is recognized to be a common and highly variable disorder [39].

Costello syndrome (CS) was first described in 1977 by Dr. Jack Costello, a New Zealand pediatrician who published in the *Australian Pediatric Journal* a report of two children with similar physical characteristics and mild intellectual handicap [40]. Although there were no further publications until 1991 [41], over the ensuing decade a number of publications defined the clinical phenotype [42–47].

Key features are developmental disability, prenatal overgrowth, and postnatal failure to thrive and short stature. Polyhydramnios, severe feeding difficulty, and congenital heart disease are common features, as are hypertrophic cardiomyopathy, and cardiac arrhythmia, predominantly atrial. Skin changes are striking, particularly increased skin over the palms and soles and the development of papilloma at moist body surfaces. The finding of excess palmar skin is part of a striking hand phenotype, comprising in addition, hyperextensibility of the small joints of the hand and a posture of flexion and ulnar deviation at the wrists.

A surprising finding was an increased risk of malignancy, particularly embryonal rhabdomyosarcoma and bladder carcinoma [48,49].

Although CS has a more severe phenotype than NS, it was recognized to have some overlapping clinical features (in particular, short stature, pattern of congenital heart disease) with both NS and cardiofaciocutaneous syndrome (CFC).

CFC was recognized from the first publications [50,51] as a disorder with overlapping features with NS, but with a more severe developmental phenotype and characteristic skin involvement, including nevi.

In 2005, mutations in *HRAS* were found to cause CS [52], the first report of mutations in an oncogene causing a developmental syndrome. Aoki and her colleagues had hypothesized that, given the phenotypic overlap between CS and NS, CS might be caused by mutations in genes of the *RAS/MAPK* cascade. Missense gain of function mutations in *PTPN11,* a *RAS/MAPK* cascade gene, encoding the tyrosine phosphatase *SHP2,* had been established as causative in around 50% of cases of NS [53].

The availability of a diagnostic test for CS permitted confirmation of the diagnosis and more precise definition of the phenotype [54,55]. An unexpected severe neonatal phenotype [56,57], and a myopathic presentation emerged [58]. A milder phenotype has been recognized [59], and a genotype–phenotype correlation has been established [60].

After the report of *HRAS* mutations as causative in CS, mutations in other pathway genes were established as causative of CFC [61,62]. The clinical phenotype of CFC has become clearer; it is now apparent that it can be associated with a very severe developmental outcome, an epilepsy that may be difficult to control, and prolonged major feeding difficulty [63]. In contrast, normal intellect has also been reported in a small number of patients, [64] and vertical transmission has been observed [65] in a family with a *MAPK2K2* mutation (previously *MEK2*).

Over the ensuing decade, a number of other pathway genes [66–68] have been found to be mutated in NS. Table 13.2 lists the currently known genes, and the clinical correlates that can be distinguished [39].

Several clinical subgroups are distinctive. LEOPARD syndrome now called *NS with multiple lentigines* was delineated on the basis of cardinal clinical findings: *L,* lentigines; *E,* electrocardiogram abnormalities; *O,* ocular hypertelorism; *P,* pulmonary stenosis; *A,* abnormal genitalia; *R,* retardation of growth; *D,* deafness. Mutations in both *PTPN11* and *RAF1* have been reported in patients described in this way. Given, however, the presence of lentigines in many RASopathy patients and the presence of all of the

TABLE 13.2 Genotype–Phenotype Correlations in Noonan Syndrome [39,66–68]

Gene	Established Correlations
PTPN11 (50%)	More PS, less ASD and HCM; more short stature, bleeding diathesis, and JMML
SOS1 (10%)	Less ASD, short stature, and intellectual disability; more ectodermal manifestations (like CFC)
RAF1 (10%)	More HCM; more nevi, café au lait patches, and lentigines
KRAS (<2%)	More severe cognitive delay; more ectodermal manifestations (like CFC)
NRAS (<1%)	No significant differences but very few cases reported
RIT1 (9%)	More HCM
RRAS	Two cases only

Key: *PS*, Pulmonary stenosis; *ASD*, Atrial septal defect; *HCM*, Hypertrophic cardiomyopathy; *JMML*, Juvenile myelomonocytic leukemia.

features in many other patients with a mutation in one of the pathway genes, the meaning of this as a particular description is unclear [69].

NS with loose anagen hair was first described as a distinctive clinical NS-like phenotype [70]. It is caused by a recurrent mutation in *SHOC2*, a positive modulator of the *RAS/MAPK* signaling cascade. In addition to loose anagen hair, patients often have dark skin with eczema or ichthyosis, a hypernasal voice, growth hormone deficiency, mitral valve dysplasia and septal defects [39].

Patients with a *CBL* mutation have an increased risk of juvenile myelomonocytic leukemia (JMML) and may have short stature and developmental disability [71,72]. The facies are highly variable, and may not be particularly suggestive of NS. A number of distinctive features have been reported: enlarged left atrium, transient chaotic ventricular dysrhythmia, cerebellar vermis hypoplasia, average adult height, aortic stenosis, and mitral valve insufficiency [39].

Although there were a number of early postmutation publications that looked carefully at differentiating clinical features across the pathway syndromes, especially between CS and CFC [73], as more patients have been studied at the molecular and clinical level, it has become apparent that the disorders of the *RAS/MAPK* pathway are on a phenotypic spectrum [74,75] with few, if any, pathognomonic features for each clinical diagnosis.

This has been reflected in the description of these conditions as "neurocardiofaciocutaneous" syndromes [76,77], or, more recently, "RASopathies" [78,79].

Clinical differentiation, particularly in early life, may be difficult. Although many patients will have facial features that are described as characteristic of one of these conditions [39,44,54], in a large number of patients, the facies are unremarkable. Making the diagnosis relies on recognizing the pattern of health and developmental problems that can arise from a *RAS/MAPK* gene mutation.

The clinical features that might in combination lead to a clinical suspicion of a RASopathy are listed in Table 13.3.

Table 13.1 includes a number of disorders that are not normally classified as "RASopathies," because of their localized nature. However, the report of a urothelial cancer in a young man with an epidermal nevus containing an *HRAS* mutation suggests that some of the risks of the RASopathies may be present in mosaic phenotypes [80].

Malignancy Risk in RASopathies

Two tumors are known to be a particular risk in NF1, optic glioma and malignant peripheral nerve sheath tumors. Less widely known is an increased risk for a wide range of malignancies [81] and a more recently established risk of breast cancer [82]. This increased risk includes childhood JMML, a rare childhood malignancy characterized by hyperactivation of the *RAS* pathway. The age of onset is, however, later, usually greater than 5 years [83].

It can therefore be hypothesized that for other RASopathies there may be a small generalized increased risk of malignancy, with or without a high risk of particular malignancies.

NS also has a strong association with JMML. A large prospective study of infant patients with *PTPN11* mutations has shown a myeloproliferative disorder (MPD) in 5.6% and with a severe disorder meeting criteria for JMML in 3% overall [84]. In this cohort, although the MPD usually resolved, JMML was associated with a high mortality in infancy. Two patients with MPD developed another malignancy in childhood.

Conversely, in a cohort of patients with JMML [85], a high proportion were found to have a germ-line mutation in *PTPN11*, with small numbers with mutations in *NRAS*, *KRAS*, or *CBL*. Only one had a clinical diagnosis of NS.

Given the impact of underlying *RAS* germ-line mutations on prognosis in JMML, early mutation analysis is important in determining prognosis and treatment [83].

Although the greatest malignancy risk in NS is for hematological malignancy [86,87], there is an increased incidence of embryonal rhabdomyosarcoma and neuroblastoma (as in CS), as well as brain tumors.

A population-based study of childhood cancer in the RASopathies has shown a relative risk for malignancy overall of 10.5: 8.1 for NS and 42.4 for CS [87].

It is hoped that the insights into cancer biology gained by the study of the germ-line disorders of the *RAS* pathway will be beneficial in understanding cancer biology in the general population. Similarly, the research on *RAS/MAPK* inhibitors

TABLE 13.3 Clinical Features Suggestive of a RASopathy

Prenatal	Infancy	Childhood
Polyhydramnios	Prematurity	Developmental disability
Macrocephaly	Severe feeding difficulty	Congenital heart disease
Macrosomia	Hypoglycemia	Cardiomyopathy
Nuchal thickening	Congenital heart disease, valvular or septal	Short stature
Hydrops/ascites	Cardiomyopathy	Eye abnormalities
Ventriculomegaly	Supraventricular tachychardia	Structural renal abnormalities
Ulnar deviation at the wrist	Tracheomalacia, bronchomalacia, both	Growth hormone deficiency
Arrhythmia	Ascites/pleural effusion lymphedema, lymphangectasia	Craniosynostosis Chiari malformation Posterior fosse compression
	Ulerythema ophryogenes, keratosis pilaris, excess palmar skin, eczema	Hyperkeratosis of the soles and palms
Advanced paternal age	Malrotation, pyloric stenosis	Hearing loss
	Monosmy 7, juvenile myelomonocytic leukemia	Malignancy, especially rhabdomyosarcoma, acute lymphoblastic leukemia, brain tumor, neuroblastoma

that has been done in pursuing effective cancer treatments may lead to treatments for other aspects of these conditions.

MAKING A DIAGNOSIS OF A GENETIC DISORDER IN THE ERA OF MOLECULAR PATHWAYS

A genetic disorder can be suspected in a number of different contexts:

- one or more congenital abnormalities, diagnosed prenatally or postnatally
- developmental difference
- stature out of step for the family, either short or tall
- family history
- unusually severe or early presentation of a condition; eg, neonatal or childhood scoliosis
- a constellation of problems that is known to be associated in a genetic syndrome; eg, see Table 13.3
- a single difference that is commonly associated with a genetic syndrome; eg, EL.

After determining the likelihood of a genetic disorder, a full medical history and physical examination are necessary as in most clinical situations. Several aspects, however, assume a great importance: family history; a growth and developmental history, including possible serial growth and head circumference data; and an assessment for minor anomalies.

Understanding the significance or otherwise of a family history requires a three-generation family tree from both sides of the family. Specific questioning to be included relates to the presence or absence of consanguinity, premature death, causes of death, pregnancy loss, neonatal or childhood death, epilepsy, special schooling, or other congenital problems affecting one or more family members. Names and dates of birth of first-degree relatives should be included if a genetic disorder is suspected and genetic investigation or referral is likely.

Many genetic syndromes are defined in part by natural history; this includes a pattern of growth. It may be unremarkable or characterized by prenatal or postnatal growth failure or both. This applies equally well to head circumference as to weight and height; serial growth measurements are very helpful in a differential diagnosis.

In assessing the significance of a congenital abnormality, a decision needs to be taken as to whether or not the abnormality under consideration is isolated or not. Most congenital abnormalities are isolated, but some, such as those affecting the posterior of the eye, heart, brain, and kidney will not be diagnosed without a specific assessment.

In deciding which abnormalities are likely to be syndromal or part of a more generalized disturbance in development, important factors in the history are family history, general health, growth, and development. Also helpful is the focused clinical examination to look for minor morphological variations that are not abnormal in themselves, but can be a clue to an underlying syndrome or an indicator of a multisystem problem. Common examples are bilateral transvers palmar creases, widely spaced eyes, and upslanting palpebral fissures. The presence of three or more minor anomalies occurs in less than 1% of infants, and when present, an underlying major malformation will be found in around 20% [88] and further investigation is warranted.

In the presence of a developmental disorder, with or without congenital abnormality, the first line of investigation is usually a chromosome microarray. A chromosome difference is the commonest cause of both developmental disability and multiple congenital abnormalities. When a clinical diagnosis of a single gene disorder is made, a chromosome investigation may still be warranted, because for some disorders, complete or partial gene deletion may be the underlying mechanism and may be detected by microarray.

There are very few gene disorders where gene testing is always abnormal and diagnostic. Gene testing will always be helpful in the disorders where the mutation is a triplet repeat (eg, fragile X syndrome, myotonic dystrophy) and in some other conditions with characteristic mutations and very distinct phenotypes (eg, achondroplasia, CS). For most genetic disorders, diagnoses remain clinical, albeit with an increasing number being able to be confirmed by gene testing.

The new era of molecular testing of multiple gene panels has been particularly helpful for very heterogeneous phenotypes like hypertrophic cardiomyopathy, where many different genes can now be tested simultaneously. Although the problems associated with attributing causality are substantial [89], this is counterbalanced by the information that can be obtained by gene testing large numbers of patients with properly documented family histories and phenotypes that maximize the likelihood of correct variant interpretation. Sharing data is equally critical to this process.

The increased capacity for gene testing has also illuminated the inaccuracies in syndrome definition on purely clinical grounds. Phenotypes associated with developmental syndromes in particular have been shown to be caused by mutations in pathways that could not be predicted clinically; eg, the fibrillinopathies and acromicric disorders. Previous diagnostic criteria relying on purely clinical examination have also been shown to be inaccurate; eg, apparently IEL, RASopathies.

Increasingly, accurate clinical diagnosis relies on consideration of patterns of differences that may reflect a perturbation in a particular cellular pathway. This includes not only the range of possible presentations, but also the need to consider potential natural history and complications in terms of those seen in other pathway disorders.

CONCLUSION

The classification of genetic disease is changing to reflect the underlying molecular pathway or perturbation at the cellular level as molecular pathology illuminates gene function and interrelationships. This has had an impact on disease nosology that is likely to increase, because understanding of the cellular basis of genetic causes of individual conditions is likely to inform diagnosis and management when the cause is unknown or multifactorial.

The increasing ability to establish the molecular cause of genetic disorders is altering the previous clinical definitions and differentiation of genetic conditions to a more pathway-based approach, where consideration of the overall pattern of disease is of higher importance.

It is likely that the approach of considering the molecular nosology will lead to more accurate diagnoses, but also to effective treatments through research that is focused on understanding the pathology at the cellular level.

REFERENCES

[1] Cassidy SB, Allanson JE, editors. Management of genetic syndromes. 3rd ed. New York: John Wiley; 2010.

[2] Firth HV, Hurst JA. Oxford desk reference: clinical genetics. Oxford: Oxford University Press; 2005.

[3] McKusick VA. On lumpers and splitters, or the nosology of genetic disease. Perspect Biol Med 1969;12(2):298–312.

[4] Neri G, Zollino M. More on the Noonan-CFC controversy. Am J Med Genet 1996;65(2):100.

[5] Superti-Furga A, Bonafe L, Rimoin DL. Molecular-pathogenetic classification of genetic disorders of the skeleton. Am J Med Genet 2001;106(4):282–93.

[6] Warman ML, Cormier-Daire V, Hall C, Krakow D, Lachman R, Le Merrer M, et al. Nosology and classification of genetic skeletal disorders: 2010 revision. Am J Med Genet Part A 2011;155:943–68.

[7] Krakow D, Vriens J, Camacho N, Luong P, Deixler H, Funari TL, et al. Mutations in the gene encoding the calcium-permeable ion channel TRVP4 produce spondylometaphyseal dysplasia, Kozlowski type and metatrophic dysplasia. Am J Hum Genet 2009;84:307–15.

[8] Andreucci E, Aftimos S, Alcausin M, Haan E, Hunter W, Kannu P, et al. TRPV4 -related skeletal dysplasias: a phenotypic spectrum highlighted by clinical and molecular findings in 21 new families. Orphanet J Rare Dis 2011;9:37.

[9] Nishimura G, Lausch E, Savarirayan R, Shiba M, Spranger J, Zabel B, et al. TRPV$-associated skeletal dysplasias. Am J Med Genet C Semin Med Genet 2012;160C(3):190–204.

[10] Miller FA, Begbie ME, Giacomini M, Ahern C, Harvey EA. Redefining disease? The nosological implications of molecular genetic testing. Perspect Biol Med 2006;49(1):99–114.

[11] Fry AM, Leaper MJ, Bayliss R. The primary cilium: guardian of organ development and homeostasis. Organogenesis 2014;10(1):62–8.

[12] Horani A, Brody SL, Ferkol TW. Picking up speed: advances in the genetics of primary ciliary dyskinesia. Pediatr Res 2014;75(1–2):158–64.

[13] Valente EM, Rosti RO, Gibbs E, Gleeson JG. Primary cilia in neurodevelopmental disorders. Nat Rev Neurol 2014;10(1):27–36.

[14] Marfan A. Un cas de déformation congénitale des quartre membres, plus prononcée aux extrémitiés, caractérisée par l'allongement des os avec un certain degré d'amincissement [A case of congenital deformation of the four limbs, more pronounced at the extremities, characterized by elongation of the bones with some degree of thinning]. In: Bulletins et memoires de la Société medicale des hôspitaux de Paris (in French), vol. 13 (3rd series); 1896. p. 220–6.

[15] Dietz HC, Cutting GR, Pyeritz RE, Maslen CL, Sakai LY, Corson GM, et al. Marfan syndrome caused by a recurrent de novo missense mutation in the fibrillin gene. Nature 1991;352:337–9.

[16] Davis MR, Summers KM. Structure and function of the mammalian fibrillin gene family: implications for human connective tissue diseases. Mol Genet Metab 2012;107:635–47.

[17] Beighton P, de Paepe A, Danks D, Finidori G, Gedde-Dahl T, Goodman R, et al. International nosology of heritable disorders of connective tissue. Am J Med Genet 1988;29:581–94.

[18] Loeys BL, Dietz HC, Braverman AC, Callewaert BL, De Backer J, Devereux RB, et al. The revised Ghent nosology for the Marfan syndrome. J Med Genet 2010;47:476–85.

[19] Chandra A, Patel D, Aragon-Martin JA, Pinard A, Collod-Beroud G, Comeglio P, et al. The revised Ghent nosology: reclassifying isolated ectopia lentis. Clin Genet 2015;87(3):284–7.

[20] Faivre L, Masurel-Paulet A, Collod-Beroud G, Callewaert BL, Child AH, Stheneur C, et al. Clinical and molecular study of 320 children with Marfan syndrome and related type 1 fibrillinopathies in a series of 1009 probands with pathogenetic FBN1 mutations. Pediatrics 2009;123(1):391–8.

[21] Détaint D, Faivre L, Collod-Beroud G, Child AH, Loeys BL, Binquet C, et al. Cardiovascular manifestations in men and women carrying a FBN1 mutation. Eur Heart J 2010;31(18):2223–9.

[22] Callewaert BL, Loeys BL, Ficcadenti A, Vermeer S, Landgren M, Kroes HY, et al. Comprehensive clinical and molecular assessment of 32 probands with congenital contractural arachnodactyly: report of 14 novel mutations and review of the literature. Hum Mutat 2009;30(3):334–41.

[23] Amado M, Maria Angelina Calado MA, Ferreira R, Lourenço T. Neonatal Marfan syndrome : a successful early multidisciplinary approach. BMJ Case Rep 2014. http://dx.doi.org/10.1136/bcr-2013-202438.

[24] Le Goff C, Cormier-Daire V. From tall to short: the role of TGFβ in growth and its disorders. Am J Med Genet C Semin Med Genet 2012;160C(3):145–503.

[25] Cecchi A, Ogawa N, Martinez HR, Carlson A, Fan Y, Penny DJ, et al. Missense mutations in FBN1 exons 42 and 42 cause Weill-Marchesani syndrome with thoracic aortic disease and Marfan syndrome. Am J Med Genet 2013;161A(9):2305–10.

[26] Hiratzka LF, Bakris GL, Beckman JA, Bersin RM, Carr VF, Casey Jr DE, et al. ACCF/AHA/AATS/ACR/ASA/SCA/SCAI/SIR/STS/SVM guidelines for the diagnosis and management of patients with thoracic aortic disease: executive summary. A report of the American College of Cardiology Foundation/American Heart Association Task Force on Practice Guidelines, American Association for Thoracic Surgery, American College of Radiology, American Stroke Association, Society of Cardiovascular Anesthesiologists, Society for Cardiovascular Angiography and Interventions, Society of Interventional Radiology, Society of Thoracic Surgeons, and Society for vascular Medicine (developed in collaboration with the American College of Emergency Physicians). J Am Coll Cardiol 2010;55:1509–44.

[27] Shores J, Berger KR, Murphy EA, Pyeritz RE. Progression of aortic root dilatation and the benefit of long-term beta-adrenergic blockage in Marfan's syndrome. NEJM 1994;330:1335–41.

[28] Sandor GG, Alghamdi MH, Raffin LA, Potts MT, Willimas LD, Potts JE, et al. A randomised, double blind pilot study to assess the effects of losartan vs. atenolol on the biophysical properties of the aorta in patients with Marfan and Loeys-Dietz syndromes. Int J Cardio 2015;179:470–5.

[29] Brooke BS, Habashi JP, Judge DP, Patel N, Loeys B, Dietz HC. Angiotensin II blockade and aortic root dilatation in Marfan's syndrome. NEJM 2008;358(26):2787–95.

[30] Lacro RV, Dietz HC, Sleeper LA, Yetman AT, Bradley TJ, Colan SD, et al. Atenolol versus losartan in children and young adults with Marfan's syndrome. N Engl J Med 2014;371:2061–71.

[31] Franken R, den Hartog A, Radonic T, Micha D, Maugeri A, van Dijk FS, et al. Beneficial outcome of losartan therapy depends on type of FBN1 mutation in Marfan syndrome. Circ Cardiovasc Genet 2015;8(2):383–8. pii:CIRCGENETICS.114.000950 [Epub ahead of print].

[32] Attenhofer Jost CH, Greutmann M, Connolly HM, Weber R, Rohrbach M, Oxenius A, et al. Medical treatment of aortic aneurysms in Marfan syndrome and other heritable conditions. Curr Cardio Rev 2014;10(2):161–71.

[33] Shimizu K, Goldfarb M, Suard Y, Perucho M, Yen L, Kamata T, et al. Three human transforming genes are related to the viral ras oncogenes (human tumor cells/molecular cloning/gene families). Proc Natl Acad Sci USA 1983;80:2112–6.

[34] Pylayeva-Gupta Y, Grabocka E, Bar-Sagi D. RAS oncogenes: weaving a tumorigenic web. Nat Rev Cancer 2013;11(11):761–74.

[35] Huson SM, Clark D, Compston D, Harper P. A genetic study of von Recklinghausen neurofibromatosis in South East Wales. I. Prevalence, fitness, mutation rate and effect on parental transmission on severity. J Med Genet 1989;26:704–11.

[36] Ekvall S, Sjörs K, Jonzon A, Vihinen M, Annerén G, Bondeson ML. Novel association of neurofibromatosis type 1-causing mutations in families with neurofibromatosis-Noonan syndrome. Am J Med Genet A 2014;164A(3):579–87.

[37] Brems H, Pasmant E, Van Minkelen R, Wimmer K, Upadhyaya M, Legius E, et al. Review and update of SPRED1 mutations causing Legius syndrome. Hum Mutat 2012;33(11):1538–46.

[38] Noonan JA, Ehmke DA. Associated noncardiac malformations in children with congenital heart disease. Midwest Soc Pediatr Res 1963;63:468–70.

[39] Roberts AE, Allanson JE, Tartaglia M, Gelb BD. Noonan syndrome. Lancet 2013;381:333–42.

[40] Costello JM. A new syndrome: mental subnormality and nasal papillomata. Aust J Paediatr 1977;13:114–8.

[41] Der Kaloustian VM, Moroz B, McIntosh N, Watters AK, Blaichman S. Costello syndrome. Am J Med Genet 1991;41:69–73.

[42] Martin RA, Jones KL. Delineation of the Costello syndrome. Am J Med Genet 1991;41:346–9.

[43] Lurie IW. Genetics of the Costello syndrome. Am J Med Genet 1994;52:358–9.

[44] Costello JM. Costello syndrome: update on the original cases and commentary. Am J Med Genet 1996:199–201.

[45] Zampino G, Mastroiacovo P, Ricci R, Zollini M, Segni G, Martini-Neri ME, et al. Costello syndrome: further clinical delineation, natural history genetic definition and nosology. Am J Med Genet 1993;47:176–83.

[46] Johnson JP, Golabi M, Norton ME, Rosenblatt RM, Feldman GM, Yang SP, et al. Costello syndrome: phenotype, natural history, differential diagnosis and possible cause. J Ped 1998;133(3):441–8.

[47] Fryns JP, Vogels A, Haegerman J, Eggermont E, Van Den Berghe H. Costello syndrome: a postnatal growth retardation syndrome with distinct phenotype. Genet Couns 1994;5(4):337–43.

[48] Kerr B, Eden OM, Dandamudi R, Shannon N, Quarrell O, Emmerson A, et al. Costello syndrome: two cases with embryonal rhabdomyosarcoma. J Med Genet 1998;35:1036–9.

[49] Gripp KW, Scott CI, Nicholson L, McDonald-McGinn DM, Ozeran JD, Jones MC, et al. Five additional Costello syndrome patients with rhabdomyosarcoma: proposal for a tumour screening protocol. Am J Med Genet 2002;108:80–7.

[50] Reynolds JF, Neri G, Herrmann JP, Blumberg B, Coldwell JG, Miles PV, et al. New multiple congenital anomalies/mental retardation syndrome with cardio-facio-cutaneous involvement–the CFC syndrome. Am J Med Genet 1986;25:413–27.

[51] Baraitser M, Patton MA. A Noonan-like short stature syndrome with sparse hair. J Med Genet 1986;24:9–13.

[52] Aoki Y, Niihori T, Kawame H, Kurosawa K, Ohashi H, Tanaka Y, et al. Germline mutations in HRAS proto-oncogene cause Costello syndrome. Nat Genet 2005;37(10):1038–40.

[53] Tartaglia M, Mehler EL, Goldberg R, Zampino G, Brunner HG, Kremer H, et al. Mutations in PTPN11 encoding the protein tyrosine phosphatase SHP-2 cause Noonan syndrome. Nat Genet 2001;29:465–8.

[54] Kerr B, Delrue M-A, Sigaudy S, Perveen R, Marche M, Burgelin I, et al. Genotype-phenotype correlation in Costello syndrome: HRAS mutation analysis in 43 cases. J Med Genet 2006;43(5):401–5.

[55] Kerr B. The clinical phenotype of Costello syndrome. In: Zenker M, editor. Noonan syndrome and related disorders. Monographs in Human Genetics, vol. 17. Basel: Karger; 2009. p. 83–93.

[56] Lo I, Brewer C, Shannon N, Shorto J, Tang B, Black G, et al. Severe neonatal manifestations of Costello syndrome. J Med Genet 2008;45:167–71.

[57] Burkitt-Wright EMM, Bradley L, Shorto J, McConnell VPM, Gannon C, Firht HV, et al. Neonatal lethal Costello syndrome and unusual dinucleotide deletion/insertion mutations in HRAS predicting p.Gly12Val. Am J Med Genet Part A 2012;158A:1102–10.

[58] Van der Burgt I, Kupsky W, Stassou S, Nadroo A, Barroso C, Diem A, et al. Myopathy caused by HRAS germline mutations: implications for disturbed myogenic differentiation in the presence of constitutive H-Ras activation. J Med Genet 2007;44:459–62.

[59] Gripp KW, Hopkins E, Sol-Church K, Stabley DL, Axelrad ME, Doyle D, et al. Phenotypic analysis of individuals with Costello syndrome due to HRAS p.G13C. Am J Med Genet A 2011;155A:706–16.

[60] McCormick EM, Hopkins E, Conway L, Catalano S, Hossain J, sol-Church K, et al. Assessing genotype-phenotype correlation in Costello syndrome using a severity score. Genet Med 2013;15(7):554–7.

[61] Niihori T, Aoki Y, Narumi Y, Neri G, Cave H, Verloes A, et al. Germline mutations in KRAS and BRAF cause cardio-facio- cutaneous syndrome. Nat Genet 2006;38:294–6.

[62] Rodriguez-Viciana P, Tetsu O, Tidyman WE, Estep AL, Conger BA, Santa Cruz M, et al. Germ-line mutations in genes within the MAPK pathway cause cardio-facio-cutaneous syndrome. Science 2006;311:1287–90.

[63] Armour CM, Allanson JE. Further delineation of cardio-facio-cutaneous syndrome: clinical features of 38 individuals with proven mutations. Lett J Med Genet 2008;45:249–54.

[64] Sarkozy A, Carta C, Moretti S, Zampino G, Digilio MC, Pantaleoni F, et al. Germline BRAF mutations in Noonan, LEOPARD, and cardiofaciocutaneous syndromes: molecular diversity and associated phenotypic spectrum. Hum Mutat 2009;30:695–702.

[65] Linden HC, Price SM. Cardiofaciocutaneous syndrome in a mother and two sons with a MEK2 mutation. Clin Dysmorph 2011;20:86–8.

[66] Aoki Y, Niihori T, Banjo T, Okamoto N, Mizuno S, Kurosawa K, et al. Gain of function mutations in RIT1 cause Noonan syndrome, a RAS/MAPK pathway syndrome. Am J Hum Genet 2013;93:173–80.

[67] Cirstea IC, Kutsche K, Dvorsky R, Gremer L, Carta C, Horn D, et al. A restricted spectrum of NRAS mutations causes Noonan syndrome. Nat Genet 2010;42:27–9.

[68] Flex E, Jaiswal M, Pantaleoni F, Martinelli S, Strullu M, Fansa EK, et al. Activating mutations in RRAS underlie a phenotype within the RASopathy spectrum and contribute to leukaemogenesis. Hum Mol Genet 2014;23(16):4315–27.

[69] Santoro C, Pacileo G, Limogelli G, Scanguetta S, Giugliano T, Piluso G, et al. LEOPARD syndrome : clinical dilemmas in differential diagnosis of RASopathies. BMC Med Genet 2014;15:44. http://dx.doi.org/10.1186/1471-2350-15-44.

[70] Mazzanti L, Cacciari E, Cicognani A, Bergamaschi R, Scarano E, Forabosco A. Noonan-like syndrome with loose anagen hair. Am J Med Genet 2003;118A:279–86.

[71] Martinelli S, De Luca A, Stellacci E, Rossi C, Checquolo S, Lepri F, et al. Heterozygous germline mutations in the CBL tumor-suppressor gene cause a Noonan syndrome-like phenotype. Am J Med Genet 2010;87(2):250–7.

[72] Niemeyer CM, Kang MW, Shin DH, Furlan I, Erlacher M, Bunin NJ, et al. Germline CBL mutations cause developmental abnormalities and predispose to juvenile myelomonocytic leukemia. Nat Genet 2010;42(9):794–800.

[73] Gripp KW, Lin AE, Nicholson L, Allen W, Cramer A, Jones KL, et al. Further delineation of the phenotype resulting from BRAF or MEK1 germline mutations helps differentiate cardio-facio-cutaneous syndrome from Costello syndrome. Am J Med Genet 2007;143A:1472–80.

[74] Nystrom A-M, Ekvall S, Berglund E, Bjorkqvist M, Braathen G, Duchen K, et al. Noonan and cardio-facio-cutaneous syndromes: two clinically and genetically overlapping disorders. J Med Genet 2008;45:500–6.

[75] Syndromes run together in the RAS pathway. Nat Genet 2006;38:267.

[76] Bentires-Ali M, Kontaridis MI, Neel BG. Stops along the RAS pathway in human genetic disease. Nat Med 2006;12:283–5.

[77] Denayer E, Legius E. What's new in the neuro-cardio-facial-cutaneous syndromes ? Eur J Pediatr 2007;166(11):1091–8.

[78] Tidyman WE, Rauen KA. The RASopathies: developmental syndromes of RAS/MAPK dysregulation. Curr Opin Genet Dev 2009;19:230–6.

[79] Zenker M. Clinical manifestations of mutations in RAS and related intracellular signal transduction factors. Curr Opin Pediatr 2011;23(4):443–51.

[80] Haffner C, Toll A, Real FE. HRAS mutation mosaicism causing urothelial cancer and epidermal nevus. NEJM Lett 2011;356:1940–2.

[81] Narrod SA, Stiller C, Lenoir GM. An estimate of the heritable fraction of childhood cancer. Brit J Cancer 1991;63:993–9.

[82] Sharif S, Moran A, Huson SM, Iddenden R, Shenton A, Howard E, et al. Women with neurofibromatosis 1 are at a moderately increased risk of developing breast cancer and should be considered for early screening. J Med Genet 2007;44:481–4.

[83] Niemeyer CM. RAS diseases in children. Haematologica 2014;99(11):1653–62.

[84] Strullu M, Caye A, Lachenaud J, Cassinat B, Gazal S, Fenneteau O, et al. Juvenile myelomonocytic leukaemia and Noonan syndrome. J Med Genet 2014;51(10):689–97.

[85] Park H-D, Lee SH, Sung KW, Koo HH, Jung NG, Cho B, et al. Gene mutations in the RAS pathway and the prognostic implication in Korean patients with juvenile myelomonocytic leukaemia. Ann Hematol 2012;91:511–7.

[86] Smpokou P, Zand DJ, Rosenbaum KN, Summar ML. Malignancy in Noonan syndrome and related disorders. Clin Genet 2015. http://dx.doi.org/10.1111/cge.12568. [e-pub ahead of print].

[87] Kratz CP, Franke L, Peters H, Kohlschmidt N, Kazmierczak B, Finckh U, et al. Cancer spectrum and frequency among children with Noonan, Costello, and cardio-facio-cutaneous syndromes. Br J Cancer March 5, 2015. http://dx.doi.org/10.1038/bjc.2015.75. [e-pub ahead of print].

[88] Méhes K. Informative morphogenetic variants in the newborn infant. Akadémiai Kiadó Budapest; 1988.

[89] MacArthur DG, Manolio TA, Dimmock DP, Rehm HL, Shendure J, Abecasis GR, et al. Guidelines for investigating causality of sequence variants in human disease. Nature 2014;508:469–76.

[90] Krantz ID. Cohesin embraces new phenotypes. Nat Genet 2014;46(11):1157–8.

[91] Skibbens RV, Colquhoun JM, Green MJ, Molnar CA, Sin DN, Sullivan BJ, et al. Cohesinopathies of a feather flock together. PLoS Genet 2013;9(12):1–6. e1004036.

[92] Al-Jassar C, Bikker H, Overduin M, Chidgey M. Mechanistic basis of desmosome-targeted diseases. J Mol Biol 2013;425:4006–22.

[93] Worman HJ, Östlund C, Wang Y. Diseases of the nuclear envelope. Cold Spring Harb Perspect Biol 2010;2:a000760.

Chapter 14

Genomic Analysis in Clinical Practice: What Are the Challenges?

A. Lucassen[1,2], I. van Langen[2]

[1]University of Southampton Medical Centre, Southampton, United Kingdom; [2]University Medical Centre Groningen and Rijksuniversiteit Groningen, Groningen, The Netherlands

Chapter Outline

Introduction	**191**	Uncertain Incidental Findings	197
Consent for Genome Testing in Clinical Practice	**191**	**The Data Interpretation Problem**	**197**
Incidental Findings	**195**	What Constitutes a Result in Genomic Medicine?	197
Management of Incidental Findings	195	Diagnostic Versus Population Screening	198
Uncertainties About for Whom the Result has Clinical		**Conclusions**	**198**
Utility, and When	196	**References**	**198**
Uncertainties About the Clinical Validity of a Test	197		

INTRODUCTION

The speed by which a person's entire genetic code (genome) can be deciphered has increased several 1000-fold over recent years, whereas the attendant costs have plummeted. Until recently, the search for a genetic diagnosis was usually a targeted approach analyzing specific genes based on particular signs, symptoms, or family history of disease. Rapid technological advances mean that this approach is shifting to untargeted sequencing of an entire genome [whole genome sequencing (WGS)], albeit with subsequent targeted analysis of the output of sequencing. Targeted approaches characteristically have a high yield for penetrant monogenic conditions; whole genome approaches have the potential to unravel a much larger proportion of genetic disease burden but require complex bioinformatic analyses to interpret the significance of individual genomic variations. Whole genome analyses promise not only to transform a clinician's diagnostic and predictive ability, but also to improve preventative therapies, surveillance regimes, and tailor treatment to an individual patient's genetic makeup. Furthermore, such approaches are likely to extend the diagnostic range from preconception to testing, through to noninvasive prenatal diagnosis to postmortem testing for the benefit of relatives.

Whole genome analyses also bring challenges for healthcare. Although the ethical, legal, and social aspects resulting from the generation of data from genome sequencing are not unique, several nuances merit serious consideration. In this chapter, we start by summarizing some of the issues we believe require consideration (see Box 14.1), and then focus on two areas that have been much debated in clinical, academic, and research communities: consent to genomic testing and what to do about incidental findings (IFs) from genomic testing. We do not discuss prenatal genomic testing, although many of the questions we raise will also be relevant here.

CONSENT FOR GENOME TESTING IN CLINICAL PRACTICE

Providing individuals with sufficient information to make decisions about investigations or interventions is a key element of good clinical practice. A discussion about the potential risks and benefits of analyzing a particular part of the genetic code (because of signs, symptoms, or a particular family history of disease) will be different to those for a whole genome analysis, where the range of possible inferences that can be made, and the degrees of certainty of many of them, are much greater. Achieving a balance between providing sufficient information but avoiding overload can be a challenge, especially for tests

Box 14.1 Summary

Some of the overlapping, such as ethical, legal, social, and practical issues that need to be addressed as genome analysis, enters clinical practice.

Complexity of Genome Outputs

Genome analysis can provide many different predictions about diagnoses, or susceptibilities to conditions with varying degrees of certainty and confidence of these predictions.

- **Epistasis**: the effect of one genetic variant being dependent on the particular genetic background it lies on, and on other (as yet unknown) genetic variants or environmental factors means the confidence intervals about clinical predictions can be very broad. This is particularly so where variants are discovered opportunistically rather than as a result of focused analysis of genetic variants associated with a personal or family history of disease.

- **Epigenetics**: the modification of gene expression rather than alteration of the genetic code itself also affects epistasis.

- **New evidence**: from larger and different population studies over time is likely to change the confidence of predictions, which may in turn need to be fed back to clinical care. Many genetic variants have been studied more in one geographical population than others and their associations with disease often do not translate to different populations [1].

Recontacting/Follow-up Policies

- Given that the clinical information accompanying a genomic result is likely to change over time as evidence is accumulated/other associations found, what obligations do health professionals have to recontact their patients in the light of evolving knowledge? How and when should someone seen several years ago be recontacted about potential new information? Who (if anyone) is liable if the patient is not made aware of new knowledge about previous test results? What responsibilities do patients have to seek updates about their genomic results or to enable health professionals to contact them? [2].

- Given that genomic testing could predict disease onset many years in the future, how should such information be recorded before the onset of clinical utility? For example, should the finding of a *BRCA1* mutation in a baby: (1) be avoided at all costs (as suggested by much professional guidance [3,4]), (2) be disclosed to the parents for future disclosure by them, or (3) be recorded in the baby's medical notes for future disclosure by a health professional and, if so, how could medical record systems best be adapted to flag such issues at appropriate times?

Familial Aspects

- Although genomic information is on the one hand very personal, it may also be relevant for others who have not (yet) sought medical advice but may be identified to be at risk through testing of another. Does knowing about at-risk relatives give the health professional involved any extra responsibilities? Are these discharged by urging intrafamilial communication, or are there times, much like contact tracing in infectious diseases, that health professionals need to be active in alerting family members? [5] These questions are particularly hard to resolve at a time when medicine is increasingly seen as personalized and where concerns about consent and confidentiality raise tensions for familial communication of risk. These questions are also raised by large-scale population WGS approaches, as recently performed in Iceland [6]. The authors used a combination of familial databases and WGS analyses of a proportion of the population to impute and infer the WGS of others.

Confidentiality

- Modern medicine often adopts an approach to confidentiality that assumes facts or inferences about a patient can and should be kept confidential at all times. Yet this is increasingly impossible to do in an age of electronic data sharing, and where interpretations of genomic variants and their effect on health is only really possible in the context of large population analyses. Given that most genomic information cannot be completely anonymized, when is confidentiality broken? Is it possible to keep the medical details about a patient confidential but share the genetic predisposition to those medical details with others in the family who may also be at risk without breaking confidentiality? How can genetic data, be they DNA sequence, diagnostic, or family history details, be appropriately shared between health professionals in different locations?

Data Management

- Developments in the collection, linking, use, and exploitation of human genomic and health data arising from advances in technology and organization raise fundamental questions about the practical exercise of privacy, autonomy, anonymity, identity, altruism, solidarity and ownership, control, and interest in data [7]. As well as the interaction between the interests of the individual data subject, other individuals, and public interest. For example, as WGS becomes more commonplace what should be stored and where? Should the DNA sample be stored, or the DNA sequence or the interpretation of the sequence? Or combinations of these? What constitutes a result? What should be part of a person's medical record, how can professionals be compliant with relevant data protection (and other) legislation? How can researchers balance better inferences from large comprehensive data sets with potential security issues around identifiable data? What role should patients have in controlling their own data; for example, should they have their sequence data on a USB stick or handheld patient records?

Research/Clinical Boundary

- Genomic technologies are rapidly evolving, so the traditional route of research to clinical evolution with a clear distinction between the two is not necessarily possible.

Box 14.1 Summary—cont'd

Patients may think they are taking part in research when in fact they are receiving a clinical service, or vice versa [8,9]. The United Kingdom's 100,000 Genome Project is explicitly a hybrid activity; patients are offered the test and any relevant results as part of their clinical diagnostic package, but they have to consent to further research on their data in order to be recruited into the program [10]. Conversely participants in United Kingdom's biobank [11] or the Netherlands' LifeLines [12] receive results of certain health investigations but not any genomic findings, nor can they choose to do so.

Public Perception of Genetics/Genomics

- Genomics is often portrayed as more clear cut, or deterministic, than there is current evidence for.
- The analytical validity of genome testing is high, but this is not the same as clinical validity or clinical utility. There is much genome variation that is either: uninterpretable, probably benign, or only pathogenic in certain circumstances; for example, in the presence of as yet unknown epistatic factors. This gap between technological advances and the interpretation of any next generation sequencing (NGS) output is neatly encapsulated by the phrase "$1000 genome; $1 million interpretation" [13]. The public discourse around genomics often conflates high analytic validity with high clinical utility. Put another way, one can be confident a particular variant exists, but not necessarily be certain of its effect on a person's health.

Consent

- Consent is only valid if the implications of [an investigation] are communicated and understood when it is given. Yet is it ever possible to give valid consent to genomic testing where the number of possible outcomes and inferences are much greater than can feasibly be discussed up front, let alone fully assimilated and their implications (both personal and familial) understood? How then can health professionals ensure their patients are adequately informed before testing without glossing over some pertinent details or facts and running the risk of being accused of returning to a bygone paternalistic approach of "doctor knows best" what test you should have? How can and should broad consent be implemented appropriately and how can appropriate shared decision making be achieved?

Incidental Findings

The greater the resolution by which the genome is analyzed, the greater the probability of finding potential abnormalities that are unrelated to the clinical question for which the test was initiated. Such findings have been called IFs, but the more that whole genome analysis becomes a routine approach, the less any finding can be truly incidental.

We will now explore some of the aspects outlined in table/summary/[whatever decided] in more detail. we start with an exploration of the law on consent with particular emphasis on UK case-based law.

---

where multiple different outcomes are possible. Individuals need to understand what genome tests can reveal, but also that some degree of uncertainty is likely.

The UK's Department of Health states "it is a general legal and ethical principle that valid consent must be obtained before starting treatment or a physical investigation." NHS Choices (a patient information arm of the National Health Service) tells us that "for consent to be valid, it must be voluntary [not due to pressure by staff, friends or family], informed, and the person consenting must have the capacity to make the decision." So, if consent is only valid if the implications of [an investigation] are communicated and understood, is it ever possible to give valid consent to genomic testing where the number of possible outcomes and inferences are much greater than can feasibly be discussed up front, or their implications (both personal and familial) understood?

Providing consent to an investigation or treatment is an important part of a person's right to self-determination. A greater focus on consent over the last few decades also reflects changing attitudes about what is viewed as acceptable practice and a desire to move away from medical paternalism. There are some clear legal landmarks in

this evolution. The *Sidaway* case [14] established that the information that should be provided to a patient would be determined by the application of the "Bolam test"; that is, what a reasonable body of medical opinion would agree would be sufficient for patients to make an informed decision. Aside from the obvious paternalistic component of such a test, the problems for such a test in a rapidly developing clinical practice such as genomics are obvious. The *Bolitho* [15] case challenged the principle of the Bolam test and instead considered information to be adequate if it is what a reasonable patient might want to know. Lord Woolfe, in the *Pearce v United Bristol Healthcare NHS Trust* [16] ruled that "the reasonable doctor must tell the patient what the reasonable patient would want to know." What does the reasonable patient want to know about genomics, and how can a reasonable patient decide at the point of testing what they would or would not want to know? What is the reasonable health professional able to tell a patient about genomics, especially when most outputs are far less deterministic than popular portrayals make them sound? More recently, the *Montgomery vs Lanarkshire* case [17] appears to refine consent into what the particular reasonable patient would want to know, potentially placing greater emphasis on health professionals to find out what the particular patient in front of them would want to know in order to be able to give valid consent. The full ramifications of this judgment are, however, not clear at the time of writing and its implications for genomic medicine even less so.

Some have responded to this by suggesting that consent to genomic testing considers broad categories of results rather than specifics. For example, a patient could be asked if they want to know only results that predict serious conditions, only if they are treatable, or if they are "actionable," etc. The thinking behind this is that it could help to maximize patient choice and empower patients in this fast-moving, complex area, but in practice it can be very difficult to implement because: (1) it can be very difficult for a patient to know what they might want to know until it is there to know, and (2) healthcare professionals (including those involved in formulating the laboratory report) can also find it difficult to decide which "bin" a particular result belongs in. For example, is a result that confers a modest increase in the chance of developing melanoma one that is serious and treatable (other than interventions that are general health advice)? Or is the discovery of an untreatable condition still "actionable" in the sense that certain actions such as family planning, or perhaps more general life planning, can be taken in the light of new knowledge? [18] Box 14.2 outlines a case history where the difficulty in placing results in particular bins is highlighted.

Although consent is clearly one of the ways in which patients can be empowered, clinical consent does not equate to asking a patient to choose whatever investigation they want. Patients can refuse medical treatments, but they cannot determine whatever treatment (or investigation) they want. Using a language of "rights" is often invoked in this setting: does the patient have a right to a test, perhaps if certain family history criteria are met? Does their relative have a "right not to know"? And how could a right to not knowing be exercised without knowing there is something to know in the first place? One important aspect to remember that rights for one person may mean responsibility for others. If a patient has a right to a test, then health professionals will have a responsibility to provide it, and this would soon turn health professionals into providers of consumer goods rather than healthcarers whose job is to try and ensure patients' choices are as informed as possible. We need to attempt to act on patients' considered choices and these might be very different from their initial choices, especially if they had understood genomics to be a much more deterministic or clear-cut investigation than it currently is.

Confusion about what consent should look like in medical practice more widely is in part reflected by the number of qualifiers or adjectives we place in front of the word: *informed* consent, *fully informed* consent, *sufficient*, and *real* are all to some extent tautologies because they are also part of the definition of consent itself; consent *is not consent unless it is informed*. Then there are adjectives like *specific*, *generic*, *open-ended*, *broad*, *presumed*, or *implied*, which might be placed before the word *consent* in an attempt to nuance the nature of consent. Douglas Winnicott, a well-known pediatrician in the 1960s, famously advocated that the "good enough" parent was best for children. At a time when no qualifier seems enough for

---

Box 14.2 The Right to Know or Not Know

William is a 46-year-old man with a diagnosis of dilated cardiomyopathy and a family history of the condition, although knowledge of his family history is somewhat scanty. His father died young from an accident so he knows little about his paternal family history. However, there is some suggestion of dementia at a relatively young age in his paternal grandmother, but no confirmation.

He is offered, and consents to, an NGS cardiomyopathy panel test (a targeted panel facilitating the screening of many genes simultaneously). As part of the consent process, he decides he does not want to know about the presenilin *(PSEN)* gene analysis, which is part of the panel test, because he knows that there are no interventions or treatments available for early-onset dementia and he would therefore rather not know if his cardiomyopathy is a result of a mutation in one of these genes.

Cardiomyopathy panels currently in use consist of about 50 genes and may include the presenilin genes *(PSEN1* and *PSEN2)*. The *PSEN* genes are primarily associated with autosomal dominant, early onset, Alzheimer disease, but these genes are also expressed in the heart and are critical to cardiac

development. Li et al. found causative *PSEN* mutations in approximately 1% of dilated familial cardiomyopathy index patients [19]. All genes are sequenced in the testing process, but William's *PSEN* genes were therefore not reported.

William's panel test result found no abnormalities. He therefore has no genetic explanation for his own and his family history of cardiomyopathy, and his three sons (aged 20–26) continue with annual cardiology surveillance. William returns to the cardiogenetic clinic some 2 years after the initial panel test to see whether cardiomyopathy genetic testing has improved sufficiently to warrant further testing. Whole exome sequencing (WES) is offered as a test that picks up a small proportion of additional diagnoses to the test William has already had. Before starting the WES, review of his previous testing shows that William has a *PSEN* mutation but that this was not disclosed because William did not want it to be. The laboratory scientist is very uncomfortable about having this information and discusses it with William's clinician, who does not want to record this information in his medical records.

consent, we should like to advocate, *good enough* consent for genomic testing. A recent study [20] reporting on clinical exome sequencing showed high satisfaction and low distress with a consent approach that we consider can best be described as *good enough*.

INCIDENTAL FINDINGS

The term *IF* was first used in radiology in cases where, for example, an asymptomatic lung tumor is detected during a chest radiograph for acute back pain. Ever since chromosome analysis has been available since the 1960s, genomics IFs have been possible, though their sporadic discovery meant the debate around the ethical issues they might raise lay dormant until the recent leaps and bounds in genomic technologies. The discovery of an XXY karyotype while screening for Down syndrome prenatally, or of misattributed paternity during cascade screening, are both IFs that have been encountered for many decades. A genomic analysis increases the chance of an IF simply because much more data are being examined, but at the same time it confuses the term *incidental*. A genome test can only have an IF if it is used to answer, for example, a particular clinical question. If the

question is "What are the abnormalities in this genome?" then there can be no IFs [21].

Consequently there has been much recent debate about whether other terms might capture the phenomenon more accurately, without an overwhelming consensus. Some of the alternative suggestions, with their associated advantages and disadvantages are summarized in Table 14.1

Different terms, or multiple terms, may need to be used by different parties involved (clinicians, researchers, patients). For example, a finding that is unexpected by a patient might be anticipatable by a clinician or researcher simply because they know their field of vision is so broad. Discussion also needs to reflect that a finding may initially have unknown or uncertain clinical significance but be reclassified as a diagnostic or incidental finding over time or after further investigations.

Management of Incidental Findings

There has been much recent debate about the management of IFs in clinical applications of whole genome technologies. The American College of Medical Genetics and Genomics (ACMG) produced guidelines that recommended the active search for particular IFs when

TABLE 14.1 Evaluation of Terms Used to Describe Incidental Genomic Findings [22]

Term	Advantages	Disadvantages
Incidental finding	Useful term when a genome test is used to investigate particular signs, symptoms or diagnoses; to indicate that findings that are incidental to such diagnoses may also be found using a broad test	The finding can provide significant information, so the term incidental may not do justice to the impact it has on the person tested, and may trivialize the significance it has.
		Furthermore, where a whole genome is investigated it could be argued that no finding can be incidental, because any finding is actively sought.
Unsought/unsolicited finding	Captures the ability of genomic technologies to generate data not necessarily related to the initial diagnostic question [23]	Findings can only be made if they are looked for or solicited. Where a clinical diagnosis/prediction is sought, other findings might be unsolicited but any whole genome analysis cannot have truly unsolicited findings. Otherwise they would not be found [24].
Unrelated finding	Highlights the notion that the finding does not explain the condition for which testing has been done	Although unrelated to the original investigation, the finding could be related to the current/future health of the tested individual and other family members.
Secondary findings/ variants	Distinguishes between findings causing the disorder for which testing was performed (primary variants) and other clinically important findings (secondary variants) [25]	The term does not capture situations where the so-called "secondary variant" is the only (significant) variant found.
		It can inaccurately suggest a temporal relationship where one finding is found first and the other the second.
		The term *variant* is often used to indicate normal genetic variation or polymorphisms. Using *variant* to describe predisposition to disease could be confusing [26].

Continued

TABLE 14.1 Evaluation of Terms Used to Describe Incidental Genomic Findings [22]—cont'd

Term	Advantages	Disadvantages
Unexpected result/ unanticipatable finding	Dividing findings into *expected/anticipatable* and *unexpected/unanticipatable* could remind clinicians that unexpected findings should be discussed with and disclosed to patients in a different way to those associated with the original reason for testing.	Since the entire genome is being examined, findings that are unrelated to the reason for the test should always be expected or anticipated [28].
	Unexpected is a term patients can easily understand. Maintaining a division between *expected* and *unexpected* helps emphasize that answers to particular clinical questions are the priority. Any other answers do not provide an explanation to the clinical question [27].	Expectations of patients, clinicians, and researchers are different from each other, so what is unexpected for one might be expected for the other.
Off-target result	Indicates that genomic tests are broader than targeted tests; what is targeted is most clearly seen, but other findings can still be found	This could give the impression that the result is not correct because it has missed a target.
Nonpertinent/ coincidental	Emphasizes that the discovery of some (coincidental) findings is unavoidable (eg, if genes are collocated with those associated with the pertinent finding)	The result may still be pertinent to an individual, just not pertinent to the original question asked.
Opportunistic findings	Emphasizes that while looking for particular genomic findings, the analyzer looks opportunistically for findings associated with unrelated conditions [29]	The opportunity the finding gives patients may be opaque to them.
Additional looked-for findings	Emphasizes that in addition to looking for specific findings other findings will also be sought; term used by Genomics England to describe specific list of extra variants that will be examined during WGS (the term is distinguished from incidental findings which Genomics England will not report) [30].	Implies that there is a predefined list of variants for which specific consent can be sought. Therefore, this term is more like opportunistic screening that incidental or unexpected findings

using a WGS/WES approach [9]. The debate that ensued was largely focused on giving patients the right to say no to certain findings, and less on whether it is really possible to provide valid consent, or indeed valid refusal, to a range of possible unknown findings. Patients should, of course, be able to say no to certain things, but this *no* has to be sufficiently informed and therefore has to be more than a tick-box exercise. Were a patient to tick the *no* box, yet a genomic finding happened to indicate a particular high risk of a sudden cardiac death for which evidence-based interventions were available, then without being certain that this *no* was sufficiently informed, the health professional might be deemed negligent for not communicating such information to the patient. If a health professional is not convinced a refusal was sufficiently informed and harm is more likely to be prevented by disclosure (through availability of surveillance or other interventions, for example) then such refusal does not have to be observed at all costs.

Uncertainties About for Whom the Result has Clinical Utility, and When

Where an evidence-based medical intervention is available (ie, a treatment, risk-reducing surgery and/or surveillance that can improve health outcomes), the clinical utility of a finding is said to be high; however, the following still need to be considered:

Box 14.3 outlines a clinical example that illustrates this difficulty. WGS results in Alice have no clinical utility for her over the next 20 years but could be very useful to the parent she has inherited the trait from so that surveillance could be started.

Do the potential benefits to adult family members outweigh the potential harms of disclosing findings that do not have immediate clinical utility for a child? The ACMG has argued that a child such as Alice does benefit by potentially preventing a severe adverse health outcome in her parent. But how can this be squared with a multitude of

---

Box 14.3 Clinical Utility of Results

Alice (age 2 years) has developmental delay and facial features suggesting a rare genetic condition. Alice is one of the first children to take part in the United Kingdom's 100,000 Genomes Project and her parents consent to a series of "additional, looked for findings." A pathogenic BRCA1 mutation is found. This has no relevance for Alice's clinical care over the next 20 years, but her mother (age 35 years) is found to also have it despite no known family history of breast and ovarian cancer. Breast surveillance and risk-reducing surgeries are discussed with her mother.

---

international guidelines that suggest children should not be tested for adult-onset genetic conditions until there is a medical benefit or until they can decide if and when to be tested? There have been many critics of this approach who opine that a child's best interest should be the only reason for testing and disclosing a result. These disagreements further evidence the tension that occurs when tests are no longer targeted to investigate specific questions. Once a result is available, clinicians might feel a need or obligation to disclose it, even if they would not have ordered that specific test in the first place.

Uncertainties About the Clinical Validity of a Test

The clinical validity of a test is the accuracy with which a particular finding predicts the presence or absence of the underlying condition. The clinical validity of an IF, both now or in the future, can be unclear if the expressivity is variable or penetrance low. The likelihood of disease can depend on other genetic, environmental, or stochastic factors that have not been, or cannot yet be, determined, leaving a degree of uncertainty. The finding may therefore give a very incomplete prediction of disease, or indeed be a risk factor rather than a diagnosis. Genetic variants predisposing to disease have been known about in the past (for example, heterozygote CHEK2 mutations) but not offered as a clinical diagnostic test because they do not predict the onset or absence of breast cancer well. Should they now be reported just because they have been found, or should the certainty by which they predict the disease in question be of a certain threshold? Disclosure might then be determined by whether or not a result meets predetermined thresholds.

Uncertain Incidental Findings

Some findings may be deemed possibly pathogenic (eg, a missense mutation in a long QT gene not previously documented) but have as yet no evidence for pathogenicity. Functional studies of the finding or familial segregation studies might assist in clarifying the clinical significance of a finding, but may equally give indefinite answers or,

because they require testing and surveillance of family members, be too difficult or resource-intensive to perform. The uncertainty and extra steps needed to find out the significance of a finding will warrant careful communication, both with the patient and, potentially, their family in turn, raising implications for clinical practice.

THE DATA INTERPRETATION PROBLEM

Much of the misperception about the diagnostic value of genome sequencing results from an oversimplification in which it is assumed there is one gene for a condition, when in fact any increase in risk conferred by a mutation may be subtle, or only manifest in the context of specific genetic background or environmental exposure. For many common diseases there are multiple risk factors and although the identification of susceptibility genes has often provided novel insights in disease biology, their clinical utility in an individual may be very low because their predictive power in isolation is very poor.

There is, however, also a risk of overinterpretation even for mutations with seemingly large effects. For affected patients where there is a strong prior probability of the gene mutation being causal because of a positive family history and or specific clinical phenotype, interpretation can be straightforward. However, if mutations are not fully penetrant, there will be carriers in the population who remain healthy. Much of our knowledge about the penetrance of mutations to date is based on family data and hence suffers from ascertainment bias [31]. Without unbiased knowledge of the effect of mutations, interpretation at the population level will be inherently problematic. Although policies to restrict genetic testing to high-risk populations were initially driven by budget restraints, and the more widespread availability of testing thought to be an advantage of declining costs, another consequence is that the interpretation of the clinical significance of a mutation is much more difficult if found without the ascertainment bias noted. That is to say, predicting the effects of a novel BRCA2 mutation in the context of a strong family history of the mutation segregating with disease in the family is far easier than when it is discovered in a population screen (Box 14.4).

What Constitutes a Result in Genomic Medicine?

In the clinical setting, certainly in the short term, diagnostic accuracy will continue to depend on additional factors such as clinical history and therefore pretest probability. The diagnostic accuracy of, for example, a germline P53 missense mutation found in a 25-year-old woman with sarcoma and a family history of breast cancer and brain tumors is much higher than finding a novel missense P53

Box 14.4 Difficulties in Clinical Interpretation of Some Findings

Beth, a 4-year-old girl, was noted by her school teachers to intermittently lose concentration in class. She was seen by a community pediatrician, who investigated her for epilepsy and when this drew a blank, referred to a pediatric cardiologist to exclude potential cardiac causes. A baseline electrocardiograph was defined as within normal limits and no family history of any possible rhythm disturbance or of sudden cardiac death was elicited. The cardiologist had been to a presentation about mainstreaming genetics and realized that long QT interval (LQT) gene carriers can be difficult to detect in childhood. An LQT1-associated mutation was identified and described by the laboratory as possibly pathogenic because it had been seen in two other independent cases and was in a highly conserved part

of the gene. A reveal device was inserted but no abnormalities in Beth's QT interval were recorded during subsequent losses of concentration, which first led to her referral. Nevertheless, it was thought appropriate to treat her with β-blockers on the basis of her genetic test result. Cascade testing of Beth's family revealed that her 8-year-old sister, mother, maternal uncle (and his son aged 3 years), and maternal grandmother all carried the same mutation. Cardiac investigations of their phenotype at rest, with exercise, and pharmacological challenge were normal or equivocal. All carriers in the family were prescribed β-blockade, given a list of potentially QT prolonging medications to avoid, and possible implantable cardiac defibrillator insertion was debated.

---

mutation in a healthy 60-year-old female with no family history. If WGS approaches are to be used to answer clinical questions, some sort of filtering of sequence output will need to take place; there are otherwise too many potential genome outputs to consider. A subtle shift has thereby occurred in the point of targeting or filtering. Medicine has long required, and practiced, the targeting of the investigations to symptoms or signs. In WGS, the targeting has moved to the output of the test. The genome, once sequenced, is filtered for particular sections or variants rather than that only certain sections are sequenced in the first place. This raises novel issues about what constitutes a result, and therefore what is recorded in a patient's medical records. Is the raw sequence a (potential) result? Can a data set be reanalyzed at any stage to answer other questions? And if so, what obligations does the healthcare or laboratory professional have to reanalyze this data at any stage? Or is it only the validated variants that were fished out of the sequence because of a particular clinical question that should be analyzed?

Diagnostic Versus Population Screening

Given the significant limitations to our current understanding of the impact of genetic variation, we believe that clinical genome sequencing should for now be focused on particular clinical presentations compatible with a genetic etiology, rather than engaging in opportunistic population screening. For example, the identification of an *APC* mutation in a person with colonic polyposis is diagnostic and highly predictive for family members. In contrast, the identification of variants such as *LQT1* (described in Box 14.4) in a population screen do not have sufficient certainty to infer as much, resulting in difficult clinical management issues. Such contextual differences may be difficult to grasp if genetics is portrayed as being clear cut, and clinical interventions may therefore be offered without sufficient evidence for their benefit.

At the time of writing, Genomics England is poised to start offering WGS to 100,000 NHS patients who either have a rare condition or a particular cancer (or infectious disease). It is hoped that by sequencing the rare-disease patient's genome, and where possible their parents' genomes, and by comparing a person's genome with the genome in their cancer, and filtering the outputs for particular diagnostic variants, genomics will start to deliver the long-promised transformation of clinical practice.

CONCLUSIONS

In this chapter we have scratched at the surface of some of the ethical (and consequently) practical issues raised as genomic medicine enters mainstream medical practice. We have raised more questions than provided answers, but hope that our questions will enable the reader to contemplate the relevant issues for their own practice and thus deliver thoughtful and sound clinical genomics.

REFERENCES

[1] Lu YF, Goldstein DB, Angrist M, Cavalleri G. Personalized medicine and human genetic diversity. Cold Spring Harb Perspect Med July 24, 2014;4(9). a008581 25059740.

[2] Otten E, Plantinga M, Birnie E, Verkerk MA, Lucassen AM, Ranchor AV, et al. Is there a duty to recontact in light of new genetic technologies? A systematic review of the literature. Genet Med December 11, 2014. http://dx.doi.org/10.1038/gim.2014.173.

[3] BSGM guidance: http://www.bsgm.org.uk/media/678741/gtoc_booklet_final_new.pdf.

[4] Lucassen A, Widdershoven G, Metselaar S, Fenwick A, Parker M. Genetic testing of children: the need for a family perspective. Am J Bioeth 2014;14(3):26–8.

[5] Michael P, Lucassen A. Genetic information: a joint account? BMJ 2004;329:165.

[6] Gudbjartsson DF, Helgason H, Gudjonsson SA, et al. Large-scale whole-genome sequencing of the Icelandic population. Nat Genet March 25, 2015. http://dx.doi.org/10.1038/ng.3247.

[7] http://nuffieldbioethics.org/wp-content/uploads/Biological_and_health_data_web.pdf.

[8] Genomics and the boundary between research and clinical care and treatment http://www.phgfoundation.org/file/16014/.

[9] Green RC, Berg JS, Grody WW, Kalia SS, Korf BR, Martin CL, et al. ACMG recommendations for reporting of incidental findings in clinical exome and genome sequencing. Genet Med 2013;15(7):565–74.

[10] http://www.genomicsengland.co.uk/clinicians-researchers-and-industry-collaborate-with-the-100000-genomes-project/ [accessed 30.03.15].

[11] http://www.ukbiobank.ac.uk/about-biobank-uk/.

[12] https://www.lifelines.nl/lifelines-research/general.

[13] Davies K. The $1,000,000 genome interpretation. Bio-IT World. 2010. October 2010.

[14] *Sidaway v Board of Governors of the Bethlem Royal Hospital* [1984] QB 493, [1984] 1 All ER 1018, CA; affd [1985] AC 871, [1985] 1 All ER 643, HL.

[15] *Bolitho v Hackney Health Authority* [1998] Ac 232, [1997] 4 All ER 771, HL, (1998) 39 BMLR 1.

[16] *Pearce v United Bristol Healthcare* NHS Trust [1999] PIQR 53; (1998) 48 BMLR 118.

[17] *Montgomery v Lanarkshire* [2015] UKSC 11.

[18] Fenwick A, Dheensa S, Crawford G, Shkedi-Rafid S, Lucassen AM. Rescue obligations and collective approaches: complexities in genomics. Am J Bioeth February 2015;15(2):23–5. http://dx.doi.org/10.1080/15265161.2014.990763.

[19] Li D, Parks SB, Kushner JD, et al. Mutations of presenilin genes in dilated cardiomyopathy and heart failure. Am J Hum Genet 2006;79:1030–8.

[20] Sie AS, Prins JB, van Zelst-Stams WAG, Veltman JA, Feenstra I, Hoogerbrugge. N. Patient experiences with gene panels based on exome sequencing in clinical diagnostics: high acceptance and low distress. Clin Genet 2015;87:319–26.

[21] Lucassen A, Houlston RS. The challenges of genome analysis in the health care setting. Genes (Basel) July 22, 2014;5(3):576–85. http://dx.doi.org/10.3390/genes5030576.

[22] Shkedi-Rafid S, Dheensa S, Crawford G, Fenwick A, Lucassen AM. Defining and managing incidental findings in genetic and genomic practice. J Med Genet Nov 2014;51(11):715–23.

[23] Parker LS. The future of incidental findings: should they be viewed as benefits? J Law Med Ethics 2008;36(2):341–51.

[24] Srebniak MI, Diderich KE, Govaerts LC, Joosten M, Riedijk S, Galjaard RJ, et al. Types of array findings detectable in cytogenetic diagnosis: a proposal for a generic classification. Eur J Hum Genet 2014;22(7):856–8.

[25] Presidential Commission for the Study of Bioethical Issues. Anticipate and communicate: ethical management of incidental and secondary findings in the clinical, research, and direct-to-consumer contexts. 2013. Available at: http://bioethics.gov/sites/default/files/FINALAnticipateCommunicate_PCSBI_0.pdf. [accessed on 20.03.15].

[26] Crawford G, Fenwick A, Lucassen A. A more fitting term in the incidental findings debate: one term does not fit all situations. Eur J Hum Genet November 20, 2013. http://dx.doi.org/10.1038/ejhg.2013.266. Published Online First.

[27] van El CG, Cornel MC, Borry P, Hastings RJ, Fellmann F, Hodgson SV, et al. ESHG Public and Professional Policy Committee: whole-genome sequencing in health care. Recommendations of the European Society of Human Genetics. Eur J Hum Genet 2013;21:S1–5.

[28] Christenhusz GM, Devriendt K, Dierickx K. Secondary variants–in defense of a more fitting term in the incidental findings debate. Euro J Hum Genet 2013;21(12):1331–4.

[29] Wright CF, Middleton A, Burton H, Cunningham F, Humphries SE, Hurst J, et al. Policy challenges of clinical genome sequencing. BMJ 2013;347:347. f6845.

[30] http://www.genomicsengland.co.uk/the-100000-genomes-project/taking-part/[accessed 29.03.15].

[31] Zollner S, Pritchard JK. Overcoming the winner's curse: estimating penetrance parameters from case-control data. Am J Hum Genet 2007;80(4):605–15.

Chapter 15

Genomic Perspective of Genetic Counseling

C.L. Gaff[1,2], I. Macciocca[3]

[1]Melbourne Genomics Health Alliance, Melbourne, Australia; [2]The University of Melbourne, Melbourne, Australia; [3]Victorian Clinical Genetics Service, Melbourne, Australia

Chapter Outline

Introduction 201
Ethos and Principles of Genetic Counseling 201
Elements of Genetic Counseling 202
Counseling Issues 203
  Uncertainty 203
  Confidentiality and Privacy of Genomic Data 203
    Data Sharing and the Undiagnosed Patient 204
    Reanalyzing Stored Genomic Data 204
    Sharing Genomic Data for Advancement of Knowledge 204
  Secondary Findings 204
  Family Communication 205

The "Healthy Genome" 205
Case Study 205
Practicalities of Counseling for Genomic Tests 206
  Genetic Information Is Key 209
  Patient Autonomy Must Be Supported 210
  Patient Emotions Make a Difference 210
  Relationship Is Integral to Genetic Counseling 210
Conclusion 210
Acknowledgment 210
References 210

INTRODUCTION

Genetic counseling is the communication process that aims to help individuals, couples, and families understand and adapt to the medical, psychological, familial, and reproductive implications of the genetic contributions to disease [1]. The term *genetic counseling* was coined well before molecular or cytogenetic testing was available [2], when a person's medical and family histories were largely the only means by which to assess the chance of disease occurrence (or recurrence); genetic counseling was predominantly education about the condition's inheritance. With the advent of tests to better diagnose genetic disease or predict future onset ("predictive testing") and the rapidly increasing number of conditions for which testing was available, pretest genetic counseling was recommended to promote informed choice. This was an opportunity for the person concerned to consider the possible medical and psychosocial consequences of the test for themselves and the wider family [3]. Posttest genetic counseling is offered for support and to facilitate adjustment [4].

The introduction of next generation sequencing tests, which can simultaneously sequence many thousands of genes at once, offers a greater level of precision than other microarray or cytogenetic analysis while providing a much greater breadth of information than Sanger gene sequencing. In this chapter, we provide an introduction to the practice of genetic counseling, with a focus on pretest counseling. We consider the counseling issues arising in genomic testing, delineating the truly novel "genomic counseling" issues from those that arise whenever new technologies for genetic diagnosis have been introduced. We address the genetic counseling for inherited (germline) mutations, not for the somatic changes found in cancer (noting that testing of cancer can identify inherited disease-causing mutations).

ETHOS AND PRINCIPLES OF GENETIC COUNSELING

From first coining of the phrase, genetic counseling was described as requiring knowledge of human genetics, respect for the sensitivities and reactions of clients, and the provision of genetic information to the full extent known (Reed, 1955 as cited in Ref.[5]). Respect for clients' emotional state and values was a response to misuses of genetics, which resulted in discrimination of those individuals and communities affected [2,6]. Decisions about prenatal testing, carrier testing, and predictive testing were recognized as personal decisions, which should be made on the basis of that person's

Medical and Health Genomics. http://dx.doi.org/10.1016/B978-0-12-420196-5.00015-0

own values, situation and preferences, rather than medical opinion. "Nondirectiveness" became a guiding principle for genetic counseling that "promotes the autonomy or self-determination and personal control of the client" [7].

The extent to which genetic counseling is a process of teaching or counseling has been debated for many years [8]; however, good communication skills are required for successful outcomes in both. McCarthy Veach et al. have developed a unifying model: the reciprocal engagement model (REM) [9]. This model defines five tenets, or principles that recognize the educative aspects of genetic counseling: (1) tenet—genetic information is key; (2) ethical stance—patient autonomy must be supported; and client-centered approach—relationship is integral to genetic counseling, patients are resilient, and patient emotions make a difference. The REM model was developed by and for trained genetic counselors, and it is not known to what extent other health professionals who also provide genetic counseling, such as medical geneticists, subscribe to these tenets. We believe the REM tenets provide a useful framework for understanding the process of genetic counseling and consequently some of the dilemmas and issues faced in genomic testing.

ELEMENTS OF GENETIC COUNSELING

The genetic counseling interview is unique, because it is neither a purely medical interview nor a purely counseling one; the combination of these approaches has its own distinctive attributes [10]. For instance, people commonly hold personal beliefs about the causes of traits or conditions in their family, which may or may not align with a medical assessment [11,12]. Genetic counseling does not necessarily replace nonscientific beliefs [13]. Ascertaining the client's understanding of why they had been referred, their own reasons for their attending and, subsequently, their own beliefs about their family history lay the basis for mutual understanding of the client's situation.

In general, genetic counseling in relation to testing encompasses four broad aspects:

1. information gathering and assessment regarding the client's personal and family medical history
2. provision of information and education about the genetic condition in question and testing being offered, including its implications for the person considering testing and their relatives
3. counseling to facilitate informed decision making regarding genetic testing and prepare them for the results by exploring the psychological impact of the genetic condition for the client and their family
4. posttest support and follow-up counseling.

Information gathering is an important component of current genetic counseling and will remain so for some time

into the genomics era. The client's physical presentation, medical history, and family history continue to be necessary for determining the inheritance pattern and diagnosis, as well as identifying relatives who may also be at risk. The likely pattern of inheritance is also used in the bioinformatics analysis of sequencing data in cases where multiple family members (eg, a trio of parents and the affected child) are tested together. The clinical information gathered before testing is also important when the variants detected by genomic sequencing[1] are being assessed for their clinical relevance (clinical interpretation).

As a psychosocial tool, drawing of the pedigree can also be used to understand the impact of the condition on the family or to understand the family narrative in relation to the condition. For example, how many family members have the condition, how they were diagnosed, and what symptoms or events are attributes to the condition or to other causes. This information, which from a client's perspective, is gathered "incidentally" as part of providing the family history, is important because research has shown that severity of the disease manifestations in the family can influence reproductive and genetic testing decisions [14]. Drawing the pedigree can also build rapport and provide an insight into family dynamics, which can affect the client's posttest experience [15]. Documenting the family history provides an opportunity to explore the nature of relationships between relatives and assess potential barriers to the communication of genetic risk information.

The provision of pretest information and counseling are both key components of informed decision making about genetic tests. As with any medical test, it is important that the person tested is aware of the utility, limitations, and risks of testing [16,17]. For those with a condition, identifying the underlying cause of that condition can bring personal and medical benefit in the form of reproductive choice, early surveillance, better management, and/or treatment. However, testing for a genetic condition presents issues that are quite distinct from those arising from other forms of medical testing, in particular, the potential to identify variants that predict future risk of disease for the patient and relatives. A person choosing to have a test therefore also faces decisions about communicating with relatives about those risks and tests to family members who may have different perceptions of the value of this information [18]. Although benefiting family members is one of the most common reasons for adults undergoing a genetic test themselves [19], most relatives do not undergo testing [20]. Other common motivations for testing are to improve health outcomes (for example, by early detection) or provide greater certainty about the future. These expectations are often not fully met.

---

1. *Genomic sequencing* is used here as a collective term for targeted next generation sequencing panels, whole exome sequencing, and whole genome sequencing.

Genetic counseling assists people to consider these possibilities before testing, on the basis that this can help people adjust posttest when they have anticipated (rather than unanticipated) experiences and emotions. Genetic counsellors also provide support and follow up care to assist adjustment.

COUNSELING ISSUES

The combination of each client's personality, coping style, expectations, and experience is of course unique. Genetic counseling aims to be patient-centered. In the context of testing, that means enabling the individual to consider the information in the light of their own circumstances and to make a decision based on these. The issues we discuss in the following section were chosen from the many that arise in genetic counseling [5] because they have particular relevance to genomic testing.

Uncertainty

Skirton and Bylund [21] describe types of uncertainty in relation to genetic conditions as including illness complexity, timescale, likelihood of a particular outcome, and nondiagnosis. Many genetic conditions have variable expression, even within one family, making it hard to predict how a condition may manifest. It can be uncertain whether a single case is the result of a de novo mutation or mosaicism of a parent's germline. Furthermore, incomplete penetrance means it is not usually possible to completely distinguish those who will develop a condition from those who will not. Much genetic information is probabilistic in nature; risk rather than certainty is being communicated.

The desire to reduce uncertainty is a known motivator for genetic testing [22,23], be it predictive testing or diagnostic testing. For adult-onset conditions, predictive testing can remove the uncertainty of inheritance if the family-specific causative mutation is known. This can relieve those who do not have the mutation from the need to undergo risk-reducing or early detection procedures. For example, a person at risk who is found to have *not* inherited the family-specific mutation causing the hereditary cancer condition of Lynch syndrome no longer requires regular colonoscopy. However, that person still has a residual risk of colorectal cancer (the same as the general population risk) so some uncertainty remains. Those who are carriers cannot know if and when a cancer will develop.

For parents of children with disabilities, diagnostic uncertainty is a major psychological stressor [24]. Parents prefer certainty of diagnosis over uncertainty, even when the prognosis is serious or life-threatening [25]. However, diagnostic testing does not always reduce uncertainty. Relatively few conditions are caused by common gene mutations, so the failure of a DNA-based test to detect a causative mutation

is an "uninformative" result, rather than truly negative one. Furthermore, knowledge of the clinical significance of much of the genome is limited. Diagnostic testing may result in the detection of variants of uncertain significance; ie, variants about which there is not sufficient evidence to prove or disprove their impact on gene function and disease. Hayeems et al. [26] describe the pervasive uncertainty experienced by parents whose children had a microarray assay for diagnosis. This was evident regardless of whether the test results were pathogenic or uncertain; pathogenic findings did not alleviate uncertainty about prognosis and informative results did not provide the hoped-for answers, leaving both groups of parents frustrated and fearful.

Uncertainty of test results is not new to genomics. The introduction of techniques that improve the diagnostic rate also inevitably increases the number of uncertain results, at least until their significance becomes clear with further experience and investigation. However, the scale of this issue is unprecedented in genomics. About 3400 genes are known to have a phenotype-causing mutation [27], a number that has increased rapidly, but many of these genes have not yet been well characterized. In our experience, on average 115 nonsynonymous or truncating rare variants are detected in Online Inheritance in Man–listed mendelian genes when single patients receive whole exome sequencing (unpublished data). Knowledge of the clinical significance of these variants will gradually increase with greater experience and functional studies. For very rare conditions and for novel genes, clinical interpretation of variants will continue to be an issue.

Confidentiality and Privacy of Genomic Data

The advent of genomic sequencing technologies has brought with it the need to store and share clinical genomic data. Many potential benefits of sharing genomic data exist for patients and the healthcare system, some of which are described in this chapter. However, to maximize these benefits, patients will need to trust that their healthcare team and policy makers will make efforts to protect their privacy and assure confidentiality [28]. A systematic review of public and research participants' views on genomic data sharing found that of the value of genomic data sharing was recognized but people concurrently held a variety of concerns regarding privacy, security, and the sensitivity of the information [29].

Before the advent of genomic sequencing technologies, a patient's single gene test result and raw sequencing traces were stored. If new genes were discovered to cause the patient's condition after the initial analysis, a new assay had to be developed and DNA sequencing or testing performed again for the new gene. The ability to store, share, and reanalyze a patient's whole exome or genome data relatively simply, without having to develop new assays, is a major advancement for patients on several fronts. First, sharing data with research studies has the potential to improve diagnosis for the diagnostically destitute patient. Second, healthcare costs may be reduced if genomic data is stored within and shared across health services in such a way that it can be reaccessed and reanalyzed for different medical purposes, throughout a patient's life. Third, sharing of clinically generated genomic data as part of international initiatives is necessary to improve understanding of human variation and consequently the interpretation of variants identified in individual patients. Each of these scenarios is discussed further in the following sections.

Data Sharing and the Undiagnosed Patient

Many patients with a phenotype that is not consistent with any known condition have benefitted enormously from the advent of genomic sequencing. Sharing of their genomic data and clinical information internationally has led to diagnosis and the discovery of new genes [30,31]. The end of these patients' diagnostic odysseys has captured the attention of the mass media, with parents describing the importance not only have having a diagnosis but also connecting with other families with the same rare condition [32,33]. Formal data sharing platforms are now in place to facilitate such sharing in an ethically responsible way to improve patient diagnosis (eg, DECIPHER [34]). Patients with rare or undiagnosed disease are often very motivated to obtain a diagnosis and therefore willing to share potentially identifying information.

Reanalyzing Stored Genomic Data

Patients without a diagnosis are likely to benefit from storage and reanalysis as further genes are discovered and clinical bioinformatics pipelines are improved to detect a broader range of variants; for example, large intragenic deletions.

The ability to reinterrogate and share a person's genomic data within or between health services after it has been analyzed for the primary indication also has the potential to improve efficiency of healthcare [35], particularly if storage costs also reduce. For example, a patient may have a genomic sequencing test because of a personal and family history of familial predisposition to cancer. Years later, the patient may return to the health service with clinical depression; a pharmacogenomics analysis could be performed on the patient's existing stored genomic data before determining which antidepressant will be effective [36,37].

Sharing Genomic Data for Advancement of Knowledge

The extent of variation within the human genome is such that international sharing of data is needed to improve understanding and distinguish normal variation from the pathogenic. Leading these efforts is the Global Alliance for Genomics and Health, and the Clinical Genome Resource

(ClinGen), among others, who have as their core mission to set up systems and policies to facilitate the sharing of genomic data to improve healthcare outcomes, including the clinical interpretation of genomic sequence variants. Sharing of individual variants is taking place (eg, ClinVar [38]), as well as sharing of whole genome/exome data (eg, the Exome Aggregation Consortium (ExAC) [39]).

Whether or not patient data can be provided to these international data-sharing initiatives, and the conditions under which sharing is possible, will vary between different jurisdictions. The Global Alliance for Genomics and Health has developed a framework for the responsible sharing of genomic and health-related data and a consent policy to guide practices that support data sharing [40]. Although sharing data is not a new concept in healthcare, genomic sequencing is drawing greater attention to this issue. For researchers and clinicians, this is a priority because the ability to detect sequence variants is exceeding our ability to interpret their meaning for patients. However, concerns about confidentiality are heightened as the data that could be shared extends beyond that related to the person's condition to their entire genome or exome, a very different situation than the sharing of genetic test information.

Secondary Findings

Whole exome and whole genome sequencing generates data on most of the genes in the exome simultaneously. Therefore data also exist on genes that have a role in the development of adult-onset conditions but are unrelated to the condition currently affecting a person. Whole exome and genome sequencing of patients therefore also presents an opportunity to test for future disease risk. Various terms have been used for these findings [41]. We use the term *secondary* here to highlight that these potential findings are not the primary indication for testing of the patient but do require laboratories to deliberately analyze and interpret variants in genes unrelated to the patient's condition.

Testing for secondary findings as part of a diagnostic procedure has proved to be contentious. Testing for secondary findings requires laboratories and health services to make decisions about the genes to be tested, the choices (or not) available to the person being tested, and whether children will be tested for secondary findings.

Policy regarding the genes to be tested for pathogenic variants has been informed by categorizing genes according to clinical validity and utility [eg, the American College of Medical Genetics and Genomics (ACMG) guidelines on incidental findings [42]]. The imperative is clearly for the return of those variants that predispose to potentially treatable or preventable conditions [43], but it is difficult is to reach agreement on the allocation of all genes into discrete categories [44,45]. The ACMG guidelines identify 56 genes as a minimum for the return of disease-causing mutations, but others apply a broader list [46].

The extent to which patients should be able to choose whether their data is analyzed for secondary findings has proved controversial, with a range of practices occurring across genomic centers [46]. Initially, ACMG guidelines suggested that all patients, irrespective of age, should be tested for the 56 genes specified. If patients did not wish to receive their results, the guidelines recommended that this should be managed by the clinician. This guidance conflicted with the ethos of autonomy and personal choice that has underpinned predictive genetic testing [47]. Many people at risk of inherited conditions choose not to learn their genetic status [19,48] and guidelines on the genetic testing of children generally do not support testing for adult-onset conditions [49]. Evidence from hypothetical discrete choice experiments suggests that more participants would choose to receive results for medically treatable, high-penetrance disorders than not receive results [50]. Yang et al. [51] reported secondary findings from clinical exome investigations to probands and made parents' results available on request. A total of 92 patients from a predominantly pediatric cohort of 2000 had secondary findings; 33 parents from 19 families then requested their own results. A diagnostic laboratory reported that only 13 people of 200 declined all findings from all of categories of secondary findings they made available [52]. Again, this was a largely pediatric cohort. Of note, parents wishing to learn their child's results had a higher acceptance rate than adults offered their own results, although numbers of adults were too small to draw firm conclusions. In response to a survey of its members, the ACMG has subsequently revised their guidelines to recommend that patients can opt out of testing for secondary findings.

Family Communication

Genetic and genomic test results can have implications for other family members. It is typical to rely on family members to communicate information about genetic risk to their relatives, and this is supported by the guidelines and position statements of international human genetics societies and bioethics bodies [53]. A systematic review found that genetic health professionals manage this issue by encourage reluctant clients to share relevant information and assisting the clients' communication by psychoeducational guidance and written information [54]. Some evidence suggests that definitive results are communicated more often than uninformative or uncertain ones [55].

We are not aware of any studies addressing family communication of whole exome or whole genome test results. On the basis that information resulting from genomic sequencing is effectively the same as that from genetic testing (ie, the presence of variants known to cause a condition with mendelian inheritance), we anticipate that the communication of a definitive genomic sequencing result is likely to be the same as a definitive genetic test result. This may be different in the future, if diagnosis or risk of disease is determined by the presence of combinations of multiple variants or the integration of information from multiple sources (eg, sequence, epigenome, transcriptome). The complexity of such information may have an impact on its communication within families.

The "Healthy Genome"

Genetic health professionals have a large body of experience with people who are affected by genetic conditions or have high-risk factors for a genetic condition (eg, family history, ethnicity) and this is reflected in the previous discussion. However, this experience may not be very informative when counseling healthy people at low risk who wish to improve their health by undergoing testing. This scenario has received considerable media attention [56]. Despite this interest, such testing is not yet widespread, in part because there remain limitations in the ability to interpret the significance of variants in the absence of a clinical phenotype and limited evidence of utility.

An evidence base will also be needed for counseling. It cannot be assumed that research findings from affected and increased-risk individuals will apply to healthy individuals at low or population risk. Early evidence comes from healthy participants in the clinical sequencing study, ClinSeq [57], and studies of people who have undergone direct-to-consumer testing. The motivations for testing in these contexts were similar to those with or at-risk of an inherited condition. The majority indicated interest in learning personal health information, some with the hope that participation would help prevent disease [58,59]. Interestingly, 8% of ClinSeq participants hoped that testing would give them a "clean bill of health" [58], suggesting that management of expectations may be important in pretest counseling. These groups may not be as attuned to perceived risk as those undergoing predictive genetic testing, who usually have a family history of the disease for which they are being tested [60].

Direct-to-consumer testing does not involve pretest genetic counseling, but participant posttest knowledge was found to be good [61]. In fact, uptake of free posttest genetic counseling was low because participants felt they understood the test result [62]. Interestingly, the discussion of results with a health professional was associated with self-efficacy [61] and completion of screening tests [63], but cause and effect cannot be determined.

CASE STUDY

A 38-year-old woman comes to her primary care physician with complaints about shortness of breath and fainting episodes. The physician organizes a range of clinical tests, including an echocardiograph, which confirms a diagnosis of dilated cardiomyopathy. She is referred to a cardiologist, who treats her symptoms. The patient asks why she has developed cardiomyopathy. The cardiologist explains there are many causes, including the possibility that her condition may have a genetic basis. The patient becomes concerned about the risks for her two children (10 and 12 years of age). The cardiologist refers her for genetic counseling to answer her concerns and questions.

Table 15.1 outlines the issues that would be discussed if this patient sought counseling in 1990, 2000, 2010, and 2015. As can be observed, many of the issues genetic counselors address are similar regardless of the point in history this patient is seen. The patient's main concern is her children; assuming the cardiomyopathy is genetically determined, most have an autosomal dominant inheritance pattern so each child has a 50% risk for inheriting the condition. Identification of a disease-causing mutation would confirm the genetic etiology of her cardiomyopathy and would make available a predictive gene test for her children. The chance of identifying a disease-causing mutation increases with advances in knowledge and testing technologies, peaking at a yield of about 30–50% in 2015 with a

genomic sequencing test. Yet even this has a high chance of being negative, that is, uninformative, because it is likely that not all the genes causing cardiomyopathy have been discovered yet. Uncertainty then remains regarding the etiology of the cardiomyopathy.

PRACTICALITIES OF COUNSELING FOR GENOMIC TESTS

Just 10 years ago, it was unimaginable that it would be possible to test hundreds or thousands of genes in a single assay with a turn-around time of months in a clinical environment. Protocols and guidance for pretest genetic counseling were developed for testing for single conditions; for example, those from EuroGentest (Box 15.1).

Guidelines and protocols aim to promote informed decision making through comprehensive pretest consideration of the test and its implication. Diagnostic genetic consultations in which single gene sequencing and cytogenetics are considered are usually an hour in length [64], whereas predictive genetic testing may be conducted over several sessions [65].

TABLE 15.1 Similarities and Differences in Genetic Counseling for Cardiomyopathy at Different Times in History

	1990	2000	2010	2015
Availability of gene test	None	Sanger sequencing test for up to 8 genes	A targeted capture, next generation sequencing panel test of 60 genes	Exome sequencing with selective analysis of 100 cardiomyopathy genes
Mutation detection rate	–	<1–2%	~30%	30–50%
Genetic Counseling Issues That Change				
Counseling regarding risk for variants of unknown significance	–	+	+++	+++
Consent and confidentiality issues regarding storage of genomic data and potential secondary uses	–	–	–	Potential secondary uses: • for reanalysis if new cardiomyopathy genes are discovered • for gene discovery research, with consent • for pharmacogenomic analysis to assist with correct medication selection and dosage for cardiac symptoms (eg, atrial fibrillation) • for opportunistic screening to determine risk for development of unrelated conditions.
Enduring Genetic Counseling Issues				
Risk to children	Could be as high as 50% if autosomal dominant form of dilated cardiomyopathy, but could also be negligible if acquired Serial (yearly) clinical screening of children recommended			
Uncertainty	Uncertainty regarding etiology of dilated cardiomyopathy and risk to children			
Limitations of the test	Negative gene test does not exclude a genetic basis			

Box 15.1 Genetic Counseling for Testing for Genetic Conditions

Persons Involved in Counseling

Counselor

Profession	Person who gives genetic counseling should be a professional specialist, as he needs to convey complex information. In common problems there is a growing need for nonphysician healthcare providers and nongenetic specialist physicians to give genetic counseling. More education in genetics is needed.
Training	Genetic counselors should receive appropriate and on-going training in genetic conditions, risk assessment, psychosocial issues, bioethics, service system, communication, and patient perspective.
Cooperation within healthcare system	Genetic counseling team consists of a team of professionals.
Duties	Genetic counselor has certain duties that include, eg, preparing for consultation, promoting public education, assisting in decision making, and being aware of the professional ethics.

Family

Common information	Genetic information is common to the whole family, which can cause conflicts. Counseling should be available to at-risk relatives. It should be considered before the test how they are contacted. Patients have a duty to contact their at-risk relatives, but this should not be a condition for test, because sometimes they have good reasons not to contact relatives. Disclosure should be agreed between the patient and the physician.
Cultural issues	When providing information, family's cultural and ethnic background should be taken into account.

Vulnerable Patients

Not able to consent	Tests should be done in the best interest of the patients who cannot give informed consent, or sometimes in the health interest of their family. Authorization of the representative is required, but the persons need to participate according to their capacity.
Children	Tests should be done in the best interest of the child, or sometimes in the health interest of their family. Authorization of parents is required, but the child needs to participate according to his or her capacity. The age at which the child is able to decide on testing should be flexible, and the older the child is, the more significance should be put on the assent. Information should be appropriate to child's capacity and needs, and the child should be involved in counseling as much as possible.

Practices of Counseling

Setting	Genetic counseling should be given in a private and comfortable place.

When to counsel

Pretest	Pretest counseling should be available.
Posttest	Posttest counseling should be available.
Discussion before counseling	It is recommended to discuss with patients before counseling to identify their needs, to increase their understanding of counseling, and to gather background information.
Periods	There should be enough of time between counseling and decision making. The waiting time should, however, not be too long; at least the test results should be disclosed as soon as possible. Patient should be informed about the time between the appointments.
Consent	Informed consent should be asked always. According to the situation, it should be either in written or verbal form; written form is often advisable. There should be a free withdrawal from the consent. Consent needs to be asked also to inform the relatives.
Psychological support	Psychological support should be given to patients when necessary to adjust them to the situation, to help them to cope with stress, and to assure their individual decision making. Patient's psychological, emotional, and social responses need to be taken into account. Sometimes it can be recommended to take a support person to the appointment.

Continued

Box 15.1 Genetic Counseling for Testing for Genetic Conditions—cont'd

Referrals	Referrals can be made to other professionals when necessary. Reason for a referral can be, eg, support, treatment, or ethical reasons.
Follow-up care	Effective systems for follow-up care are needed.
Counseling integral part of testing	Counseling is an integral part of genetic testing and more links between laboratories and clinical services are needed.

Information-Giving

Content of information	Information should consist of medical facts of the disorder, risk figures, benefits and harms of testing, limitations of the test, reliability of the test, implications of testing, familial implications, probabilities of inheritance, prevention and treatment, information on available support, and alternative choices.
What kind of information	Information should be appropriate, balanced, honest, objective, full, sufficient, accurate, age-appropriate, and based on up-to-date knowledge.
Understanding	Patient's understanding needs to be ensured. Clear language should be used and interpreters invited when needed. Written summary of information is advisable.
Ways to give information	It is sometimes useful to use alternative ways to distribute information, such as leaflets, videos, and visual presentation.

Test-type–Based Information Giving

Predictive test	Appropriate counseling should be given. It should include advantages and disadvantages of knowing the result, psychological and social issues, complex appearance of the disease and difficulties to quantify probabilities, possibilities to treatment, and planning of life.
Prenatal test	Appropriate counseling should be given. Options, limitations, and risks of test should be discussed. Psychological aspects need to be taken into account. Medical facts and child's future should be discussed.
Preimplantation test	Appropriate counseling should be given.
Carrier test	Appropriate counseling should be given and reproductive options discussed.

Values

Nondirectiveness	Nondirectiveness needs to be provided to the patient. It does not mean abandoning the patient, but personal conviction of the counselor should not be involved. Nondirectiveness is, however, difficult in practice and it needs to be considered case-specifically.
Autonomy	Gene test should always be voluntary; only in rare cases can compulsory testing be accepted. Counseling needs to respect individual's values and culture. No pressure should be put on the person receiving counseling, and every decision should be equally accepted. Independent decision making should be encouraged, but the counselor can assist the patient.
Confidentiality	Confidentiality needs to be assured to the patient. Patient's private information needs to be protected from third parties. In the situations that nondisclosure would cause severe harm, confidentiality can be breached.
Right to know and to not know	Right to know and to not know genetic information are both important, but sometimes, if it is in the patient's interest, the right to know overweighs the right not to know.
Well-being	In counseling, patient needs to be taken care of. Good should always overweigh harm.
Equal access	Everyone should have equal access to genetic testing and counseling.
Individual's interests	Interests of individual should prevail over the interests of the society.

Box 15.1 Genetic Counseling for Testing for Genetic Conditions—cont'd

Problems

Practical problems	Varying practices related to genetic testing and counseling is a problem, because the services should be equal for everyone. Lack of trained professionals, inaccuracy of test results, risks involved in testing, and costs of counseling are also problems to be solved.

Ethical Problems

Discrimination	There is a chance that there will be discrimination on the basis of genes in the society. Testing can also stigmatize on the basis of one's ethnic origins. Testing for sex-selection should not be allowed in order to avoid discrimination on the basis of sex. Testing creates attitudes toward disabled people, and in counseling the quality of disabled people's lives is often defined without their contribution. Eugenics needs to be seen as a threat; a rich society includes all kinds of people.
Conflict between values	There is often conflict between confidentiality and duty to warn and provide care and between the right to know and the right to not know. These need to be handled sensitively.
Disclosure	It is sometimes difficult to decide what information to disclose; eg, unexpected findings and information not related to disease.
Society	In genetic testing the beneficence of both the individual and the society needs to be considered. Increasing range of options that gene tests bring changes society. One of these options is commercial genetic testing, in connection with which proper counseling should be guaranteed.
Determinism	The use of genetic information may lead to deterministic thinking, and science may be considered as godhead.
Cultural norms	There are different cultural norms on health. Therefore dialogue is needed, especially because of the cross-border patients.

Relation to Other Healthcare

Genetic testing is different	Genetic information differs from other health information because it has familial and generational implications, it is predictive, and the emphasis is not on treatment; it is severe, private, and complex, and deals with special ethical dilemmas.
Exceptionalism is not good	It is not good to think that genetic testing is too exceptional. This leads to deterministic thinking. Genetics should rather be integrated into mainstream healthcare.

Adapted from EuroGen Test. Summary of the Guidelines for Genetic Counselling. http://www.eurogentest.org/index.php?id=675 [accessed 03.09.15].

Genomic sequencing is now a part of clinical practice. Targeted next generation sequencing for disease-specific genes is well established, but already being replaced by whole exome sequencing. Early examples of the use of whole genome sequencing for clinical care of healthy and affected individuals also exist in the literature [66,67]. The list of issues to be covered in pretest counseling (see Box 15.1) remains just as relevant for these tests. However, incorporating them all into clinical sessions/protocols is a challenge. The REM provides a useful framework to highlight these challenges, which are considered below through the lens of the tenets of the REM.

Genetic Information Is Key

The tenet that *genetic information is key* recognizes the importance of information to informed decision making. The person offered testing should have a good understanding of the test, its purpose, reliability, and limitations. Guidelines also recommend that the person should have an understanding of the condition being tested for: its inheritance, natural history, and management before testing [68]. When one or two genes are to be sequenced, it is feasible to discuss the relevant conditions, the test, and the potential clinical implications. However, when next generation sequencing is performed, because there are many differential diagnoses or a diagnostic conundrum, the magnitude and complexity of the information to be conveyed and understood is amplified. When some 3400 genes for mendelian conditions are to be sequenced, counseling challenges include:

- Providing accurate information about the disease. A molecular diagnosis could disprove the suspected clinical diagnosis, and/or the causative mutation(s) for the suspected condition may have unexpected consequences [69]. Some people have been found on genome sequencing to have two conditions simultaneously [51].

- Conveying the limitations of testing, in particular that the failure to detect a causative mutation, is uninformative and does not exclude the presence of a mutation that could not be detected by the test.
- Explaining the limitations of current knowledge, specifically the possibilities that variants of uncertain significance may be detected, and the possibility that new information may result in reclassification of variants over time.
- Describing the service or laboratory's policy on secondary findings and, where relevant, ensuring that the patient is able to make an informed choice about receiving the results of opportunistic genetic screening.
- Explaining that data is stored and may be reanalyzed for clinical purposes, including what might lead to reanalysis and whether the patient will be contacted.
- Exploring the patient's willingness for data sharing to be shared, including what might be shared, for what purposes, and whether or not any information will be returned.

Patient Autonomy Must Be Supported

More than simply ensuring there is no coercion to undergo testing, patient autonomy is best supported when patients actively consider their options and preferences. It has been estimated that counseling before a genomic test could take 2–6 hours, if all possibilities are covered comprehensively [70,71]. Bowdin et al. find that counseling sessions of 30–45 minutes are necessary to avoid information overload, with consent needing to be an iterative process occurring over several sessions [72]. This is a challenge for busy services with limited resources and lengthy waiting times for appointments.

Patient Emotions Make a Difference

Hope for positive change is evident in patients' motivations for testing, be it for their own health or those of the next and future generations. Yet, as indicated, uncertainty is rife in genomic testing. Furthermore, some patients expect genomic tests will provide greater information about prognosis and clinical course than is usual for genetic tests, because "all" the genetic information is sequenced (G. Brett, personal communication). Patients preferred face-to-face consultations [73] and appreciated thorough consent processes but found the time required was too long [71].

Relationship Is Integral to Genetic Counseling

Ormond et al. [74] commented on genetic counseling in the context of whole genome sequencing "This process will be difficult, lengthy and expensive, but how the consent process could be meaningfully undertaken in any less intensive way is hard to imagine." There is debate about the best response

to this genomic uncertainty [26]. Austin et al. [75] point out that the psychological dimensions of genetic counseling practice are associated with more positive client outcomes. As the provision of genetic information does not alleviate uncertainty or the resultant anxiety that can arise, Austin and colleagues propose that helping clients adapt to these uncertainties is a more effective response. Trepanier and Wickland [76] observe, however, that the volume of information to be conveyed may reinforce the use of the education-based approaches, with more time spent on conveying these at the cost of exploration of the psychosocial impact.

CONCLUSION

The advent of genomic technologies in clinical practice provides new opportunities for better healthcare, while presenting genetic counseling challenges. Although the issues are fundamentally familiar, the difficulties of ensuring a process that truly enables a person to make an informed choice about genomic testing are manifold.

It is not only patients who are faced with uncertainty; the issues highlighted here remain largely unresolved for professionals. Nevertheless, practitioners are of necessity adopting strategies that reflect their clinical experiences and health service policy [46]. Practice will naturally evolve as experience grows, but these different practices are also a chance for evaluation and therefore an evidence base for effective genetic counseling. More emphasis may be placed on posttest counseling; this may be particularly the case for people choosing to have "healthy genome" investigation.

The focus of germline genomic testing is largely on diagnosis with the hope that this will lead to improved care. At some point, evidence for treatment or prevention will drive genomic testing. This could well change the dynamics of informed consent and require another reconsideration of genetic counseling approaches.

ACKNOWLEDGMENT

We would like to thank Melissa Martyn for assistance with the preparation of this chapter.

REFERENCES

[1] Human Genetics Society of Australasia. What is genetic counselling? http://www.hgsa.org.au/education-training/genetic-counselling/genetic-counselling-training-and-accreditation [accessed 01.09.15].

[2] Resta RG. Eugenics and nondirectiveness in genetic counseling. J Genet Couns 1997;6(2):255–8.

[3] International Huntington Association and the World Federation of Neurology Research Group on Huntington's Chorea. Guidelines for the molecular genetics predictive test in Huntington's disease. J Med Genet 1994;31(7):555–9.

[4] Resta R, Biesecker BB, Bennett RL, Blum S, Hahn SE, Strecker MN, et al. A new definition of genetic counseling: National Society of Genetic Counselors' Task Force report. J Genet Couns 2006;15(2):77–83.

[5] McCarthy Veach P, Le Roy BS, Bartels DM. Facilitating the genetic counselling process: a practice manual. NY: Springer; 2003.

[6] Markel H. The stigma of disease: implications of genetic screening. Am J Med 1992;93(2):209–15.

[7] Biesecker BB. Goals of genetic counseling. Clin Genet 2001;60:323–30.

[8] Kessler S. Psychological aspects of genetic counseling. IX. Teach Couns J Genet Couns 1997;6(3):287–95.

[9] Veach PM, Bartels DM, Leroy BS. Coming full circle: a reciprocal-engagement model of genetic counseling practice. J Genet Couns 2007;16(6):713–28.

[10] Evans C. Genetic counselling: a psychological approach. Cambridge: Cambridge University Press; 2006.

[11] Chapple A, May C, Campion P. Lay understanding of genetic disease: a British study of families attending a genetic counseling service. J Genet Couns 1995;4(4):281–300.

[12] McAllister M. Personal theories of inheritance, coping strategies, risk perception and engagement in hereditary non-polyposis colon cancer families offered genetic testing. Clin Genet 2003; 64:179–89.

[13] Weil J. Mothers postcounseling beliefs of cause of their childrens genetic disorders. Am J Hum Genet 1991;48(1):145–53.

[14] Henneman L, Bramsen I, van der Ploeg H, Ader HJ, van der Horst HE, Gille JJP, et al. Participation in preconceptional carrier couple screening: characteristics, attitudes, and knowledge of both partners. J Med Genet 2001;38(10):695–703.

[15] Galvin KM, Young M. Family systems theory. In: Gaff C, Bylund CL, editors. Family communication about genetics. New York: Oxford University Press; 2010. p. 102–19.

[16] National Health and Medical Research Council. Ethical aspects of human genetic testing: an information paper. 2000.

[17] Eurogentest. Recommendations for genetic counselling related to genetic testing. 2006.

[18] Gaff CL, Collins V, Symes T, Halliday J. Facilitating family communication about predictive genetic testing: probands' perceptions. J Genet Couns 2005;14(2):133–40.

[19] Bleiker EM, Esplen MJ, Meiser B, Petersen HV, Patenaude AF. 100 years Lynch syndrome: what have we learned about psychosocial issues? Fam Cancer 2013;12(2):325–39.

[20] Hodgson J, Metcalfe S, Gaff C, Donath S, Delatycki MB, Winship I, et al. Outcomes of a randomised controlled trial of a complex genetic counselling intervention to improve family communication. Eur J Hum Genet 2015;24(3):356–60.

[21] Skirton H, Bylund CL. Management of uncertainty. In: Gaff C, Bylund CL, editors. Family communication about genetics. New York: Oxford University Press; 2010. p. 136–51.

[22] Braithwaite D, Sutton S, Steggles N. Intention to participate in predictive genetic testing for hereditary cancer: the role of attitude toward uncertainty. Psychol Health 2002;17(6):761–72.

[23] Wolff K, Nordin K, Brun W, Berglund G, Kvale G. Affective and cognitive attitudes, uncertainty avoidance and intention to obtain genetic testing: an extension of the theory of planned behaviour. Psychol Health 2011;26(9):1143–55.

[24] Lenhard W, Breitenbach E, Ebert H, Schindelhauer-Deutscher HJ, Henn W. Psychological benefit of diagnostic certainty for mothers of children with disabilities: lessons from Down syndrome. Am J Med Genet Part A 2005;133A(2):170–5.

[25] Makela NL, Birch PH, Friedman JM, Marra CA. Parental perceived value of a diagnosis for intellectual disability (ID): a qualitative comparison of families with and without a diagnosis for their child's ID. Am J Med Genet Part A 2009;149A(11):2393–402.

[26] Hayeems RZ, Babul-Hirji R, Hoang N, Weksberg R, Shuman C. Parents' experience with pediatric microarray: transferrable lessons in the era of genomic counseling. J Genet Couns 2015;25(2):298–304.

[27] O.M.I.M. [accessed 30.08.15].

[28] McGuire AL, Fisher R, Cusenza P, Hudson K, Rothstein MA, McGraw D, et al. Confidentiality, privacy, and security of genetic and genomic test information in electronic health records: points to consider. Genet Med 2008;10(7):495–9.

[29] Shabani M, Bezuidenhout L, Borry P. Attitude of research participants and the general public towards genomic data sharing: a systematic literature review. Expert Rev Mol Diagn 2014;14(8):1053–65.

[30] Enns GM, Shashi V, Bainbridge M, Gambello MJ, Zahir FR, Bast T, et al. Mutations in NGLY1 cause an inherited disorder of the endoplasmic reticulum-associated degradation pathway. Genet Med Off J Am Coll Med Genet 2014;16(10):751–8.

[31] Taft RJ, Vanderver A, Leventer RJ, Damiani SA, Simons C, Grimmond SM, et al. Mutations in DARS cause hypomyelination with brain stem and spinal cord involvement and leg spasticity. Am J Hum Genet 2013;92(5):774–80.

[32] Kaminsky L. Decoding Massimo Damiani's rare genetic disease. The Weekend Australian; 2013.

[33] Mnookin S. One of a kind: what do you do if your child has a condition that is new to science?. The New Yorker; 2014.

[34] Chatzimichali EA, Brent S, Hutton B, et al. Facilitating collaboration in rare genetic disorders through effective matchmaking in DECIPHER. Hum Mutat 2015;36(10):941–9.

[35] Hoffman JM, Haidar CE, Wilkinson MR, Crews KR, Baker DK, Kornegay NM, et al. PG4KDS: a model for the clinical implementation of pre-emptive pharmacogenetics. Am J Med Genet Part C, Seminars Med Genet 2014;166C(1):45–55.

[36] Hall-Flavin DK, Winner JG, Allen JD, Carhart JM, Proctor B, Snyder KA, et al. Utility of integrated pharmacogenomic testing to support the treatment of major depressive disorder in a psychiatric outpatient setting. Pharmacogenet genomics 2013;23(10):535–48.

[37] Porayette P, Flockhart D, Gupta SK. One size fits one: pharmacogenetics in gastroenterology. Clin Gastroenterol Hepatol Off Clin Pract J Am Gastroenterol Assoc 2014;12(4):565–70.

[38] Landrum MJ, Lee JM, Riley GR, Jang W, Rubiinstein WS, Church DM, et al. ClinVar: public archive of relationships among sequence variation and human phenotype. Nucleic Acids Res 2014;42(Database issue):D980–5.

[39] Exome Aggregation Consortium (ExAC). http://exac.broadinstitute. org [accessed 03.09.15].

[40] Global Alliance for Genomics & Health. http://genomicsandhealth. org [accessed 03.09.15].

[41] Weiner C. Anticipate and communicate: ethical management of incidental and secondary findings in the clinical, research, and direct-to-consumer contexts (December 2013 report of the Presidential Commission for the Study of Bioethical Issues). Am J Epidemiol 2014;180(6):562–4.

[42] Green RC, Berg JS, Grody WW, Kalia SS, Korf BR, Martin CL, et al. ACMG recommendations for reporting of incidental findings in clinical exome and genome sequencing. Genet Med Off J Am Coll Med Genet 2013;15(7):565–74.

[43] Turbitt E, Wiest MM, Halliday JL, Amor DJ, Metcalfe SA. Availability of treatment drives decisions of genetic health professionals about disclosure of incidental findings. Eur J Hum Genet 2014;22(10):1225–8.

[44] Green RC, Berg JS, Berry GT, Biesecker LG, Dimmock DP, Evans JP, et al. Exploring concordance and discordance for return of incidental findings from clinical sequencing. Genet Med Off J Am Coll Med Genet 2012;14(4):405–10.

[45] Grove ME, Wolpert MN, Cho MK, Lee SS, Ormond KE. Views of genetics health professionals on the return of genomic results. J Genet Couns 2014;23(4):531–8.

[46] Berg JS, Amendola LM, Eng C, Van Allen E, Gray SW, Wangle N, et al. Processes and preliminary outputs for identification of actionable genes as incidental findings in genomic sequence data in the Clinical Sequencing Exploratory Research Consortium. Genet Med Off J Am Coll Med Genet 2013;15(11):860–7.

[47] Burke W, Antommaria AH, Bennett R, Botkin J, Clayton EW, Henderson GE, et al. Recommendations for returning genomic incidental findings? We need to talk!. Genet Med Off J Am Coll Med Genet 2013;15(11):854–9.

[48] Christiaans I, Birnie E, Bonsel GJ, Wilde AA, van Langen IM. Uptake of genetic counselling and predictive DNA testing in hypertrophic cardiomyopathy. Eur J Hum Genet 2008;16(10):1201–7.

[49] Borry P, Fryns JP, Schotsmans P, Dierickx K. Carrier testing in minors: a systematic review of guidelines and position papers. Eur J Hum Genet 2006;14(2):133–8.

[50] Regier DA, Peacock SJ, Pataky R, van der Hoek K, Jarvik GP, Hock J, et al. Societal preferences for the return of incidental findings from clinical genomic sequencing: a discrete-choice experiment. Can Med Assoc J 2015;187(6):E190–7.

[51] Yang Y, Muzny DM, Xia F, Niu Z, Person R, Ding Y, et al. Molecular findings among patients referred for clinical whole-exome sequencing. Jama 2014;312(18):1870–9.

[52] Shahmirzadi L, Chao EC, Palmaer E, Parra MC, Tang S, Gonzalez KD. Patient decisions for disclosure of secondary findings among the first 200 individuals undergoing clinical diagnostic exome sequencing. Genet Med Off J Am Coll Med Genet 2014;16(5):395–9.

[53] Forrest LE, Delatycki MB, Skene L, Aitken M. Communicating genetic information in families: a review of guidelines and position papers. Eur J Hum Genet 2007;15(6):612–8.

[54] Mendes A, Paneque M, Sousa L, Clarke A, Sequeiros J. How communication of genetic information within the family is addressed in genetic counselling: a systematic review of research evidence. Eur J Hum Genet 2015;24(3):315–25.

[55] Patenaude AF, Dorval M, DiGianni LS, Schneider KA, Chittenden A, Garber JE. Sharing BRCA1/2 test results with first-degree relatives: factors predicting who women tell. J Clin Oncol Off J Am Soc Clin Oncol 2006;24(4):700–6.

[56] Cadwalladr C. What happened when I had my genome sequenced. The Guardian; 2013.

[57] Biesecker LG, Mullikin JC, Facio FM, Turner C, Cherukuri PF, Blakesley RW, et al. The ClinSeq Project: piloting large-scale genome sequencing for research in genomic medicine. Genome Res 2009;19(9):1665–74.

[58] Facio FM, Brooks S, Loewenstein J, Green S, Biesecker LG, Biesecker BB. Motivators for participation in a whole-genome sequencing study: implications for translational genomics research. Eur J Hum Genet 2011;19(12):1213–7.

[59] Goldsmith L, Jackson L, O'Connor A, Skirton H. Direct-to-consumer genomic testing: systematic review of the literature on user perspectives. Eur J Hum Genet 2012;20(8):811–6.

[60] Taber JM, Klein WM, Ferrer RA, Lewis KL, Biesecker LG, Biesecker BB. Dispositional optimism and perceived risk interact to predict intentions to learn genome sequencing results. Health Psychol Off J Div Health Psychol Am Psychol Assoc 2015;34(7):718–28.

[61] Carere DA, Kraft P, Kaphingst KA, Roberts JS, Green RC. Consumers report lower confidence in their genetics knowledge following direct-to-consumer personal genomic testing. Genet Med Off J Am Coll Med Genet 2015;18(1):65–72.

[62] Darst BF, Madlensky L, Schork NJ, Topol EJ, Bloss CS. Perceptions of genetic counseling services in direct-to-consumer personal genomic testing. Clin Genet 2013;84(4):335–9.

[63] Bloss CS, Wineinger NE, Darst BF, Schork NJ, Topol EJ. Impact of direct-to-consumer genomic testing at long term follow-up. J Med Genet 2013;50(6):393–400.

[64] Cooksey JA, Forte G, Flanagan PA, Benkendorf J, Blitzer MG. The medical genetics workforce: an analysis of clinical geneticist subgroups. Genet Med 2006;8(10):603–14.

[65] MacLeod R, Tibben A, Frontali M, Evers-Kiebooms G, Jones A, Martinez-Descales A, et al. Recommendations for the predictive genetic test in Huntington's disease. Clin Genet 2013;83(3):221–31.

[66] Ashley EA, Butte AJ, Wheeler MT, Chen R, Klein TE, Dewey FE, et al. Clinical assessment incorporating a personal genome. Lancet 2010;375(9725):1525–35.

[67] Saunders CJ, Miller NA, Soden SE, Dinwiddie DL, Noll A, Alnadi NA, et al. Rapid whole-genome sequencing for genetic disease diagnosis in neonatal intensive care units. Sci Transl Med 2012;4(154):154–235.

[68] EuroGen Test. Summary of the Guidelines for Genetic Counselling. http://www.eurogentest.org/index.php?id=675 [accessed 03.09.15].

[69] Smith KR, Damiano J, Franceschetti S, Carpenter S, Canafoglia L, Morbin M, et al. Strikingly different clinicopathological phenotypes determined by progranulin-mutation dosage. Am J Hum Genet 2012;90(6):1102–7.

[70] Mayer AN, Dimmock DP, Arca MJ, Bick DP, Verbsky JW, Worthey EA, et al. A timely arrival for genomic medicine. Genet Med Off J Am Coll Med Genet 2011;13(3):195–6.

[71] Tabor HK, Stock J, Brazg T, McMillin MJ, Dent KM, Yu JH, et al. Informed consent for whole genome sequencing: a qualitative analysis of participant expectations and perceptions of risks, benefits, and harms. Am J Med Genet Part A 2012;158A(6):1310–9.

[72] Bowdin SC, Hayeems RZ, Monfared N, Cohn RD, Meyn MS. The SickKids Genome Clinic: developing and evaluating a pediatric model for individualized genomic medicine. Clin Genet 2015;89(1):10–9.

[73] Meiser B, Storey B, Quinn V, Rahman B, Andrews L. Acceptability of, and information needs regarding, next-generation sequencing in people tested for hereditary cancer: a qualitative study. J Genet Couns 2015;25(2):218–27.

[74] Ormond KE, Wheeler MT, Hudgins L, Klein TE, Butte AJ, Altman RB, et al. Challenges in the clinical application of whole-genome sequencing. Lancet (London, Engl) 2010;375(9727):1749–51.

[75] Austin J, Semaka A, Hadjipavlou G. Conceptualizing genetic counseling as psychotherapy in the era of genomic medicine. J Genet Couns 2014;23(6):903–9.

[76] Wicklund C, Trepanier A. Adapting genetic counseling training to the genomic era: more an evolution than a revolution. J Genet Couns 2014;23(4):452–4.

Chapter 16

Genetics and Genomics of Reproductive Medicine and Health

D. Kumar[1,2]

[1]The University of South Wales, Pontypridd, Wales, United Kingdom; [2]Cardiff University School of Medicine, University Hospital of Wales, Cardiff, United Kingdom

Chapter Outline

Introduction	213
Female Infertility	213
Congenital Anomalies of the Reproductive Tract	213
Disorders of Abnormal Ovulation	214
Premature Ovarian Failure	215
Disorders of the Endometrium	217
Endometriosis	218
Male Factor Infertility	218
Congenital Anomalies of the Male Reproductive Tract	218
Molecular Genetics of Male Factor Infertility	219
Epigenetics and Epigenomics in Male Factor Infertility	219
Mitochondrial Genes in Male Factor Infertility	221
New Genomic Applications in Male Factor Infertility	221
Genetic Factors in In Vitro Fertilization	221
Preimplantation Genetic Testing and Screening	222
Preimplantation Genetic Diagnosis	222
Recent Advances in Prenatal Diagnosis	223
Aneuploidy (Trisomy 21) Pregnancy	223
Cell-Free Fetal DNA	223
Summary	224
References	224

INTRODUCTION

The fields of reproductive medicine and clinical obstetrics cover a very broad range of disease conditions, as well as natural physiological processes such as pregnancy, parturition, and the menstrual cycle. Unique in this field of study is the need to consider the relevance of multiple genomes: mother and fetus (in pregnancy) or mother and father (in fertilization and infertility). This chapter provides a clinically oriented overview on the available genetic and genomic information relevant to the practice of reproductive medicine and health.

FEMALE INFERTILITY

In all clinical settings, investigations for unexplained infertility predominantly include detailed clinical assessment of the woman. However, male factor infertility is encountered in around 15% cases. Essentially investigations focus on three categories of the female reproductive system disorders: congenital anomalies of the female genital tract, disorders associated with ovarian function, and structural and functional disorders of the endometrium. There is plenty of evidence available to implicate genetic and genomic factors in all these three disease categories.

Congenital Anomalies of the Reproductive Tract

Investigations for infertility should include a thorough clinical assessment supported by relevant internal exploration and imaging to exclude congenital anatomical anomalies. This would apply to both male and female partners. In most cases, this could also include laboratory investigations to assess adequate sperm quantity and quality, endocrine profile for satisfactory ovulation, chromosome analysis with new molecular techniques [fluorescence in situ hybridization (FISH) and array comparative genomic hybridization (aCGH)], and, if applicable, relevant molecular genetic analysis (eg, cystic fibrosis).

Developmental anomalies of the female reproductive tract per se may diminish the natural potential of achieving a successful and viable pregnancy. The majority of female genital tract congenital anomalies are isolated and reflect unequal and abnormal embryological differentiation of the müllerian tract. However, it is likely that abnormal müllerian canal differentiation could be associated with other malformations, predominantly of mesodermal origin. There are several recognizable multiple malformation syndromes that include müllerian as well as cloacal developmental anomalies (Table 16.1). Several genes with regulatory

Medical and Health Genomics. http://dx.doi.org/10.1016/B978-0-12-420196-5.00016-2

TABLE 16.1 Selection of Syndromes with Abnormal Female Reproductive Tract Anatomy

Syndrome	Gynecological Anomaly	OMIM	Gene (Locus)	Gene Product
Adams-Oliver	Vaginal atresia	100300	3q13; 3q21	*ARHGAP31; RBP2*
Antley-Bixler	Hypoplastic labia majora Vaginal atresia	207410		*FGFR2 FGFR3 POR*
Apert	Vaginal atresia	101200	10q25-q26	*FGFR2*
Bardet-Biedl	Vaginal atresia	209900	Several	*BBS1-12*
Fraser	Bicornate uterus Vaginal atresia	219000	4q21; 13q13	*FRAS1 FREM2*
Fryns	Vaginal atresia Bicornate uterus	229850	1q41	?
Gorlin	Bicornate uterus	109400	9q22-q31	*PTCH1 SUFU*
Hand-foot-genital	Absent/small uterus	140000	7p15	*HOXA13*
Johanson-Blizzard	Bicornate uterus	243800	15q14-q21	*UBR1*
McKusick-Kaufman	Vaginal/cervical atresia Hydrometrocolpos	236700	20p12	*MKKS*
MURCS association	Vaginal atresia	601076	12q14; 14q31	??
Roberts-SC phocomelia	Bicornate uterus	268300	8p12-8p21	*ESCO2*
Schinzel-Giedion	Bicornate uterus	269150	18q21	*SETBP1*
Sienomelia sequence	Vaginal atresia	182940	?	?
Urorectal septum	Vaginal atresia	?	10q25-10q26?	

MURCS, Müllerian duct aplasia, renal aplasia, and cervicothoracic somite dysplasia; *OMIM,* Online Mendelian Inheritance in Man.

sequences and polymorphisms are now assigned to some of these dysmorphic conditions. Diagnostic applications of the molecular pathology in some of these conditions offer reliable tools for confirmation of the diagnosis and reproductive counseling. The scope of this chapter, however, limits the space for a detailed description of the key dysmorphic conditions with major female reproductive tract anomalies.

In addition to the malformation syndromes listed in Table 16.1, isolated anomalous development of the female reproductive tract is likely to be part of the clinical spectrum of a chromosomal disorder; for example, Turner syndrome resulting from X chromosome aneuploidy (45,X) or structural X chromosome abnormalities (deletion, inversion, or isochromosome). Thus it is essential that detailed cytogenetic analysis be carried out in any such case. New molecular cytogenetic techniques (FISH and aCGH) and new genomic tools (targeted/whole exome sequencing and whole genome sequencing) are invaluable for elucidating the causative underlying structural genomic abnormality. It is important that any case be investigated at a dedicated tertiary pediatric gynecology or infertility unit equipped with

the necessary molecular and imaging diagnostic facilities, supported by a skilled multidisciplinary team (pediatric gynecologist, developmental pediatrician, pediatric surgeon, clinical geneticist, reproductive medicine clinician, and clinical psychologist).

Disorders of Abnormal Ovulation

Biological mechanisms governing oogenesis and ovulation are complex and beyond the scope of this chapter. Series of hormonal and molecular processes are involved. Primary genetic endocrine diseases, for example, panhypopituitarism, may include extremely low levels of gonadotrophins [the follicle stimulating hormone (FSH) and luteinizing hormone (LH)], leading to hypogonadotropic hypogonadism. Primary amenorrhea and anovulation are major reproductive features in this group of disorders. In contrast, abnormal ovulation may also result in ovarian disorders with normal or elevated gonadotrophins: hypergonadotropic hypogonadism. This chapter reviews genetic and genomic aspects of two major group of conditions in the latter category:

premature ovarian failure (POF) and polycystic ovary syndrome. The interested reader may refer to dedicated texts on reproductive medicine for other less common conditions.

Premature Ovarian Failure

POF, or premature menopause, refers to the development of amenorrhea as a result of the cessation of ovarian function before the age of 40 years. The diagnosis is based on elevated FSH levels in the menopausal range (usually above 40 IU/l) detected on at least two occasions a few weeks apart [1]. Women with POF suffer from anovulation and hypoestrogenism and have primary or secondary amenorrhea, infertility, sex steroid deficiency, and elevated gonadotrophins. The condition affects approximately 1% of women, occurring in 10–28% of women with primary amenorrhea and in 4–18% of those with secondary amenorrhea. Early loss of ovarian function has significant psychosocial sequelae and major health implications, with nearly twofold age-specific increase in mortality [2].

A wide spectrum of pathogenic mechanisms may lead to the development of POF, including chromosomal, genetic, autoimmune, metabolic (galactosemia), infectious (mumps), and iatrogenic (anticancer treatments) causes. However, in a large proportion of POF cases no cause is found, and they are classified as idiopathic or karyotypically normal spontaneous ovarian failure. In one series of idiopathic POF, up to 30% of cases were ascertained with an autoimmune cause [3].

Genetics of Premature Ovarian Failure

Most cases of POF are sporadic, with a small number of familial cases indicating the role of genetic factors in its pathogenesis [4]. Genetic mechanisms include specific X and autosomal chromosomal abnormalities and a few uncommon single-gene disorders (Table 16.2). There is insufficient evidence for single nucleotide polymorphisms (SNPs) and genome-wide copy number variations (CNVs) to be causally linked to POF. Various genetic mechanisms implicated in the pathogenesis of POF include reduced haploinsufficiency (gene dosage effect) and nonspecific chromosome effects that impair meiosis. These can lead to ovarian failure by causing decrease in the pool of primordial follicles, increased atresia of the ovarian follicles resulting from apoptosis, or failure of follicle maturation.

Familial Premature Ovarian Failure

The large majority of POF cases are sporadic, with a small proportion having a reliable family history. The overall incidence of familial cases is around 4%. However, various studies indicate that up to one-third of cases are familial or inherited POF [4]. The variation between reported incidences might be explained by differences in the definition of POF and the idiopathic form, by differences in population

recruitment, and by selection and recall bias. Pedigree studies on affected families show autosomal dominant sex-limited transmission or X-linked inheritance with incomplete penetrance. An adequate family history can distinguish between familial and sporadic POF. The risk of female relatives developing POF may be higher in familial POF than in sporadic cases. Early diagnosis of familial predisposition permits prediction of impending menopause, and susceptible women can be guided to achieve their reproductive goals by timely planning of pregnancy.

In contrast to generally neutral effects of balanced autosomal translocations, balanced X/autosomal translocations often lead to POF, with more than 100 cases of postpubertal women with X/autosomal balanced translocations reported [7]. Molecular investigations in women with POF and experiments on transgenic animal models have led to the identification of a number of candidate genes for POF; mutations in these genes have been identified in 10% of POF cases; however, functions of many of these genes are not known [8].

Fragile X Syndrome Type A and *FMR1*

Fragile X syndrome type A (FRAXA) is one of the common inherited mental retardation conditions, occurring predominantly in the male. The inheritance pattern does not follow the classic mendelian pattern, and the condition is often described as an X-linked semidominant condition. The critical gene is called *FMR1*, located on Xq27.3, outside the Xq POF-critical region. Pathological expansion of the trinucleotide repeats located at its 5′ untranslated region result in abnormal expression of *FMR1*. Four types of alleles are identified based on the number of repeats: normal (6–40), intermediate (41–60), premutated (61–200), and fully mutated (>200). The full mutation is associated with the phenotype of FRAXA. The *FMR1* gene product is expressed in oocytes and encodes an RNA-binding protein involved in translation.

POF was first noted as an unexpected phenotype among heterozygous carriers of the fragile X premutation in the early 1990s. Subsequent studies showed that *FMR1* trinucleotide expansions in the premutation range of 50–200 repeat units, but not full mutations, are associated with POF [9]. The underlying mechanism for this association is unclear. Presently, there exists firm evidence for a significant association between fragile X premutation carrier status and premature menopause, as shown by both the analysis of women carrying the premutated allele and the screening of women affected by POF. The results of an international collaborative study examining premature menopause in 760 women from fragile X families showed that 16% of the 395 premutation carriers had experienced menopause before the age of 40 years, compared with none of the 238 full-mutation carriers, and one (0.4%) of the 237 controls [10].

TABLE 16.2 Genetic Aspects of Premature Ovarian Failure

Categories	Chromosome	Gene	Gene Locus
Mutations identified	X chromosome genes	FMR1	Xq27.3
		FMR2	Xq28
		BMP15	Xp11.2
	Autosomal genes	FOXL	3q22-q23
		FSHR	2p21-p16
		LH receptor	2p21
		FSHβ variant	11p13
		LHβ	19q13.32
		Inhibin A	2q33-q36
		GALT	9p13
		AIRE	21q22.3
		EIF2B2, EIF2B4, and EIF2B5	14q24.3, 2p23.3, 3q27
		NOGGIN	17q22
		POLG	15q25
Candidate genes	X chromosome genes	DIAPH2	Xq22
		DFFRX	Xp11.4
		XPNPEP2	Xq25
		ZFX	Xp22.3-p21.3
		FSHPRH1	Xq22
		XIST	Xq13.2
	Autosomal genes	WT1	11p13
		ATM	11q22.3
Mutations not identified	X chromosome genes	AT2	Xq22-q23
		c-kit	4q12
		SOX3	Xq26-q27
	Autosomal genes	MIS	19p13.3–13.2

Adapted from Gutierrez-Mateo C, Benet J, Wells D, Colls P, Bermudez MG, Sanchez-Garcia JF, et al. Aneuploidy study of human oocytes first polar body comparative genomic hybridization and metaphase II fluorescence in situ hybridization analysis. Hum Reprod 2004;19(12):2859–68.

The incidence of FRAXA premutations has been shown to vary among women with POF, depending on the proportion of sporadic and familial cases. Thirteen percent of pedigrees with the familial POF and 3% of women with the sporadic form of POF have been found to carry FRAXA premutations, compared with an expected prevalence of one in 590. It is likely that carriers who received the premutation from their fathers were at a higher risk of POF (28%) than those who received the premutation from their mothers (4%), suggesting that POF may be limited to permutations that are paternally inherited [11]. Nevertheless, from the practical point of view, FRAXA premutations are certainly worth seeking in those with familial POF in order to enable genetic counseling and, hopefully, limit the transmission of fragile X syndrome to future generations. Some units might also consider screening for FRAXA premutations in sporadic cases.

In patients who have the cytogenetic changes of fragile X syndrome but who are FMR1-mutation negative, a second site of fragility (FRAXE) is seen, lying approximately 150–600 kb distal to the FRAXA site at Xq28, and similarly folate-sensitive. An excess of small alleles with fewer than 11 repeats at the FRAXE locus were found in women with POF. It is believed that microdeletions within FMR2 may be a significant cause of POF, being found in 1.5% of women with the condition, and in only 0.04% of the general female population [12].

Mutations in the bone morphogenetic protein 15 gene (*BMP15*), mapped to Xp11.2, are associated with POF. Bone morphogenetic proteins are extracellular signaling proteins belonging to the transforming growth factor-β superfamily, which also includes growth differentiation factors (GDFs). *BMP15* is an oocyte-specific GDF that stimulates folliculogenesis and granulosa cell growth and is expressed in oocytes during early folliculogenesis. *BMP15* lies within the Xp POF-critical region [13].

Autosomal Abnormalities in Premature Ovarian Failure

Autosomal translocations are uncommon in women with POF. Most reports of translocations document X/autosome balanced translocations, with no common autosomal breakpoint. Reciprocal translocation between chromosomes 2 and 15 [46; XX,t [2,15] (q32.3; q13.3)] was recorded in a woman with POF [14]. In addition, it is well known that women with trisomy 21 (Down syndrome) are twice as likely to develop POF as are their corresponding age-related, chromosomally normal cohort, implying possible etiological associations of the autosomal genes on chromosome 21.

Several autosomal loci and genes are now known to exist that are causally associated with POF. Notable examples include blepharophimosis epicanthus inversus syndrome (BPES) (OMIM 110100) and autoimmune polyendocrinopathy-candidiasis-ectodermal dystrophy (OMIM 240300). BPES, an autosomal dominant genetic condition, is characterized by complex eyelid malformation and inverted epicanthus. Two forms are described: type I with POF in affected females, and type II not associated with POF [15]. POF-related infertility is inherited as an autosomal dominant sex-limited trait. BPES type I associated with POF maps to 3q22–q23, as does type II; the putative gene is the winged helix/forkhead transcription factor gene (*FOXL2*), and is mutated in both BPES types I and II [16]. The FOXL2 protein appears predominantly in the ovary in adult humans, and its corresponding gene is the first human autosomal gene in which dominant mutations have been implicated in ovarian maintenance and differentiation. More than 100 intragenic mutations and variants of *FOXL2* are listed, including uncommonly pathogenic association with nonsyndromic POF.

Polycystic Ovarian Syndrome

Polycystic ovarian syndrome (PCOS) is the most common endocrine condition afflicting women, occurring in 5% of the population. It is characterized by enlarged ovaries with multiple small follicles, anovulation, and an androgen-secreting stroma, presumably from the theca cells of the follicles. Recent genomic studies have improved our understanding of the molecular mechanisms underlying PCOS [17]. Analysis of the isolated and cultured theca cells showed an increase in gene transcription for aldehyde dehydrogenase 6 and retinol dehydrogenase 2. It is likely that many other genomic studies on hormone and growth factor responses in fibroid smooth muscle cells will be performed in the near future. If genetic pathways can be elucidated that differ between fibroids and myometrium, it may be possible to create medical treatments that selectively inhibit fibroid growth without interfering in myometrial function. This is particularly important in premenopausal women whose families have not yet been completed.

Disorders of the Endometrium

The endometrium is a dynamic tissue that responds to multiple stimuli, depending on physiological and environmental conditions, including steroid hormones, an implanting conceptus, withdrawal of steroid hormones, contraceptive steroids, selective steroid hormone-receptor modulators, infection, transient cell populations, and metaplastic and neoplastic agents. Microarray gene-expression profiling allows large numbers of genes to be investigated simultaneously, and the resulting gene-expression patterns can then be correlated with different structural, functional, or clinical parameters [18].

Endometrial gene-expression profiling is vastly different when comparing data obtained using endometrial cell cultures from profiling derived from endometrial biopsies, because cells may have different responses in vitro than in vivo. These differences may be the result of the processing of the cells, the culture medium used, and the loss of paracrine interactions with other cells in situ. Especially important when dealing with endometrial tissue is careful clinical characterization of the subjects donating the tissue, their hormonal stage, their age, any history of endometrial pathologies, and other medical conditions; it is also important to minimize intersubject variability and to maximize data reliability.

There are only few studies reporting on endometrial gene-expression profiling.

In an ambitious project, the transcriptional profiling of human endometrium during the menstrual cycle through a seven-phase time-course analysis was attempted [19]. In addition to identifying 425 genes that showed at least twofold upregulation in at least one stage of the menstrual cycle, the study also demonstrated prediction of the menstrual cycle stages based on the transcriptional profile of the samples, and identified different patterns of expression for groups of the genes associated with different endometrial biological processes such as implantation and menstruation. It is also suggested that discordant patterns of gene expression may help identify endometrial samples with subtle abnormalities not readily apparent in routine histopathology [19].

Endometriosis

Endometriosis is defined as the presence of endometrial tissue in ectopic locations outside the uterine cavity, typically on the pelvic peritoneal surfaces of the uterus and its ligaments, on the ovaries, and in the rectovaginal septum [20]. Between 5% and 10% of women have endometriosis, which causes a range of symptoms from pain to infertility.

The uterine (eutopic) endometrium from women with endometriosis may have endogenous abnormalities that promote the establishment of the disease. Matrix metalloproteinases *(MMPs)* 7 and 9 are normally expressed in endometrium during menstruation and are inhibited by progesterone during the secretory phase of the cycle. However, in women with endometriosis, these enzymes are constantly expressed during the secretory phase. This may allow retrograde endometrial tissue to occupy the peritoneal surface. Other studies also found that eutopic endometrium from women with endometriosis expressed higher levels of *MMP2*, membranous type 1 *MMP, MMP3,* and urokinase-type plasminogen activator *(uPA),* and lower levels of *TIMP2* than endometrium from normal women [21].

Up to 50% of women with infertility have endometriosis. Endometriosis-related infertility may be caused by toxicity to sperm and embryos by peritoneal fluid in women with endometriosis, and implantation failure resulting from abnormal expression of several genes (that either regulate or inhibit implantation) in women with endometriosis compared with normal women. Eutopic endometrium from women with endometriosis may be histologically normal but biochemically abnormal. During the window of implantation, several genes are aberrantly expressed in women with endometriosis compared to controls, including integrin αVβ3, *MMPs (MMP2, MMP3, MMP7,* and *MMP11),* transcription factors such as hepatocyte nuclear factor and endometrial bleeding factor, enzymes involved in steroid hormone metabolism (aromatase, 17β-hydroxysteroid dehydrogenase), leukemia inhibitory factor *(LIF), HOX* genes, and progesterone receptor isoforms [21]. Few groups have used microarrays to compare the transcriptional profile of ectopic and eutopic endometrium [22].

Knowledge of gene expression in normal endometrium and how it changes during the menstrual cycle is a prerequisite to an understanding of gene-expression data in endometrial disorders. Studying gene expression of normal endometrium via the high-throughput approach of DNA microarrays also allows the identification of new signature genes or biomarkers of normal physiological functions of endometrium such as implantation and menstruation. It is likely that some biomarker genes will be useful in molecular

or immunohistochemical diagnostics, or as molecular targets for drug discovery and therapeutic intervention. One of the most promising and clinically relevant applications of global transcript analysis with microarray is the gene-expression profiling of endometrial disorders to establish a genome-based diagnosis. That promise has not been fulfilled, although remarkable progress has been made. The data generated with microarrays certainly contribute to our understanding of endometrial disorders.

MALE FACTOR INFERTILITY

Male factor infertility is implicated in about half of all infertile couples. This includes wide-ranging potential causes that require thorough evaluation for accurate detection and management. Couples with a component of male factor infertility need a systematic evaluation directed at the apparently normal-looking healthy male partner to maximize their reproductive potential. A systematic approach with a complete medical history in conjunction with targeted investigations is necessary in all cases of suspected male factor infertility. The physical examination should be focused on signs of an associated endocrine disorder; the presence of gynecomastia; examination of the penis, scrotum, and testis; and rectal examination for prostatic enlargement. The semen analysis is mandatory in all cases, but it is by no means sufficient to determine the cause or decide on management plan therapy. Limited imaging investigations may also assist in evaluating male factor infertility.

Genetic factors implicated in male factor infertility include chromosomal abnormalities (47,XXY and structural Y chromosome abnormalities), single-gene disorders interfering in the sperm's motility or propagation (cystic fibrosis), spermatogenesis (mutations in *SRY* and 5-α reductase genes), and mutations in the *DAZ* gene cluster [23]. The prognosis for any given couple depends, in large part, on the cause of the infertility. Without a firm understanding of the genetics, anatomy, and physiology, and their interactions necessary to permit full functioning of the male reproductive system, the evaluation becomes an inefficient exercise that often fails to elucidate the precise cause of infertility. Treatment success relies not just on the clinical diagnosis, but on determining the cause of male factor infertility as well.

Congenital Anomalies of the Male Reproductive Tract

Exclusion of a congenital anatomical anomaly of the male reproductive tract is the essential first step in managing male factor infertility. Most congenital anomalies of the male reproductive tract are sporadic, without any appreciable recurrent genetic basis. However, genetic factors play a significant role in the causation of congenital anomalies; for example, cystic fibrosis transmembrane conductance regulator *(CFTR)*–related congenital bilateral absence of the vas deferens. Genetic factors underlying a congenital anomaly of the male reproductive tract include both chromosomal and mendelian conditions (Table 16.3). There are limited clinically relevant genomic data or information available on these. It is nevertheless likely that low-risk alleles, gene polymorphisms, and genome-wide CNVs could have some causal and/or functional roles in male factor infertility. Such studies are important, but they might be difficult to conduct and interpret because of the phenotypic heterogeneity of congenital anomalies of the male reproductive tract.

Molecular Genetics of Male Factor Infertility

Mutations in several Y-linked genes and interstitial deletions of the Y chromosome may cause impaired spermatogenesis and anomalous development of male gonads. Y chromosome microdeletions, predominantly of the long arm, are a common cause of infertility in males (Fig. 16.1) [24]. A particular area of interest on Yq is the azoospermia factor region (*AZF* region), which contains genes involved in the growth and development of sperm (*AZFa, AZFb,* and *AZFc*), more prevalent in men who are azoospermic and severely oligozoospermic. It is essential to look for the three *AZF* deletions when planning assisted reproductive technology (ART), because microdeletions are always passed on to the male offspring, and fertilization and pregnancy rates are not affected by microdeletions on the *AZFc* region when using intracytoplasmic sperm inoculation (ICSI). Multiple gene deletions in the *AZFb* and *AZFc* areas can produce a wide range of infertile phenotypes [25].

It is critical that azoospermic and severely oligozoospermic men be tested for microdeletions, both for accurate diagnosis and for genetic counseling, before performing ART

[26]. Initial screen for Y chromosome microdeletions is crucial before employing more expensive and technically challenging testing methods. Several autosomal and X-linked mendelian disorders are known to complicate male factor infertility. It is important to investigate men with the clinical suspicion of these disorders (Table 16.4). Diagnostic molecular genetic methods are available for confirmation of the diagnosis. A referral to the clinical genetic service for investigations and genetic counseling is important and should be included in the management plan.

Epigenetics and Epigenomics in Male Factor Infertility

Essentially, *epigenetic phenomenon* refers to an alteration in gene functioning without any DNA sequence changes in the coding regions of the gene. Epigenetic and epigenomic factors are now increasingly discussed and implicated in a number of biological processes (see chapter: Epigenetics and epigenomics in human health and disease). Since spermatogenesis involves complex series of events, it is vulnerable to an accumulation of genetic and epigenetic errors that can severely affect the spermatogenic process. There is evidence to suggest that epigenetics and epigenomics are among the several genetic factors in male factor infertility [27].

Histone markers signify DNA-imprinting control regions during the formation of spermatozoa. The transcriptional control of gene expression is regulated by the addition of acetyl, methyl, ubiquitin, and phosphate groups to histones. Imprinting, the methylation of DNA, determines which genes from the paternal and maternal genomes are expressed in the embryo and is critical for normal development. The imprinted regions of DNA are reset every reproductive cycle, which allows novel parental imprints to be established on the germ cells. Imprinting is achieved by the

TABLE 16.3 Prevalence and Phenotypes of Common Chromosomal Abnormalities Associated with Male Infertility

Genetic Abnormality	Phenotype	Prevalence, %
Chromosomal abnormalities	Azoospermia to normozoospermia	5 (total infertile population), 15 (azoospermia)
Klinefelter syndrome	Azoospermia to severe oligozoospermia	5 (severe oligozoospermia), 10 (azoospermia)
Robertsonian translocation	Azoospermia to normozoospermia	0.8 (total infertile population), 1.6 (oligozoospermia), 0.09 (azoospermia)
Y Chromosome microdeletions	Azoospermia to oligozoospermia	10–15 (azoospermia), 5–10 (oligozoospermia)
AZFa deletion	Azoospermia, Sertoli cell–only syndrome	1.5–1.0
AZFb deletion	Azoospermia, spermatogenic arrest	1.5–1.0
AZFc deletion	Severe oligozoospermia to nonobstructive azoospermia	6–12
Partial *AZFc* deletions	Azoospermia to normozoospermia	3–5

Note: Prevalence listed refers to listed phenotype, unless noted otherwise.
Adapted with permission from O'Flynn O'Brien KL, Varghese AC, Agarwal A. The genetic causes of male factor infertility: a review. Fertil Steril 2010;93(1):1–12.

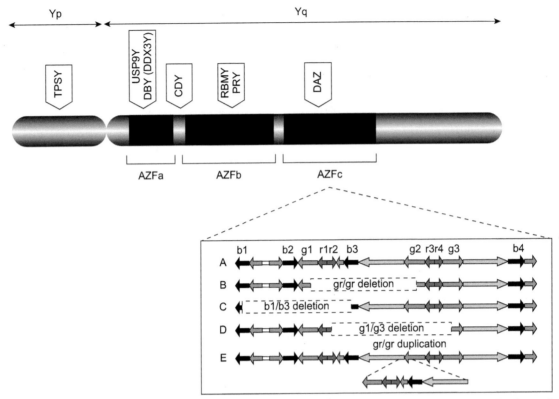

FIGURE 16.1 Y chromosome abnormality/deleted in azoospermia *(DAZ)* spectrum. Image of Y chromosome displaying AZF regions and associated genes. Enlarged portion of AZFc region highlights discussed microdeletions (A) Normal AZFc region; (B) gr/gr deletion; (C) b1 /b3 deletion; (D) g1/g3 deletion; (E) gr/gr duplication. *Adapted with permission from O'Flynn O'Brien KL, Varghese AC, Agarwal A. The genetic causes of male factor infertility: a review. Fertil Steril 2010;93(1):1–12.*

TABLE 16.4 Genes on Y Chromosome with Suspected Involvement in Male Factor Infertility

Gene	Location	Reasons for Investigation
USP9Y	AZFa	Involved in efficiency of spermatogenesis; deletion or shortening may cause azoospermia, oligozoospermia, or oligoasthenozoospermia
DBY	AZFa	Involved in premeiotic germ cell development
RBMY	AZFb	RNA-binding protein/testis-specific splicing factor; reduced expression in azoospermic men
PRY	AZFb	Regulation of apoptosis
DAZ	AZFc	Regulation of translation, meiosis, and germ cell population; codes for RNA-binding proteins; reduced expression in azoospermic men; partial deletions related to oligozoospermia
CDY	Yq	Involved in histone replacement
TSPY	Yq	Regulates timing of spermatogenesis; greater copy number in infertile patients

Adapted with permission from O'Flynn O'Brien KL, Varghese AC, Agarwal A. The genetic causes of male factor infertility: a review. Fertil Steril 2010;93(1):1–12.

differential marking of DNA regions with histone modifications, methylation, or possibly both, to allow only one copy of a gene to remain active. In fertile men with a normal ejaculate, paternal differentially methylated regions (DMRs) of the DNA should be methylated, and the maternal DMRs should be unmethylated. Approximately 14% of infertile male patients had abnormalities in the DMRs of the paternal

imprint, and 21% had abnormalities in the maternal imprint. Most patients with abnormalities in both imprint regions were oligozoospermic. Additionally, men with abnormally imprinted DMRs had low success rates with ART. It was also discovered that oligozoospermic men probably had a higher risk of transmitting imprinting errors to their children [28]. ART could have negative consequences on the imprinting of sperm because it may use sperm that are not yet fully mature, and, consequently, whose epigenetic code is not established. If the sperm are too immature or abnormal, it is more likely that the offspring could be born with an imprinting disorder.

The correlation between the incidence of imprinting disorders and ART in men with abnormal sperm is a controversial topic. Several researchers have asserted that ART such as ICSI causes imprinting disorders like Angelman syndrome and Beckwith-Wiedemann syndrome [29]. One study reports the incidence of Beckwith-Wiedemann syndrome was almost 5% in children conceived by ART in comparison with an incidence of less than 1% in the general population [24].

Mitochondrial Genes in Male Factor Infertility

Mitochondrial DNA (mtDNA) inheritance may also have an impact on male factor infertility (see chapter: High-performance Fire Retardant Polyamide Materials). Abnormal mitochondria are known to interfere with sperm motility because of aberrations in the mitochondrial sheath. There is concern about passing mutated mtDNA to offspring using ICSI in ART, because an entire sperm is injected into the oocyte and the mtDNA is conserved. In normal fertilization, the sperm mitochondria are lost along with the loss of the sperm tail. In addition, it is also known that oligospermic males have a relatively higher prevalence of mtDNA mutations, and various studies have demonstrated a correlation between abnormal mtDNA and dysfunctional sperm. However, data are conflicting as to the functional impact of mtDNA injected during ICSI. It is widely believed that the paternal mtDNA is degraded after fertilization, and therefore the functional clinical impact of mitochondrial abnormality in children conceived through ICSI/ART is not confirmed.

New Genomic Applications in Male Factor Infertility

Incorporating techniques such as genomics, proteomics, and metabolomics into infertility research could assist us in creating a complete portrait of the genes involved in infertility, and would enable improvements in ART for the development of more precisely targeted solutions. Microarrays are valuable tools for the identification of gene-expression profiles of infertile phenotypes. Examining the simultaneous expression of genes allows geneticists to determine molecular signatures related to infertile phenotypes. Microarray technology is also useful in the examination of spermatogenesis. An analysis of gene expression over time could be performed to determine the genes that are involved in each stage of the process. Genomic analysis can also be used to determine differentially transcribed genes. An enhanced understanding of transcription regulation could help geneticists discover how different expression patterns

affect a patient's fertility. Additionally, microarrays can be used to study the effect of hormones or growth factors on gene-expression profiles. Some advantages of using a microarray are that it is a noninvasive test, and it is very effective in studying germ cells because they express 4% of the genome. Disadvantages of genomics are that gene expression can vary between two different samples, and infertile patients might have pockets of gene expression that are difficult to detect using microarrays.

Metabolomics is another emerging area of research in the evaluation of the role of genetic factors in male factor infertility. Metabolomics involves measuring the expression of metabolites, small biomarkers that indicate the functionality of a cell, and characterizing them for certain diseases or physiological states. The identification of the human metabolome will reveal the functional phenotype of the system being studied, whether it is a single cell or an entire organism. Mass spectroscopy, nuclear magnetic resonance spectroscopy, and other chromatography methods can be used to create profiles of metabolites. Pathway or cluster analysis is used to determine subsets of metabolites that can be used to quantitatively characterize patients for diagnosis [30]. By identifying differences in the expression of metabolites in infertile phenotypes, new methods of diagnosis and treatment of male factor infertility can be developed that are inexpensive and noninvasive. Future clinical applications of metabolomics may include gamete selection (assessing the best sperm to use for ART) and functional genomic testing (screening for aneuploidy and other genetic conditions). Next, efficient clinical methods must be developed to compare standardized metabolomic signatures with patients' personal metabolomic profiles for the creation of individualized fertility care. These novel technologies hold promise for advances in the ways in which information about genetic profiles can aid infertility patients [31].

GENETIC FACTORS IN IN VITRO FERTILIZATION

A key problem limiting success during in vitro fertilization (IVF) is a diminished reserve of ovarian follicles, with increasing maternal age being the greatest factor influencing success. Unfortunately, there are no reliable predictors of ovarian reserve to inform patients of the chances of success with treatment, which would be information that could help individuals make decisions regarding participation in often-expensive treatment. Mice-model experiments have shown numerous genes with altered expression in ageing oocytes, including genes involved in genome stability, oxidative stress, and mitochondrial function. By comparing patterns of gene expression from women with normal and with reduced ovarian reserves, the investigators found preliminary evidence of a molecular expression pattern that may predict poor ovarian reserve.

It has been shown that up to 61% of first-trimester miscarriages after IVF have chromosomal abnormalities detectable by aCGH, a laboratory technique allowing scanning of whole genomes for structural changes such as either missing or extra pieces of chromosomes in a cell or tissue. Genomic screening of single cells taken from embryos before transfer, with tools such as aCGH, is a possible way of reducing miscarriage rates. Successful use of aCGH in women with recurrent implantation failure is reported. The feasibility of using aCGH on single cells taken from embryos has been proven in principle, with reported detection of trisomy 13 and 15 with amplified DNA from a single cell [32].

Some investigators are employing microarrays to study male factors in infertility. Gene-expression profiles from ejaculated spermatozoa of normal fertile men have been characterized [24]. A "normal" gene-expression profile in normal fertile men is important as a baseline to compare with that of infertile or subfertile men. However, this might not clearly identify or predict specific individual genes with dysregulated activity in infertile males. In an attempt to identify genes with a role in spermatogenesis, gene-expression profiles of human adult and fetal testis have been compared, as well as spermatozoal messenger RNA (mRNA) with adult testis mRNA [33]. Presumably, genes upregulated in adult testis and spermatozoa in these experiments, respectively, are increasingly likely to have an important role in the spermatogenic process.

Preimplantation Genetic Testing and Screening

Developments in molecular cytogenetic and genomic techniques have offered prospects of selecting an embryo before implantation using IVF-based ARTs. This approach allows successful reproduction outcomes in managing recurrent reproductive failure and allowing the couple the choice of selective reproduction when faced with increased genetic risk in future offspring. This includes two distinct approaches: preimplantation genetic screening (PGS) and preimplantation genetic diagnosis (PGD). PGS involves the selection of embryos before transfer into the uterus to increase the success of assisted reproduction. Genetic analysis is carried out on one or two blastomeres that are microsurgically removed from the embryo on day 3 of the culture. Embryos with abnormal chromosomal complements are discarded, and the selected embryos can be transferred on day 5 or frozen for future transfer. FISH is the main technique, involving the use of fluorescently labeled DNA probes to paint fetal DNA in interphase nuclei. Recent reports involve the use of aCGH, but this is not yet available for clinical use. Indications for PGS include advanced maternal age, repeated implantation failure, and idiopathic recurrent pregnancy loss, and in order to improve pregnancy rates in single-embryo transfers.

In most cases, the results are helpful in making an appropriate embryo selection. However, because of the high level of chromosomal mosaicism in the cleavage stages of embryonic development, the interpretation of findings might be difficult, leading to reanalysis. Secondly, because the contemporary FISH methods do not capture the full complement of chromosome material, the extent to which PGS is useful in improving pregnancy rates and outcomes is debated. Consequently, there is a move toward employing other, new methods in PGS. The current FISH technology is not recommended for the indications noted. Analysis of polar bodies may yield improved pregnancy outcomes by detecting maternal genetic abnormalities in eggs, including meiotic errors that result in aneuploidy. Newer array-based methods, including 24-chromosome SNP arrays or virtual karyotyping, will probably replace FISH because they provide more genetic information [24].

Preimplantation Genetic Diagnosis

PGD allows the selection of disease-free embryos for transfer into the uterus. Genetic analysis is usually carried out as described for PGS. FISH is used to detect sex chromosomes and specific chromosomal abnormalities, or polymerase chain reaction (PCR) is used to amplify DNA for molecular diagnosis. The first births after PGD of structural chromosomal abnormalities with the use of aCGH and microarray analyses were recently reported [34]. The first and second polar bodies can be analyzed to determine the presence of maternal genetic contributions (ie, X-linked diseases and autosomal dominant diseases), including carrier states for Duchenne muscular dystrophy, incontinentia pigmenti, and neurofibromatosis type 2. PGD has been successfully used in a number of monogenic dominant, recessive, and sex-linked diseases (www.hfea.gov.uk/preimplantation-genetic-diagnosis.html).With current molecular genetic methods and the rapid inclusion of the next generation genomic methods, the diagnosis of mendelian disorders is highly accurate, with a false-positive rate of less than 1%. Misdiagnosis has been attributed to laboratory error, including transfer of the wrong embryo; contamination by extraembryonic material; allele dropout (when one of the alleles is not amplified on PCR); use of the wrong probes or primer sets; and chromosomal mosaicism.

PGD is increasingly available in the United States and Europe. However, in most countries, its practice is relatively unregulated. In the United Kingdom, legal requirements and guidelines are provided by the statutory Human Embryology and Fertilisation Authority (www.hfea.gov.uk/preimplantation-genetic-diagnosis.html). In addition, professional societies (eg, the American Society for Reproductive Medicine and the European Society of Human Reproduction and Embryology) have issued guidelines and recommended the accreditation of laboratories performing such genetic diagnoses [24].

RECENT ADVANCES IN PRENATAL DIAGNOSIS

Laboratory techniques that are routinely employed in analyzing the aborted material, fetal tissue or cells, and parental bloods include tissue culturing, conventional karyotyping, FISH, quantitative fluorescence PCR (QF-PCR), and, increasingly, the new technique of aCGH. Newer techniques are being rapidly introduced with the increasing use of new generation genomic methods (exome sequencing, Sanger sequencing, whole-genome sequencing, and deep sequencing) in clinical practice; however, they are limited by technical and cost factors [35].

Aneuploidy (Trisomy 21) Pregnancy

Trisomy 21 (Down syndrome) is the most common cause of genetic mental retardation, occurring in one in 700 live births. Testing for Down syndrome is offered as a part of routine antenatal care for women in most developed nations, and has been increasingly introduced in developing countries. Unfortunately, noninvasive screening tests of maternal serum screening markers are insufficient to detect a Down syndrome pregnancy. The panel of serum tests used in antenatal screening for Down syndrome includes free β human chorionic gonadotropin, unconjugated estriol, pregnancy-associated plasma protein A *(PAPPA),* and dimeric inhibin A. Recently, *ADAM12,* a novel serum marker with biological properties similar to those of *PAPPA,* has been shown to improve detection rates. Several studies have advocated improved performance of serum screening when combined with first-trimester ultrasonographic measurement of nuchal translucency [36]. However, it is difficult to implement this as a population-based screening program because of the shortage of adequately trained sonographers and problems with quality assurance. The objective of antenatal Down syndrome screening is detecting high-risk pregnancies among younger women (under 35 years of age), who are then offered one of the two invasive tests for fetal chromosome analysis. The invasive tests required (amniocentesis or chorionic villous sampling of amniotic liquor) entail a miscarriage risk of 0.5–1.0%.

Traditionally, fetal chromosome analysis involves cell culture and full karyotyping. This is slow and expensive but assures a comprehensive analysis, minimizing the risk of failure to detect any other chromosomal abnormality. However, the rapid introduction of molecular cytogenetic methodology led to the increased use of commercially available fluorescent hybridization kits for selective cytogenetic analysis for chromosomal aneuploidies of 13, 18, and 21 chromosomes. The technique has improved to employ QF-PCR. It is claimed that a selective, but fast and less expensive, cytogenetic analysis using QF-PCR is more cost-effective when used in a carefully selected cohort of

17,500 women, using a combined screening strategy of serum testing with increased nuchal translucency (>4 mm) [36]. Microarray studies of trisomy 21 pregnancies include an attempt to identify new maternal serum biochemical markers for Down syndrome pregnancy by complementary DNA microarrays. Other approaches for faster detection of chromosomal aneuploidies include the use of cell-free fetal DNA from the amniotic fluid to form fluorescent probes for hybridization to aCGH.

Cell-Free Fetal DNA

Most prenatal genetic diagnosis (PND) is based on fetal tissue obtained by invasive methods involving chorionic villus biopsy (>11 weeks), amniocentesis (>15 weeks), and fetoscopy (>18 weeks). All invasive PND methods involve additional risk of miscarriage (1–3%). However, with high-resolution ultrasound imaging, precise localization has considerably reduced fetal risks. Efforts to develop noninvasive PND (NIPND) led to the consideration of testing circulating fetal cells in the maternal circulation. It has long been recognized that nucleated fetal cells reach the maternal circulatory system, but attempts to isolate these rare cells from maternal blood (which typically number one to six cells per milliliter of maternal blood) and use them for genetic testing have been disappointing because of low sensitivity. Cell-free fetal DNA is currently the material of choice for NIPND. It represents 3–6% of circulating cell-free DNA in maternal plasma, and it can be detected in the first trimester of pregnancy, increasing in abundance as the placenta grows. Cell-free fetal DNA fragments are much smaller than cell-free maternal DNA, which facilitates DNA sequence analysis. Although fetal DNA is detectable at 5 weeks of gestation, current methods of analysis are unreliable before 7 weeks of gestation [37]. Cell-free fetal RNA and DNA are not released from the fetus but from apoptotic placental trophoblast cells. These hold greater promise for genetic testing as a result of advances in DNA-sequencing methods and informatics [38].

Because the presence of the Y chromosome defines the male sex, its detection or lack thereof in maternal blood can be used to determine the fetal sex. A recent review and meta-analysis of fetal sex determination [39] with the use of maternal cell-free fetal DNA reported very good sensitivity, but the greatest sensitivity and specificity in the use of Y chromosome sequences to determine sex are obtained after 20 weeks of gestation, at which time ultrasonography is the preferred (and inexpensive) method.

In addition to sex determination, the detection of paternal genomic contributions to cell-free fetal DNA can be used to determine fetal rhesus D factor (RhD) status with high accuracy in the pregnancy of an RhD-negative woman. This approach can also be used to detect paternally transmitted dominant single-gene disorders, including Huntington disease, achondroplasia, and myotonic dystrophy. Carrier status for cystic fibrosis, hemoglobinopathies, and 21-hydroxylase deficiency has also been determined.

Recent reports on sequencing of the cell-free fetal DNA indicate that it has allowed the detection of underrepresentation or overrepresentation of chromosome-specific sequences, leading to applications in diagnosing trisomies of chromosomes 13, 18, and 21 [40]. Sequencing-based measurements of the proportion of small DNA fragments derived from chromosome 21 that exceed a threshold value relative to sequences from euploid reference samples have been reported to have a positive predictive value of 96.6% and a negative predictive value of 100%. This approach is based on "shotgun sequencing" of the small cell-free fetal DNA fragments and has the potential of identifying less common, more complex aneuploid states resulting from unbalanced chromosomal rearrangements or partial chromosome duplication. Other potential applications of genetic analysis of cell-free fetal DNA, including whole genome sequencing, include detection of a fetal microdeletion syndrome and genomic CNV. If these approaches became technically feasible, it is not clear whether they would be used as a screening method or as a diagnostic test? They would need to be cost-effective, with sufficiently rapid reporting of results in order to have a meaningful effect on parental decision-making.

SUMMARY

Recent additions to our scientific knowledge and technological developments have considerably revolutionized the contemporary practice of reproductive medicine. Utilization of medical genetic expertise and the application of conventional genetic laboratory methods are now routinely employed in managing major referrals to a modern reproductive medicine facility in a tertiary unit. Major areas of concern that have benefitted include POF, gynecological conditions like endometriosis, male factor infertility, preimplantation genetic screening, assisted reproduction techniques, early recurrent pregnancy loss, prenatal diagnosis, preeclampsia, and preterm birth. The use of microarrays has permeated most areas of study within the reproductive, obstetrics, and gynecology fields. The number of diseases and conditions within this broad field that are currently undergoing genomic studies is too large to be summarized in this chapter. The rate at which microarray studies and next generation genomic methods are being published makes it certain that this chapter will always be short of information. Nonetheless, this chapter provides an overview of the way that genetic and genomic input and studies are important in a number of areas in the current practice of reproductive medicine.

REFERENCES

[1] Conway GS. Premature ovarian failure. Br Med Bull 2000;56(3): 643–9.
[2] Kalantaridou SN, Davis SR, Nelson LM. Premature ovarian failure. Endocrinol Metabolism Clin N Am 1998;27(4):989–1006.
[3] Conway GS, Conway E, Walker C. Mutation screening and isoform prevalence of the follicle stimulating hormone receptor gene in women with premature ovarian failure, resistant ovary syndrome and polycystic ovary syndrome. Clin Endocrinol 1999;51(1):97–9.
[4] Wieacker P. Genetic aspects of premature ovarian failure. J Reprod Med Endokrinol – J Reprod Med Endokrinol 2009;6(1):17–8.
[5] Deleted in review.
[6] Deleted in review.
[7] Loughlin SA, Redha A, MicIver J, Boyd E, Carothers A, Connor JM. Analysis of the origin of Turner's syndrome using polymorphic DNA probes. J Med Genet 1991;28(3):156–8.
[8] Harris SE, Chand AL, Winship IM, Gersak K, Aittomäki K, Shelling AN. Identification of novel mutations in FOXL2 associated with premature ovarian failure. Mol Hum Reprod 2002;8(8):729–33.
[9] Allingham-Hawkins DJ, Babul-Jirji R, Chitayat D, Holden JJ, Yang KT, Lee C, et al. Fragile X premutation is a significant risk factor for premature ovarian failure: the international collaborative POF in fragile X study, preliminary data. Am J Med Genet 1999; 83(4):322.
[10] Uzielli ML, Guarducci S, Lapi E, Cecconi A, Ricci U, Ricotti G, et al. Premature ovarian failure (POF) and fragile X premutation females: from POF to fragile X carrier identification, from fragile X carrier diagnosis to POF association data. Am J Med Genet 1999;84(3):300–3.
[11] Hundscheid RDL, Sistermans EA, Thomas CMG, Braat DDM, Straatman H, Kiemeney LALM, et al. Imprinting effect in premature ovarian failure confined to paternally inherited fragile X premutations. Am J Hum Genet 2000;66(2):413–8.
[12] Murray A, Webb J, Dennis N, Conway G, Morton N. Microdeletions in FMR2 may be a significant cause of premature ovarian failure. J Med Genet 1999;36(10):767–70.
[13] Di Pasquale E, Beck-Peccoz P, Persani L. Hypergonadotropic ovarian failure associated with an inherited mutation of human bone morphogenetic protein-15 (BMP15) gene. Am J Hum Genet 2004;75(1):106–11.
[14] Kawano Y, et al. Premature ovarian failure associated with a Robertsonian translocation. Acta Obstet Gynecol Scand 1998;77(4):467–9.
[15] Zlotogora J, Sagi M, Cohen T. The blepharophimosis, ptosis, and epicanthus inversus syndrome: delineation of two types. Am J Hum Genet 1983;35(5):1020–7.
[16] Crisponi L, Deiana M, Loi A, Chiappe F, Uda M, Amati P, et al. The putative forkhead transcription factor FOXL2 is mutated in blepharophimosis/ptosis/epicanthus inversus syndrome. Nat Genet 2001;27(2):159–66.
[17] Jansen E, Laven JS, Dommerholt HB, Polman J, van Rijt C, van den Hurk C, et al. Abnormal gene expression profiles in human ovaries from polycystic ovary syndrome patients. Mol Endocrinol 2004;18(12):3050–63.
[18] Giudice LC. Elucidating endometrial function in the post-genomic era. Hum Reprod Update 2003;9(3):223–35.
[19] Ponnampalam AP, Weston GC, Trajstman AC, Susil B, Rogers PA. Molecular classification of human endometrial cycle stages by transcriptional profiling. Mol Hum Reprod 2004;10(12):879–93.

[20] Giudice LC, Kao LC. Endometriosis. Lancet 2004;364(9447):1789–99.

[21] D'Hooghe TM, Debrock S, Meuleman C, Hill JA, Mwenda JM. Future directions in endometriosis research. Ann. N. Y Acad Sci 2004;1034:316–25.

[22] Kao LC, Germeyer A, Tulac S, Lobo S, Yang JP, Taylor RN, et al. Expression profiling of endometrium from women with endometriosis reveals candidate genes for disease-based implantation failure and infertility. Endocrinology 2003;144(7):2870–81.

[23] Krausz C, Forti G, McElreavey K. The Y chromosome and male fertility and infertility. Int J Androl 2003;26(2):70–5.

[24] O'Flynn O'Brien KL, Varghese AC, Agarwal A. The genetic causes of male factor infertility: a review. Fertil Steril 2010;93(1):1–12.

[25] Nuti F, Krausz C. Gene polymorphisms/mutations relevant to abnormal spermatogenesis. Reprod Biomed Online 2008;16(4):504–13.

[26] Sadeghi-Nejad H, Farrokhi F. Genetics of azoospermia: current knowledge, clinical implications, and future directions. Part II: Y chromosome microdeletions. Urol J 2007;4(4):192–206.

[27] Lawrence LT, Moley KH. Epigenetics and assisted reproductive technologies: human imprinting syndromes. Semin Reprod Med 2008;26(2):143–52.

[28] Hartmann S, Bergmann M, Bohle RM, Weidner W, Steger K. Genetic imprinting during impaired spermatogenesis. Mol Hum Reprod 2006;12(6):407–11.

[29] Cox GF, Burger J, Lip V, Mau UA, Sperling K, Wu BL, et al. Intracytoplasmic sperm injection may increase the risk of imprinting defects. Am J Hum Genet 2002;71(1):162–4.

[30] He Z, Chan WY, Dym M. Microarray technology offers a novel tool for the diagnosis and identification of therapeutic targets for male infertility. Reproduction 2006;132(1):11–9.

[31] Deepinder F, Chowdary HT, Agarwal A. Role of metabolomic analysis of biomarkers in the management of male infertility. Expert Rev Mol Diagn 2007;7(4):351–8.

[32] Gutierrez-Mateo C, Benet J, Wells D, Colls P, Bermudez MG, Sanchez-Garcia JF, et al. Aneuploidy study of human oocytes first polar body comparative genomic hybridization and metaphase II fluorescence in situ hybridization analysis. Hum Reprod 2004;19(12):2859–68.

[33] Montjean D, De La Grange P, Gentien D, Rapinat A, Belloc S, Cohen-Bacrie P, et al. Sperm transcriptome profiling in oligozoospermia. J Assist Reprod Genet 2012;29(1):3–10.

[34] Alfarawati S, Fragouli E, Colls P, Wells D. First births after preimplantation genetic diagnosis of structural chromosome abnormalities using comparative genomic hybridization and microarray analysis. Hum Reprod 2011;26(6):1560–74.

[35] Bodurtha J, Strauss 3rd JF. Genomics and perinatal care. N Engl J Med 2012;366(1):64–73.

[36] Chitty LS, Kagan KO, Molina FS, Waters JJ, Noclaides KH. Fetal nuchal translucency scan and early prenatal diagnosis of chromosomal abnormalities by rapid aneuploidy screening: observational study. BMJ 2006;332(7539):452–5.

[37] Larrabee PB, Johnson KL, Pestova E, Lucas M, Wilber K, LeShane ES, et al. Microarray analysis of cell-free fetal DNA in amniotic fluid: a prenatal molecular karyotype. Am J Hum Genet 2004;75(3):485–91.

[38] Go AT, van Vugt JM, Oudejans CB. Non-invasive aneuploidy detection using free fetal DNA and RNA in maternal plasma: recent progress and future possibilities. Hum Reprod Update 2011;17(3):372–82.

[39] Devaney SA, Palomaki GE, Scott JA, Bianchi DW. Noninvasive fetal sex determination using cell-free fetal DNA: a systematic review and meta-analysis. JAMA 2011;306(6):627–36.

[40] Chiu RW, Akolekar R, Zheng YW, Leung TY, Sun H, Chan KC, et al. Non-invasive prenatal assessment of trisomy 21 by multiplexed maternal plasma DNA sequencing: large scale validity study. BMJ 2011;342:c7401.

Chapter 17

Stratified and Precision Medicine

D. Kumar[1,2]

[1]The University of South Wales, Pontypridd, Wales, United Kingdom; [2]Cardiff University School of Medicine, University Hospital of Wales, Cardiff, United Kingdom

Chapter Outline

Introduction	227	Challenges for Stratified Medicine	232
Molecular, Genetic, and Genomic Revolutions in Medicine	227	The Future of Stratified Medicine	233
Personalized Medicine	228	Summary	234
Stratified Medicine	230	References	234

INTRODUCTION

Recent innovations and new developments in molecular biology and biotechnology have revolutionized the sophisticated diagnosis of most complicated and rare conditions. This is evident from overwhelming "gene-specific" and "genome-driven" diagnoses in many inherited and genetic disorders. However, the success on the therapeutic sector has not been encouraging and continues to face many hurdles and limitations. Most therapeutic regimens are designed for the "average model patient." Although "one-size-fits-all-approach," treatments can be very successful for some patients, they are not for many others. However, this has changed with rapid advances made in evidence-based personalized medicine. The whole process involves a stepwise approach in building the holistic picture referred to as *stratified medicine* with the ultimate aim of individualized or personalized therapeutic interventions. This chapter briefly summarizes scope and limitations of genetic and genomic applications and new exciting opportunity for stratified and personalized medicine.

During the last decade, new genomic diagnostic tools and molecular innovations have led to the emergence of precision medicine, an innovative approach to disease prevention and treatment that takes into account individual differences in people's genes, environments, and lifestyles. Precision medicine gives clinicians tools to better understand the complex mechanisms underlying a patient's health, disease, or condition, and to better predict which treatments will be most effective. Advances in precision medicine have already led to powerful new discoveries and several new treatments that are tailored to specific characteristics of individuals, such as a person's genetic makeup or the genetic profile of an individual's tumor. This is leading to a transformation in the way we can treat diseases such as cancer. Patients with breast, lung, and colorectal cancers, as well as melanomas and leukemia, for instance, routinely undergo molecular testing as part of patient care, enabling physicians to select treatments that improve chances of survival and reduce exposure to adverse effects.

The potential for precision medicine to improve care and speed the development of new treatments has only just begun to be exploited. Translating initial successes to a larger scale will require a coordinated and sustained global effort. Through collaborative public and private partnerships, stratified and precision medicine initiatives will harness advances in genomics and biotechnology for accelerating biomedical discoveries. This grand ambition is shared by the US President Barack Obama in his State of the Union address:

> *Doctors have always recognized that every patient is unique, and doctors have always tried to tailor their treatments as best they can to individuals. You can match a blood transfusion to a blood type—that was an important discovery. What if matching a cancer cure to our genetic code was just as easy, just as standard? What if figuring out the right dose of medicine was as simple as taking our temperature?*

MOLECULAR, GENETIC, AND GENOMIC REVOLUTIONS IN MEDICINE

After the phenomenal discovery of the structure of the nucleic acids (DNA and RNA) and subsequent sequencing of the human genome, rapid progress is continually made in understanding the molecular bases of human disease.

Medical and Health Genomics. http://dx.doi.org/10.1016/B978-0-12-420196-5.00017-4

However, for around 4 decades the focus was largely on the molecular basis; for example, deciphering the structure of hemoglobin to unravel the complexities of a number of inherited and acquired blood diseases. This period was the hallmark of molecular medicine. To a large extent, developments in this field were also supported by understanding the causation of human disease in the context of genes, inheritance patterns, and making clear diagnoses using the gene–molecule approach, however limited [1]. However, these advances and evolving trends in medicine had to wait for relevant laboratory research before applications in the clinic. The practice of molecular medicine in genetic and genomic terms is now a reality and most clinicians are engaged with this, however discreetly or indirectly [2]. This juncture changed dramatically with the near complete sequencing of the entire human genome, opening new horizons for clinical medicine and even extending into domains of public and population health [3].

Although 99.5% of the human genome sequence is similar in any two individuals, the structural variation, however, in the remainder of the genome could be functionally relevant to an individual's unique genetic constitution or genomic signatures [4]. This "personalized" sequence variation is undeniably important and is agreeably the fundamental basis of genomic medicine or health genomics. Functional annotation for individual sequence variation, when complete, will be crucial in diagnosing and selecting appropriate therapeutic agents [5]. This is likely to be vastly improved with the availability of targeted sequencing of selected genes, exons, or promoters. There are fewer variants in protein coding than noncoding sequences. Variants that cause amino acid changes, and thus altered protein product, are in general dissimilar (nonsynonymous) compared with those that lack such an association (synonymous). If an excess of nonsynonymous substitution is observed for one particular coding region, then this can be taken as an indicator of diversifying (positive) selection. With the help of next generation sequencing technologies, more and more variants are being characterized and sequence annotations made available. It is envisaged that ultimately a fuller picture will emerge of the variants that alter genome function, and will enable selection of those that contribute to health and disease in a particular individual [6].

Since 2005, rapid and unprecedented progress has been made in applied and translational genomic research, leading to practical and dynamic utilizations in clinical medicine. Complicated laboratory techniques of genome sequencing [whole exome sequencing (WES), targeted deep capture, and whole genome sequencing (WGS)] are no longer confined to research settings [7]. With the advent of next generation sequencing, the speed of generating enormous genome sequencing data is considerably greater at successively lower costs [8]. An individual may obtain a personal genome sequence at just under US $1000. Requests are being generated for WES and WGS from the offices of nonspecialist medical practitioners or even direct from the consumer [9]. Sincere efforts are put in place, both in public and private sectors, for establishing the role of genome sequencing in clinical medicine and public health. Higher state-level commitments are declared to set standards and guidelines for the genome sequencing in both clinical and research settings. The recent 100,000 Genomes Project of the United Kingdom government is a good example. This chapter reviews the current position of genomics in medical and health perspectives, particularly personalized genome sequencing, and examines strengths and weaknesses of the genome sequencing for medical purposes and positive health.

In the context of personalized medicine, genetic and genomic laboratory techniques are now increasingly applied to select patients based on specific genetic and molecular signatures that might influence the outcome of a particular therapeutic regimen. This approach has gained recognition and momentum, leading to the emergence of stratified medicine. For these emerging concepts, a new term, *precision medicine*, is now increasingly used. Fundamentally, stratified medicine (or precision medicine) implies individual patient's clinical care based on specific risk of disease or response to therapy by using diagnostic tests or techniques, whether conventional or genetic [10]. This approach has far-reaching implications. Patients and healthcare providers both benefit from more targeted and effective treatments, whereas industry benefits from the potential for more efficient therapeutic development, as well as the market expansion for these new treatments.

The development of stratified medicine or precision medicine is being pursued globally as its benefits are increasingly recognized. To most medical and health practitioners, the concept and philosophy behind stratified medicine are not unfamiliar. However, this approach is now remarkably strengthened with increasing accuracy and sophistication of genomic and molecular medicine. It is widely acknowledged that the power of personalized and stratified medicine is considerably enhanced with the availability of individual genome sequence information [11].

PERSONALIZED MEDICINE

The success of genomic medicine will depend upon the ability to sequence an individual's full genome. With the benefit of new technologies, it is possible to generate gigabases of data as short sequence reads, and to assemble the data accurately using the finished sequence as a template. This will provide the essential database of human genome variation for a given population. Comparison of these data sets will provide a full profile of common genome variation along each chromosome. Detection of each variant will help in estimating the recombination rates and correlation along each chromosome. This approach could give important baseline information on healthy tissue compared with pathological tissue. For example, a comparison of the cancer genome sequences could allow for monitoring DNA changes on a genome-wide basis for cancer development. A similar approach could also be applied in other diseases. This genomic information on both healthy and diseased tissue could be used in screening an individual's disease risk and devising appropriate therapy and medical advice, paving the way forward for personalized medicine.

As the human genome functional annotation becomes available, the prospects of personalized medicine will improve. A hypothetical scenario is described [12] where variation in the *PPAR-γ* gene, one of the susceptibility genes in type 2 diabetes mellitus, is employed in the selection of the most appropriate oral hypoglycemic agent (Fig. 17.1).

The chromosome 3 region (A) 12,300–12,450 kb, numbering as in build 34 (http://www.ensembl.org), contains the *PPAR-γ* gene structure [dark blue (dark gray in print versions)] with an alternative promoter [light blue (lighter gray in print versions)], hypothetical noncoding functional variants [green shaded boxes (light gray in print versions)], and functional variants [red (gray in print versions)]. Magnification of the variant segments (B) shows the translated sequence with nucleotide changes [functional variants highlighted in blue (light gray in print versions)] and amino-acid changes [pink (gray in print versions), C]. The amino-acid variant results in variation in the protein molecule as confirmed from a linked database. This variant protein molecule contains a specific drug binding site [blue (light gray in print versions)] for antidiabetic thiazolidinedione, an oral hypoglycemic agent. A number of biological consequences,

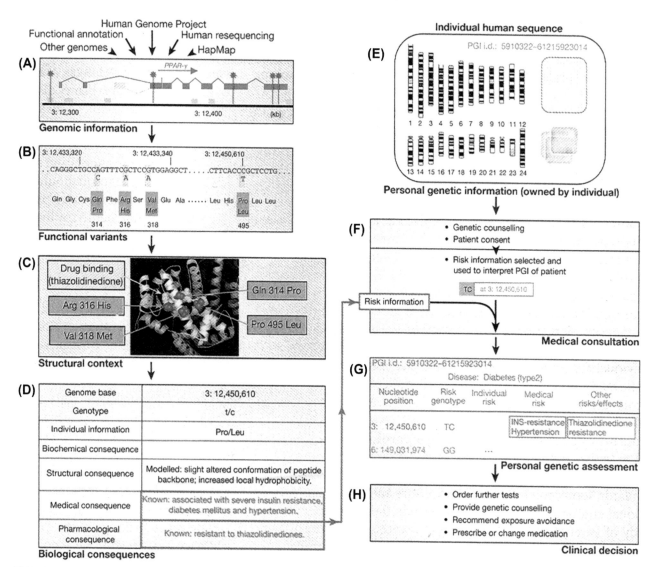

FIGURE 17.1 A hypothetical model of personalized medicine. *PGI*, Personal genetic information. *Adapted from Chapter 19, Figure 19.4, Kumar D. Genetic and genomic approaches to clinical medicine, In: Kumar D, Eng C, editors. Genomic Medicine-Principles and Practice, NY: Oxford University Press; 2014.*

such as biochemical, medical, and pharmacological, can be predicted using linked database information (D). This information can be regularly updated and curated allowing a detailed listing of the likely consequences. A small subset of this information would define the disease or drug outcome or side effect associated with each variant, and would enable the clinician to provide specific risk information in clinical consultation. This information could be made available on the public domain subject to stringent review and including only those data for which medical relevance was established.

The use of personal genetic information in a clinical setting could be requested and consent given by the individual concerned. The individual sequence acquired could be restricted to one or two genotypes or as much as a complete genome sequence. The information thus acquired would be exclusive and private and wholly owned by the individual. It could be stored electronically, protected by a high-security code requiring unique personal identifiers, such as used for storing multiple fingerprint or iris-pattern, for access only with the consent of the individual (E). The information might be taken either before consultation or afterward, and in either case the individual would be subject to counseling by the medical practitioner and must give consent.

The clinical consultation could initiate a specific investigation (F). The personal annotated genetic information of the individual patient, such as a set of gene mutations or variants for cardiovascular disease, would be made available for interpretation with respect to the clinical phenotype. The clinician would use the available risk information concerning each variant to provide a genetic assessment for the individual (G). In the case illustrated, the individual has the heterozygous genotype TC at position 3:12,450,610. This corresponds to having both Pro 495 and Ala 495 forms of the protein PPAR-γ. This genotype confers an increased risk of insulin-resistant diabetes mellitus on the individual, and also resistance to the thiazolidinedione class of antidiabetic drugs. Combining this with risk information for other genotypes would help to make informed subsequent clinical decisions (H).

Thus with easy access to a well-annotated human genome and availability of cheap, accurate WGS technology, an individual could acquire either a specific or complete genetic health profile, including risk and resistance factors. The information could then be used to improve and guide important medical decisions, to assess the risk of possible future exposures, and to select preventive treatments for improved health [12].

In brief, the practice of personalized or specifically individualized medicine will become the central focus of the future practice of clinical medicine. However, this will demand lots of commitment, perseverance, and investment at personal, family, community and public, or state levels. Inevitably and understandably, this approach will raise several ethical and social concerns for the fear of inequity, discrimination (primarily because of enormous costs and affordability), and potential misuse or abuse (malpractice). The practice of personalized medicine should not be allowed to develop without relevant professional and statutory safeguards. This approach should be one of the other major ingredients of the clinical practice pathway, what is often referred to *the 4 P's of Medicine*; medicine that will be more predictive, personalized, preemptive, and participatory [13]. To this list, the new fifth term of *precise* is increasingly used.

To reach these key long-term goals, the US National Institutes of Health, the UK National Institute of Health Research, and many other organizations are actively pursuing and promoting research in the aforementioned areas. These organizations are strategically investing in research to further our understanding of the fundamental causes of diseases at their earliest genetic, genomic, and molecular stages. The central theme of the personalized medicine is based on the simple basic concept that individuals respond differently to environmental factors including therapeutic interventions, according to their genetic/genomic endowment and their own behavior and lifestyle. In the future, applied and translational genomic and molecular research will allow us to predict how, when, and in whom a disease will develop [14]. We can envision a time when we will be able to precisely target or stratify treatment on a personalized (individualized) basis to those who need it, avoiding treatment to those who do not (Fig. 17.2). Ultimately, this individualized approach will allow us to preempt disease before it occurs, utilizing the participation of individuals, communities, and healthcare providers in a proactive and preparatory fashion, as early as possible, and throughout the natural cycle of a disease process [15].

STRATIFIED MEDICINE

Stratified medicine is the grouping of patients based on risk of disease or response to therapy by using diagnostic tests or techniques [10]. Patients and healthcare providers both benefit from more targeted and effective treatments, whereas industry benefits from the potential for more efficient therapeutic development, as well as the market expansion for these new treatments. The development of stratified medicine is being pursued globally as its benefits are increasingly recognized. The concept and philosophy behind stratified medicine are not unfamiliar. However, this approach is now remarkably strengthened with increasing accuracy and sophistication of the genomic and molecular medicine. [16] Stratified approaches to therapy are expected to become the standard for the management of a whole range of diseases (for example, chronic heart failure), provided that these match certain criteria as recommended by leading clinicians and scientists [17]. The Academy of

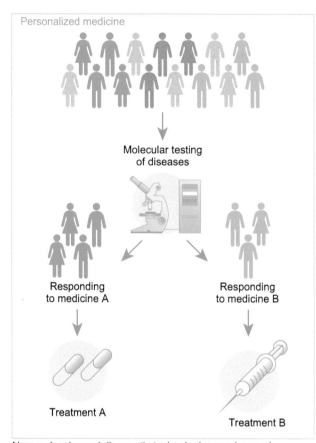

Personalized medicine

Molecular testing
of diseases

Responding
to medicine A

Responding
to medicine B

Treatment A

Treatment B

New molecular and diagnostic technologies can be used
to match select groups of patients with treatments that may
give them the best results

FIGURE 17.2   Personalized medicine: comparison of outcomes of treatment A and treatment B. *Adapted with permission-http://www.pfizer.ie/personalized_med.cfm.*

TABLE 17.1 Criteria for Stratified Medicine

1. Continued research to understand the genetic and molecular bases of diseases
2. Development and use of increasingly sophisticated and powerful informatics technology
3. Improvement and standardization of clinical data collection and linkage with genomic and other databases
4. Increased collection of tissues for biomarker research and evaluation, and its organization in national and international biobanks
5. Greater efficiency and productivity in the development of therapeutics and diagnostics
6. The introduction of flexible and novel approaches for the regulatory assessments of innovative stratified medicine products
7. Improved flexibility in pricing for stratified medicine products, both for the diagnostic process and for the associated therapy, to ensure cost-effectiveness for payers while encouraging innovation.

The Academy of Medical Sciences, UK. Realizing the potential of stratified medicine; 2013.

Medical Sciences in the United Kingdom (www.acmedsci.ac.uk) has recommended criteria for Stratified Medicine (Table 17.1) [18].

Several programs and incentives are now operational for stratified medicine to enable partnership across academia, industry, healthcare systems, regulatory/pricing authorities, research funders, and patient groups. The progress toward stratified medicine, increasingly confused with personalized medicine relies fundamentally upon data, which is central to the applied and translational research to understand the molecular basis of disease; development of targeted interventions; effective regulation, health technology assessment, and valuation of stratified medicine products; and the stratification of treatment by physicians (Table 17.2) [19].

Among many examples of stratified approaches in planning and executing treatment for common cancers, the case for non–small cell lung cancer (NSCLC) is noteworthy; probably it is the best paradigm in the context of stratified medicine (Figs. 17.3 and 17.4).

Side (A) of Fig. 17.4 represents the heterogeneity in patients with adenocarcinoma of the lung according to

driver oncogenes that are crucial for selecting targeted drugs for treatment. The number of people reflects approximate incidence. Heterogeneity in patients with *EGFR* mutations (B), resulting in heterogeneous response to EGFR-TKI. Most (~90%) *EGFR* mutations are known to be a predictor of response to EGFR-TKI; however, the *T790M* mutation can be detected in 0.5% or 38% of TKI-naive patients by direct sequencing or by the polymerase chain reaction–based sensitive method, respectively. Low *IκB* expression which occurs in approximately 30% of patients, low *BIM* expression in approximately 20% or 38% of patients, or high *CRKL* expression in approximately 3% of patients are reported to be associated with shorter progression-free survival outcomes upon EGFR-TKI treatment. On the left of Fig. 17.4 (C), there is heterogeneity in individual patients treated using targeted therapy that results from selection pressure and the microenvironment of the tumor. Even in a single patient there can be heterogeneity in the resistance mechanisms activated. Further treatment would likely result in even greater heterogeneity.

It has been known for some time that mutations in *KRAS* were associated with squamous cell lung cancer. Further research indicated that mutations in the epidermal growth factor receptor gene (*EGFR*) could be used in targeting the treatment, notably those of exons 19 and 20 [20]. Similarly, the *EML4–ALK* mutation can be used as an example of how molecular understanding accompanied by targeted medicines has transformed the treatment of patients with NSCLC [21]. In 2007, research demonstrated that approximately 5% of NSCLC cases involved this mutation. Within 3 years, targeted therapies were developed and demonstrated dramatic efficacy; now patients with lung cancer can have biopsy tissue sent for genetic analysis to ascertain their suitability

TABLE 17.2 Stratified Medicine in Current Medical Practice [18]

		Case Study		US approval		EU approval	
		Drug (Rx)	Companion Diagnostic (CDx)	Rx	CDx	Rx	CDx
1	Breast cancer	**Herceptin (trastuzumab)** Roche/Genentech	**HercepTest** Dako	Sep. 1998	Sep. 1998	Aug. 2000	Yes
2	HIV	**Ziagen (abacavir)** GSK/ViiV Healthcare	*HLA-B*57:01* **screening assay**	Dec. 1998	N/A: unbranded test	Jul. 1999	N/A: unbranded test
3	Breast cancer	N/A: Dx only	**Oncotype Dx** Genomic Health	N/A: Dx only	Not FDA approved: use supported by literature	N/A: Dx only	2007
4	Colorectal cancer	**Vectibix (panitumumab)** Amgen	**EGFR pharmDx kit** Dako	Sep. 2006	Sep. 2006	Sep. 2007	Yes
			therascreen:KRAS RGQ PCR kit Qiagen		Jul. 2012		Yes
5	Melanoma	**Zelboraf (vemurafenib)** Roche/Plexxikon	**cobas 4800 BRAF V600 mutation test** Roche	Aug. 2011	Aug. 2011	Feb. 2012	Yes
6	Non–small cell lung cancer	**Xalkori (crizotinib)** Pfizer	**Vysis ALK Break Apart FISH probe kit** Abbott Molecular Diagnostics	Aug. 2011	Aug. 2011	Jul. 2012 (conditional marketing authorization)	Sep. 2011
7	Cystic fibrosis	**Kalydeco (ivacaftor)** Vertex Pharmaceuticals/ Cystic Fibrosis Foundation Therapeutics Inc.	*G551D* **mutation test**	Jan. 2012	N/A: unbranded test	Jul. 2012	N/A: unbranded test
8	Melanoma	**BRAF/MEK inhibator (trametinib and dabrafenib)** GSK	**BRAF mutation kit (v600E & K)** bioMérieux	In development			

for this treatment and they receive an accurate, genetically derived diagnosis in 7–10 days [22]. These developments have transformed therapy for those 5% with NSCLC driven by the *EML4–ALK* mutation, meaning that by simply taking two capsules per day, the cancer shrinks or disappears for more than one in every two people treated, rather than for one in every 10 as was the case with traditional chemotherapy [23]. Although this dramatic response is not always sustained over time, it is highly beneficial to patients.

Challenges for Stratified Medicine

There are several challenges and obstacles to realizing the full potential of benefits of the substantial progress in genomic and molecular research in pursuit of stratified approaches to clinical medicine:

- Standardization of genome sequencing platforms is necessary to avoid laboratory-to-laboratory variability complicating the analysis of combined datasets.
- High levels of enrollment for sequencing are required to benefit from the accumulation of whole genome sequence data, which will require that privacy and data protection concerns be addressed.
- Because of the complexity, capital expense of equipment, and size of datasets, progress in molecular medicine is increasingly requiring collaboration between many academic groups, public institutions, and industry, often across countries.
- Genomic information on its own, although useful, is only part of the story. Greater knowledge is gained when such genetic information is linked to clinical outcomes. Thus there remains a major hurdle to link

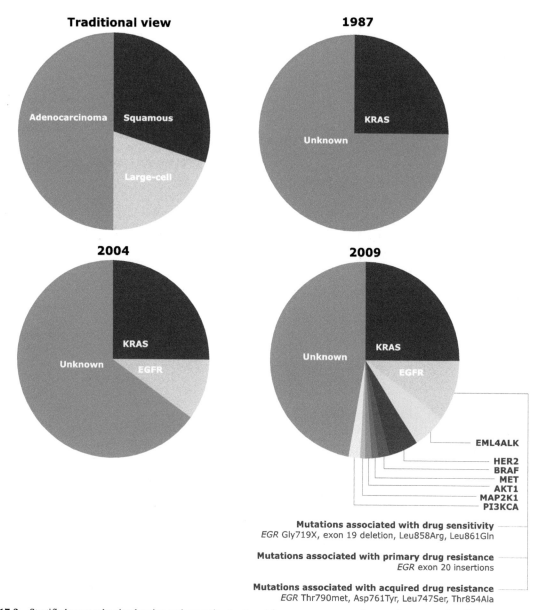

Traditional view

Adenocarcinoma

Squamous

Large-cell

1987

KRAS

Unknown

2004

KRAS

Unknown

EGFR

2009

Unknown

KRAS

EGFR

EML4ALK

HER2
BRAF
MET
AKT1
MAP2K1
PI3KCA

Mutations associated with drug sensitivity
EGR Gly719X, exon 19 deletion, Leu858Arg, Leu861Gln

Mutations associated with primary drug resistance
EGR exon 20 insertions

Mutations associated with acquired drug resistance
EGR Thr790met, Asp761Tyr, Leu747Ser, Thr854Ala

FIGURE 17.3   Stratified approaches in planning and executing treatment for common cancers, the case for non–small cell lung cancer (NSCLC) [18].

genome databases to healthcare records, which need to be electronic for this to be done efficiently.

- Research is still required so that genetic variations are not only correlated to diseases, but causal links are established, if the underlying molecular mechanisms of disease are to be understood.
- Correlation of genetic variation and disease may sometimes not transcend ethnic groups. The Pharmacogenetics for Every Nation initiative has been set up to address this issue (www.pgeni.org).
- The effect of epigenetic variations on drug response, pharmacoepigenomics, needs further research [24]. Epigenetic variations are inheritable and affect gene expression levels and therefore phenotype, yet do not result from changes in the DNA sequence [25].

The Future of Stratified Medicine

There are multiple factors that will determine the development and adoption of stratified approaches to medicine. There are "pull" factors, in that the healthcare system needs to become increasingly effective and sustainable, in particular the economic policies for investment and cost reimbursement. There are also "push" factors from recent advances in medical science, informatics, and the pharmaceutical industry requiring substantial improvements in research and development productivity to remain a viable sector in the long term [26]. These factors accelerate the momentum of stratified medicine and can be transformative in the provision of care. Detailed discussion on this aspect of stratified medicine is beyond the scope and remit of this chapter.

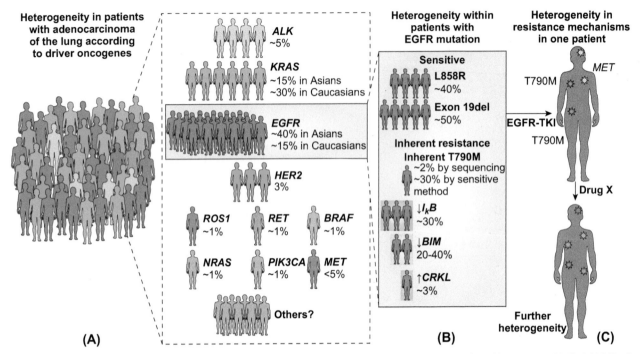

FIGURE 17.4 Stratified approach to planning and selection for targeted therapy in adenocarcinoma of lungs. *Adapted from Tetsuya M, Kenichi S, Yatabe Y. Surgery for NSCLC in the era of personalized medicine. Nat Rev Clin Oncol 10:235–244. http://dx.doi.org/10.1038/nrclinonc.2013.22.*

However, the following major areas are important [18] to consider for planners and developers of stratified medicine:

- effective and sustainable healthcare systems
- scientific and technological advances
- diagnostic applications to accommodate new disease categories
- challenges facing the pharmaceutical industry
- role of the regulatory and statutory agencies.

SUMMARY

An important milestone in the history of medical science is the recent completion of the human genome sequence. The progress on identification of approximately 23,000 genes and their regulatory regions provides the framework for understanding the molecular basis of disease. This advance has also laid the foundation for a broad range of genomic tools that can be applied to medical science. These developments in gene and gene product analysis across the whole genome have opened the way for targeted molecular genetic testing in a number of medical disorders. This is destined to change the practice of medicine; future medical practice will be more focused and individualized, often referred to as *personalized medicine*. However, despite these exciting advances, many practicing clinicians perceive the role of molecular genetics, in particular that of medical genomics, as confined to the research arena with limited clinical applications. Genomic medicine applies the knowledge and understanding of all genes and genetic variation in human

disease. This chapter introduces genomics-based advances in personalized disease susceptibility screening, diagnosis, prognostication, a stratified approach for genomics-led therapeutics, and prediction of treatment outcome in various areas of medicine.

Finally, the art and science of the practice of medicine at all times are true reflections of dynamic adjustment of the physical state of the human body and environmental pressures. In this context, the innate characteristics conferred by the genetic and genomic constitution provide the framework on which a range of life-time environmental experiences and pressures would act and manifest in either positive or morbid (disease) states. This was echoed clearly over 100 years ago across the medical community in one of the classic Harveian Orations of the Royal College of Physicians in London, England:

It was in Padua that medicine, long degraded and disguised, was now to prove her lineage as the mother of natural science, and the truth of the saying of Hippocrates, that to know the nature of man one must know the nature of all things.

Sir Clifford Allbutt, Regius Professor of Physic, Harveian Oration (1900) [27].

REFERENCES

[1] Harper PS. A short history of medical genetics, vol. 57. USA: Oxford University Press; 2008.
[2] Kumar D. Clinical medicine in the genome era: an introduction. Genomics Clin Med 2008;(53):145.

[3] Research, W.H.O.A.C.o.H. Genomics and world health: report of the Advisory Committee on Health Research 2002. World Health Organization; 2002.

[4] Cooper DN, Chen JM, Ball EV, Howells K, Mort M, Phillips AD, et al. Genes, mutations, and human inherited disease at the dawn of the age of personalized genomics. Hum Mutat 2010;31(6): 631–55.

[5] Hamburg MA, Collins FS. The path to personalized medicine. N. Engl J Med 2010;363(4):301–4.

[6] Green ED, Guyer MS. Charting a course for genomic medicine from base pairs to bedside. Nature 2011;470(7333):204–13.

[7] Snijders AM, Nowak N, Segraves R, Blackwood S, Brown N, Conroy J, et al. Assembly of microarrays for genome-wide measurement of DNA copy number. Nat Genet 2001;29(3):263–4.

[8] Berg JS, Khoury MJ, Evans JP. Deploying whole genome sequencing in clinical practice and public health: meeting the challenge one bin at a time. Genet Med 2011;13(6):499–504.

[9] Bloss CS, Schork NJ, Topol EJ. Effect of direct-to-consumer genomewide profiling to assess disease risk. N. Engl J Med 2011;364(6):524–34.

[10] Bieber T. Stratified medicine: a new challenge for academia, industry, regulators and patients. Stratif Med 2013:3.

[11] Ziegler A, Koch A, Krockenberger K, Grosshennig A. Personalized medicine using DNA biomarkers: a review. Hum Genet 2012;131(10): 1627–38.

[12] Bentley DR. Genomes for medicine. Nature 2004;429(6990): 440–5.

[13] Hood LE, Galas DJ. P4 medicine: personalized, predictive, preventive, participatory a change of view that changes everything. 2009.

[14] Xu LH, Zheng H, Sedmak DD, Sadee W. The re-emerging concept of personalized healthcare. 2008.

[15] Yang YT, Wiley E, Leppard J. Individualized medicine and pharmacogenomics: ethical, legal and policy challenges. J Med Pers 2011;9(2):48–57.

[16] Trusheim MR, Berndt ER, Douglas FL. Stratified medicine: strategic and economic implications of combining drugs and clinical biomarkers. Nat Rev Drug Discov 2007;6(4):287–93.

[17] Silver MA, Maisel A, Yancy CW, McCullough PA, Burnett Jr JC, Francis GS, et al. BNP Consensus Panel 2004: a clinical approach for the diagnostic, prognostic, screening, treatment monitoring, and therapeutic roles of natriuretic peptides in cardiovascular diseases. Congest Heart Fail 2004;10(Suppl. 5):1–30.

[18] T.A.o.M.S. (UK). Realizing the potential of stratified medicine. 2013.

[19] Khoury MJ, Coates RJ, Fennell ML, Glasgow RE, Scheuner MT, Schully SD, et al. Multilevel research and the challenges of implementing genomic medicine. JNCI Monogr 2012;(44):112–20.

[20] Pao W, Miller VA. Epidermal growth factor receptor mutations, small-molecule kinase inhibitors, and non–small-cell lung cancer: current knowledge and future directions. J Clin Oncol 2005;23(11):2556–68.

[21] Harris TJR, McCormick F. The molecular pathology of cancer. Nat Rev Clin Oncol 2010;7(5):251–65.

[22] Tsuboi M, Ohira T, Saji H, Miyajima K, Kajiwara N, Uchida O, et al. The present status of postoperative adjuvant chemotherapy for completely resected non–small cell lung cancer. Ann Thorac Cardiovasc Surg 2007;13(2):73.

[23] Azzoli C.G. Dx/Rx: lung cancer: lung cancer. Jones & Bartlett Publishers.

[24] Baer-Dubowska W, Majchrzak-Celińska A, Cichocki M. Pharmacoepigenetics: a new approach to predicting individual drug responses and targeting new drugs. Pharmacol Rep 2011;63(2):293–304.

[25] Jaenisch R, Bird A. Epigenetic regulation of gene expression: how the genome integrates intrinsic and environmental signals. Nat Genet 2003;33:245–54.

[26] Tambuyzer E. Towards a framework for personalized healthcare: lessons learned from the field of rare diseases. Pers Med 2010;7(5):569–86.

[27] Allbutt TC. The Harveian Oration on physiological darkness before Harvey: delivered before the Royal College of Physicians on October 18th. Br Med J 1900;2(2078):1271.

Chapter 18

Teaching and Training Medicine in Genomic Era

B. Korf

University of Alabama at Birmingham, Birmingham, AL, United States

Chapter Outline

Introduction	**237**
Integration of Genomics into Medical Practice	**237**
The Practice of Genomic Medicine	238
Prevention	238
Diagnosis	239
Therapy	239
The Roles of the Geneticist and Nongeneticist Health Providers	
	240
Competencies Required in Genomic Medicine	241
Genomics and the Medical Education Landscape	**242**

Premedical Education	242
Medical School	242
Residency Education	242
Postgraduate Education	243
Approaches to Medical Genomics Education	**243**
Medical School	243
Residency Education	244
Continuing Medical Education	244
Conclusions and Final Comments	**244**
References	**244**

INTRODUCTION

The promise of genomic medicine is that new approaches will be available to prevent, diagnose, and treat both rare and common disease. Most practicing physicians, however, trained before the human genome was sequenced and are unfamiliar with the principles of genomic medicine. Multiple surveys and studies bolster this view, indicating a lack of preparedness on the part of the healthcare community for the integration of genomics into medical practice [1–6]. It is clear that there is a need for education of the healthcare workforce if genomic medicine is to fulfill its promise. This need, however, must be viewed against a background in which the practice of medicine is becoming more and more complex. As a result, there are increasing demands on professional school and postgraduate curricula, although the time available to teach has not changed. This means that genomic medicine must be taught to the current and future generations of health providers, but that the approaches to providing this education must be efficient and innovative. This chapter will provide an overview of the challenge of medical education in the genomics era and highlight some approaches that may be helpful in meeting this challenge. We will first look at how genomics may be integrated in

medicine in the coming years, thereby providing the rationale for genomics education. Next we will examine the current structure of medical education, focusing on the US system, with which the author is most familiar. We will next explore some approaches that have been used, and then make some final comments on the future of genomic medical education.

INTEGRATION OF GENOMICS INTO MEDICAL PRACTICE

Before one can specify how medical professionals should be educated in genomic medicine, it is necessary to ask how they will actually use genomics in their practice. This question raises one of the central dilemmas in medical genomics education: namely that it is hard to motivate the community to alter the educational approach to prepare trainees for the future when no one is sure what the future practice of genomic medicine will look like. This section will provide a broad outline of how genomics might integrate into medicine in the next several years (not decades), the distinction between roles of genomic medicine professionals and the general medical community, and the kinds of competencies that will be needed for both.

Medical and Health Genomics. http://dx.doi.org/10.1016/B978-0-12-420196-5.00018-6

The Practice of Genomic Medicine

It is sometimes speculated that the day may come when everyone has his or her genome sequenced, producing data that would be available for a lifetime to guide medical decision making. This would profoundly alter the landscape of medical practice, but it is debatable whether it is either feasible or desirable. It is safe to say that, if this ever happens, it will not be in the next few years. Genomics is already being integrated into the present-day practice of medicine, and for this discussion we will assume a gradual integration of genomics into medical practice for the foreseeable future. We will consider three domains of application of genomics to medicine: prevention, diagnosis, and treatment.

Prevention

Genetic approaches to prevention of disease have been practiced for 50 years, in the form of newborn screening. Newborn screening began with the ability to perform an inexpensive test to detect elevation of phenylalanine in newborn blood spots and the recognition that treatment before onset of symptoms could prevent irreversible damage to the nervous system. There has been a gradual increase in the repertoire of conditions that are included for newborn screening in the United States. Each state decides individually what to include, but there is now an organized national process to assess the evidence supporting inclusion of specific tests on the panel and there is wide acceptance of a minimum list of disorders appropriate for screening [7–9]. Tests are based on detection of biochemical intermediates by tandem mass spectrometry, or other approaches such as hemoglobin electrophoresis for globin disorders, otoacoustic emissions for deafness, and immune reactive trypsinogen for cystic fibrosis. None of these is a genetic test; ie, a test that begins with analysis of DNA to search for a gene mutation. It has been suggested, however, that genomic sequencing at birth would provide a universal newborn screen for almost all genetic disorders (not all, because some types of pathological variants are not easily detected by sequencing). This notion raises significant issues regarding consent, identification of adult-onset disorders in a newborn, finding variants of unknown significance, and identification of variants that predict disease for which there is no preventative strategy [10–12]. In one survey, many parents indicated their interest in having access to their newborn's genome sequence, but the questions were answered hypothetically, so it is uncertain how this would translate to real practice [13]. The matter is now under study in several research studies funded by the US National Institutes of Health.

The workflow in newborn screening is instructive, even if the current approach is a genetic, rather than genomic one. Blood is taken from a heel stick in the nursery, mandated by law and not requiring parental consent. In most states, testing is done in a public health laboratory. Abnormal results are called to the pediatrician identified as being responsible for the care of the baby upon discharge from the hospital, who is expected to institute immediate therapy and arrange consultation with a metabolic disease expert for confirmation of the diagnosis and management. The pediatrician is also expected to notify the baby's parents about the possible diagnosis and next steps.

Although pediatricians are likely to be familiar with the concept of newborn screening, most are probably unfamiliar with care of some of the rare inborn errors of metabolism that may be identified by newborn screening. The American College of Medical Genetics and Genomics (ACMG) has instituted the use of "ACT sheets" that provide essential information on the immediate care of the child to bridge the gap from recognition of the abnormal screen result to consultation with a specialist [14] (http://www.ncbi.nlm.nih. gov/books/NBK55827/). This approach provides the pediatrician with information just when it is needed, making it unnecessary to remember how to manage these disorders.

There are two additional examples of preventative strategies for genetic conditions. One is preconception screening for couples planning pregnancy, based on detection of carrier status for autosomal recessive conditions. This has been practiced for many years in populations at high risk, such as Tay-Sachs disease screening for Ashkenazi Jewish couples or hemoglobinopathy screening for those of African, Mediterranean, and Asian descent [15]. Testing panels have gradually expanded as the cost of testing has come down. It is now possible to obtain panels that test for over a 100 disorders, including some that are extremely rare. Testing is usually arranged by obstetrician-gynecologists who are part of prepregnancy planning, or through community-based programs. Couples found to be at risk are usually offered genetic counseling, provided by a board-certified genetic counselor or by the ordering physician.

The other approach to the prevention of rare genetic disease involves detection of mendelian traits that predispose to cancer or other conditions such as hypercholesterolemia or rare inherited forms of diabetes or hypertension. Usually the clue is family history, or in some cases an unusual presentation, such as early age of onset. There has been much discussion of the value of taking a family history as a component of routine medical care, including concern that physicians may not have the time, let alone the training, to obtain a thorough family history [16–18]. There has also been effort to encourage the public to obtain family history information on their own; for example, at family get-togethers such as Thanksgiving, and to share the information with their physicians [19]. Family history can be a powerful tool in the recognition of conditions such as hereditary breast and ovarian cancer, but not all individuals at risk are aware of their family history, and some have small families with no known affected individuals.

One of the major hopes for genomic medicine is that it will enable prevention of common multifactorial disorders, such as hypertension or Alzheimer disease. Genome-wide association studies have revealed numerous genetic markers that are associated with common disorders; some of these may be involved in the pathogenesis of the conditions, whereas others are in linkage disequilibrium with unknown genetic traits that contribute to risk of the disorder [20]. In spite of major progress in this area, for the most part only a small proportion of the estimated heritability of these traits are accounted for by currently known genetic markers [21]. Hence genetic testing reveals only slight changes in the odds of being affected, in most cases too small to be of value in prevention. Furthermore, for most conditions, definitive preventative strategies are not available, other than life style changes such as weight loss, which can be recommended without genomic testing. Genomic tests that offer prediction of risk of common disease have been offered, though the US Food and Drug Administration has prevented use of this approach in providing medical advice. The future of genomic testing for risk of common disease remains uncertain, depending largely on the degree to which the testing is truly predictive and the effectiveness of prevention strategies.

Diagnosis

Use of genetic and genomic approaches in diagnosis of disease have more than a 50-year history, beginning with the identification of trisomy 21 as the cause of Down syndrome. Technological advances have gradually improved the precision of genetic diagnosis and expanded to a repertoire of genetic conditions for which diagnostic testing can be provided. Beginning in the 1980s, as the genes that underlie mendelian disorders began to be discovered, molecular genetic testing has enabled precise diagnosis of symptomatic individuals, those at risk based on family history, and prenatal testing of the fetus. Some genetic tests are ordered by generalist physicians, such as pediatricians, who arrange chromosomal analysis for patients suspected of having an abnormality such as trisomy 21. Genetic tests are also commonly ordered by specialists, such as neurologists who order tests for muscular dystrophy or spinocerebellar ataxia, or cardiologists who order tests for cardiomyopathy. Sometimes these are done in consultation with a genetics professional, such as a medical geneticist or genetic counselor, but often the testing is ordered and interpreted by the nongeneticist specialist.

During the past decade, genomic approaches have been introduced that are dramatically changing the landscape of diagnostic testing. The first was the application of cytogenomic microarrays, which permit detection of copy number variants below the limit of detection with the light microscope. The second was the introduction of genomic sequencing—so far mostly exome sequencing, although now also whole genome sequencing. This permits detection of pathological variants without the need to know in advance the specific gene to test; it is most helpful in short circuiting the "diagnostic odyssey" in which a patient seeks a diagnosis, often for years and with the assistance of multiple physicians, yet is frustrated by the lack of definitive results [22]. Genomic tests such as microarrays and genomic sequencing may be ordered by generalist physicians or nongeneticist specialists, although it is common for these to be done in collaboration with a genetics professional. In some institutions this is a requirement, given the high costs of testing, and some insurance companies may require participation of a geneticist for the same reason. Involvement of a geneticist can be especially helpful in the interpretation of results; for example, in instances of finding a variant of unknown significance or incidental findings indicating risk of a condition other than the one for which testing was done.

Genomic testing is also beginning to see wide use in diagnosis of cancer. Identification of cancer-specific rearrangements or genetic variants can be used to obtain a specific diagnosis and guide therapy [23]. Testing is often done as part of the pathological analysis of a tumor specimen. Detailed reports of the results of sequencing can help guide the oncologist to know the significance of genomic findings and to use this information to decide on a therapeutic regimen.

Therapy

Treatment of genetic disorders also has a long history, mainly through the treatment of inborn errors of metabolism. As noted previously, this is usually managed by an expert in biochemical genetics working together with a primary care physician. As the pathophysiology of other genetic conditions has come to be understood, additional disorders are amenable to treatment; for example, cystic fibrosis [24], Duchenne muscular dystrophy [25], and lysosomal storage disorders [26]. In most cases, treatment is managed by a specialist who is most familiar with the disorder and its natural history; for example, a neurologist with expertise in neuromuscular disease would treat a patient with Duchenne muscular dystrophy. It is likely that this approach will continue to apply as treatments are developed for other genetic conditions; it will also be the case if genetic studies reveal new treatments for multifactorial disorders, although in some cases the treatments are likely to be used by primary care physicians. The important knowledge here is not that of genetics and genomics, but rather the principles of management of the condition itself.

Another area where genetic and genomic testing will influence therapeutics is in pharmacogenetics. Testing for specific polymorphisms can reveal individual patterns of drug metabolism or risk of idiosyncratic reactions that can guide choice of medication and dosaging [27]. For most such applications there is still debate about issues such as cost-effectiveness, clinical utility, the best way to ensure timely testing, and interpretation of results to guide treatment decisions. Haga et al. [28] surveyed primary care physicians and noted that most were aware of the possibility of pharmacogenetic testing, but few felt prepared to use it in practice. It may not, however, be necessary for the physician to learn which pharmacogenetic test to order or how to use the information to guide treatment. Rather, alerts may be built into the electronic prescribing system, providing point-of-care decision support; pharmacogenetic information might be incorporated into the system, offering the physician a suggested dosage or a warning if the medication is not appropriate to use in a specific patient. Testing for pharmacogenetic polymorphisms might be done with quick turnaround at the point of care, or patients might be tested for important polymorphisms one time in life, with the results available any time a clinical decision needs to be made. The pharmacogenetic information might then reside in the electronic health record, in some central cloud-based storage location, or with the patient herself.

The Roles of the Geneticist and Nongeneticist Health Providers

Through this discussion, reference has been made to a role for both geneticists and nongeneticist providers. There are several types of training paths for medical geneticists (Table 18.1). Physicians can obtain residency training in medical genetics and genomics and be eligible for certification by the American Board of Medical Genetics and Genomics (ABMGG). Medical genetics is recognized in the United States by the American Board of Medical Specialties. Similar training and recognition for medical geneticists exists in many other countries. Laboratory geneticists can be trained in clinical cytogenetics, clinical biochemical genetics, and clinical molecular genetics, and are also eligible for ABMGG certification. Pathologists can receive training in molecular genetic pathology, also leading to board certification. Genetic counselors complete a 2-year master's degree program and are eligible for certification by the American Board of Genetic Counseling (ABGC). There is also a training program for nurses who receive special competency as advanced practice nurse geneticists.

There are fewer than 2000 board-certified medical geneticists, and perhaps twice that number of genetic counselors, but far too few to serve the needs of the population, particularly if genomic testing becomes widespread. Therefore nongeneticist providers will need to take substantial responsibility for implementation of genomic medicine. This is indeed compatible with the picture of a collaborative model for care that was described previously.

The preventative strategy of newborn screening has always involved collaboration of a pediatrician and a specialist, typically a biochemical geneticist for inborn errors of metabolism. Given the complexity of these

TABLE 18.1 Various Training Paths in Medical Genetics

Title	Description	Training	Certification
Medical genetics	Physicians specializing in medical genetics	2-year residency after at least 2 years in an ACGME-accredited residency	ABMGG
Genetic counseling	Specialists in genetic risk assessment, coordination of care, and counseling	2-year master's degree program	ABGC
Clinical cytogenetics	Laboratory specialist in cytogenetics (chromosomal analysis)	2-year fellowship after MD or PhD	ABMGG
Clinical biochemical genetics	Laboratory specialist in biochemical genetics (inborn errors of metabolism)	2-year fellowship after MD or PhD	ABMGG
Clinical molecular genetics	Laboratory specialist in molecular genetic testing	2-year fellowship after MD or PhD	ABMGG
Medical biochemical genetics	Physician specializing in care of patients with biochemical genetic disorders	1-year fellowship after medical genetics training	ABMGG
Molecular genetic pathology	Physician (usually pathologist) training in molecular genetic testing	1-year fellowship (usually after pathology training)	ABMGG/American Board of Pathology
Advanced practice nursing in genetics	RN specializing in genetics	Postgraduate genetics training and experience	International Society of Nurses in Genetics

ABGC, American Board of Genetic Counseling; *ABMGG*, American Board of Medical Genetics and Genomics; *ACGME*, Accreditation Council for Graduate Medical Education.

disorders and the risks of severe morbidity or death if a child is not treated correctly, it is likely that this type of care will always require specialists. As noted previously, point-of-care information such as the ACMG ACT sheets provide the pediatrician with important guidance when and where it is needed. Preconceptional screening typically is done by obstetrician-gynecologists, often with the assistance of genetic counselors. Screening for individuals at high risk of cancer can be done with a few pointed personal and family history questions by a primary care physician, and can also be facilitated by public education. Genetic counselors, working in partnership with a medical geneticist, typically provide access to testing and interpretation of results. In some cases, the counselor may work with a specialist such as an oncologist, although providing such counseling requires special training in the interpretation of genetic information, which goes beyond typical oncology training.

Genetic diagnostic testing is often done by specialists who see patients with specific medical problems, such as neurological or cardiac disorders. These specialists should be familiar with the interpretation of genetic tests in their specialty, although some may employ a genetic counselor to work directly with families. The genetic counselor, in turn, ideally works in collaboration with a medical geneticist who can assist in the interpretation of complex results. Genomic testing, including microarray and genomic sequencing, may be best done with involvement of a medical geneticist, often working together with a genetic counselor. It is possible for a nongeneticist specialist, or even a generalist, to order genomic testing and provide interpretation to a patient, but the time required to do this and the need to explain complex results require a special set of skills.

Therapeutic approaches are likely to be implemented by the specialists who are most familiar with the specific conditions. Medical geneticists often take primary responsibility for care of patients with multisystem disorders or inborn errors of metabolism, and are likely to remain involved in the care of these patients. As noted above, pharmacogenetic test interpretation is likely to be done semiautomatically by decision support built into the electronic prescribing system with advice from the laboratory geneticist and others, such as pharmacists.

Competencies Required in Genomic Medicine

The previous sections make the case that competency in genomic medicine will be required by a broad array of health practitioners, including generalists (such as pediatricians, family practitioners, and internists), specialists, and geneticists. Efforts have been made to define the skills needed by these various groups in order to incorporate genomics into their practice. In the United States, the National Coalition for Health Professional Education in Genetics issued a

set of competencies of health providers [29], and a parallel effort has been conducted in Europe [30,31].

The issues raised by genomic medicine have been considered by several medical professional societies. The National Human Genome Research Institute has convened a group of representatives of these societies to form the Intersociety Coordinating Committee for Practitioner Education in Genomics (ISCC) [32]. Several working groups have been formed, including one to define a framework for competencies of nongeneticist physicians. This framework identified a set of "entrustable professional activities"; that is, activities that a physician would be expected to be able to carry out without supervision [33]. The framework is summarized in Table 18.2. It was expected that this framework could be used by a specialty group to customize a set of competencies relevant to their area of practice.

For the physician geneticist, the ACMG has developed a set of competencies that are intended to guide training programs [34]. Aside from areas traditionally included in training, such as dysmorphology, genetic risk assessment, and interpretation of genetic testing, specific competencies were included for genomic testing, including, for example, use of genomic sequencing. These competencies formed the backbone for formulation of milestones for medical genetics residency training required by the Accreditation Council for Graduate Medical Education [35]. Efforts are now underway to do the same for training of laboratory geneticists, spearheaded by the Association of Professors of

TABLE 18.2 Entrustable Professional Activities in Genomic Medicine Proposed by ISCC Working Group on Competencies [33]

Entrustable Professional Activity	Description
Family history	Elicit, document, and act on relevant family history pertinent to the patient's clinical status.
Genomic testing	Use genomic testing to guide patient management.
Patient treatment based on genomic results	Use genomic information to make treatment decisions.
Somatic genetics	Use genomic information to guide the diagnosis and management of cancer and other disorders involving somatic genetic changes.
Microbial genomic information	Use genomic tests that identify microbial contributors to human health and disease, as well as genomic tests that guide therapeutics in infectious diseases.

ISCC, Intersociety Coordinating Committee for Practitioner Education in Genomics.

Human and Medical Genetics (APHMG). The ABGC has used a formal analysis of current practice to inform the content of the certifying examination and thereby the training program curricula [36].

GENOMICS AND THE MEDICAL EDUCATION LANDSCAPE

Any effort to address the educational needs of health practitioners to obtain competency in genomic medicine must take account of the context for medical education. This is critical to ensure that approaches are pragmatic and sustainable in an era when the demands on the medical curriculum are already straining the system. In this section we will focus on physician education in the United States, considering the vector from premedical education, through medical school, residency, and postgraduate education.

Premedical Education

In the United States, students must complete a 4-year college degree before matriculating into medical school. The system is different in Europe and many other parts of the world, where medical education begins immediately after completion of secondary school. The latter system offers the opportunity to tailor the curriculum, but training may occur over 6 years rather than 8 years, offering less time to incorporate genomics into the curriculum. In the US system, the medical profession can exert indirect control over the undergraduate curriculum, through admissions standards and the Medical College Admission Test (MCAT).

Medical schools in the United States expect a student to have completed a broad-based undergraduate education, although most require some emphasis on scientific content, especially chemistry and biology, as well as exposure to mathematics. Most do not have explicit requirements to complete course work in genetics or genomics, although there may be an expectation that some genetics would have been included in biology courses. The MCAT has recently been revised [37] and includes coverage of genetics as a fundamental concept. In 2009, the Association of American Medical Colleges (AAMC) and the Howard Hughes Medical Institute (HHMI) published a document, "Scientific Foundations for Future Physicians," which included scientific competencies both for premedical and medical education [38]. Principles of genetics as well as DNA structure and function were included as competencies for the premedical curriculum.

Medical School

Medical schools have emphasized scientific principles underlying medical practice since the Flexner report resulted in the reform of US medical education in 1910 [39].

The Carnegie Foundation for the Advancement of Teaching, which commissioned the Flexner report, conducted a second analysis 100 years later [40]. This report emphasized the need to individualize approaches to learning and to instill in students skills that promote lifelong learning. These are critical to success in a rapidly evolving area such as genomic medicine, although the report does not single out this area.

U.S. Medical schools are accredited under the auspices of the Liaison Committee on Medical Education (LCME). The LCME does not prescribe the details of the curriculum, however. It does include genetics among basic sciences that should be covered, but the details of how to do so are left to the individual schools. The previously mentioned AAMC/HHMI report did include principles of genetics as a competency to be achieved by medical students, although this document only provided very high-level descriptions of the various competencies in basic science, including genetics. The APHMG, in contrast, has published detailed curricular guidelines for genetics education in medical schools (http://media.wix.com/ugd/3a7b87_7064376a9eb346cfa1b85bc2f137c48f.pdf), although the degree to which these are followed is not known.

One force for standardization of the medical curriculum is the US Medical Licensing Examination (USMLE), administered by the National Board of Medical Examiners. The USMLE consists of three steps. The first, typically taken at the end of the second year of medical school, is focused on the basic sciences. The second is taken toward the end of the 4 years of medical school and tests clinical knowledge and clinical skills. The third, taken usually during residency, is intended to assess an individual's ability to practice without supervision. The APHMG and ACMG have conducted reviews of the USMLE content to determine coverage of genetics over the years, and have noted a gradual increase in genetics coverage from when this was first done in the 1990s [41].

Residency Education

Residency education in the United States is governed by the ACGME and certification of residency graduates by various specialty boards operating under the American Board of Medical Specialties. The residency in each specialty is governed by a residency review committee (RRC) that reviews the curriculum as well as procedures and policies, and grants accreditation to each residency program. Each RRC has published a set of milestones that define the competencies expected of residents at various stages of their training, with the expectation that these will be the basis for monitoring progress of a resident through the training experience. As noted previously, milestones for medical genetics have been identified based on the ACMG competencies document.

Postgraduate Education

The mainstay of postgraduate education has been continuing medical education (CME). CME programs are overseen by the Accreditation Council for Continuing Medical Education, which accredits providers of CME. Participation in CME is required by state medical licensing boards and hospital credentialing committees, although the specific content is at the discretion of the practitioner (other than an expectation that CME be in the practitioner's area of practice).

In recent years, another opportunity for postgraduate education has arisen in the form of maintenance of certification (MOC) [42]. In the past, once a physician achieved board certification in an area of specialty that certification was valid for life. Since institution of MOC, however, specialty certifications are time-limited, requiring participation in MOC to renew certification every 10 years. The MOC process includes adherence to professional standards, maintaining currency through participation in CME and other specialty-specific mechanisms, passing a recertification examination every 10 years, and ongoing assessment of performance in practice. The specific approaches to MOC differ for each specialty, but collectively the MOC process provides an opportunity to include new content, including genomic medicine, at the discretion of the specialty.

APPROACHES TO MEDICAL GENOMICS EDUCATION

In this section we will briefly review examples of how genetics and genomics have been incorporated into educational programs in medical school, residency, and continuing education. This review will rely on approaches that have been described in the medical literature; there are undoubtedly other unpublished initiatives that will not be covered here.

Medical School

Among the major trends in medical education, three stand out: integration of the curriculum, problem-based learning, and hands-on teaching. All three have been the subject of genomic education initiatives.

The concept behind integrated education is that students need to recognize the connections between the basic sciences and clinical practice. In the past, the curriculum for the first 2 years of medical school was typically run by individual basic science departments; currently the curriculum is managed by a curriculum committee and there is an effort to meld basic science and clinical applications. Plunkett-Rondeau et al. [43] surveyed medical school genetics directors and found that the majority had integrated genetics into the basic science curriculum but only 26% included formal genetics teaching in the clinical years. In 75% of cases, genetics was taught as part of an integrated curriculum rather than as a stand-alone course. This represents an increase in the proportion that used an integrated approach reported in an earlier survey by Thurston et al. [44].

Some schools have included genetics as a major theme of the curriculum. One published example is the Genes to Society curriculum instituted at Johns Hopkins School of Medicine in 2009. The curriculum emphasizes the biological, environmental, and social factors that contribute to the risk of disease and its management in an individual. It was heavily influenced by the book by Johns Hopkins faculty member Barton Childs [45], which emphasized this perspective. The Double Helix curriculum at the University of Rochester School of Medicine weaves genetics through the 4 years [46]. Baylor College of Medicine has developed a genetics track curriculum for students with a special interest in genetics [47]. The curriculum provides a variety of didactic and hands-on experiences, including clinical rotations in genetics and a community-based experience providing interaction of students and patients and families with genetic disorders.

Problem-based learning is an approach to integrate basic science and clinical medicine through the use of clinical cases. Students work in groups to apply their knowledge of basic science to solve diagnostic puzzles and/or formulate a plan of management. One example is a curriculum (since revised) at Harvard Medical School that included a genetics course with a set of problem-based learning exercises [48]. The University of Chicago has utilized a case-based individual learning exercise that incorporates use of Internet resources in genetics [49]. Emory University has employed case-based exercises to prepare students to interpret genetic and genomic laboratory studies in a "virtual diagnostic laboratory" [50].

Hands-on approaches to learning put students into real or simulated exercises to enhance learning of basic science and/or clinical principles and skills. University of Alabama at Birmingham has employed a role-play exercise to expose students to genetic counseling [51]. Students were assigned to various clinical disorders (hereditary breast and ovarian cancer, fragile X syndrome, or autosomal recessive retinitis pigmentosa) and provided mock genetic counseling sessions regarding hypothetical test results. Several schools have incorporated exercises that involve students in the interpretation of genomic tests, either done on themselves or using simulated data. At Stanford, students were offered the chance to undergo personal genomic testing through a direct-to-consumer company [52,53]. Those who underwent testing demonstrated a greater increase in genomic knowledge based on pretest and posttest evaluations. A group of 19 students at Mount Sinai School of Medicine were offered whole genome sequencing free of charge and were provided instruction in interpretation of the data [54].

Surveys of the students indicated an increase in knowledge, and, with the exception of one student who received a false-positive result for a potentially life-threatening disorder, low levels of psychological stress.

Residency Education

The competency and milestone-based efforts in medical genetics residency education have already been mentioned. Riegert-Johnson et al. [55] provided an outline for incorporation of genetics into internal medicine residency education. Haspel et al. [56] discussed the importance of familiarizing pathology residents with genomics and have defined a curriculum. Genetics is mentioned in the ACGME milestones for some specialties, such as family medicine and neurology, but this does not mean that it is not included in the curricula for other specialties. Hoop et al. [57] surveyed psychiatry residency faculty and residents about the status of genetics in their training; nearly half reported a paucity of faculty prepared to teach genetics to psychiatry residents. Altshuler et al. [58] have incorporated objective structured clinical examinations (OSCEs) that focus on genetic issues into residency training. OSCEs are increasingly used in both medical student and residency education, using either real or simulated patients to enhance acquisition and testing of clinical skills.

Continuing Medical Education

Several surveys of practicing physicians have indicated an interest in genetics education, especially in areas such as the genetics of common disorders, psychosocial and counseling issues, recognizing individuals at risk for a genetic disorder, and indications for referral to a geneticist [3,59–63]. There have been two national projects aimed at education of health practitioners in genetics. The Genetics in Primary Care Initiative teamed geneticists with primary care providers in a train-the-trainer model. They developed teaching materials for both in-residency education and continuing medical education [64,65]. The American Academy of Pediatrics has developed a Genetics in Primary Care Institute (https://geneticsinprimarycare.aap.org), including a set of webinars and other educational materials to facilitate the incorporation of genetics into pediatric practice [66].

Gaff et al. [67] designed an educational approach in Victoria, British Columbia, Canada, called the *Genetics File* that was targeted at general practitioners and was based on adult learning theory. The approach consisted of creating and distributing a genetics manual and providing workshops to practitioners. Carroll et al. [68] described an interactive workshop offered to primary care providers that, based on a survey, increased self-reported knowledge of genetics and confidence in approaching genetic issues. Online learning systems may also be an effective way to reach primary care providers to offer genetics education [69].

CONCLUSIONS AND FINAL COMMENTS

Genetics and genomics are rapidly moving from having a peripheral role in medical practice, important mostly in the care of rare disorders, to an increasingly central role. It is widely recognized that health providers will need to become more literate in the use of genetics and genomics if the full promise of genomic medicine is to be realized. Achieving this will require a multifaceted approach that engages heath professionals along the continuum from undergraduate education through postgraduate continuing education. In spite of what might seem to be a daunting task, however, I will end this chapter with a note of optimism. This is not the first time in recent history that health providers have been presented with a need to acquire skills in a new area. It has happened before, with the advent of antibiotic therapy, with advanced imaging, and with a constant influx of new medications, to name a few instances. It is especially challenging to mount a major educational initiative for changes that are anticipated but have not yet arrived; nevertheless when new approaches prove their value, the health profession has shown that it can adapt and learn what is needed to advance care into new areas.

REFERENCES

[1] Hayflick SJ, Eiff MP, Carpenter L, Steinberger J. Primary care physicians' utilization and perceptions of genetics services. Genet Med November 1998;1(1):13–21.
[2] Lapham EV, Kozma C, Weiss JO, Benkendorf JL, Wilson MA. The gap between practice and genetics education of health professionals: HuGEM survey results. Genet Med July 2000;2(4):226–31.
[3] Metcalfe S, Hurworth R, Newstead J, Robins R. Needs assessment study of genetics education for general practitioners in Australia. Genet Med March 2002;4(2):71–7.
[4] Baars MJH, Scherpbier AJJA, Schuwirth LW, Henneman L, Beemer FA, Cobben JM, et al. Deficient knowledge of genetics relevant for daily practice among medical students nearing graduation. Genet Med May 2005;7(5):295–301.
[5] Burke S, Stone A, Bedward J, Thomas H, Farndon P. A "neglected part of the curriculum" or "of limited use?" Views on genetics training by nongenetics medical trainees and implications for delivery. Genet Med February 2006;8(2):109–15.
[6] Harvey EK, Fogel CE, Peyrot M, Christensen KD, Terry SF, McInerney JD. Providers' knowledge of genetics: a survey of 5915 individuals and families with genetic conditions. Genet Med 2007;9(5):259–67. Nature Publishing Group.
[7] Watson MS, Mann MY, Lloyd-Puryear MA. Newborn screening: toward a uniform screening panel and system, executive summary. Pediatrics 2006;117.
[8] Watson MS. Current status of newborn screening: decision-making about the conditions to include in screening programs. Ment Retard Dev Disabil Res Rev 2006;12.
[9] Dougherty D, Boyle C, Watson M. Committee report: method for evaluating conditions nominated for population-based screening of newborns and children. Genet Med 2010;12.
[10] Ethical issues in DNA sequencing in the neonate. Clin Perinatol December 2014;41(4):993–1000.

[11] Ulm E, Feero WG, Dineen R, Charrow J, Wicklund C. Genetics professionals' opinions of whole-genome sequencing in the newborn period. J Genet Couns October 28, 2014;24.

[12] Newborn screening by whole-genome sequencing: ready for prime time? Clin Chem September 2014:1243–4.

[13] Parents are interested in newborn genomic testing during the early postpartum period. Genet Med December 4, 2014.

[14] Howell RR, Engelson G. Structures for clinical follow-up: newborn screening. J Inherit Metab Dis August 10, 2007;30(4):600–5.

[15] Grody WW, Thompson BH, Gregg AR, Bean LH, Monaghan KG, Schneider A, et al. ACMG position statement on prenatal/preconception expanded carrier screening. Genet Med 2013;15(6):482–3. Nature Publishing Group.

[16] Teng K, Acheson LS. Genomics in primary care practice. Prim Care June 2014;41(2):421–35.

[17] Rubinstein WS, Acheson LS, O'Neill SM, Ruffin MT, Wang C, Beaumont JL, et al. Clinical utility of family history for cancer screening and referral in primary care: a report from the Family Healthcare Impact Trial. Genet Med November 2011;13(11):956–65.

[18] Berg AO, Baird MA, Botkin JR, Driscoll DA, Fishman PA, Guarino PD, et al. National Institutes of Health State-of-the-Science Conference statement: family history and improving health. Ann Intern Med December 15, 2009;151(12):872–7.

[19] Carmona RH, Wattendorf DJ. Personalizing prevention: the U.S. Surgeon General's family history initiative. Am Fam Physician January 1, 2005;71(1):36–9.

[20] Welter D, MacArthur J, Morales J, Burdett T, Hall P, Junkins H, et al. The NHGRI GWAS catalog, a curated resource of SNP-trait associations. Nucleic Acids Res January 2014;42(Database issue):D1001–6. Oxford University Press.

[21] Manolio TA, Collins FS, Cox NJ, Goldstein DB, Hindorff LA, Hunter DJ, et al. Finding the missing heritability of complex diseases. Nature October 10, 2009;461(7265):747–53. England.

[22] Yang Y, Muzny DM, Reid JG, Bainbridge MN, Willis A, Ward PA, et al. Clinical whole-exome sequencing for the diagnosis of mendelian disorders. N Engl J Med October 17, 2013;369(16):1502–11.

[23] Wang L, Wheeler DA. Genomic sequencing for cancer diagnosis and therapy. Annu Rev January 14, 2014;65(1):33–48. http://dx.doi.org/10.1146/annurev-med-120811-171056.

[24] Ramsey BW, Davies J, McElvaney NG, Tullis E, Bell SC, Dřevínek P, et al. A CFTR potentiator in patients with cystic fibrosis and the *G551D* mutation. N Engl J Med November 3, 2011;365(18):1663–72. United States.

[25] Hoffman EP, Connor EM. Orphan drug development in muscular dystrophy: update on two large clinical trials of dystrophin rescue therapies. Discov Med November 2013;16(89):233–9.

[26] Klein AD, Futerman AH. Lysosomal storage disorders: old diseases, present and future challenges. Pediatr Endocrinol Rev November 2013;11(Suppl. 1):59–63.

[27] Scott SA. Personalizing medicine with clinical pharmacogenetics. Genet Med December 2011;13(12):987–95. United States.

[28] Haga SB, Burke W, Ginsburg GS, Mills R, Agans R. Primary care physicians' knowledge of and experience with pharmacogenetic testing. Clin Genet October 2012;82(4):388–94. Blackwell Publishing Ltd.

[29] Jenkins J, Blitzer M, Boehm K, Feetham S, Gettig E, Johnson A, et al. Recommendations of core competencies in genetics essential for all health professionals. Genet Med April 2001;3(2):155–9.

[30] Challen K, Harris HJ, Julian-Reynier C, Kate Ten LP, Kristoffersson U, Nippert I, et al. Genetic education and nongenetic health professionals: educational providers and curricula in Europe. Genet Med May 2005;7(5):302–10.

[31] Skirton H, Lewis C, Kent A, Coviello DA. Members of Eurogentest Unit 6 and ESHG Education Committee. Genetic education and the challenge of genomic medicine: development of core competences to support preparation of health professionals in Europe. EJHG September 2010;18(9):972–7.

[32] Manolio TA, Murray MF. Inter-Society Coordinating Committee for Practitioner Education in Genomics: the growing role of professional societies in educating clinicians in genomics. Genet Med August 2014;16(8):571–2.

[33] Korf BR, Berry AB, Limson M, Marian AJ, Murray MF, O'Rourke PP, et al. Framework for development of physician competencies in genomic medicine: report of the competencies working group of the Inter-Society Coordinating Committee for Physician Education in Genomics. Genet Med April 24, 2014;16.

[34] Korf BR, Irons M, Watson MS. Competencies for the physician medical geneticist in the 21st century. Genet Med November 2011;13(11):911–2. United States.

[35] Korf BR. The medical genetics residency milestones. J Graduate Med Educ March 2014;6(1 Suppl. 1):87–90.

[36] Hampel H, Grubs RE, Walton CS, Nguyen E, Breidenbach DH, Nettles S, et al. Genetic counseling practice analysis. J Genet Couns June 2009;18(3):205–16. Springer US.

[37] Kirch DG, Mitchell K, Ast C. The new 2015 MCAT: testing competencies. JAMA December 4, 2013;310(21):2243–4. American Medical Association.

[38] Committee A-H. Scientific foundations for future physicians. 2009. Washington.

[39] Flexner A. The Flexner report on medical education in the United States and Canada. New York: Carnegie Foundation; 1910.

[40] Cooke M, Irby D, O'Brien BC. Educating physicians: A call for reform of Medical School and Residency. 2010.

[41] Desnick RJ, Korf B, Blitzer MG, Saul R. Summary of the Association of Professors of Human and Medical Genetics Fourth Annual Workshop. Am J Med Genet 2000:169–72.

[42] Miller SH. American Board of Medical Specialties and repositioning for excellence in lifelong learning: maintenance of certification. J Contin Educ Health Prof June 1, 2005;25(3):151–6. Wiley Subscription Services, Inc., A Wiley Company.

[43] Plunkett-Rondeau J, Hyland K, Dasgupta S. Training future physicians in the era of genomic medicine: trends in undergraduate medical genetics education. Genet Med February 12, 2015;17.

[44] Thurston VC, Wales PS, Bell MA, Torbeck L, Brokaw JJ. The current status of medical genetics instruction in US and Canadian medical schools. Acad Med May 2007;82(5):441–5.

[45] Childs B. Genetic medicine: A logic of disease. Baltimore: Johns Hopkins Press; 2003.

[46] Robinson DM, Fong C-T. Genetics in medical school curriculum: a look at the University of Rochester school of medicine and dentistry. J Zhejiang Univ Sci B January 2008;9(1):10–5. Zhejiang University Press.

[47] Dhar SU, Alford RL, Nelson EA, Potocki L. Enhancing exposure to genetics and genomics through an innovative medical school curriculum. Genet Med January 2012;14(1):163–7.

[48] Korf BR. Integration of genetics into clinical teaching in medical school education. Genet Med 2002;4(6 Suppl):33S–8S.

[49] Waggoner DJ, Martin CL. Integration of internet-based genetic databases into the medical school pre-clinical and clinical curriculum. Genet Med June 2006;8(6):379–82.

[50] Bean LJ, Fridovich-Keil J, Hegde M, Rudd MK, Garber KB. The virtual diagnostic laboratory: a new way of teaching undergraduate medical students about genetic testing. Genet Med November 2011;13(11):973–7.

[51] McIlvried DE, Prucka SK, Herbst M, Barger C, Robin NH. The use of role-play to enhance medical student understanding of genetic counseling. Genet Med October 2008;10(10):739–44.

[52] Salari K, Karczewski KJ, Hudgins L, Ormond KE. Szecsi PB, editor. Evidence that personal genome testing enhances student learning in a course on genomics and personalized medicine. PLoS One 2013;8(7):e68853.

[53] Vernez SL, Salari K, Ormond KE, Lee SS-J. Personal genome testing in medical education: student experiences with genotyping in the classroom. Genome Med 2013;5(3):24. BioMed Central Ltd.

[54] Sanderson SC, Linderman MD, Zinberg R, Bashir A, Kasarskis A, Zweig M, et al. How do students react to analyzing their own genomes in a whole-genome sequencing course? Outcomes of a longitudinal cohort study. Genet Med January 29, 2015;17.

[55] Riegert-Johnson DL, Korf BR, Alford RL, Broder MI, Keats BJ, Ormond KE, et al. Outline of a medical genetics curriculum for internal medicine residency training programs. Genet Med 2004;6(6):543–7.

[56] Haspel RL, Arnaout R, Briere L, Kantarci S, Marchand K, Tonellato P, et al. A call to action: training pathology residents in genomics and personalized medicine. Am J Clin Pathol June 2010;133(6):832–4. American Society for Clinical Pathology.

[57] Hoop JG, Savla G, Roberts LW, Zisook S, Dunn LB. The current state of genetics training in psychiatric residency: views of 235 U.S. educators and trainees. Acad Psychiatry March 2010;34(2):109–14. Springer-Verlag.

[58] Altshuler L, Kachur E, Krinshpun S, Sullivan D. Genetics objective structured clinical exams at the Maimonides Infants & Children's Hospital of Brooklyn, New York. Acad Med November 2008;83(11):1088–93.

[59] Calefato JM, Nippert I, Harris HJ, Kristoffersson U, Schmidtke J, Kate Ten LP, et al. Assessing educational priorities in genetics for general practitioners and specialists in five countries: factor structure of the Genetic-Educational Priorities (Gen-EP) scale. Genet Med. 2008 ed. February 2008;10(2):99–106.

[60] Julian-Reynier C, Nippert I, Calefato JM, Harris H, Kristoffersson U, Schmidtke J, et al. Genetics in clinical practice: general practitioners' educational priorities in European countries. Genet Med. 2008 ed. February 2008;10(2):107–13.

[61] Houwink EJ, van Luijk SJ, Henneman L, van der Vleuten C, Dinant GJ, Cornel MC. Genetic educational needs and the role of genetics in primary care: a focus group study with multiple perspectives. BMC Fam Pract February 16, 2011;12(1):5. BioMed Central Ltd.

[62] Houwink EJF, Henneman L, Westerneng M, van Luijk SJ, Cornel MC, Dinant JG, et al. Prioritization of future genetics education for general practitioners: a Delphi study. Genet Med March 2012;14(3):323–9.

[63] Li J, Xu T, Yashar BM. Genetics educational needs in China: physicians' experience and knowledge of genetic testing. Genet Med December 11, 2014;17.

[64] Burke W, Acheson L, Botkin J, Bridges K, Davis A, Evans J, et al. Genetics in primary care: a USA faculty development initiative. Community Genet November 30, 2001;5(2):138–46.

[65] Laberge AM, Fryer-Edwards K, Kyler P. Long-term outcomes of the "Genetics in Primary Care" faculty development initiative. Fam Med 2009;41.

[66] Saul RA. Genetic and genomic literacy in pediatric primary care. Am Acad Pediatr 2013:S198–202.

[67] Gaff CL, Aitken M, Flouris A, Metcalfe SA. A model for the development of genetics education programs for health professionals. Genet Med July 2007;9(7):451–7.

[68] Carroll JC, Rideout AL, Wilson BJ, Allanson JM, Blaine SM, Esplen MJ, et al. Genetic education for primary care providers: improving attitudes, knowledge, and confidence. Can Fam Physician December 2009;55(12):e92–9. The College of Family Physicians of Canada.

[69] Houwink EJF, van Teeffelen SR, Muijtjens AMM, Henneman L, Jacobi F, van Luijk SJ, et al. Sustained effects of online genetics education: a randomized controlled trial on oncogenetics. EJHG March 2014;22(3):310–6.

Chapter 19

Genomics, New Drug Development, and Precision Medicines

D. McHale[1], M. Penny[2]

[1]UCB, Braine L'Alleud, Belgium; [2]Biogen, Cambridge, MA, United States

Chapter Outline

Introduction	**247**	**Applying Pharmacogenetics to Drug Development**	**253**
The Drug Discovery and Development Process	248	Pharmacodynamic Variability	253
Preclinical Testing	248	Pharmacokinetic Variability	254
Clinical Development	249	**Predicting Safety**	**255**
Phase I	249	Predicting Type B Adverse Events	255
Phase II	249	Predicting Type A Adverse Events	255
Phase III	250	Individualized Therapy: An Integrated Response	256
Phase IV	250	Improving Disease Classification: Stratified Medicines	256
Applying Genomics to Drug Discovery	**250**	Adverse Drug Reactions	257
Choosing the Best Drug Targets	250	**Summary**	**257**
Complex Trait Genetics	251	**References**	**258**
Single Gene Disorders/Traits	251		
Drug-Specific Targets Approach	252		
Effect of Genetic Variation on Compound Screening	252		

INTRODUCTION

Pharmaceutical companies have historically focused their drug discovery and development programs on finding therapies for broad use in large disease populations, the so-called "blockbuster business model." A blockbuster drug is usually defined as one with peak annual sales of greater than $1 billion and is generally developed for long-term use to treat common complex chronic disorders in the general population. The strategy to identify and develop blockbuster drugs has been the response to the high cost of drug discovery and development. A survey of the drug development costs of 68 new compounds from 10 pharmaceutical companies estimated that the cost to develop a new drug in 2000 was $802 million [1]. The high cost of developing drugs can be attributed to two main factors: the large size and duration of the clinical trials required to provide the data to show safety and efficacy of the compound, and the high rate of attrition of compounds in clinical development. Fewer than 10% of compounds entering phase I clinical development reach the market, the majority failing in clinical development because lack of efficacy in phase II. This

lack of recent research and development success in finding blockbuster drugs combined with financial pressure caused by patent expiration and downward pressure on pricing has led to a shift in strategy for many companies within the biopharmaceutical industry. Companies are shifting toward the discovery and development of stratified medicines. A stratified medicine is one that is targeted at a subgroup of a traditionally classified disease; eg, trastuzumab (Herceptin) for the treatment of *Her2* overexpressing breast cancer. Stratified medicines offer significant opportunity to the industry because they have an increased probability of success and the potential of smaller programs, to the regulators as the benefit-risk profiles of these medications are greater than unselected medications, to the payers because they are more cost-effective, and most importantly to patients because they are more effective and safer therapies. Genomics has a large role to play in the development of stratified medicines because many of the tools used to stratify the patient populations are genomic; eg, epidermal growth factor receptor (*EGFR*) mutation status and gefitinib, *KRAS* mutation status and cetuximab (Erbitux) and panitumumab (Vectibix), *ALK4* mutation status and crizotinib.

Pharmacogenomics, the investigation of variations of DNA and RNA characteristics (germline or tumor) as related to drug response in individual patients or groups of patients, is one of a number of initiatives employed by the pharmaceutical industry to stratify patient populations.

A major cause of the attrition of drugs for lack of efficacy is the heterogeneity of the diseases we currently classify as single entities. Most would be better referred to as *syndromes* rather than single diseases. The disease classification currently used is based on phenotypic consequences of disease processes rather than the underlying pathological mechanisms. This has led to the clustering of heterogeneous disease syndromes based on symptoms rather than based on molecular pathology. Genomics will be an important tool in reclassifying diseases into a new molecular taxonomy of human disease. Oncology is one therapeutic area where this is most advanced because the scientific evidence base for tumor etiology is more advanced than in other area. The majority of drug development programs in oncology are now stratifying patient populations based on molecular changes in the tumor. It is widely expected that this approach will expand across other therapeutic areas as our understanding of disease biology improves.

The Drug Discovery and Development Process

The generation of an idea that a particular protein might be a suitable therapeutic target for the treatment of a disease sets in motion what is often depicted as a linear process known as the drug discovery and development pipeline, in which new medicines follow a set route from early discovery and preclinical stages through a set of clinical development processes to the marketplace. In reality the process is generally far from linear, but for the purposes of describing the component parts we will consider it a sequential process.

The ultimate aim of the drug discovery process is to find a chemical (eg, small molecule) or biological reagent, such as an antibody, which has the potential to be a drug that can be moved into preclinical and then clinical testing. In order to start the process of identifying a potential drug, a biological assay testing interactions with the drug target must be developed. This assay is generally based on a cloned and expressed form of the drug target and will be converted into a format that will allow high-throughput testing, as millions of chemicals may need to be screened in the assay. The need to screen millions of chemicals means that it is usually only feasible to screen one protein variant of the target in the high-throughput screen. It is therefore vital to screen the "right" variant. In the situation where there may be more than one form of the protein that can be included in the screen, it is important to know that the most biologically relevant and/or the most

common variant is screened, and it may be necessary to screen the chemical matter against more than one form of the protein. This is not always the most common form of the protein; for example, verumafenib, a novel drug for the treatment of malignant melanoma, was identified by specifically screening against the V600E mutated form of the *BRAF* protein to ensure it only blocked signaling of the pathogenic form.

The high-throughput screens generally identify several potential "hits," which need to be tested in more rigorous biological assays to determine the type of interaction and the effects. Promising "leads" are then developed by a series of minor chemical changes to the original lead, and the final candidate is chosen based on the selectivity and potency criteria required for the drug candidate. This candidate is then taken forward into preclinical testing.

The final testing phase is usually based on in vivo testing of the compound in animal models that are demonstrated to have some translatability to the target human disease or a range of ex vivo models of human tissue recapitulating components of the disease. The predictability and translatability of these models to humans varies with different diseases and is the focus of biomedical research in many therapeutic areas.

Preclinical Testing

Once a drug candidate has been made, it goes into a range of preclinical toxicology testing that includes in vitro screening tests to identify potential pharmacological effects at other receptors that could lead to adverse events, and genetic toxicology testing, which evaluates mutagenicity and pathogenicity. Only if these are satisfactory does animal testing begin. The animal testing is done in two species and is staged to ensure that as few animals as possible are used and that major problems are picked up early. Toxicology studies to evaluate long-term exposure, reproductive toxicological effects, juvenile toxicity, and carcinogenicity are generally only performed once the data have been obtained from short-term human studies that support safety and efficacy. To date, toxicology induced by new chemicals is identified and classified by standard phenotypic and histological changes. Although this picks up the majority of potential toxic effects, it can be insensitive to subtle changes and can identify species-specific effects that can be difficult to interpret. A greater understanding of the molecular changes after drug administration could identify more subtle effects and species-specific effects. The applicability of animal models of disease could also be assessed by evaluating molecular changes rather than probably misleading phenotypic similarities. Adverse events can be caused by unexpected consequences of the primary pharmacology or by unexpected interactions with off-target proteins. Understanding the mechanism of the toxicological effects is important, because this allows a more quantitative evaluation of the risk of the event happening in humans. Genomics can be used to identify interactions with off-target proteins, because transcription changes induced in the organ damaged by the compound can point to the mechanism of the toxicity. This is often referred to as *toxicogenomics*. Multiple consortia, eg, the Predictive Safety Testing Consortium and SafeSciMET, are working to identify genomic biomarkers that are more sensitive than current histopathological scores, allowing early detection of toxicology and the demonstration of species-specific toxic effects. Similarly where specific organ toxicity is expected because of the mechanism of action of the compound or known off-target effects, then transcription changes can offer a more sensitive assay to detect early organ damage.

Clinical Development

Once the initial in vitro testing and acute animal toxicology studies (generally 14 days) have been performed, then it is possible to start testing the candidate drug in humans. The human studies have traditionally been split into four phases (I–IV), each with specific aims (Box 19.1).

Phase I

The first time a novel compound (or biological therapy) is tested in humans, a broad range of dosages and dosing strategies are tested starting at very low exposures to minimize any risks to the clinical trial participants. Although these initial studies have generally been performed on healthy volunteers, there is an increasing trend toward incorporating patients as early as possible. The dose is escalated over several weeks starting at between 10- and 100-fold below the expected pharmacological exposure levels to a maximum tolerable level or several-fold beyond the expected maximum clinical dosage (whichever is reached sooner). The aim is to identify common adverse events and the relationship with plasma exposure as well as to establish the basic pharmacokinetic (PK) parameters. As drug development continues, more studies are performed to understand the effects of multiple dosing, specific drug–drug interactions, and food effects. The aim of these studies is to provide a more comprehensive understanding of the PKs and

significant causes of variability in PK profiles. Collections of pharmacogenomic samples in phase I clinical protocols allow the assessment of the impact of genetic variation on drug metabolism and transport.

There is a growing trend for performing some of these very early studies in patients and these are often referred to as *phase Ib studies*. The primary intent of these studies is still to establish safety and PKs of the compound, but the use of patients allows early indicators of target engagement and biomarkers of efficacy to provide evidence that the compound is modulating the proposed mechanism. Where it is possible to biopsy disease tissue in these studies, then white cell transcription analysis can provide some evidence that the target pathway is being modulated. This is generally restricted to some tumor types and dermatological conditions; eg, psoriasis where it is possible to obtain high-quality tissue samples.

Phase II

Phase II is traditionally divided into phase IIa, where the aim is to demonstrate the safety and PK parameters in patients, and IIb, where the aim is to establish efficacy and delineate the dose–response curve. However, most companies now endeavor to generate some biomarker data in the phase IIa studies to provide some evidence of efficacy and confidence to progress into the more expensive and larger phase IIb dose ranging study. This is a critical time, because up to 50% of all drug candidates will fail in phase II. If preclinical data or data from translational medicine studies have identified a patient population more likely to respond to the mechanism, eg, *BRAF* activating mutation-positive melanoma tumors for MEK inhibitors, then the studies can be restricted to this patient population to increase the likelihood of seeing an efficacy signal. Even when there is no strong a priori hypothesis then samples collected in phase II studies for pharmacogenomic analysis are useful for testing less validated hypotheses on the impact of genetic variation with respect to drug response, particularly for genes with large effects, because these studies are limited in that they comprise relatively small numbers of patients (50–100). Samples for these pharmacogenomic studies may be collected with specific consent for genotyping of named genes within the protocol, which can be correlated with clinical data collected in the trial.

Box 19.1 Human Studies Have Traditionally Been Split into Four Phases (Phases I–IV)

- Phase I: pharmacokinetic (PK) and safety profiles in healthy volunteers
- Phase II: safety and efficacy in patients and the establishment of the dose response
- Phase III: safety and efficacy at the chosen dosage
- Phase IV: postapproval studies to answer specific safety or efficacy questions and to support commercial strategies

Phase III

Phase III trials form the basis of the regulatory approval, and they are large studies evaluating the safety and efficacy of the candidate at the clinical dose and in the population where the drug will ultimately be used. The cost of this phase of development is significantly more than the others and so failure at this point has a major impact on the company. The larger numbers of patients included in these studies provide more power for pharmacogenomic analysis. In addition, these samples also provide a useful resource for more disease-focused phenotype–genotype correlations, and often samples are collected with broad consent for genotyping that allows the investigation of many candidate genes.

The patient population studies in the phase III program form the basis of the population approved to use the drug once it is launched. Therefore if a genetically defined patient population is used in these studies, then the drug will only be approved for use in that group of patients. However, it is often necessary to include at least one study where all patient groups are included to ensure that there is not an unexpected benefit in the nonselected population and also to provide a safety database for that group should they be prescribed the drug once it is approved. The inclusion of a prospectively stratified "all comers" strategy also allows a more robust evaluation of the positive and negative predictive value of the test and, importantly, differentiation between a predictive pharmacogenomics test where the test identifies subjects who differentially respond to the drug from a prognostic test where the test-differentiated subjects have a more severe prognosis from the disease regardless of treatment paradigm.

Phase IV

Drug testing does not stop with regulatory approval, and phase IV studies are run after the drug has been approved. Sometimes there are clinical studies required by regulatory authorities as a postapproval commitment. These generally test a specific question around safety and efficacy or are used to generate data to support commercial strategies. Studies conducted after the regulatory approval of the drug represent an excellent resource for the implementation of a pharmacogenomics strategy because of the availability of larger sample sets. The potential to collect genomic samples from thousands of individuals recruited into large phase IV clinical studies presents the opportunity to link genomic data to quality clinical data, biomarker data, and, in many cases, long-term follow-up monitoring. An area where postmarket pharmacogenomic surveillance can have a great impact is in addressing safety issues. The availability of large numbers of patients on active treatments not only provides the material to look for pharmacogenomic effects but is also a valuable resource for understanding the molecular basis for

disease, which in turn feeds back into the idea generation in the early discovery section of the pipeline.

The studies performed within drug development programs are still classified according to this system, but, increasingly, companies are looking to generate potential signals of efficacy data in the early phase I and IIa studies (sometimes called the *learn phase*) to provide confidence that the compound will work before investing in the more expensive phase IIb and III studies (sometimes called the *confirm phase*).

APPLYING GENOMICS TO DRUG DISCOVERY

Choosing the Best Drug Targets

One key area where genetics has impacted the drug discovery and development process is in target selection. A significant number of compounds fail in development because the target and hence mechanism of action of the drug is not linked to the pathogenesis of the disease to which they are directed. Taking the view that the more you know about a drug target early in the discovery process, the less likely it is to fail in development caused by lack of confidence in rationale (CIR), many companies are now investing up front in understanding the molecular genetics of the complex diseases we treat and using genetics to identify novel targets and prioritize target selection from candidate gene lists for drug development programs. The advances in DNA sequencing, bioinformatics, and genetic analysis are offering great opportunity to use human genetics to identify novel targets.

Before 1990, pharmaceutical companies had worked on approximately 500 potential drug targets with around 100 of these mechanisms having produced marketed drugs [2]. Initial analysis of the final draft of the Human Genome Project suggested that the total number of drug targets with small chemicals might increase to 5000 [3]. However, not all of these targets will be relevant to disease and therefore current estimates are that there are 600–1500 drug targets in the human genome [2]. This expansion of potential targets in concert with the rising cost of drug development means that the choice of targets is increasingly important.

Given the length of time it takes to get from an idea to a compound to the market, there are few prospective examples of marketed compounds where genomics has provided a new drug target or supported its initial CIR, and thus there is insufficient data to show that having genetic or genomic CIR from complex traits has significantly increased candidate survival in the drug development pipeline. Human genetics is a simple and effective way of beginning to assess the molecular evidence and provide the CIR for establishing a drug development program for a particular target. It is possible to retrospectively identify positive genetic associations between drug targets and incidence or severity of disease for drugs that are widely prescribed; for example, angiotensin-converting enzyme inhibitors and hypertension [4,5], β-agonists and asthma [6,7], and serotonin reuptake inhibitors and depression [8,9]. However, this is not always the case, because the proton-pump inhibitors, used to treat gastroesophageal reflux disease (GERD) are one of the most commonly prescribed classes of drugs worldwide, but very little is known about the molecular genetics of GERD and there is no reported association between the genes encoding the α and β subunits of the drug target hydrogen/potassium adenosine triphosphatase (ATPase) and the disease [10]. Knockout mouse data also provides evidence relevant to the function of target on the phenotype [11]. The CIR for the statins, one of the most successful drug classes to be developed for the lowering of low-density lipoprotein cholesterol, was derived from biochemistry; the 3-hydroxy-3-methylglutaryl–coenzyme A (HMG-CoA) reductase knockout mouse is lethal, and there are very few published genetic association studies on HMG-CoA reductase [12].

Complex Trait Genetics

The ability to carry out large-scale whole genome studies in well-characterized populations extends the candidate gene approach, and has increased the potential to identify novel targets and new pathways that are relevant to disease. Linkage studies have had some success in identifying genetic variants associated with complex diseases; examples include phosphodiesterase 4D (*PDE4D*) and stroke [13], organic cation transporter (OCTN) and discs large homologue 5 (*Drosophila*) (*Dlg5*) genes with inflammatory bowel disease, [14,15] and 5-lipoxygenase-activating protein (*FLAP*) and myocardial infarction and stroke [16]. These studies have provided some supporting evidence for the link between potential drug targets and disease, but rarely as the only evidence supporting this link. This is owing to the fact that the reproducibility of early genetic association studies was poor with many false-positives reported, the identification of the causative variant is often challenging, and hence the prediction of whether the genetic variant is causing an increase or decrease in protein function can be a challenge. Three major advances have occurred in the last decade that has positively impacted the use of complex trait genetics. The first was the publication of the Wellcome Trust Case Control Consortium, which clearly demonstrated the need for larger sample sizes and rigorous quality control procedures [17]. The second has been the rapid development of DNA sequencing, which in 2013 is reaching a point where it is possible to sequence large cohorts of subjects, allowing the evaluation of rare variants as well as the common variants covered by the whole genome association studies [18]. The final advance is that the development of

bioinformatics and genetic analysis is allowing the combining of the genetic variations into pathway maps looking for dysregulated pathways rather than just individual single nucleotide proteins (SNPs) [19]. This allows for the identification of optimal intervention points in pathways and the design of functional experiments, confirming the direction of the dysregulation and hence whether an agonist or antagonist approach is required.

Single Gene Disorders/Traits

One key approach to increase the predictivity of genetic data is to use rare genetic syndromes to identify drug targets with high confidence that pharmacological approaches will mimic the human phenotype seen in the family. The last 5 years have seen the first cohort of drugs reach approval or late-stage clinical development where human genetics either identified the target or provided significant confidence in the approach. Examples of these drugs are included in Table 19.1 but include maraviroc and *CCR5* (HIV), tofacitinib and the *JAK* kinases (rheumatoid arthritis), romosozumab and sclerostin (*SOST*) (postmenopausal osteoporosis), and vemurafenib and *BRAF* (melanoma).

The identification of *CCR5* as a potential therapeutic target for HIV infection came from the identification of *CCR5* and the coreceptor for HIV and a genetic study of individuals, who, despite multiple high-risk exposures, did not become infected with the virus. The association between a common mutation in the gene encoding CCR5 that resulted in a nonfunctional protein and resistance to HIV infection identified the *CCR5* receptor as a coreceptor used by HIV to infect cells in the majority of primary infections. Individuals who were homozygous for this mutation (*CCR5Δ32*) and therefore had no functional CCR5 protein

TABLE 19.1 Drugs with Targets Defined by Human Genetics

Drug	Gene	Phenotype	Indication
Maraviroc	*CCR5*	HIV resistance	HIV
Tofacitinib	*JAK 3*	Severe combined immunodeficiency	Rheumatoid arthritis
Romosozumab	Sclerostin	Sclerosteosis	Osteoporosis
Clopidogrel	*P2yR*	Congenital bleeding	Ischemic heart disease
Alirocumab	*PCSK9*	Hypercholesterolemia	Ischemic heart disease
In development	*Nav 1.7*	Insensitivity to pain	Pain

were apparently healthy and resistant to infection by HIV [20]. Subsequent candidate gene studies have shown that the *CCR5Δ32* mutation is associated with slower progression to AIDS [21]. Recent data have shown that a genetic polymorphism in the promoter of the *CCR5* gene, resulting in increased CCR5 expression, is more common in individuals rapidly progressing to AIDS [22]. Thus within 7 years of the publication of genetic evidence that *CCR5* would be a valid target in HIV therapy, clinical validation of this drug target was achieved, with both Pfizer and Schering-Plough publishing data showing significant viral load drops in patients with HIV infection treated with the potent CCR5 antagonists maraviroc and Schering C (vicriviroc), respectively [23].

The discovery of Janus kinase (JAK) and the identification of causative mutations in the *Jak3* gene and severe combined immunodeficiency (SCID) highlighted the key role of this target in cytokine signaling and lymphocyte development and function and provided CIR for the development of a selective *Jak3* antagonist for the treatment of rejection in renal transplantation and rheumatoid arthritis. As with CCR5 above, the fact that individuals with the mutations only have the very specific effects of immunodeficiency and no other apparent deleterious phenotype means that these genetic data also provide confidence in safety (CIS) for the therapeutic approach [24].

Sclerosteosis is a rare genetic condition seen in only a small number of families in the world. A key aspect of the disease phenotype of sclerosteosis is bone overgrowth. This bone overgrowth is seen in the heterozygotes when they have generalized increase in bone density and mass, and the homozygotes when they have increased bone growth and density that can lead to nerve entrapment syndromes causing deafness and visual problems. The gene for sclerosteosis was identified in 2005 and the disease is cause by the absence of a protein called *sclerostin* [25]. Sclerostin is a secreted protein which is highly amenable to a biologics approach and where reduction in circulating sclerostin will lead to increase in one density. This led to collaboration between UCB Celltech and Amgen to produce an antibody to sclerostin for the treatment of postmenopausal osteoporosis. This antibody has now been tested in phase IIb and shown to increase bone mineral density to a greater extent than current therapies.

Drug-Specific Targets Approach

An alternative strategy to the single gene and whole genome approaches is to carry out association studies in a large subset of specific drug target genes. Several companies have taken this approach to explore genetic association with as many targets as possible in many indications. Oxagen is a biopharmaceutical company specializing in understanding the genetic basis of common human diseases. One of the main areas of interest for the company is in G-protein coupled receptors (*GPCRs*); 20–30% of marketed drugs are targeted to the products of this class of genes. There are over 750 *GPCR* genes, thus Oxagen applied a filtering process to select the best targets for further analysis based on expression profiling, known biology, whether they have a known drug targeted to them, or are likely to be chemically tractable before high-throughput genetic analysis [26]. The Structural Genomics Consortium has focused on kinases (the kinome). This consortium is funded by private and public sources and focuses on the identification of crystal structures of novel kinases and then the development of chemical tools. In concert with this, there have been considerable efforts to identify kinases and their role in disease. Much of this has focused on the use of genetic mutations of kinases in cancer and genetic associations in conditions such as rheumatoid arthritis.

With the increasing use of genetics to drive target identification in well-defined patient populations comes the dilemma of knowing which of all the targets identified is the best to take forward. Many of the positive genetic associations with disease from linkage disequilibrium based whole genome association studies (LD-WGA) are likely to occur in noncoding regions of the genome and the basis for a strong association, if replicated, will be unknown. Recent data investigating noncoding parts of the genome have revealed the importance of these regions in regulating gene expression [27].

The application of whole genome technologies to understanding common complex disease has increased the number of potential targets.

Effect of Genetic Variation on Compound Screening

Regardless of the original source of the target, genetic analyses are important in understanding how to move forwards in the drug discovery process. Undertaking a comprehensive analysis of the genetic variation that exists in putative drug targets will provide information that has the potential to impact drug discovery processes downstream. In an internal study within Pfizer comparing coding SNP (cSNP) frequency, a selection of 111 genes encoding potential druggable targets and 160 genes considered to be "nondruggable" targets identified that 15% (26/111) of the putative targets were not polymorphic at the amino acid level, whereas 40% (45/111) had one or two cSNPs. There are also well-documented differences in the frequencies of specific polymorphisms between ethnic groups. Prior knowledge of any polymorphisms in a target can be incorporated into target validation, lead optimization, and inform preclinical projects supporting the development of the compound. The effect of genetic variation can be assessed through in vitro assays that incorporate a comparison of polymorphic targets either by using cells or biological reagents obtained from donors of known genotypes where available, or by site-directed mutagenesis. This will facilitate early assessment of the potential impact of genetic variation on the activity of compounds and offer the potential to choose candidates that are least likely to be influenced by the target polymorphism [28].

Gaining an early understanding of the impact of genetic variation can increase confidence in chemistry (CIC). For example, chemokine receptor 5 (*CCR5*) has been shown to be the second coreceptor required for primary HIV infection. As such, it was a very attractive drug target for the treatment of HIV, because blockade of *CCR5* should reduce HIV entry into cells and hence lower viral turnover. There have been multiple polymorphisms reported in the *CCR5* gene, and some of these have been associated with effects on HIV infection rates and/or progression from infection to AIDs. A key question that had to be asked was what the functional effects of these polymorphisms were and whether they would impact the effectiveness of the therapy. Preclinical experiments demonstrated that the predominant effect of the functional polymorphisms was to alter receptor expression rather than structure, and hence that the variability could be managed by identifying a dosage that could effectively inhibit viral entry across a wide range of receptor expression levels.

The pharmacogenomic studies included in the preclinical phase of drug discovery that provide CIR and CIC and support nomination of a candidate drug for development are not intended to replace any of the clinical studies required for exploratory drug development or predict response in patient populations. The preclinical strategy will produce data to inform the pharmacogenomic plan for compounds in exploratory and full development. The challenge facing pharmacogenomic specialists in the pharmaceutical industry is to use the available genomic data to improve the efficiency of clinical trials.

APPLYING PHARMACOGENETICS TO DRUG DEVELOPMENT

Despite relatively early introduction, pharmacogenetics (see chapter: Content and Variation of the Human Genome) remained a relatively small field for the next 40 years because of the fact that although it was well recognized that all drugs exhibited significant interindividual variability in response, the genetic tools to examine this variability were not available. Apart from a few standard approaches, for example, renal impairment studies and gender differences, there was limited investigation of this phenomenon during drug development. The approach of the drug companies and regulators alike was to ensure that all compounds had a sufficiently good therapeutic index that the average benefit significantly outweighed the potential risk. This has led to the withdrawal or termination of development of a number of compounds with good efficacy but an insufficient

population-based safety profile, which can often be driven by a small number of potentially serious adverse events. These events can be categorized into those that are expected based on an understanding of the pharmacological action of the drug (type A) and those that correlate with plasma exposure levels or idiosyncratic (type B) [29]. The mechanism of idiosyncratic reactions are generally unknown and do not have a clear dose–response relationship. There are two basic pharmacological implications of underlying genetic factors: pharmacodynamic (PD) variability (drug absorption and delivery to desired target), and PK variability (drug metabolism and excretion).

Pharmacodynamic Variability

The importance of being able to predict drug response is highlighted by the fact that it has been estimated that approximately 30% of prescriptions written do not benefit the patient and even in highly controlled environments, such as clinical trials, it is rare to get response rates significantly above 70% [30]. If we assume that subjects take the medication in the prescribed manner, then lack of efficacy may result from inadequate exposure to the drug (PK variability), an inability to respond to the therapy because of genetic variation in the target and/or downstream effectors (PD variability), or because the pharmacological intervention does not alter the underlying pathophysiological process (disease heterogeneity). Whereas some commentators have suggested that differences in disease genetics (disease heterogeneity) should be considered as separate from pharmacogenetics, at a practical level, understanding this genetic variation will result in the same outcome; for example, understanding increased or decreased likelihood of response to therapy. Therefore this group will be included in the PD variability subgroup. There are now multiple examples of the use of pharmacogenetics to predict drug response. The majority of these are in oncology where tumor mutations have been shown to drive PD response in multiple areas. The best known examples of this are trastuzumab and imatinib (Gleevec). In the case of trastuzumab, amplification of the *Her2* gene leads to upregulated Her2 expression in approximately 25% of all breast cancers. These tumors are responsive to trastuzumab, whereas tumors with very low levels of expression of Her2 do not respond. Imatinib is a treatment for Philadelphia chromosome–positive chronic myeloid leukemia (CML) specifically designed to target the BCR ABl fusion protein generated from this chromosomal translocation. It also is active in tumors with mutated *KIT* genes; eg, *GIST*. Table 19.2 contains a list of antitumor therapies aimed at genotypically defined tumors.

Vemurafenib is a very exciting example, because this compound was screened using the common V600E mutation of the *BRAF* gene. The mutation is present in 35–45% of melanoma cases. A counter screen of nonmutated *BRAF*

TABLE 19.2 Selected Drugs: Genotype–Phenotype Correlations

Drug	Indication	Gene
Imatinib	Gastrointestinal stromal tumor	*KIT*
Gefitinib	Non–small cell lung cancer	*EGFR*
Erlotinib	Non–small cell lung cancer	*EGFR*
Cetuximab	Colorectal cancer	*KRAS*
Panitumumab	Colorectal cancer	*KRAS*
Crizotinib	Non–small cell lung cancer	*Alk4*
Vemurafenib	Melanoma	*BRAF*

was also run, ensuring the identified compound was specific for the mutated allele. This drug is highly effective in V600E-positive tumors and has a very good safety profile because it does not bind to the nonmutated protein, hence only working within the tumor cells.

Although the majority of examples are in oncology, there are exemplars in other therapeutic areas as well. One of the clearest examples is in the treatment of hepatitis C subjects who have the AA polymorphism in their interferon gene have a greater chance of responding to interferon therapy. Other examples exist particularly in the rare disease in which therapies are directed at specific genetic disorders and, in this case, it is disease genetics rather than PKs. Despite the success observed over the last 5 years, most therapies tested do not appear to have a clear pharmacogenetic signature. It may be that the current approaches are unable to identify the correct genetic variation or, more likely, the combination of variants that can predict response or that genetic variation is not a major cause of the heterogeneity of drug response.

Pharmacokinetic Variability

Interindividual variation in drug metabolism is now a well-documented phenomenon, but it was not until Mahgoub et al., in 1977 [31] described the polymorphic metabolism of debrisoquine that significant interest grew in the genetic contribution. The cytochrome P450 (CYP) enzyme family protects the body from xenobiotic agents and is the major route of metabolism of many drugs [32]. Several of these enzymes (for example, 2D6, 2C9, and 2C19) are known to have functional genetic polymorphisms that result in significant reductions or increases in function [33,34]. Genetic variation in cytochrome P450 2D6 (CYP2D6) is well characterized and approximately 10% of Caucasians make no

2D6 enzyme. Experiments with the antihypertensive agent debrisoquine were the first proven examples of a pharmacogenetic effect. Debrisoquine is metabolized by the CYP2D6 enzyme. An individual who makes no 2D6 and takes a standard dose of debrisoquine will suffer a profound hypotensive event resulting from high plasma exposure levels caused by an inability to metabolize the drug [35]. Approximately 20% of all drugs are metabolized by 2D6 and subjects who are unable to make this enzyme are at increased risk of developing adverse events when taking one of these compounds [36].

Interindividual variation in drug metabolism is well documented. Approximately 20% of drugs are metabolized by the CYP2D6 enzyme [36]. The incorporation of genetic testing for CYP2D6 or related enzymes in clinical trials has the potential to identify, prospectively, subjects who are likely to have adverse events because of poor metabolism or those who may have limited response through inadequate exposure because of ultrarapid metabolism.

Many drug-metabolizing enzymes have genetic variants, leading to reduced or increased function with consequent impact on the PK variability. Despite this knowledge, there are no drugs for which pharmacogenetic tests are routinely applied, and only recently has it become accepted best practice to test for the presence of variation in the gene encoding the thiopurine methyltransferase (TPMT) enzyme before prescription of azathioprine and 6-mercaptopurine. Approximately one in 300 individuals are homozygous for mutations in the gene encoding TPMT [37]. If treated with a standard dose of azathioprine (6-mercaptopurine), these individuals have a substantially increased risk of developing the potentially fatal complication of red cell aplasia [37]. Suitable dosage reduction decreases this risk. The recent decision by the clinical pharmacology division of the US Food and Drug Administration (FDA) to recommend that subjects be tested for TPMT enzyme status (either phenotypically or genotypically) before receiving a dose of 6-mercaptopurine is evidence of the increasing awareness of the value of understanding interindividual variation in drug metabolism. Similarly, the recently approved drug atomoxetine (Strattera) from Eli Lilly provides safety data for CYP2D6-poor and CYP2D6-extensive metabolizers, and the availability of a suitable test to distinguish these two groups is also included on the label, although there is no recommendation about using the test and adjusting the dosage according to genotype.

As the clinical value of these tests becomes established and is translated into practice, so will the acceptability of requiring a metabolizing enzyme diagnostic test before dispensation of the drug. Clear demonstration of the advantages of prospectively using a diagnostic test versus clinical management of drug dosaging will also be vital if these tests are to be used in clinical practice. This will also allow the development of chemicals with narrow therapeutic windows and predominantly metabolized by a polymorphic enzyme. Many of these compounds have historically been terminated, as the risk of adverse events caused by high plasma exposures outweighed the potential benefit. A clinically acceptable way of managing this risk would make safe use of these compounds possible.

PREDICTING SAFETY

Predicting Type B Adverse Events

The last 5 years has demonstrated that pharmacogenetics can be used to predict some rare adverse events. Extreme PD adverse responses to drugs have been described in the past; eg, malignant hyperthermia and inhaled anesthetics, and succinyl choline deficiency and prolonged paralysis. More recently an immunogenetic explanation for rare hypersensitivity reactions was discovered. Abacavir (Ziagen) was a key drug in highlighting the role of *HLA* variation and drug hypersensitivity. Two retrospective studies have identified the *HLA*-B*57:01 allele of the major histocompatibility complex (*MHC*) class I B gene as a genetic determinant of hypersensitivity to abacavir [38,39]. The availability of a relatively large patient population led to the identification of the *HLA*-B*57:01-Hsp70-Hom variant haplotype in 94.4%

of cases compared with only 0.4% of controls. Analysis in different ethnic groups, however, showed that *HLA*-B*57:01 alone would not be sufficiently predictive of hypersensitivity in diverse patient populations, suggesting that other genetic determinants of hypersensitivity remain to be identified. Additional *HLA* associations with adverse drug reactions (ADRs) have been described. Chung et al., in 2004 [40] described an association between *HLA*-B*1502 and Stevens-Johnson syndrome in Han Chinese. Again, this association appears to be confined to the Han Chinese. Additional *HLA* associations and immune mediated adverse events have since been confirmed (Table 19.3).

Predicting Type A Adverse Events

ADRs are a major cause of morbidity, leading to approximately 5% of all hospital admissions, and severe ADRs are a leading cause of death in young adults. Despite initial optimism, pharmacogenetics has had limited impact in reducing this morbidity and mortality. Genetic variation can influence our risk of developing type A adverse events by either increasing our exposure to the active agent or altering the PD effects of the drug. Warfarin is one of the best understood examples of how genetic variation can influence risk of adverse events. Bleeding events

TABLE 19.3 Drug Response Modification Associated with Genetic Polymorphisms in "Disease-Modifying" or "Treatment-Modifying" Genes

Gene or Gene Product	Disease or Drug Effect	Medication	Influence of Polymorphism
Adducin	Hypertension	Diuretics	Myocardial infarction or stroke
Apolipoprotein E (*APOE*)	Atherosclerosis, ischemic cardiovascular events	Statins (simvastatin)	Enhanced survival
Apolipoprotein E (*APOE*)	Alzheimer disease	Tacrine	Clinical improvement
HLA	Toxicity	Abacavir	Hypersensitivity reaction
Cholesterol ester transfer protein (*CETP*)	Progression of atherosclerosis	Statins (pravastatin)	Slowing of atherosclerosis
Ion channels (*HERG, KvLQT1, Mink MiRP1*)	Congenital long QT syndrome	Erythromycin, cisapride, terfenadine, clarithromycin, quinidine	Increased risk of drug-induced torsade de pointes
Methylguanine methyltransferase (*MGMT*)	Glioma	Carmustine	Response of glioma
Parkin	Parkinson disease	Levodopa	Clinical improvement and levodopa-induced dyskinesias
Prothrombin and factor V	Deep-vein thrombosis and cerebral vein thrombosis	Oral contraceptives	Increased risk of deep-vein and cerebral-vein thrombosis
Stromelysin-1	Atherosclerosis progression	Statins (pravastatin)	Reduction in cardiovascular events: death, myocardial infarction, stroke, angina; reduction in risk of angioplasty

Adapted from Evans WE, McLeod HL. Pharmacogenomics: drug disposition, drug targets, and side effects. N Engl J Med February 6, 2003;348(6):538–549.

while taking warfarin is one of the most common adverse events resulting in significant morbidity. Underlying genetic variation accounts for at least 50% of the risk of developing a bleeding event. This risk is predominantly driven by two key genes: the drug-metabolizing enzyme *CYP2C19* and the gene encoding the vitamin K receptor. Studies by Lane et al. [41] have shown that poor metabolizers' status of cytochrome *CYP2C19* have a approximately four fold increase in plasma exposure of S-warfarin (the active moiety). The increase in exposure results in a five fold increase in bleeding risk caused by PK variability. The vitamin K receptor is the target for warfarin and required for the production of vitamin K–dependent clotting factors. A common variant in this receptor results in a decrease in vitamin K receptor function. Whereas this normally causes no significant sequelae, it does impact response to warfarin. Individuals who are homozygous for the rare allele have an increase in bleeding risk of approximately two fold when taking warfarin. By combining the results of these genotypes, it is possible to refine an individual's risk of developing a bleeding adverse event if they are given a standard dose of warfarin. Prospective trials are now ongoing to determine the utility of using genotype results to adjust the starting dosage of warfarin.

Individualized Therapy: An Integrated Response

In real life, the response of an individual is based on both the plasma exposure and how that affects the various physiological processes in the target organs. Evans and Relling generated a hypothetical graph representing the PK and PD variation in concert [42].

Variation in drug metabolizing enzymes can dramatically impact plasma exposure levels. However, it is not until we integrate this with variation in genes affecting PD response in the right hand column that we start to get a real understanding of the impact on response for the individual. It is important to realize that dosage-related adverse events are observed in extensive metabolizers as well as poor metabolizers, but the incidence is dependent upon the frequency of variation in the genes affecting PD response. As the frequency of variation in genes affecting PD response approaches 0.5, the predictive power of a test solely looking at drug metabolism decreases. Similarly, the predictive power of a test evaluating variation in genes impacting PD response will vary depending upon PK variability. Most published pharmacogenetic studies concentrate on single genes or small numbers of candidate genes, which are likely to impact either PK or PD variability. It is unsurprising that these studies fail to demonstrate high positive or negative predictive information for drug response that is in general caused by a combination of both of these factors. As we move forward, a more holistic approach to the examination of genetic factors impacting drug response should lead to the identification of sets of SNPs with higher predictive values, leading to improved prescribing (Table 19.1).

Improving Disease Classification: Stratified Medicines

The need to accurately and precisely characterize the disease under investigation has important implications in drug development. The current disease classification system has changed little in the last 100 years and is based on the phenotypic clustering. That is, diseases that present with similar symptoms have been classified as having the same condition. These diseases are therefore more like syndromes and do not necessarily reflect a common underlying pathology. Likewise, there may be conditions with similar pathological mechanisms that are classified as different diseases because the phenotypic features are not similar enough. A very clear example of this is in oncology, where many mechanisms are represented in subsets of organs classified tumors; eg, *EGFR* mutations are present in multiple tumor types. The knowledge from the outset of a drug discovery program that there are molecular subtypes of disease means that appropriate preclinical experiments can be developed early to predict the likelihood of a pharmacogenomic effect, and this information can be used advantageously in the drug development program. Combining genotype data with other genomic data provides valuable information related to disease subtype. Integration of genotyping data with gene expression has identified subtypes of obesity phenotypes in a mouse model [44]. Using similar approaches and including microRNA, epigenetic, proteomic, and metabonomic analyses in well-defined patient cohorts will provide powerful tools to aid the dissection of the phenotype of disease in humans in order to drive the development of targeted therapies based on molecular subclassification. This reclassification of disease has become the focus of several cross-academic/industry consortiums, and the next decade could see the development of new disease taxonomies reflecting the true molecular mechanisms of the pathology rather than the end consequences.

One therapeutic area where using genetic and genomic technologies has undoubtedly had a major and measurable impact on understanding the molecular subtypes of disease is oncology. The advances in understanding the molecular mechanisms predisposing a patient to cancer have seen the number of oncology compounds in clinical development rise from 10 to over 400 in 10 years. The majority of the new compounds now being tested are classed as targeted biotech medicines. Imatinib and trastuzumab were the first two such targeted compounds approved. Trastuzumab is a therapy targeting the *HER2/neu* receptor in breast cancer. The rationale for this therapy was based on a sound understanding of the underlying molecular pathology. It was known that only 20–30% of breast tumors overexpress this protein and it was demonstrated in the drug development program that response to trastuzumab was limited to subjects whose tumors overexpressed the target [45]. Similarly, imatinib is a therapy targeting the fusion protein product resulting from the Philadelphia chromosomal translocation observed in most cases of CML [46]. This therapy provided dramatic efficacy in cases of CML with the chromosomal translocation and was rapidly approved by the FDA.

Since the rapid approval and success of imatinib and trastuzumab, many other targeted cancer therapies have entered clinical trials, thus highlighting the absolute requirement to continue to investigate and understand the underlying molecular mechanisms that are associated with disease. Gefitinib (Iressa) was the first of class-selective EGFR inhibitors to receive accelerated approval based on preliminary data from phase II studies in non–small cell lung carcinoma (NSCLC) patients. Activating mutations and overexpression of EGFR were known to occur in many cancers, providing CIR for development of an EGFR inhibitor for cancer treatment. Inactivation of the *EFGR* gene in mice did not cause any major phenotypic effects, which, in turn provided CIS with respect to pharmacological inhibition of this target [47]. However, tumor response to treatment in the clinical trials was only observed in 9–19% of patients. Subsequent analysis to predict factors that would indicate good response to gefitinib identified that female gender, nonsmoking status, and specific histological subtype of tumor was associated with better response to therapy. Investigation of biological and markers of response failed to show an association with EGFR expression levels. However, somatic mutations in the ATP-binding site of the tyrosine kinase domain of EGFR were observed more often in the tumors of patients who responded to gefitinib. The *EGFR* mutations are located close to the putative binding site for compounds like gefitinib and lead to increased signaling in the growth factor pathway, and thus tumors harboring these mutations are more susceptible to treatment with an EGFR inhibitor [48]. This highlights the importance of defining the molecular subtypes of disease and understanding the impact on response to therapy. Had the molecular profile of NSCLC been identified before testing in humans, it may have been possible to design preclinical cell-based assays to determine whether the genetic profile of the tumor would influence response to therapy and then inform clinical trial design.

The majority of oncology programs now in development are focusing on stratified populations based on genetic or genomic classifications of tumor type.

Adverse Drug Reactions

In a recent study of ADRs, 5% of hospital admissions in the United Kingdom were identified as being the result of ADRs. Over 70% were considered avoidable, and whereas drug interactions accounted for the majority of the ADRs and older drugs were implicated in the hospital admission, there is still a need to understand the underlying causes of all ADRs [49]. It is difficult to detect rare adverse events in the confines of a clinical trial and the current system for monitoring ADRs has been suggested to be too disparate. A move to a more comprehensive epidemiological approach to monitoring drug safety has been proposed. The inclusion of pharmacogenomic analyses within this approach would allow the systematic assessment of the contribution of genetic determinants to ADRs. Pharmacogenomic surveillance in large phase IV trials of approved compounds will have a great impact in addressing safety issues.

One therapeutic area where detailed pharmacosurveillance, including pharmacogenomic analyses and postapproval, is not new is in the antiretroviral treatment of HIV infection. Viral resistance and drug toxicity are common and often lead to treatment failure. Determination of HIV genetic sequences and viral load are constantly monitored to assess viral resistance to highly active antiretroviral therapy. Polymorphisms in drug transporters and drug metabolizing enzymes have also been monitored in HIV therapy. Two retrospective studies have identified the *HLA*-B*57:01 allele of the *MHC* class I B gene as a genetic determinant of hypersensitivity to abacavir [38,39]. Analysis in different ethnic groups, however, showed that *HLA*-B*57:01 alone would not be sufficiently predictive of hypersensitivity in diverse patient populations, suggesting that other genetic determinants of hypersensitivity remain to be identified. Implementation of pharmacogenetic postapproval will have a role in increasing the CIS of new products.

SUMMARY

The genomic revolution has offered the pharmaceutical industry the potential of improving the efficiency of drug development by reducing the current high failure rate through better choice of targets and improved understanding of drug response early in development. To the healthcare providers, it offers the potential to reduce the burden of adverse events by identifying those subjects at increased risk and offering them alternative therapies, as well as targeting its resources to use newer, more expensive treatments on subjects who will derive most benefit. Finally, and most importantly, it offers to the patient the opportunity with their physician to identify from the range of available therapeutic options the one most suited to them. Although pharmacogenetic testing is unlikely to be able to guarantee that the therapy will work and will not cause an adverse event, it will increase the probability that a drug will work and reduce uncertainty around adverse events and provide a rational way of choosing between therapies.

As our understanding of genomics improves, so will our ability to determine key factors involved in variability of drug response. The quest for precision medicines will start at the beginning of the drug discovery process with more comprehensive understanding of the molecular basis of the disease, molecular stratification, and the role of the drug target in the pathological process. Significant PK variability will be explained by systematic evaluation of all the relevant metabolizing enzymes and transport proteins. The drug candidates will only be tested in patients with suitable variants of the drug target. Drugs will be approved with variable dosage levels dependent upon underlying genotypes affecting drug response variation at the desired drug target. Finally, genetic- and genomic-led drug development and evaluation will not stop with the approval, but postmarketing (phase IV) research will endeavor to identify the causes of uncommon adverse events, leading to continuous refinement of how we use drugs throughout their life cycle.

REFERENCES

[1] DiMasi JA, Hansen RW, Grabowski HG. The price of innovation: new estimates of drug development costs. J Health Econ 2003;22(2):151–85.

[2] Drews J. Drug discovery: a historical perspective. Science 2000;287(5460):1960–4.

[3] Hopkins AL, Groom CR. The druggable genome. Nat Rev Drug Discov 2002;1(9):727–30.

[4] Zee RY, Lou YK, Griffiths LR, Morris BJ. Association of a polymorphism of the angiotensin I-converting enzyme gene with essential hypertension. Biochem Biophys Res Commun 1992;184(1):9–15.

[5] Province MA, Kardia SL, Ranade K, Rao DC, Thiel BA, Cooper RS, et al. A meta-analysis of genome-wide linkage scans for hypertension: the National Heart, Lung and Blood Institute Family Blood Pressure Program. Am J Hypertens 2003;16(2):144–7.

[6] Turki J, Pak J, Green SA, Martin RJ, Liggett SB. Genetic polymorphisms of the beta 2-adrenergic receptor in nocturnal and nonnocturnal asthma: evidence that Gly16 correlates with the nocturnal phenotype. J Clin Invest 1995;95(4):1635–41.

[7] Santillan AA, Camargo Jr CA, Ramirez-Rivera A, Delgado-Enciso I, Rojas-Martinez A, Cantu-Diaz F, et al. Association between β2-adrenoceptor polymorphisms and asthma diagnosis among Mexican adults. J Allergy Clin Immunol 2003;112(6):1095–100.

[8] Ogilvie AD, Battersby S, Bubb VJ, Fink G, Harmar AJ, Goodwim GM, et al. Polymorphism in serotonin transporter gene associated with susceptibility to major depression. Lancet 1996;347(9003):731–3.

[9] Golimbet VE, Alfimova MV, Shchebatykh TV, Abramova LI, Kaleda VG, Rogaev EI. Serotonin transporter polymorphism and depressive-related symptoms in schizophrenia. Am J Med Genet 2004;126B(1):1–7.

[10] Post JC, Ze F, Ehrlich GD. Genetics of pediatric gastroesophageal reflux. Curr Opin Allergy Clin Immunol 2005;5(1):5–9.

[11] Zambrowicz BP, Sands AT. Knockouts model the 100 best-selling drugs: will they model the next 100? Nat Rev Drug Discov 2003;2(1):38–51.

[12] Tong Y, Zhang S, Li H, Su Z, Kong X, Liu H, et al. 8302A/C and (TTA)n polymorphisms in the HMG-CoA reductase gene may be associated with some plasma lipid metabolic phenotypes in patients with coronary heart disease. Lipids 2004;39(3):239–41.

[13] Gretarsdottir S, Thorleifsson G, Reynisdottir ST. The gene encoding phosphodiesterase 4D confers risk of ischemic stroke. Nat Genet 2003;35(2):131–8.

[14] Peltekova VD, Wintle RF, Rubin LA, Amos CI, Huang Q, Gu X, et al. Functional variants of OCTN cation transporter genes are associated with Crohn disease. Nat Genet 2004;36(5):471–5.

[15] Stoll M, Corneliussen B, Costello CM, Waetzig GH, Mellgard B, Koch WA, et al. Genetic variation in DLG5 is associated with inflammatory bowel disease. Nat Genet 2004;36(5):476–80.

[16] Helgadottir A, Manolescu A, Thorleifsson G. The gene encoding 5-lipoxygenase activating protein confers risk of myocardial infarction and stroke. Nat Genet 2004;36(3):233–9.

[17] Burton PR, Clayton DG, Cardon LR, Craddock N, Deloukas P, Duncanson A, et al. Genome-wide association study of 14,000 cases of seven common diseases and 3000 shared controls. Nature 2007;447:661–78. http://www.nature.com/nature/journal/v447/n7145/full/nature05911.html. The Wellcome Trust Case Control Consortium.

[18] Hinds DA, Seymour AB, Durham LK, Banerjee P, Ballinger DG, Milos PM, et al. Application of pooled genotyping to scan candidate regions for association with HDL cholesterol levels. Hum Genomics 2004; Nov;1(6):421–34.

[19] John S, Shephard N, Liu G, Zeggini E, Cao M, Chen W, et al. Whole-genome scan, in a complex disease, using 11,245 single-nucleotide polymorphisms: comparison with microsatellites. Am J Hum Genet 2004;75(1):54–64.

[20] Samson M, Libert F, Doranz BJ, Rucker J, Liesnard C, Farber CM, et al. Resistance to HIV-1 infection in Caucasian individuals bearing mutant alleles of the CCR-5 chemokine receptor gene. Nature 1996;382(6593):722–5.

[21] Michael NL, Louie LG, Rohrbaugh AL, Schultz KA, Dayhoff DE, Wang CE, et al. The role of CCR5 and CCR2 polymorphisms in HIV-1 transmission and disease progression. Nat Med 1997;3(10):1160–2.

[22] Salkowitz JR, Bruse SE, Meyerson H, Valdez H, Mosier DE, Harding CV, et al. CCR5 promoter polymorphism determines macrophage CCR5 density and magnitude of HIV-1 propagation in vitro. Clin Immunol 2003;108(3):234–40.

[23] Feinberg J. Meeting notes from the 43rd Interscience Conference on Antimicrobial Agents and Chemotherapy (ICAAC): new CCR5 antagonist shows antiretroviral effect. AIDS Clin Care 2003;15(11):94–5.

[24] O'Shea JJ, Husa M, Li D, Hofmann SR, Watford W, Roberts JL, et al. JAK3 and the pathogenesis of severe combined immunodeficiency. Mol Immunol 2004;41(6–7):727–37.

[25] Brunkow ME, Gardner JC, Van Ness J, Paeper BW, Kovacevich BR, Proll S, et al. Bone dysplasia sclerosteosis results from loss of the SOST gene product, a novel cystine knot-containing protein. Am J Hum Genet 2001;68:577–89.

[26] Allen MJ, Carey AH. Target identification and validation through genetics. Drug Discov Today 2004;3(5):183–91.

[27] ENCODE Project Consortium. Identification and analysis of functional elements in 1% of the human genome by the ENCODE pilot project. Nature 2007;447(7146):799–816.

[28] Penny MA, McHale D. Pharmacogenomics and the drug discovery pipeline: when should it be implemented? Am J Pharmacogenomics 2005;5(1):53–62.

[29] Rawlins M, Thompson J. Mechanisms of adverse drug reactions. In: Davies D, editor. Textbook of adverse drug reactions. Oxford: Oxford University Press; 1991. p. 18–45.

[30] Silber BM, editor. Pharmacogenomics, biomarkers and the promise of personalised medicine. Pharmacogenetics-pharmacogenomics. 2000.

[31] Mahgoub A, Idle JR, Dring LG, Lancaster R, Smith RL. Polymorphic hydroxylation of Debrisoquine in man. Lancet September 17, 1977;2(8038):584–6.

[32] Danielson PB. The cytochrome P450 superfamily: biochemistry, evolution, and drug metabolism in humans. Curr Drug Metab 2002;3:561–97.

[33] Lee CR, Goldstein JA, Pieper JA. Cytochrome P450 2C9 polymorphisms: a comprehensive review of the in-vitro and human data. Pharmacogenetics 2002;12(3):251–63.

[34] Shimizu T, Ochiai H, Asell F, Shimizu H, Saitoh R, Hama Y, et al. Bioinformatics research on inter-racial difference in drug metabolism. I. Analysis on frequencies of mutant alleles and poor metabolizers on CYP2D6 and CYP2C19. Drug Metab Pharmacokinet 2003;18(1):48–70.

[35] Idle JR, Mahgoub A, Lancaster R, Smith RL. Hypotensive response to debrisoquine and hydroxylation phenotype. Life Sci 1978;22(11):979–83.

[36] Cascorbi I. Pharmacogenetics of cytochrome P4502D6: genetic background and clinical implication. Eur J Clin Invest 2003;33(Suppl. 2):17–22.

[37] Evans WE. Pharmacogenetics of thiopurine S-methyltransferase and thiopurine therapy. Ther Drug Monit 2004;26(2):186–91.

[38] Hetherington S, Hughes AR, Mosteller M. Genetic variations in HLA-B region and hypersensitivity reactions to abacavir. Lancet 2002;359(9312):1121–2.

[39] Mallal S, Nolan D, Witt C, Masel G, Martin AM, Moore C, et al. Association between presence of *HLA*-B*5701, *HLA*-DR7, and *HLA*-DQ3 and hypersensitivity to HIV-1 reverse-transcriptase inhibitor abacavir. Lancet 2002;359(9308):727–32.

[40] Chung WH, Hung SI, Hong HS, Hsih MS, Yang LC, Ho HC, et al. Medical genetics: a marker for Stevens–Johnson syndrome. Nature 2004;428:486.

[41] Lane S, Al-Zubiedi S, Hatch E, Matthews I, Jorgensen AL, Deloukas P, et al. The population pharmacokinetics of R- and S-warfarin: effect of genetic and clinical factors. Br J Clin Pharmacol January 2012;73(1):66–76.

[42] Evans WE, Relling MV. Pharmacogenomics: translating functional genomics into rational therapeutics. Science 1999;286(5439):487–91.

[43] Evans WE, McLeod HL. Pharmacogenomics: drug disposition, drug targets, and side effects. N Engl J Med February 6, 2003;348(6):538–49.

[44] Schadt EE, Lamb J, Yang X, Zhu J, Edwards S, Guhathakurta D, et al. An integrative genomics approach to infer causal associations between gene expression and disease. Nat Genet 2005;37(7):710–7.

[45] Vogel CL, Cobleigh MA, Tripathy D, Gutheil JC, Harris LN, Fehrenbacher L, et al. Efficacy and safety of trastuzumab as a single agent in first-line treatment of *HER2*-overexpressing metastatic breast cancer. J Clin Oncol 2002;20(3):719–26.

[46] Deininger MW, Goldman JM, Lydon N, Melo JV. The tyrosine kinase inhibitor CGP57148B selectively inhibits the growth of *BCR-ABL*-positive cells. Blood 1997;90(9):3691–8.

[47] Wong RW. Transgenic and knock-out mice for deciphering the roles of *EGFR* ligands. Cell Mol Life Sci 2003;60(1):113–8.

[48] Lynch TJ, Bell DW, Sordella R, Gurubhagavatula S, Okimoto RA, Brannigan BW, et al. Activating mutations in the epidermal growth factor receptor underlying responsiveness of non-small-cell lung cancer to gefitinib. N Engl J Med 2004;350(21):2129–39.

[49] Pirmohamed M, James S, Meakin S, Green C, Walley TJ, Farrar K, et al. Adverse drug reactions as cause of admission to hospital: prospective analysis of 18,820 patients. BMJ 2004;329(7456):15–9.

Chapter 20

Cancer Genetics and Genomics

J. Whitworth, E. Maher

University of Cambridge, Cambridge, United Kingdom

Chapter Outline

Introduction	261	Inherited Cancer Genes	266
Inherited Cancers: Germ Line and Somatic	261	Phenotypic Effects of Mutations in Cancer Predisposition Genes	266
Oncogenes, Tumor Suppressor Genes, and the Identification of Cancer Predisposition Genes	261	Genotype–Phenotype Correlation	280
Identifying Cancer Predisposition Genes	262	Genetic Testing and Mainstreaming	281
High, Medium, and Smaller Risks	263	Genetic Testing and Next Generation Sequencing	281
Genetic Testing in Cancer Predisposition Syndromes	264	Reasons for Nondetection of Causative Genetic Changes Through Clinical Testing	282
Risk Information as Therapy	264	Conclusion	283
Cancer Surveillance	265	References	283
Prophylactic Surgery	265		
Pharmacological Management	266		

INTRODUCTION

In the 19th century, microscopic analysis led to the observation that chromosome aberrations can occur in malignant cells [1]. Boveri was one of the first to put forward the idea that such aberrations might be a key causative factor in tumorigenesis. By studying abnormal mitoses in sea urchin embryos, he hypothesized that abnormal cellular properties, including malignancy, were conferred by an unbalanced chromosome complement. His work included assertions regarding "inhibiting chromosomes"; ie, those that normally act to suppress cell division, and "stimulatory chromosomes," which change a cell's relationship with its external environment to encourage a proliferative state. These ideas were remarkably prophetic of the proto-oncogenes and tumor suppressor genes (TSGs; see Oncogenes, Tumor Suppressor Genes, and the Identification of Cancer Predisposition Genes) that later became familiar [2]. It was not until the 1960s that a specific chromosomal abnormality was associated with a particular tumor, when the Philadelphia chromosome (resulting from a translocation of chromosomes 9 and 22) was identified as present in all studied chronic myeloid leukemia patients [3].

Abnormal chromosomes may be a result of the genomic instability caused by tumorigenic processes but may also, as in the case of the Philadelphia chromosome, be important initiating events. With the development of gene sequencing techniques, it became possible to study the molecular consequences of such events at the gene level and also define causative genetic abnormalities, not visible by chromosome analysis, that occur at the nucleotide level.

INHERITED CANCERS: GERM LINE AND SOMATIC

Oncogenes, Tumor Suppressor Genes, and the Identification of Cancer Predisposition Genes

A crucial step in understanding how genetic changes in tumors led to cancer was the development of the concept of the oncogene, a gene that has undergone change rendering it contributory to tumor development. These changes result in enhanced or altered function of the given gene's normal counterparts (ie, proto-oncogenes). Proto-oncogenes are involved in a variety of cellular processes pertinent to cell growth/proliferation, including cell cycle regulation and growth signaling. Oncogenes were initially identified through analysis of tumor cells with malignant properties that had been induced by a retrovirus. The tumorigenic potential of such viruses was found to be caused by one component gene of the virus, designated the *oncogene* [4]. Further research revealed that orthologs of the viral oncogenes were present in normal cells and were labeled *proto-oncogenes* [4]. The discovery of the other main class of gene significant in the development of cancer, that of the TSG, was to provide the key step in the understanding of inherited tumors. In its normal state a TSG often functions to inhibit cell proliferation, but loss of function mutations compromise this role and thereby promote tumorigenesis. In some cases, inactivation of a TSG directly leads to cellular attributes that favor malignant transformation but in other cases (eg, inactivation of a TSG involved in normal DNA repair mechanisms with resulting failure to repair mutations in TSGs or proto-oncogenes) the protumorigenic process is indirect.

Although some genetic changes appear critical in conferring tumor-defining properties to cells, a cell's transition from normal to malignant is a multistep process. A source of much debate, often based on epidemiological evidence, has been how many changes are required for the completion of this process. Work by Nordling observed cancer mortality correlating with age and estimated that, on average, six mutational events in a given cell were required for a cancer to occur [5]. The work only studied certain cancer types and observed that many malignancies did not conform to this model.

Seminal work by Knudson et al. suggested that at least one cancer type (childhood retinoblastoma) could be caused by two rate-limiting mutational events affecting different alleles at the same locus. By observing tumor prevalence in presumed inherited and sporadic cases and estimating the mutation rate at that relevant theoretical locus, a model was proposed whereby those who inherit a mutated TSG allele from a parent require only a single mutation event ("hit") in the other allele for a tumor to be initiated. Sporadic cases, on the other hand, must acquire both hits postconception [6]. This model explains the very high risk of retinoblastoma in individuals with a germline mutation and the later onset of the tumor in sporadic than in inherited cases. Knudson's work and the subsequent identification of the *RB1* TSG [7] was a key step in the development of inherited cancer genetics (Box 20.1). Apart from highlighting the role of TSGs in cancer pathogenesis, it demonstrated that genetic factors causing cancer predisposition were identifiable through the study of affected families and that inherited cancer genes can be implicated in sporadic cases.

The discovery of *RB1* prompted an extensive and ongoing search for further cancer predisposition genes that has yielded findings of relevance to affected individuals and also tumors occurring outside of the familial/germline context. Identified genes may not conform to the two-hit TSG model and a number of constitutionally activated proto-oncogenes have been found to be associated with familial cancers [eg, those in the *RET* gene leading to multiple endocrine neoplasia type 2 (MEN2) [10,11]]. The search for inherited cancer genes has focused on individuals with

> **Box 20.1 *RB1*: The Discovery of a Cancer Predisposition Gene**
>
> Retinoblastoma is a rare childhood cancer affecting the eye. Bilateral tumors occur in the familial form and unilateral tumors usually represent a lack of an inherited factor, namely a genetic aberration affecting the *RB1* gene. This gene was identified through analysis of retinoblastomas with Knudson's two-hit hypothesis in mind. There was previous evidence that a section of chromosome 13 was the area affected by the hypothesized second hit. Some retinoblastoma patients had been shown to harbor a constitutional deletion in this area [8], and acquired loss or partial deletion of this region of chromosome 13 had been shown in retinoblastoma tumor cells [9]. Identification of the *RB1* gene within the target region was followed by demonstrating that inherited cases had an inactivating germline mutation and a second hit (deletion, mutation, or other change) in the tumor cells, whereas tumors from noninherited sporadic cases harbored inactivating hits in both *RB1* alleles that were not present in the patient's normal cells (so both events were somatic) [7].

clinical characteristics of inherited cancer predisposition (eg, family history of cancer, multiple tumors, phenotypic marker, and/or young age at diagnosis) but advances in genetic technology are providing the means for large-scale mutation screening in individuals without classical features of an inherited cancer syndrome.

Identifying Cancer Predisposition Genes

A variety of approaches have been employed to identify inherited cancer genes, including candidate gene analysis, positional cloning, positional candidate gene analysis, and comprehensive sequencing with next generation sequencing (NGS) technologies (Box 20.2). Earlier studies concentrated on large families with high penetrance cancer phenotypes and used genetic linkage studies to identify regions that segregated with cancer predisposition in the family. In some cases (eg, *RB1*; see Box 20.1) the putative localization of the inherited cancer gene was supported by deletion/allele loss analysis in tumor material. Having defined an interval of interest, all genes within the region were then sequenced until the gene containing germline mutations was identified. As illustrated by *RB1*, some inherited cancer genes are targets for somatic mutations in sporadic tumors. A typical candidate gene approach would be to take a gene known to be somatically mutated in a sporadic cancer and test whether inherited cases harbor germline mutations in that gene (eg, *PBRM1* mutations in familial renal cell carcinoma [12]). A positional candidate gene approach has been popular, particularly since the advent of detailed maps of the human genome, because after the identification of a candidate region of linkage, the genes within the target

Box 20.2 *POT1*: An Example of Cancer Predisposition Gene Identification Using NGS Techniques

Researchers investigating genetic susceptibility to melanoma identified a novel predisposition gene (*POT1*) by the application of NGS techniques to families where multiple members were affected.

Whole exome or whole genome sequencing (WGS) was performed in individuals belonging to 28 families where two or more melanomas had occurred and there were three or more available members for analysis. This latter criterion increased the ability of the researchers to exclude potentially causative variants that were not present in affected cases (or were present in unaffected cases).

The large number of variants identified through sequencing were filtered using bioinformatic techniques to exclude those that were unlikely to be the cause of melanoma. Variants were excluded if they were common and therefore unlikely to explain the relatively rare phenomenon of familial melanoma. They were also not retained if they were unlikely to affect the function of the corresponding protein product.

In the families that had variants fitting the criteria, 320 genes were found. Crucially, only five of these were shared by more than one of the families. The research team focused on these and demonstrated that the variants in one of them (*POT1*) could be found in more families with familial melanoma and disrupted biological function of the gene [13].

interval can be prioritized according to their relevance to cancer and sequenced accordingly. High-throughput NGS techniques have enabled germline exome/genome sequencing in cases in which genetic linkage study are not possible (eg, DNA may only be available from a single affected family member) and, although the challenges of correctly interpreting large numbers of rare genetic variants should not be underestimated, can greatly facilitate the identification of inherited cancer genes.

Methodologies as described in Box 20.2 generally begin without a hypothesis in terms of which gene may harbor the causative mutation. Initial analysis, particularly for NGS approaches, is likely to generate gene lists for further investigation, and selection of putative causative genes at this stage is aided by candidature provided by existing sources of evidence. This might be expression of the gene in the organ of interest or involvement of it in a cellular process relevant to cancer (eg, DNA repair). Observation of somatic alterations in a cancer type corresponding to that seen in the studied family may also prompt candidature. An example of this in the identification of *RB1* is illustrated in Box 20.1 and the knowledge that *TP53* was recurrently mutated in many cancer types was helpful in the genetic characterization of Li-Fraumeni syndrome [14]. Molecular characterization of tumors has become more extensive through cancer genome projects and is likely to further assist with identification of cancer predisposition genes in this way. The cBioPortal, for

example, contains information from over 20,000 sampled tumors from various projects [15].

High, Medium, and Smaller Risks

Although NGS technologies and improved candidature have enhanced the rate of novel cancer predisposition gene discovery, mutations in many of these have been estimated to cause lower cancer risks than some of the earlier discoveries such as *APC*, *VHL*, and *TP53*. The majority of high-risk cancer predisposition genes affecting a large numbers of families may have been discovered, leaving more recent and future discoveries falling into a number of categories according to the risk that mutations affecting them confer.

Newly identified high-risk genes are more likely to be rare and account for a smaller proportion of overall cancer burden. Current sequencing and bioinformatic techniques applied to small numbers of families (see Box 20.2) are well placed to identify them. Their clinical utility will be significant for affected families and can provide insights into nonfamilial tumors of similar type.

Mutations in genes leading to moderate levels of cancer risk are unlikely to cause a characteristic cancer in all carrying members of families where they are present. Consequently, study designs relying on analysis of which variants are shared between affected family members may be less informative. Studies comparing variants seen in large series of patients who have a particular cancer with control series can reveal significant association of variant with tumor without necessarily reflecting a very high risk of that tumor developing. Examples of predisposition genes identified in this way include *BRIP1* and *PALB2*, reported to confer a relative breast cancer risk of 2% and 2.3%, respectively [16,17]. Interestingly, further observation of *PALB2* mutation carriers has revised this estimated risk to a much higher level [18].

Identification of the genetic basis of highly penetrant cancer predisposition phenotypes can make a major difference to the management of affected families (see Genetic Testing in Cancer Predisposition Syndromes) but only impact on small minority of patients with the relevant cancer. Genome-wide association studies (GWASs) of large cohorts of cancer patients have been employed to identify common genetic variants that predispose to specific cancers. The identification of such susceptibility alleles can provide pointers as to molecular pathways significant to particular tumors but generally have not been translated into clinical practice because the increased risk associated with each variant is small. Ultimately, clinical utility might be provided by identifying the small number of individuals who harbor multiple predisposing variants that combine to put the individual at a significantly increased risk. One analysis to assess this potential combined risks derived from 77 variants previously associated with breast cancer in a GWAS to produce a combined risk score. The score was used to stratify over 30,000 breast cancer cases and controls into quintiles. It was found that in those without a family history, the highest scoring quintile, had significantly a higher lifetime breast cancer risk (5.2%) than those in the lowest scoring quintile (16.6%). This difference was more pronounced in those with a first-degree relative with breast cancer (8.6% vs. 24.4%) [19]. Risk estimates such as those assigned to the higher risk group approach those deemed sufficient for intervention in the cancer genetic clinic.

GENETIC TESTING IN CANCER PREDISPOSITION SYNDROMES

Identification of cancer predisposition genes through the study of affected individuals and families has provided the opportunity to extend genetic analysis to large numbers of individuals, often without a family history, who are considered to be at risk of an inherited cancer syndrome (eg, at least a quarter of patients with pheochromocytoma/paraganglioma without a relevant family history have a germline mutation). The case to perform germline genetic testing may be particularly compelling where rarer tumors cluster within the same family or even individual, because alternative causes are less likely. The picture is more ambiguous where common tumors cluster. Such a scenario may represent inherited predisposition or be the result of higher population incidences of particular tumors, leading to the occurrence in multiple family members. Sporadic tumors that appear consistent with a constitutional genetic cause may be termed *phenocopies* and are more likely to be present where environmental factors commonly lead to a particular tumor type. However, this is not a simple relationship; assessment of the proportion of cancer cases attributable to 14 common preventable environmental exposures has shown relatively low figures for many of the most common tumors occurring in the population; eg, breast (26.8%) and colorectal (54.4%) cancers. [20] In many cases, clinical genetic testing is able to distinguish between clustering of common tumors caused by mutations in cancer predisposition genes and other mechanisms.

Genetic testing is considered diagnostic in an individual who has previously been diagnosed with a cancer and where a genetic explanation is sought. Predictive testing generally interrogates the genetic status of an unaffected person for a causative variant previously identified in a family member, although advances in testing technology and knowledge of cancer predisposition genes is likely to lead to much more predictive testing where such a mutation has not formally been found in a relative. In the ideal scenario, results from such analysis can provide accurate risk assessment and potentially insight into tumor biology in the tested individual. Clinical utility may therefore be derived in a number of ways.

Risk Information as Therapy

Risk information based on genetic test results may enhance management by providing a clear explanation for cancers that are of early onset and affect multiple family members. Negative results can provide reassurance but often leave open the possibility of unidentified pathogenic variants such that a genetic predisposition cannot be definitively excluded. A genetic diagnosis, although potentially increasing perceived future cancer risk, often comes with a defined spectrum of associated tumors. These specific risks, which can be managed, may be preferable to a generalized increased cancer risk that may be perceived by the patient before testing. A notable exception to this is Li-Fraumeni syndrome, where the cancer risks are broad and there is little clinical surveillance demonstrated to be of benefit. Uptake of predictive testing for this condition, however, is broadly similar to other commonly performed predictive tests where better management can be offered [21].

A systematic review of psychological outcomes in women with a family history of breast cancer undergoing genetic testing found a significant reduction in psychological distress in women receiving negative results and little change in those receiving positive results [22]. A study of individuals undergoing predictive testing for *BRCA1* or *BRCA2* mutations found that 92% would recommend the process to others in the same situation [23]. Assessment of benefits such as this should be seen in the context of testing uptake, which has been shown to be around half of individuals eligible for predictive testing in the most commonly seen conditions in the genetics clinic [24,25]. Those not pursuing testing may not have gone on to have an equally positive experience.

Individuals consulting clinical services for assessment for a possible predisposition syndrome often do so in an attempt to provide a genetic diagnosis in the family. This gives the opportunity for relatives to predict and manage their risks and may be of benefit to the consulting patient even if their own prognosis is poor. An assessment of motivations for diagnostic testing in a series of patients undergoing testing showed greater importance placed on this than a desire to increase certainty as to whether or not an increased risk of colorectal cancer was present [26].

Finally, identification of mutations in cancer predisposition genes in a potential parent may facilitate reproductive decisions or lead to testing for the variant in a fetus in utero [prenatal diagnosis (PND)] or preimplantation embryos resulting from in vitro fertilization [preimplantation genetic diagnosis (PGD)]. Test results can affect decisions as to whether to continue a pregnancy or which embryos to select for implantation. This is a model largely used in severe (mainly noncancer) childhood-onset disorders and is less commonly applied to most cancer predisposition syndromes because of their generally later onset and more manageable manifestations. However, a number of adult-onset cancer syndromes appear on the Human Fertilization and Embryo Authority's approved list for PGD [27] and high levels of agreement with PND for Lynch syndrome among individuals at risk of Lynch syndrome [28] suggest that this may change in future.

Cancer Surveillance

The current mainstay of management after the diagnosis of a cancer predisposition syndrome is interventions designed to prevent cancers occurring or to diagnose them at a more treatable stage. The latter is achieved through regular clinical surveillance of at-risk tissue. This may be via a number of modalities, depending on the tissue and syndrome in question. Frequency and age at which surveillance investigations are performed are guided by observational evidence from series of affected cases. The quality of such guidance is often compromised by the rarity of a condition and/or ascertainment biases influencing which patients are included in the series. The gold standard of a prospective study of mutation carriers to observe may be difficult to achieve for these reasons. Collaboration between centers is likely to lead to more accurate assessments of risks and also screening outcomes.

The effectiveness of surveillance programs is uncertain for most predisposition syndromes. In more common conditions, larger cohorts that can be assembled in one or few centers can provide greater clarity on this issue. In Lynch syndrome, a number of studies comparing screened and unscreened groups have been published, allowing a systematic review. The multiple studies showing reduction in colorectal cancer incidence and related mortality in screened patients (with regular colonoscopy) provide convincing evidence for the utility of this strategy.

In rarer conditions, inference may have to be made from less direct sources of information. Von Hippel-Lindau disease (VHL) is one such condition that predisposes affected individuals to central nervous system hemangioblastoma, pheochromocytoma, and renal cell carcinoma. Protocols for surveillance are widely used but no prospective follow-up data comparing screened with unscreened patients exists. However, life expectancy in VHL patients has been observed by identifying cases through genetics services and cancer registries. Whereas a reduced life expectancy was noted (52.5 years), the mean survival rate increased by 16.3 years in patients diagnosed after 1990 when the genetic service registry, and therefore increased systematic surveillance protocols, was introduced [29].

Surveillance programs for conditions with diverse tumor risks can lead to the proposal of complex screening programs that may have poor acceptability for patients and whose outcomes are difficult to assess. An appealing approach is that of a single modality screening test that can address these issues. A candidate for a modality like this is a whole body magnetic resonance imaging scan, which is being investigated for Li-Fraumeni syndrome. Crucial outcomes will include the rate of potentially significant findings on scan and effects of resulting interventions. This is true for any screening test, but a whole-body approach combined with the range of tumor risks in Li-Fraumeni syndrome make them particularly pertinent.

Clinical surveillance has more potential to do harm (eg, through unnecessary surgery) where the penetrance of a mutation in a given cancer predisposition gene is not high. It should be noted that penetrance and cancer risks can vary according to how cases are ascertained; eg, different results for *BRCA1* cancer risks have been reported in cases ascertained from high-risk cancer genetic clinics and those detected from population-based screening programs. In hereditary leiomyomatosis and renal cell carcinoma (HLRCC) caused by *FH* mutations, only 15–20% of mutation carriers develop kidney cancer but of those that do, many are at an advanced stage, resulting in a poor prognosis [30]. As well as assessing screening programs in as extensive series as possible, therefore, an important area of research is risk stratification within inherited cancer syndromes. This might be based on the particular variant in the causative gene (see Genotype–Phenotype Correlation) or through germline variants in other genes that may influence cancer risk (modifier genes). Alternatively, acceptability, specificity, and sensitivity of screening tests might be improved for those individuals at low risk by exploiting the phenomena of circulating tumor DNA. Identification of specific genetic markers of tumor cell origin in a predisposition syndrome such as HLRCC could facilitate an effective surveillance program based on blood sampling.

Prophylactic Surgery

In some predisposition syndromes, prophylactic surgery may represent the most effective preventative strategy. Utility and uptake of this can depend on a number of factors such as level of risk reduction from tissue removal, function (and loss thereafter) of the tissue in question, and likelihood of complications after the procedure. These factors need to be considered against the efficacy of surveillance strategies as an alternative. Prophylactic surgery can result in dramatic reduction in tumor risk (eg, thyroidectomy in MEN2B/2A). In hereditary breast and ovarian cancer caused by *BRCA1* and *BRCA2* mutations, bilateral mastectomy is estimated to reduce the risk of breast cancer by around 90% [31]. Utility of surgery in other syndromes may be more difficult to estimate owing to rarity of the condition and/or lack of an adequate control (no surgery performed) group with which to compare survival rates. In familial adenomatous polyposis, the risk of colorectal cancer [32] has been estimated

TODO

at a level sufficient to warrant colectomy in all diagnosed cases, leaving a low number of cases with an intact colon for further study.

Pharmacological Management

In cancer predisposition syndromes, the benefits of chemotherapeutic preventative strategies are more likely to outweigh disadvantages resulting from side effects or economic cost. Use of preventative agents may be based on observations related to tumors occurring outside of the familial context. Observations of lower colorectal cancer rates in individuals taking long-term aspirin [33] led to a trial of its use in Lynch syndrome, which leads to high risks of that malignancy. A significant reduction in bowel cancer incidence was observed, leading to the recommendation that this medication should be discussed with affected individuals [34].

Recent years have brought an increasing focus on the use of a patient's germline mutation status to guide medical therapy to both treat and prevent tumor development. This knowledge can provide insight into the biology of the tumor sufficient to prompt clinical trials. Though such therapies are limited in number, they are a source of wider optimism in cancer predisposition syndromes. Strategies generally involve the inhibition of an abnormally active/upregulated gene product or cellular pathway (Box 20.3). Notable exceptions to this exist such as the example of PARP inhibitors in *BRCA1/2*-related cancers, which target and inhibit a DNA repair mechanism (base excision repair) in tumor cells already known to be deficient in a different repair process (double-stranded DNA repair by homologous recombination) [35].

More extensive definition of tumor phenotypes associated with predisposition syndromes by improved molecular analysis should continue to yield abnormalities targetable by therapeutic agents. Sequencing of breast cancers with whole genome sequencing (WGS) has previously revealed a distinct mutational signature in germline *BRCA1*- or *BRCA2*-related cancers that is consistent with the known DNA repair deficit in these patients [40]. Similar work may be rewarding with regard to tumors from patients with other syndromes, perhaps where the function of the relevant gene is less well characterized. This should provide opportunities for stratification and treatment akin to those produced by the study of sporadic cancers.

INHERITED CANCER GENES

The canon of genes where germline mutation leads to cancer susceptibility currently stands in three figures, although in practice, defining such a gene may be difficult. For every gene where mutation carriers have high risks, there are others where mutation carriers are more likely to remain unaffected by the cancer they are at risk from in comparison with the general population. Many genetic variants have been associated with increased risk of particular cancers through GWASs but the risk conferred by these is generally not at a level prompting the management strategies described previously. A useful definition, then, might be one that can assign patients to risk groups (based on genotype) amenable to risk-management strategies.

A comprehensive review of cancer predisposition genes was published by Rahman in 2014 and included genes where rare mutations conferred a doubling of relative risk of cancer and lead to 5% of carriers being affected with cancer [41]. An adapted version of the list, with illustrative intent, is included at the end of the chapter and includes tumors that are primarily associated with mutations in each gene (Table 20.1). Referral to the original article is advised for a comprehensive approach.

Phenotypic Effects of Mutations in Cancer Predisposition Genes

Predisposition genes are involved in an array of cellular processes where aberrant function can lead to cancer-defining phenomena such as genomic instability, disrupted cell cycle regulation, or increased proliferation. Mutations in these genes can lead to phenotypic expression consistent with various models of inheritance seen in high-penetrance genetic conditions. The majority of associated phenotypes are inherited in an autosomal dominant manner, often with unaffected mutation carriers seen in the family. Mutations in a gene causing a dominantly inherited syndrome may be

---

Box 20.3 Vismodegib: Pharmacological Use from Germline Genetic Insight

Gorlin syndrome (basal cell nevus syndrome) is an autosomal dominant condition associated, among a wide range of other features, with large numbers of early-onset basal cell carcinomas. The study of affected families demonstrated that it was caused by mutations in *PTCH1*, the gene product of which performs an inhibitory function in the hedgehog signaling pathway [36,37]. This pathway is a key regulator of cellular development in early life but is abnormally active in Gorlin-related basal cell carcinoma cells because of a second hit in the wild-type *PTCH1* allele [38]. *PTCH1* was subsequently shown to be mutated in most sporadic basal cell carcinomas (BCCs), and a trial of an agent previously known to inhibit the hedgehog pathway was shown to be efficacious. This agent was vismodegib, and subsequent trials have also shown reduction in BCC occurrence in Gorlin patients [39]. The example illustrates the development of therapy based on germline mutation status and also how identification of relevant mutations in rare syndromic tumors can inform knowledge and treatment of their sporadic counterparts.

TABLE 20.1 Selected Cancer Predisposition Genes Arranged by Phenotypic Association

Gene Symbol	Gene Name	Alternative Symbols	Chromosome Location	Associated Cancer Syndrome(s)	Major Associated Tumor Types	Loss or Gain of Function Leads to Cancer Risks	State in Which Associated With Cancer Risk	Year of Association	Reference (PubMed ID) for First Association
Breast and/or Ovarian Cancer									
BRCA1	Breast cancer 1, early onset	BRCC1, PPP1R53, RNF53	17q21	Hereditary breast-ovarian cancer	Breast cancer Ovarian cancer	Loss	Biallelic	1994	7545954
BRCA2	Breast cancer 2, early onset	FANCD1	13q12	Hereditary breast-ovarian cancer Fanconi anemia (D1) (biallelic mutations)	Biallelic mutations: Myeloid hemato-logical malignancy (medulloblastoma) Wilms tumor Monoallelic mutations: Breast cancer Ovarian cancer Prostate cancer Pancreatic cancer	Loss	Monoallelic or biallelic	1995	8524414
PALB2	Partner and localizer of BRCA2	FANCN	16p12.1	Fanconi anemia (N) (biallelic mutations)	Biallelic mutations: Myeloid hemato-logical malignancy Medulloblastoma Neuroblastoma Wilms tumor Monoallelic mutations: Breast cancer Pancreatic cancer	Loss	Monoallelic or biallelic	2007	17200671 17200672 17200668 17287723
CHEK2	Checkpoint kinase 2	CDS1, CHK2	22q12.1		Breast cancer	Loss	Biallelic	2002	11967536 12094328
RAD51C	RAD51 homolog C (Saccharomyces cerevisiae)	RAD51L2	17q25.1	Fanconi anemia (O) (biallelic mutations)	Monoallelic mutation: Ovarian cancer	Loss	Monoallelic or biallelic	2010	20400964

Continued

TABLE 20.1 Selected Cancer Predisposition Genes Arranged by Phenotypic Association—cont'd

Gene Symbol	Gene Name	Alternative Symbols	Chromosome Location	Associated Cancer Syndrome(s)	Major Associated Tumor Types	Loss or Gain of Function Leads to Cancer Risks	State in Which Associated With Cancer Risk	Year of Association	Reference (PubMed ID) for First Association
RAD51D	RAD51 homolog D (S. cerevisiae)	RAD51L3	17q11		Ovarian cancer	Loss	Biallelic	2011	21822267
BRIP1	BRCA1 interacting protein C-terminal helicase 1	FANCJ, BACH1	17q22	Fanconi anemia (J) (biallelic mutations)	Biallelic mutations: Myeloid hematological malignancy Squamous cell carcinoma (SCC) (head and neck, esophagus, genital tract) Monoallelic mutations: Breast cancer Ovarian cancer	Loss	Monoallelic or biallelic	2005	16153896 16116424 16116423
Cartilaginous Cancer									
EXT1	Exostosin 1	LGCR, LGS	8q24.11		Chondrosarcoma	Loss	Biallelic	1995	7550340
EXT2	Exostosin 2	SOTV	11p12-p11		Chondrosarcoma	Loss	Biallelic	1996	8782816
Central Nervous System Tumors									
NF1	Neurofibromin 1	NFNS, VRNF, WSS	17q12	Neurofibromatosis type 1	Glioma Malignant peripheral nerve sheath tumor	Loss	Biallelic	1990	2134734 1694727
NF2	Neurofibromin 2 (merlin)	ACN, BANF, SCH	22q12.2	Neurofibromatosis type 2	Vestibular schwannoma Meningioma Ependymoma	Loss	Biallelic	1993	8453669 8379998
SMARCE1	SWI/SNF-related, matrix associated, actin dependent regulator of chromatin, subfamily e, member 1	BAF57	17q21.2		Mengingioma	Loss	Biallelic	2013	23377182

Gene	Gene name	Alias	Location	Syndrome	Cancer	Loss	Allelic	Year	Reference
SUFU	Suppressor of fused homolog (*Drosophila*)	*SUFUH, SUFUXL*	10q24.32		Medulloblastoma meningioma	Loss	Biallelic	2002	12068298
Gastrointestinal Cancer									
APC	Adenomatous polyposis coli	*DP2.5*	5q21	Familial adenomatous polyposis	Colorectal cancer Hepatoblastoma Desmoid tumor	Loss	Biallelic	1991	1651174 1651562 1651563 1678319
MUTYH	*mutY* Homolog (*Escherichia coli*)	*MYH*	1p34.1		Colorectal cancer	Loss	Monoallelic	2002	11818965
MLH1	*mutL* Homolog 1, colon cancer, nonpolyposis type 2 (*E. coli*)	*COCA2, HNPCC2*	3p21.3	Mismatch repair (MMR) deficiency syndrome (biallelic mutations). Lynch syndrome/hereditary nonpolyposis colon cancer (monoallelic mutations)	Biallelic mutations: Brain tumors Hematological malignancy Embryonal tumors Monoallelic mutations: Colorectal cancer Endometrial cancer Ovarian cancer	Loss	Monoallelic or biallelic	1994	8128251 8145827
MSH2	*mutS* Homolog 2, colon cancer, nonpolyposis type 1 (*E. coli*)	*COCA1, HNPCC1*	2p21	MMR deficiency syndrome (biallelic mutations) Lynch syndrome/hereditary nonpolyposis colon cancer (monoallelic mutations)	Biallelic mutations: Brain tumors Hematological malignancy Embryonal tumors Monoallelic mutations: Colorectal cancer Endometrial cancer Ovarian cancer Sebaceous adenoma, carcinoma, epithelioma	Loss	Monoallelic or biallelic	1993	8252616 8261515

Continued

TABLE 20.1 Selected Cancer Predisposition Genes Arranged by Phenotypic Association—cont'd

Gene Symbol	Gene Name	Alternative Symbols	Chromosome Location	Associated Cancer Syndrome(s)	Major Associated Tumor Types	Loss or Gain of Function Leads to Cancer Risks	State in Which Associated With Cancer Risk	Year of Association	Reference (PubMed ID) for First Association
MSH6	mutS Homolog 6 (E. coli)	GTBP	2p16	MMR deficiency syndrome (biallelic mutations) Lynch syndrome/ hereditary nonpolyposis colon cancer (monoallelic mutations)	Biallelic mutations: Brain tumors Hematological malignancy Embryonal tumors Monoallelic mutations: Colorectal cancer Endometrial cancer Ovarian cancer	Loss	Monoallelic or biallelic	1997	9354786
PMS2	PMS2 postmeiotic segregation increased 2 (S. cerevisiae)	PMSL2, HNPCC4	7p22	MMR deficiency syndrome (biallelic mutations) Lynch syndrome/ hereditary nonpolyposis colon cancer (monoallelic mutations)	Biallelic mutations: Brain tumors Hematological malignancy Supratentorial primitive neuroectodermal tumors Monoallelic mutations: Colorectal cancer Endometrial cancer Ovarian cancer	Loss	Monoallelic or biallelic	1994	8072530
POLD1	Polymerase (DNA directed), delta 1, catalytic subunit	POLD, CDC2, CRCS10, MDPL	19q13.33	Polymerase proofreading—associated polyposis (PPAP)	Colorectal cancer Endometrial cancer	Loss	Biallelic	2013	23263490
POLE	Polymerase (DNA directed), epsilon, catalytic subunit	FILS, POLE1, CRCS12	12q24.33	PPAP	Colorectal cancer	Loss	Monoallelic or biallelic	2013	23263490

Gene	Gene name	Aliases	Location	Syndrome	Cancer		Biallelic	Year	PubMed ID
SMAD4	SMAD family member 4	MADH4, DPC4	18q21.1	Juvenile polyposis syndrome	Colorectal cancer	Loss	Biallelic	1998	9582123
BMPR1A	Bone morphogenetic protein receptor, type IA	ALK3, CD292	10q22.3	Juvenile polyposis syndrome	Colorectal cancer Gastric cancer Hamartoma	Loss	Biallelic	2001	11381269
KIT	v-kit Hardy-Zuckerman 4 feline sarcoma viral oncogene homolog	PBT, C-Kit, CD117, SCFR	4q12		Gastrointestinal stromal tumor (GIST)	Gain	Biallelic	1998	9697690
PDGFRA	Platelet-derived growth factor receptor, alpha polypeptide	PDGFR2, RHEPDGFRA, CD140a, CD140A, PDGFRA/BCR Fusion	4q12		GIST	Gain	Biallelic	2004	14699510
CDH1	Cadherin 1, type 1, E-cadherin (epithelial)	UVO, CD324	16q22.1	Hereditary diffuse gastric cancer	Breast cancer (lobular) Gastric cancer (diffuse)	Loss	Biallelic	1998	9537325
RHBDF2	Rhomboid 5 homolog 2 (Drosophila)	RHBDL6	17q25.1		Esophageal cancer	Gain	Biallelic	2012	22265016
STK11	Serine/threonine kinase 11	LKB1	19p13.3	Peutz-Jeghers syndrome	Colorectal cancer Gastric cancer Breast cancer Sex cord–stromal tumor	Loss	Biallelic	1998	9425897 9428765
Haematological Cancer									
CEBPA	CCAAT/enhancer binding protein (C/EBP), alpha	CEBP	19q13.1		Myeloid hematological malignancy	Loss	Biallelic	2004	15575056

Continued

TABLE 20.1 Selected Cancer Predisposition Genes Arranged by Phenotypic Association—cont'd

Gene Symbol	Gene Name	Alternative Symbols	Chromosome Location	Associated Cancer Syndrome(s)	Major Associated Tumor Types	Loss or Gain of Function Leads to Cancer Risks	State in Which Associated With Cancer Risk	Year of Association	Reference (PubMed ID) for First Association
GATA2	GATA binding protein 2	NFE1B	3q21.3	Emberger/monocytopenia and mycobacterial infection (monoMAC) syndrome	Myeloid hematological malignancy	Loss	Biallelic	2011	21892158 21892162 21765025 21670465
RUNX1	Runt-related transcription factor 1	AML1, CBFA2	21q22.3		Myeloid hematological malignancy (leukemia)	Loss	Biallelic	1999	10508512
GBA	Glucosidase, beta, acid	GLUC, GBA1, GCB	1q21	Gauchersyndrome type 1	Myeloma Lymphoma Hepatocellular carcinoma	Loss	Monoallelic	1987	2880291
SBDS	Shwachman-Bodian-Diamond syndrome	SDS	7q11	Schwachman-Bodian-Diamond syndrome	Myeloid hematological malignancy	Loss	Monoallelic	2003	12496757
DKC1	Dyskeratosis congenita 1, dyskerin	DKC, NOLA4, NAP57, XAP10, Dyskerin, CBF5, DKCX,	Xq28	Dyskeratosis congenita	Acute myeloid leukemia SCC (head and neck, anorectal)	Loss	Monoallelic (X-linked recessive inheritance)	1998	9590285
Hepatobiliary System Cancer									
HMBS	Hydroxymethylbilane synthase	PBGD, UPS, PORC, PBG-D	11q23.3	Porphyria (AI)	Hepatocellular carcinoma	Loss	Biallelic	1989	2563167
HFE	Hemochromatosis	HLA-H, HFE, MVDC7, HH, HLAH, TFQTL2	6p22.2	Hemochromatosis	Hepatocellular carcinoma Cholangiocarcinoma	Loss	Monoallelic	1996	8696333

Gene	Gene name	Alias	Locus	Syndrome	Cancer				
SERPINA1	Serpin peptidase inhibitor, clade A (alpha-1 antiproteinase, antitrypsin), member 1	*PRO0684, A1A, A1AT, AAT, PI, PI1, PRO2275, alpha1AT*	14q32.13	α1 Antitrypsin deficiency	Hepatocellular carcinoma	Loss	Monoallelic	1982	3485248 7045697
Lung Cancer									
EGFR	Epidermal growth factor receptor	*ERBB1*	7p12		Non-small cell lung cancer	Gain	Biallelic	2005	16258541
DIS3L2	DIS3 mitotic control homolog (S. cerevisiae)-like 2	*FAM6A*	2q37.1	Perlman syndrome	Wilms tumor	Loss	Monoallelic	2012	22306653
Phaeochromocytoma/Paraganglioma and/or Renal Cell Carcinoma									
VHL	von Hippel-Lindau tumor suppressor, E3 ubiquitin protein ligase	*VHL1*	3p25	von Hippel-Lindau syndrome	Renal cell cancer Pheochromocytoma Neuroendocrine tumor (pancreas) Hemangioblastoma (central nervous system, retina)	Loss	Biallelic	1993	8493574
FLCN	Folliculin	*BHD*	17p11.2	Birt-Hogg-Dube syndrome	Renal cell cancer Oncocytoma	Loss	Biallelic	2002	12204536
MAX	MYC associated factor X	*bHLHd4*	14q23	Familial paraganglioma-pheochromocytoma syndrome	Paraganglioma Pheochromocytoma	Loss	Biallelic	2011	21685915
MET	Met proto-oncogene (hepatocyte growth factor receptor)	*HGFR, c-met*	7q31		Renal cell cancer (papillary carcinoma)	Gain	Biallelic	1997	9140397

Continued

TABLE 20.1 Selected Cancer Predisposition Genes Arranged by Phenotypic Association—cont'd

Gene Symbol	Gene Name	Alternative Symbols	Chromosome Location	Associated Cancer Syndrome(s)	Major Associated Tumor Types	Loss or Gain of Function Leads to Cancer Risks	State in Which Associated With Cancer Risk	Year of Association	Reference (PubMed ID) for First Association
SDHAF2	Succinate dehydrogenase complex assembly factor 2	SDH5, PGL2	11q12.2	Familial paraganglioma-pheochromocytoma syndrome	Paraganglioma Pheochromocytoma	Loss	Biallelic	2009	19628817
SDHB	Succinate dehydrogenase complex, subunit B, iron sulfur (Ip)	SDH1	1p36.1-p35	Familial paraganglioma-pheochromocytoma syndrome	Paraganglioma Pheochromocytoma Renal cell cancer	Loss	Biallelic	2001	11404820
SDHC	Succinate dehydrogenase complex, subunit C, integral membrane protein, 15 kDa	PGL3	1q21	Familial paraganglioma-pheochromocytoma syndrome	Paraganglioma Pheochromocytoma GIST	Loss	Biallelic	2000	11062460
SDHD	Succinate dehydrogenase complex, subunit D, integral membrane protein	PGL1	11q23	Familial paraganglioma-pheochromocytoma syndrome	Paraganglioma Pheochromocytoma GIST	Loss	Biallelic	2000	10657297
TMEM127	Transmembrane protein 127		2q11.2		Pheochromocytoma	Loss	Biallelic	2010	20154675
FH	Fumarate hydratase	Fumarase	1q42.1	Hereditary leiomyomatosis and renal cell carcinoma	Renal cell cancer Leiomyosarcoma (uterus)	Loss	Monoallelic or biallelic	2001	11865300

Gene	Protein	Alternative symbols	Location	Syndrome	Cancer	Loss/Gain	Allelic	Year	PMID
SDHA	Succinate dehydrogenase complex, subunit A, flavoprotein (Fp)	FP, PGL5, SDH1, SDH2, SDHF, CMD1GG	5p15.33	Carney-Stratakis syndrome	Paraganglioma Pheochromocytoma GIST	Loss	Monoallelic or biallelic	2010	20484225
RET	ret Proto-oncogene	HSCR1	10q11.2	Multiple endocrine neoplasia 2A/2B Familial medullary thyroid carcinoma	Medullary thyroid cancer Pheochromocytoma	Gain	Biallelic	1993	8099202
Skin Cancer									
PTCH1	Patched 1	PTCH, NBCCS, BCNS	9q22.3	Nevoid basal cell carcinoma syndrome Gorlin syndrome	Basal cell carcinoma Medulloblastoma	Loss	Biallelic	1996	8658145 8681379
CDK4	Cyclin-dependent kinase 4	PSK-J3	12q14		Melanoma	Gain	Biallelic	1996	8528263
CDKN2A	Cyclin-dependent kinase inhibitor 2A	p16, p14ARF	9p21		Melanoma (p16 and p14ARF) Pancreas cancer (p16) Astrocytoma (p14ARF)	Loss	Biallelic	1994	7987387 7987388
POT1	Protection of telomeres 1	CMM10, HPOT1	7q31.33	Familial melanoma	Melanoma	Loss	Biallelic	2014	24686849
XPA	Xeroderma pigmentosum, complementation group A	XP1, XPAC	9q22.3	Xeroderma pigmentosum (A)	Basal cell carcinoma SCC Melanoma	Loss	Monoallelic	1990	2234061
XPC	Xeroderma pigmentosum, complementation group C	RAD4, XPCC	3p25	Xeroderma pigmentosum (C)	Basal cell carcinoma Squamous cell carcinoma Melanoma	Loss	Monoallelic	1993	8298653

Continued

TABLE 20.1 Selected Cancer Predisposition Genes Arranged by Phenotypic Association—cont'd

Gene Symbol	Gene Name	Alternative Symbols	Chromosome Location	Associated Cancer Syndrome(s)	Major Associated Tumor Types	Loss or Gain of Function Leads to Cancer Risks	State in Which Associated With Cancer Risk	Year of Association	Reference (PubMed ID) for First Association
COL7A1	Collagen, type VII, alpha 1	Long-chain collagen, EBD1, EBDCT, EBR1	3p21.31	Epidermolysis bullosa	Squamous cell carcinoma (skin)	Loss	Monoallelic or biallelic	1993	8513326
Wilms Tumor									
WT1	Wilms tumor 1	AWT1, EWS-WT1, GUD, NPHS4, WAGR, WIT-2, WT33	11p13	Wilms tumor, anirida, genitourinary anomalies, and intellectual disability (WAGR) syndrome Denys-Drash syndrome Frasier syndrome	Wilms tumor Gonadoblastoma	Loss	Biallelic	1991	1673293
Pediatric Syndromic Presentatiobs With Associated Cancer Risks									
BLM	Bloom syndrome, RecQ helicase-like	BS, RECQ2, RECQL3	15q26.1	Bloom syndrome	Lymphoma and acute lymphoblastic leukemia hematological malignancy Myeloid hematological malignancy SCC Gastric, colorectal cancers	Loss	Monoallelic	1995	7585968

Gene	Name	Aliases	Locus	Syndrome	Cancer		Allelic	Year	ID
NBN	Nibrin	NBS1	8q21	Nijmegen breakage syndrome	Lymphoma Medulloblastoma Glioma Rhabdomyosarcoma	Loss	Monoallelic	1998	9590180 9620777
GPC3	Glypican 3	SDYS, DGSX	Xq26.1	Simpson-Golabi-Behmel syndrome	Wilms tumor Hepatoblastoma, hepatocellular carcinoma Neuroblastoma Gonadoblastoma	Loss	Monoallelic (X-linked recessive inheritance)	1996	8589713
ATM	Ataxia telangiectasia mutated	TEL1, TELO1	11q22.3	Ataxia/telangiectasia (biallelic mutations)	Biallelic mutations: Lymphoid hematological malignancy (leukemia, lymphoma) Monoallelic mutations: Breast cancer	Loss	Monoallelic or biallelic	1995	7792600
FANCA	Fanconi anemia, complementation group A	FANCH	16q24.3	Fanconi anemia (A)	Myeloid hematological malignancy SCC (head and neck, esophagus, genital tract)	Loss	Monoallelic	1996	8896564 8896563
HRAS	v-Ha-ras Harvey rat sarcoma viral oncogene homolog	HRAS1	11p15.5	Costello syndrome	Rhabdomyosarcoma Neuroblastoma Transitional cell carcinoma (bladder)	Gain	Biallelic	2005	16170316
PTPN11	Protein tyrosine phosphatase, nonreceptor type 11	CFC, NS1, SHP2, BPTP3, PTP2C, PTP-1D, SH-PTP2, SH-PTP3	12q24.13	Noonan syndrome	Juvenile myelomonocytic leukemia Neuroblastoma	Gain	Biallelic	2001	11704759 12717436
SOS1	Son of sevenless homolog 1 (Drosophila)	GF1, GGF1, GINGF, HGF, NS4	2p22.1	Noonan syndrome	Rhabdomyosarcoma	Gain	Biallelic	2007	17143285

Continued

TABLE 20.1 Selected Cancer Predisposition Genes Arranged by Phenotypic Association—cont'd

Gene Symbol	Gene Name	Alternative Symbols	Chromosome Location	Associated Cancer Syndrome(s)	Major Associated Tumor Types	Loss or Gain of Function Leads to Cancer Risks	State in Which Associated With Cancer Risk	Year of Association	Reference (PubMed ID) for First Association
SMARCA4	SWI/SNF related, matrix associated, actin dependent regulator of chromatin, subfamily a, member 4	BAF190, BAF190A, BRG1, MRD16, RTPS2, SNF2, SNF2L4, SNF2LB, SWI2, hSNF2b	19p13.2	Rhabdoid predisposition syndrome	Rhabdoid tumor	Loss	Biallelic	2010	20137775
SMARCB1	SWI/SNF related, matrix associated, actin dependent regulator of chromatin, subfamily b, member 1	INI1, SNF5L1	22q11	Rhabdoid predisposition syndrome	Rhabdoid tumor (renal, extrarenal) Central primitive neuroectodermal tumor	Loss	Biallelic	1999	10521299
Syndromes With Broad Phenotypic Spectrum With Tumors as Main Presenting Feature									
MEN1	Multiple endocrine neoplasia 1	Menin	11q13	Multiple endocrine neoplasia type 1	Parathyroid, pituitary adenoma Neuroendocrine tumor Carcinoid tumor Adrenocortical carcinoma	Loss	Biallelic	1997	9103196
BAP1	BRCA1-associated protein-1 (ubiquitin carboxy-terminal hydrolase)	UCHL2	3p21.31-p21.2		Melanoma (cutaneous, uveal) Mesothelioma Meningioma Lung cancer (adenocarcinoma)	Loss	Biallelic	2011	21874003
DICER1	Dicer 1, ribonuclease type III	DCR1, Dicer, HERNA, MNG1	14q32.13	DICER1 syndrome	Pleuropulmonary blastoma Cystic nephroma Ovarian sex cord tumor	Loss	Biallelic	2009	19556464

Gene	Full name	Aliases	Locus	Syndrome	Tumors		Allelic	Year	PubMed
PRKAR1A	Protein kinase, cAMP-dependent, regulatory, type I, alpha	TSE1	17q23-q24	Carney complex	Myxoma (cardiac, cutaneous, breast) Thyroid cancer Sex cord-stromal tumor	Loss	Biallelic	2000	10973256
PTEN	Phosphatase and tensin homolog	BZN, MHAM	10q23.3	Cowden syndrome PTEN hamartoma tumor syndrome	Breast cancer Thyroid cancer Endometrial cancer	Loss	Biallelic	1997	9140396
TP53	Tumor protein p53	LFS1	17p13.1	Li-Fraumeni syndrome	Breast cancer Sarcoma Adrenocortical carcinoma Astrocytoma	Loss	Biallelic	1990	1978757
TSC1	Tuberous sclerosis 1	LAM, hamartin	9q34	Tuberous sclerosis 1	Renal cell cancer, angiomyolipoma Subependymal giant cell astrocytoma Rhabdomyoma (cardiac)	Loss	Biallelic	1997	9242607
TSC2	Tuberous sclerosis 2	TSC4, tuberin	16p13.3	Tuberous sclerosis 2	Renal cell cancer, angiomyolipoma Subependymal giant cell astrocytoma Rhabdomyoma (cardiac)	Loss	Biallelic	1993	8269512
RB1	Retinoblastoma 1	OSRC, RB	13q14.2		Retinoblastoma Pinealoma Sarcoma Melanoma	Loss	Biallelic	1987	2885916

Original table appears in Rahman N. Realizing the promise of cancer predisposition genes. Nature 2014;505:302–8.

embryonically lethal in the rare scenario that one is inherited from both parents. A number of recessive inherited cancer syndromes have also been described, most notably colonic polyposis and colorectal cancer caused by biallelic mutations in *MUTYH* [42]. There are other intriguing examples where the inheritance of a mutation in a cancer predisposition gene has contrasting effects, depending on whether it is inherited in the monoallelic or biallelic state. This might lead to a phenotype differing qualitatively or in terms of severity. Heterozygous *SDHB* mutations cause pheochromocytoma and paragangliomas [43], whereas the inheritance of a mutation on both chromosomes leads to a neurodevelopmental disorder [44]. The tumor risks in biallelic cancer predisposition mutations may still be present but are often unable to manifest themselves because of poor life expectancy. Monoallelic *ATM* mutations are associated with an increased risk of breast cancer. Ataxia telangiectasia caused by biallelic inheritance causes a number of features such as ataxia and hematological cancers, but breast cancer has also been reported in longer surviving cases [45].

Some inherited cancer syndromes such as Li-Fraumeni syndrome caused by *TP53* mutations are associated with an increased risk of a wide range of cancer types. but most conditions are known to lead to increased risk of a small number of specific tumors. Even Li-Fraumeni–related cancers are among a set of four core malignancy types in 70% of cases [46]. The reason for this specificity is yet to be elucidated in most cases, although biological explanations include the restriction of gene expression/action to particular tissues and aberrant cellular mechanisms rendering cells susceptible to further mutation through environmental exposures only relevant to particular organs.

Some phenotypic specificity may be explained by ascertainment biases influencing the study of cancer-predisposition genes and their associated tumor risks. Identification of such genes has usually been through the preferential study of families where there are a number of occurrences of the same tumor or group of tumors. Furthermore, more likely to be included in studies are those cases where the phenotype is more severe; eg, earlier age of tumor diagnosis. The identification of novel cancer predisposition genes in these scenarios is likely to underestimate the range of tumors associated with mutations in that gene and overestimate the clinical severity of harboring them. These effects may be contributed to by the effect of clinical criteria to guide access to genetic testing (Boxes 20.4 and 20.5).

Genotype–Phenotype Correlation

When a variant in a cancer predisposition gene is detected by clinical testing, laboratory and clinical teams seek to assess that variant's pathogenicity by assessing aspects such as the likely effect on the gene product and consistency with the observed histopathological characteristics of

---

Box 20.4 Lynch Syndrome and Ascertainment Bias

Lynch syndrome is a cancer predisposition syndrome conferring susceptibility to a variety of cancers, primarily colorectal. It is caused by heterozygous mutations in mismatch repair genes such as *MLH1*.

Colorectal cancer is a common condition and clinical criteria have previously been used to identify those families likely to have tumors resulting from Lynch syndrome as opposed to an alternative cause. The Amsterdam criteria [47] were developed to provide consistency of reporting of suspected Lynch syndrome families and require a severe family history in order to be fulfilled. The Bethesda criteria [48] are designed to prompt the initiation of molecular investigations for Lynch syndrome and incorporate a wider range of families while still requiring relatively strong evidence of a predisposition syndrome.

Where genetic investigations were/are less available, rationing of analysis is likely to be based on such criteria on the basis that those fulfilling them are more likely to harbor a causative mutation and represent a more efficient use of resources. This may lead to an overestimate of the tumor risks associated with identified mismatch-repair mutations, because those families in whom the effect is less severe are less likely to been eligible for testing. Those who received testing (and tested positive) may also have greater risks conferred by other modifying genetic variants and not only because of the identified mutation, overestimating its effect.

Indeed, earlier studies estimated a higher risk of colorectal cancer than has more recently been reported. A large 1999 analysis of registry-recorded Finnish mutation carriers showed a cumulative incidence of colorectal cancer of 82% by age 70, 68 times higher than the population from which the cases were drawn [49]. However, a 2009 assessment of mutation carriers identified through genetics clinics and corrected for ascertainment bias estimated a lower cumulative incidence of 66% to age 70 [50].

---

tumors in the family. If deemed deleterious, the patient is commonly managed according to the tumor risks assigned to all deleterious variants in that gene. It is well recognized, however, that individual variants in the same gene can have contrasting clinical effects. In MEN2, caused by activating missense mutations in the *RET* proto-oncogene, a variety of tumors are observed, including medullary thyroid cancer, pheochromocytoma, and parathyroid hyperplasia/adenoma [55]. The level of risk for each of these tumors is influenced by the *RET* codon affected by the mutation to the extent that mutation status is an integral part of clinical management guidelines. For example, mutations in codon 634 are associated with an increased risk of pheochromocytoma and it is recommended that biochemical screening should start at age 8 years if detected rather than age 20 years as for many other mutations [56]. They are also associated with cutaneous lichen amyloidosis, which is not reported for other variants [57]. The p.Met918Thr variant is only associated

Box 20.5 *BAP1* and an Expanding Tumor Phenotype

BAP1 is a recently described cancer predisposition gene associated with susceptibility to an increasing array of tumors. Its discovery illustrates the potential utility of somatic mutation databases to provide gene candidature. Subsequent association with further tumor types demonstrates the ability of next generation sequencing (NGS) techniques to aid rapid definition of a broader phenotype than that described by the original association.

Acting on observational data that suggested hereditary predisposition to uveal melanoma (UM) in a proportion of cases, Abdel Rahman et al. sequenced *BAP1* in a series of UM patients with a clinical indication of hereditary susceptibility. Candidature of *BAP1* had been suggested by a number of lines of evidence, including previous study showing that around half of UM's had a somatic *BAP1* mutation [51]. *BAP1* is located on chromosome 3 and monosomy of this chromosome is commonly observed in these tumors [52]. One in 53 probands was found to have a truncating *BAP1* mutation. Their UM demonstrated loss of the wild-type allele and reduced protein product on immunohistochemistry, as did a lung adenocarcinoma diagnosed in the proband and a meningioma from a mutation-carrying relative [53].

Since this discovery, mutations in this gene have been associated with a variety of other cancers, notably renal cell carcinoma (RCC). Popova et al. identified a splice-site mutation in a family with four individuals affected with RCC and subsequently validated this finding by showing mutations in 11 of 60 families with aggregations of RCC in addition to tumors previously associated with *BAP1* variants [54]. The initial finding was obtained through whole exome sequencing (WES) where the pertinent gene was not proposed as an initial candidate, indicating that the non–hypothesis-based analysis of multiple genes simultaneously has identified more tumor associations at an earlier stage.

with the MEN2B subtype, which includes some additional features such as gastrointestinal ganglioneuromatosis [58]. Even the finding of a truncating mutation may not imply that the function of the gene product is lost and will produce a risk profile similar to other patients with only one functional allele of a given gene. The *BRCA2* c.9976A>T variant introduces a premature stop codon, but this is toward the 3′ end of the gene and is not considered to significantly increase breast and ovarian cancer risk [59].

GENETIC TESTING AND MAINSTREAMING

Individuals with suspected cancer predisposition syndromes (eg, with a strong family history of a particular tumor) are generally referred to a clinical cancer genetics service for assessment. Such services have traditionally acted as gatekeepers to testing of the single gene that is most likely to explain the phenotype of the patient consulting their service (after extensive pretest counseling relevant to that gene).

This model has provided extensive information to patients and opportunities for risk management but has a number of limitations. First, access to testing is restricted to those patients and families who conform to a phenotypic definition known to be characteristic of mutations in a given gene. This risks nondetection of significant mutations in patients whose cancers may not fit with the presumed "typical phenotype" and serves to maintain potentially inaccurate tumor estimates established because of ascertainment biases (see previous discussion). Second, sequential testing of candidate genes may be costly in terms of laboratory resources per patient and lead to a lengthy wait for results. Third, the benefits of genetic testing rely on referral from a general practitioner or specialist, which may not always occur (even when referral criteria are fulfilled).

Genetic Testing and Next Generation Sequencing

In recent years, the development of NGS technologies has begun to challenge the standard model because of the possibility of lower cost and higher throughput analysis. Using these innovations to bring the initiation of testing more into the realm of the clinician dealing with the presenting problem is likely to help address the aforementioned issues. In this alternative model, the clinical cancer geneticist's primary role may be to use specialist knowledge to advise on ambiguous results and initiate further testing or analysis in suitable undiagnosed cases.

A methodology already in widespread use is that of the gene panel. In these assays, multiple genes potentially relevant to the patient's phenotype can be sequenced simultaneously. The number of genes on a panel may vary according to the number of genes associated with the relevant phenotype and purposes of the laboratory using it.

With analysis of increasing numbers of genes comes increasing probability of identifying variants of uncertain significance, which presents difficulties to the clinician in advising patients and making management decisions. This is particularly true for most gene panels because variants in most or all of the genes analyzed potentially explain the phenotype of the tested patient. Pretest counseling for this possibility is an important consideration in the use of NGS applications for diagnostic purposes.

Results generated from greater access to gene panel testing have the potential to reduce the biases resulting from testing criteria as more mutations are found in patients with phenotypes previously considered uncharacteristic for the gene in question. It increases the knowledge base pertaining to the effects of mutations in cancer predisposition genes, particularly if clinical and variant information from these (often rare) cases is shared among clinicians and researchers through online databases.

However, panel testing still restricts the number of genes analyzed based on those most likely to be mutated in the phenotype for which the panel was designed. Mutations in genes hitherto thought to be unrelated will therefore not be detected through this method. The likelihood of this reduces as the number of tested genes increases, and some panels aim to comprehensively cover all known predisposition genes.

A yet more agnostic approach is that of whole exome sequencing (WES), which aims to provide coverage of all coding regions of the genome through selective amplification and capture of those areas before sequencing. Data relating to candidate genes can be selectively analyzed in a "virtual panel" technique, with the remainder stored for future interrogation should new candidates come to light. Alternatively, all genes can be analyzed using bioinformatic and experimental techniques to identify likely causative variants. The latter approach is largely restricted to research studies. WGS, also largely used in academic settings, does not rely on selective amplification and capture, and generates sequence for all coding and noncoding regions. This can produce better coverage of coding regions than exome sequencing and has the additional advantages of detecting chromosomal translocations and noncoding variants that may be significant in causing tumors in the patient whose sample is analyzed. A further result of WES and WGS is the potential to produce incidental findings; ie, variants potentially affecting health in ways that did not prompt the test, such as carrier status for a recessive condition. These may be desirable for patients and families to receive but should receive attention in counseling procedures before these assays.

Reasons for Nondetection of Causative Genetic Changes Through Clinical Testing

The discussed techniques are all well placed (assuming the correct gene is targeted) to detect single nucleotide changes leading to amino acid substitution or protein truncation. These mutations make up the majority of deleterious variants in cancer predisposition genes but a number of other potential mechanisms may cause tumor susceptibility and not reliably be detected by all standard methods.

Constitutional chromosome abnormalities can predispose to tumor development. This may be the result of deletion, leading to an individual possessing only one functional tumor suppressor allele, a duplication disrupting the function of a TSG, or an increase in the gene dosage of a protooncogene. Translocations may disrupt TSGs. Single gene and gene panel tests do not detect chromosome abnormalities or whole exon deletion/duplications, and additional laboratory techniques such as multiplex ligation-dependent probe amplification are required for comprehensive mutation detection. Tools exist for deletion/duplication detection from WES data, although this is more difficult than where WGS is used [60]. The latter technique also has the advantage of being able to detect and define translocation breakpoints, because sequence incorporating two genomic regions normally at distant chromosomal locations can be identified in the data.

The phenomenon of somatic mosaicism is well recognized as a cause of tumor predisposition that may evade detection by conventional genetic testing. Neurofibromatosis type 2 is a condition associated with various central nervous system tumors, particularly vestibular schwannomas. It is caused by mutations in the *NF2* gene, and mosaicism for a cell population containing them is estimated to account for around one-third of cases [61]. Mosaicism has significant implications aside from influencing mutation detection in the laboratory. It can lead to attenuated phenotypes that may prevent further investigation for the condition in question and is of reassurance to other family members, because mosaic mutations occur postconception rather than being inherited (notwithstanding the small possibility of germline mosaicism where the cell population with the mutation is present in ovaries or testes and can be inherited by offspring).

The detection of mosaicism by blood sampling depends on cells carrying mutations making up at least a proportion of circulating nucleated cells. If this is the case, the probability of detecting them will be enhanced by a greater number of independent molecular inquires in the analyzed DNA sample for a given base of interest. Conventional sequencing of single genes may reveal mosaicism, but this is not reliable and is easy to put down to an imperfect sequence readout. NGS techniques have the capability to "call" a particular base hundreds of times in a sample and consequently can reveal mutations that are present in only a proportion of cells from which DNA was extracted.

Cell populations containing mutations in cancer predisposition genes may not be represented in blood and present obvious difficulties with detection, and more examples of this situation are emerging. Identical *HIF2A* mutations were found in a patient with a paraganglioma and somatostatinoma and were felt to explain both tumors' formation. The mutation was not detected, however, in blood or other samples including urine, buccal cells, and nails [62]. In the not uncommon scenario where multiple tumors occur in the same patient [63], it may be advantageous to perform mutation analysis on both tumors. The finding of a common mutation that is not present in blood would be reassuring for family members (because a hereditary condition becomes much less likely) and may guide treatment. Such analysis may become more widespread as NGS technologies are applied in surgical and oncological settings.

Cancer predisposition genes that do not contain nucleotide variants in their coding sequence or regulatory regions may still be disrupted by epigenetic phenomena such as methylation that influences the expression of that gene. This manner of disruption, termed *epimutation*, is not detected by any of the methods discussed and requires additional assays not routinely used in clinical laboratories. An example of a condition where epimutations are known to be relevant is Lynch syndrome, where deletions of *EPCAM* are causative in an estimated 1–3% of affected families. [64] A functional *EPCAM* gene product is not necessary for DNA mismatch repair but deletion of a 3′ section of this gene that normally terminates its transcription can lead to disruption of the adjacent promoter of *MSH2* via continued transcription and consequent hypermethylation. [65] Assessment of the extent of epimutations in undiagnosed cancer predisposition syndromes will require further work in this vein that considers other factors in addition to a particular gene's DNA sequence itself.

CONCLUSION

Identification and further study of constitutional genetic alterations leading to tumor predisposition has characterized a number of susceptibility syndromes, forming the basis of an area of clinical practice, namely clinical cancer genetics. This has been based on core ideas, including the role of the genetic healthcare professional as gatekeeper for a small number of tests, and clinical criteria to ensure relatively high likelihood of a positive result before testing. Management has involved risk prediction based on genetic status and consequent risk management with either surveillance or prophylactic surgery. This model continues to be relevant, but high-throughput NGS testing strategies will have a profound impact on clinical cancer genetics and on both laboratory and clinical practice.

The number of genes tested for by a clinical cancer geneticist has traditionally been small in number, allowing pretest counseling covering aspects of the hypothesized causative gene in detail. Negative results prompt consideration of further genes or management on the basis of family history (ie, risk based on empiric observations) alone. NGS technologies have accelerated the discovery of novel predisposition genes and also made it more feasible, both economically and technically, to test them simultaneously in the clinic. These advances raise the possibility or performing much more extensive genetic testing for multiple conditions for patients fulfilling less rigid clinical criteria. A candidate to initiate this is the clinician treating the patient for their presenting tumor.

NGS tests also generate a far greater amount of data than traditional single gene tests and are consequently more likely to produce difficult-to-interpret results, requiring reference to clinical, bioinformatic, and biological information. An emerging role for the clinical cancer geneticist will be to translate those results into a clinical decision and to provide posttest counseling to guide a patient through their results.

The advances are also leading to greater opportunity to influence clinical outcomes with management strategies. This may be through better stratification of risk based on the exact mutation observed (rather than only the gene mutated) or the presence of variants in other genes that might modify risk. Knowledge of both these aspects will be improved by an increase in available data acquired through more extensive testing in cancer patients. It is hoped that the genetic status of a patient will also be much more extensively used to guide drug therapy developed on the basis of cellular pathways in which the gene in question is involved. It may become apparent that to limit access to genetic testing is to limit access to personalized/precision medicine (see chapter: Genomic Analysis in Clinical Practice: What Are the Challenges?)

REFERENCES

[1] Wunderlich V. JMM: past and present. J Mol Med 2002;80:545–8.

[2] Boveri T. Concerning the origin of malignant tumours. Translated and annotated by Henry Harris J Cell Sci 2008;121(Suppl):1–84.

[3] Nowell PC. Review series personal perspective discovery of the Philadelphia chromosome: a personal perspective. J Clin Invest 2007;117:2033–5.

[4] Bishop M. Cellular oncogenes. 1983.

[5] Nordling CO. A new theory on cancer-inducing mechanism. Br J Cancer 1953;7:68–72.

[6] Knudson AG. Mutation and cancer: statistical study of retinoblastoma. Proc Natl Acad Sci USA 1971;68:820–3.

[7] Fung YT, et al. Structural evidence for the authenticity of the human retinoblastoma gene. Science 1986;236:1657–61.

[8] Yunis JJ, Ramsay N. Retinoblastoma and subband deletion of chromosome 13. Am J Dis Child 1978;132:161–3.

[9] Balaban G, Gilbert F, Nichols W, Meadows AT, Shields J. Abnormalities of chromosome #13 in retinoblastomas from individuals with normal constitutional karyotypes. Cancer Genet Cytogenet 1982;6:213–21.

[10] Mulligan LM, et al. Germ-line mutations of the *RET* proto-oncogene in multiple endocrine neoplasia type 2A. Nature 1993;363:458–60.

[11] Donis-Keller H, et al. Mutations in the *RET* proto-oncogene are associated with MEN 2A and FMTC. Hum Mol Genet 1993;2:851–6.

[12] Benusiglio PR, et al. A germline mutation in *PBRM1* predisposes to renal cell carcinoma. J Med Genet 2015. http://dx.doi.org/10.1136/jmedgenet-2014-102912.

[13] Robles-Espinoza CD, et al. *POT1* loss-of-function variants predispose to familial melanoma. Nat Genet 2014. http://dx.doi.org/10.1038/ng.2947.

[14] Malkin D, et al. Germ line p53 mutations in a familial syndrome of breast cancer, sarcomas, and other neoplasms. Science 1990;250:1233–8.

[15] cBioPortal for Cancer Genomics. at:http://www.cbioportal.org/.

[16] Seal S, et al. Truncating mutations in the Fanconi anemia J gene *BRIP1* are low-penetrance breast cancer susceptibility alleles. Nat Genet 2006;38:1239–41.

[17] Rahman N, et al. *PALB2*, which encodes a *BRCA2*-interacting protein, is a breast cancer susceptibility gene. Nat Genet 2007;39:165–7.

[18] Antoniou AC, et al. Breast-cancer risk in families with mutations in *PALB2*. N Engl J Med 2014;371:497–506.

[19] Mavaddat N, et al. Prediction of breast cancer risk based on profiling with common genetic variants. JNCI 2015;107. djv036–djv036.

[20] Boyd L, Parkin DM. The fraction of cancer attributable to lifestyle and environmental factors in the UK in 2010. J Epidemiol Community Health 2011;65:A143.

[21] Lammens CRM, et al. Genetic testing in Li-Fraumeni syndrome: uptake and psychosocial consequences. J Clin Oncol 2010;28:3008–14.

[22] Butow PN, Lobb EA, Meiser B, Barratt A, Tucker KM. Psychological outcomes and risk perception after genetic testing and counselling in breast cancer: a systematic review. Med J Aust 2003;178:77–81.

[23] Metcalfe KA, et al. An evaluation of needs of female *BRCA1* and *BRCA2* carriers undergoing genetic counselling. J Med Genet 2000;37:866–74.

[24] Barrow P, et al. Improving the uptake of predictive testing and colorectal screening in Lynch syndrome: a regional primary care survey. Clin Genet 2015. http://dx.doi.org/10.1111/cge.12559.

[25] Meijers-Heijboer EJ, et al. Presymptomatic DNA testing and prophylactic surgery in families with a *BRCA1* or *BRCA2* mutation. Lancet 2000;355:2015–20.

[26] Esplen MJ, et al. Motivations and psychosocial impact of genetic testing for HNPCC. Am J Med Genet 2001;103:9–15.

[27] Human Fertilisation and Embryology Authority, S. and I. D. W. team., PGD conditions licensed by the HFEA – testing and screening. at:http://guide.hfea.gov.uk/pgd/.

[28] Dewanwala A, et al. Attitudes toward childbearing and prenatal testing in individuals undergoing genetic testing for Lynch syndrome. Fam Cancer 2011;10:549–56.

[29] Wilding a, et al. Life expectancy in hereditary cancer predisposing diseases: an observational study. J Med Genet 2012;49:264–9.

[30] Menko FH, et al. Hereditary leiomyomatosis and renal cell cancer (HLRCC): renal cancer risk, surveillance and treatment. Fam Cancer 2014:637–44. http://dx.doi.org/10.1007/s10689-014-9735-2.

[31] Rebbeck TR, et al. Bilateral prophylactic mastectomy reduces breast cancer risk in *BRCA1* and *BRCA2* mutation carriers: the PROSE study group. J Clin Oncol 2004;22:1055–62.

[32] Jasperson KW, Burt RW. APC-associated polyposis conditions. 2014. at: http://www.ncbi.nlm.nih.gov/books/NBK1345/.

[33] Schuler M. Aspirin and the risk of colorectal cancer in women. N Engl J Med 1996;334:121–2.

[34] Burn J, et al. Long-term effect of aspirin on cancer risk in carriers of hereditary colorectal cancer: an analysis from the *CAPP2* randomised controlled trial. Lancet 2011;378:2081–7.

[35] Farmer H, et al. Targeting the DNA repair defect in *BRCA* mutant cells as a therapeutic strategy. Nature 2005;434:917–21.

[36] Johnson RL, et al. Human homolog of *PATCHED*, a candidate gene for the basal cell nevus syndrome. Science 1996;272:1668–71.

[37] Hahn H, et al. Mutations of the human homolog of Drosophila *patched* in the nevoid basal cell carcinoma syndrome. Cell 1996;85:841–51.

[38] Gailani MR, et al. Developmental defects in Gorlin syndrome related to a putative tumor suppressor gene on chromosome 9. Cell 1992;69:111–7.

[39] Tang JY, et al. Inhibiting the hedgehog pathway in patients with the basal-cell nevus syndrome. N Engl J Med 2012;366.

[40] Nik-Zainal S, et al. Mutational processes molding the genomes of 21 breast cancers. Cell 2012;149:979–93.

[41] Rahman N. Realizing the promise of cancer predisposition genes. Nature 2014;505:302–8.

[42] Al-Tassan N, et al. Inherited variants of *MYH* associated with somatic G:C-->T:A mutations in colorectal tumors. Nat Genet 2002;30:227–32.

[43] Astuti D, et al. Gene Mutations in the Succinate Dehydrogenase Subunit SDHB Cause Susceptibility to Familial Pheochromocytoma and to Familial Paraganglioma. AJHG 2001;69:49–54.

[44] Alston CL, et al. Recessive germline *SDHA* and *SDHB* mutations causing leukodystrophy and isolated mitochondrial complex II deficiency. J Med Genet 2012;49:569–77.

[45] Gatti R. Ataxia-telangiectasia. 2010. at: http://www.ncbi.nlm.nih.gov/books/NBK26468/.

[46] Schneider K, Zelley K, Nichols K, Garber J. Li-Fraumeni syndrome. GeneReviews [Internet]. Seattle: University of Washington; 2013. at: http://www.ncbi.nlm.nih.gov/books/NBK1311/.

[47] Vasen HF, Mecklin JP, Khan PM, Lynch HT. The International Collaborative Group on Hereditary Nonpolyposis Colorectal Cancer (ICG-HNPCC). Dis Colon Rectum 1991;34:424–5.

[48] Umar A, et al. Revised Bethesda Guidelines for hereditary nonpolyposis colorectal cancer (Lynch syndrome) and microsatellite instability. J Natl Cancer Inst 2004;96:261–8.

[49] Aarnio M, et al. Cancer risk in mutation carriers of DNA-mismatch-repair genes. Int J Cancer 1999;81:214–8.

[50] Stoffel E, et al. Calculation of risk of colorectal and endometrial cancer among patients with Lynch syndrome. Gastroenterology 2009;137:1621–7.

[51] Harbour JW, et al. Frequent mutation of *BAP1* in metastasizing uveal melanomas. Science 2010;330:1410–3.

[52] Horsman DE, White VA. Cytogenetic analysis of uveal melanoma: consistent occurrence of monosomy 3 and trisomy 8q. Cancer 1993;71:811–9.

[53] Abdel-Rahman MH, et al. Germline *BAP1* mutation predisposes to uveal melanoma, lung adenocarcinoma, meningioma, and other cancers. J Med Genet 2011;48.

[54] Popova T, et al. Germline *BAP1* mutations predispose to renal cell carcinomas. Am J Hum Genet 2013;92:974–80.

[55] Giusti F, Marini F, Brandi ML. In: Pagon RA, Adam MP, Bird TD, et al., editors. GeneReviews [Internet]. Seattle: University of Washington; 2013. at: http://www.ncbi.nlm.nih.gov/books/NBK1257/.

[56] Kloos RT, et al. Medullary thyroid cancer: management guidelines of the American Thyroid Association. Thyroid 2009;19:565–612.

[57] Seri M, et al. A Cys634Gly substitution of the *RET* proto-oncogene in a family with recurrence of multiple endocrine neoplasia type 2A and cutaneous lichen amyloidosis. Clin Genet 1997;51:86–90.

[58] Hofstra RM, et al. A mutation in the *RET* proto-oncogene associated with multiple endocrine neoplasia type 2B and sporadic medullary thyroid carcinoma. Nature 1994;367:375–6.

[59] NHGRI: Breast Cancer Information Core. at:https://research.nhgri.nih.gov/projects/bic/.

[60] Tan R, et al. An evaluation of copy number variation detection tools from whole-exome sequencing data. Hum Mutat 2014;35:899–907.

[61] Moyhuddin a, et al. Somatic mosaicism in neurofibromatosis 2: prevalence and risk of disease transmission to offspring. J Med Genet 2003;40:459–63.

[62] Zhuang Z, et al. Somatic *HIF2A* gain-of-function mutations in paraganglioma with polycythemia. N Engl J Med 2012;367:922–30.

[63] Rosso S, et al. Multiple tumours in survival estimates. Eur J Cancer 2009;45:1080–94.

[64] Kuiper RP, et al. Recurrence and variability of germline *EPCAM* deletions in Lynch syndrome. Hum Mutat 2011;32:407–14.

[65] Ligtenberg MJL, et al. Heritable somatic methylation and inactivation of *MSH2* in families with Lynch syndrome due to deletion of the 3′ exons of *TACSTD1*. Nat Genet 2009;41:112–7.

Chapter 21

The Provision of Medical and Health Genetics and Genomics in the Developing World

N. Sirisena, D. Sumathipala, K. Wettasinghe, V.H.W. Dissanayake
University of Colombo, Colombo, Sri Lanka

Chapter Outline

Introduction   285
Public Health Programs in Genetics and Genomics   287
Medical Services Incorporating Genetics and Genomics   290
Summary   293
References   293

INTRODUCTION

The main goal of provision of health interventions in genetics and genomics is the prevention and care of diseases with a significant underlying genetic contribution [1]. Medical genetics traditionally concerns itself with inherited single gene disorders and genetic diagnostics, accompanied by nondirective counseling, to help patients make informed decisions based on their genetic risk profile. A genomics approach has applications far beyond simply genetic disorders; it can lead to an in-depth understanding of the function of multiple genes and gene–environment interactions, thereby shedding more light on the common complex and multifactorial diseases, and resulting in the development of numerous diagnostic/prognostic biomarkers and targeted therapeutics.

Compared with developed nations, developing countries are burdened by high levels of poverty and their limited resources to promote health and combat disease are scarce. Disease profiles in these countries are characterized by a higher proportion of nutritional, infectious, and social conditions leading to ill health and death. Superimposed on this reality, there is an increasing visibility of diseases traditionally associated with developed countries (eg, noncommunicable diseases), especially in countries that are achieving the middle income status. Facing this hard reality, developing countries must carefully assess their disease burden, set their priorities, and use their scarce health budgets wisely, making sure that the prevention and care of genetic disorders is not neglected

and finds an appropriate place among other health priorities. Therefore achieving health benefits will depend on how healthcare service providers ensure that genetic and genomic technologies are used effectively, efficiently, equitably, and responsibly for the population as a whole.

It is said that the differences in the goals and strategies for genetic services between the developed and the developing world have more to do with epidemiology, economics, and education than with differences in culture, traditions, and/or religion. For example, epidemiological factors determine the priorities of health services (ie, balance between communicable versus noncommunicable diseases), the economic factors determine the availability of funds and their allocation, and the level of education of health professionals and the general public influence the promotion and uptake of different community-based prevention programs [2].

The Human Genome Project, with its completion in 2003, has been the mainstay of biomedical research, influencing the future of medical and heath genetics [3]. Optimism about the potential of genomics has been fueled by the discovery of the molecular basis of inherited diseases, some of which has led to novel therapeutic targets. Advances have contributed to the gain of knowledge in basic sciences, as well as health sciences, in the diagnosis and management of disease and pharmacogenomic testing before the administration of certain medications. Although genomics has begun to improve diagnostics and therapeutics, profound improvement in the provision of healthcare is dependent on the concomitant improvement in policy frameworks and their implementation.

Medical and Health Genomics. http://dx.doi.org/10.1016/B978-0-12-420196-5.00021-6

The essential components of genetic services and public health genetic programs that are applicable in the context of developing countries have been described [1]. They include:

1. Genetic services and public health genetic programs that combine prevention as well as care for affected individuals and families

   The commonest monogenic disorders are the hemoglobinopathies (thalassemias and sickle cell disease and its variants), conditions that have a particularly high incidence in sub-Saharan Africa, the Mediterranean region, the Middle East, the India subcontinent, and throughout southeast Asia [4]. Simple, inexpensive tests have been developed and are now available for carrier screening of these diseases [5]. The implementation of premarital/preconceptional carrier testing, accompanied by genetic counseling and prenatal genetic testing, provide critical information for making informed choices about marriage, reproduction, and lifestyle. Different programs for prenatal diagnosis for hemoglobinopathies have been implemented, at different levels of healthcare in countries like Nigeria, Pakistan, Cuba, and India [6]. Prenatal diagnosis using rapid and inexpensive nucleic-acid amplification technology, such as the polymerase chain reaction, has been found to be useful in the diagnosis of sickle cell anemia, a condition associated with a high level of mortality and morbidity in most parts of sub-Saharan Africa [1].

2. Genetic services and public health genetic programs that prioritize simple, low-cost technology and cost-effective activities according to the prevalence and severity of existing genetic diseases and the expectations and cultural traditions of the community focusing on genetic risk detection and subsequent referral

   Basically, there are two broad categories of genetic risk: (1) personal risk of developing a genetically determined disorder, and (2) reproductive risk of transmitting an inherited disorder to one's offspring. Implementation of genetic diagnostic services should always be accompanied by appropriate genetic counseling and support services with the aim of informing patients about their genetic risk and protecting them from discrimination. An example of a success story is the implementation of a program for spinocerebellar atrophy type 2 in the province of Holguin, Cuba, which is home to hundreds of affected individuals and thousands of at-risk individuals. A public health program of presymptomatic testing of at-risk individuals, followed by genetic counseling and the offer of prenatal diagnosis, which has been put in place by the public health system there, has resulted in a drastic reduction of cases [7]. Similarly, genetic services in Cyprus grew out of the need for prevention and control of β-thalassemia, because 16% of the population were carriers. A plan for prevention with health education, community involvement, carrier detection in the general population, and genetic counseling was put in place in

the early 1970s. Soon after, a comprehensive prenatal diagnosis program followed. Over the ensuing years, the annual birth incidence of thalassemia homozygotes dropped 97% as a result of public education, community participation, genetic counseling, and voluntary prenatal diagnosis followed by the option of pregnancy termination. The success of this program was attributed to the fact that it was based on public education and was developed with consideration for the local cultural practices [8].

3. Genetic services rooted in primary healthcare and linked with regional secondary and tertiary levels of care

   Health services in Cuba are centrally planned and supported by the state according to national goals. In the early 1980s, the National Center of Medical Genetics was created and staffed with medical geneticists trained abroad. This core staff trained several teams of pediatricians and obstetricians for the major hospitals in every Cuban province, where genetic services were established. Prenatal prevention programs addressed mainly sickle cell disease, congenital malformations, and chromosome abnormalities. All components of the Cuban genetic program are regionalized with family physicians as the first line of care at the primary level and a referral network of regional genetic clinics is in place. Currently all provinces in Cuba have genetic clinics staffed by clinical geneticists and nurses are being trained in genetic counseling. The goals and methods of the Cuban genetic program have been well accepted by the Cuban population and there has been a significant reduction in the birth prevalence of sickle cell anemia, neural tube defects, severe congenital heart defects, and other serious malformations. The Cuban approach to the control and management of genetic disorders and birth defects is based on the identification of main epidemiological priorities, taking advantage of preexistent health programs, proper regionalization, and development of appropriate technology. The key factors in the program's success were a clear statement of goals and policies, high-level political support, and an adequate organization of existing resources [9].

   The use of information and communication technology to centers with genetic counseling services to patients at other centers seems to be mainly confined to developed countries [10]. In Sri Lanka, a pilot project aimed at offering genetic counseling services available at the only clinical genetic center in the country in Colombo, the country's capital, to patients in distant parts of the country using Skype technology had been implemented in 2006 with limited success [11].

4. Prenatal genetic services aimed at detecting genetic reproductive risks through carrier screening for prevalent recessive conditions and prenatal screening for neural tube defects and chromosomal abnormalities that is adequately regionalized in tertiary care centers to ensure cost-efficiency and quality of service

Experience shows that these programs are generally successful when couples identified as "at risk" have access to diagnostic tests, free access to genetic counseling, and the option to terminate affected pregnancies as observed in a number of developing countries, such as China, Cuba, Cyprus, India, South Africa, and important localities in Mexico [12]. However, an important limitation to the successful development of these services in some developing countries is the legal prohibitions that prevent selective abortion for the reason of fetal abnormalities [1].

To provide these genetic services, primary healthcare personnel should be properly trained to perform defined tasks, such as recording and interpreting a family history, detecting reproductive genetic risks, and detecting signs and symptoms that require referral to the next level of care. This is an area where there are many challenges to overcome [13].

In addition, application of genetics and genomics in healthcare should be accompanied by research appropriate to answer questions pertinent to specific populations because, after all, each population would have its own unique genetic background [14]. Thus epidemiological research should be stimulated to provide accurate information on the prevalence and types of birth defects, genetic diseases, and genetic predisposition to common diseases existing at the country level. Issues of population structure such as consanguinity, founder effects, and cultural and geographic isolation should be investigated to determine their influence on geographical clusters of genetic diseases. It is also important to assess the health beliefs, cultural traditions, and social expectations of communities before setting program goals and other interventions [2].

It is interesting to note that Mexico's National Institute for Genomic Medicine has described in a recent publication a nine-point strategy for the adoption of genomic medicine [15]. This includes:

a. building an innovative organizational design
b. establishing the initial infrastructure
c. initiating nationwide strategic alliances
d. conducting research and development in genomic medicine
e. applying genomic technology to common health problems
f. reaching excellence in teaching and training programs
g. supporting academic programs in genomic medicine
h. addressing ethical, social, and legal issues
i. translating genomic knowledge into products and services.

The provision of health interventions in genetics is basically organized under two main categories:

1. public health programs such as public education and genetic screening aimed at populations with the main aim of prevention

2. medical genetic services as part of general healthcare services, primarily addressing the care and management of individuals and families affected with or at risk for genetic disorders.

PUBLIC HEALTH PROGRAMS IN GENETICS AND GENOMICS

A number of World Health Organization (WHO)–convened advisory groups have formulated the criteria for applying genetics and genomics to the health problems of developing countries. The following are the main recommendations for developing countries:

1. The need to recognize the burden imposed by genetic disorders and birth defects
   The three major categories of disorders with a genetic component include chromosomal abnormalities, monogenic (single gene) disorders, and complex multifactorial disorders. Chromosome imbalances resulting from numerical or structural aberrations contribute to about 10–20% of the public health burden of spontaneous abortions, congenital anomalies, mental retardation, and developmental disorders worldwide. The average birth prevalence of chromosomal abnormalities in liveborns is 1 in 200 [1]. Some other sources suggest that the birth prevalence of chromosome disorders is higher in developing than in developed nations, and may reach 3 per 1000 for Down syndrome and about 6 per 1000 for all chromosome abnormalities. These figures are consistent with the higher proportion of births to women of advanced age and the limited access to family planning, prenatal diagnosis, and selective pregnancy termination in most developing countries [2].

   Although individually rare, the highly penetrant monogenic disorders, eg, thalassemia, sickle cell disease, cystic fibrosis, phenylketonuria, the hemophilias, muscular dystrophies, and Huntington disease, collectively affect about 1% of the population worldwide [16]. The hemoglobinopathies (sickle cell disease and thalassemias) constitute the major public health problem posed by genetic conditions, particularly in the developing world.

   The multifactorial diseases caused by complex interactions between environmental factors such as nutrition, infections, exposure to toxins, and lifestyle factors, and multiple gene variants of minor effect constitute the major bulk of the global burden of disease. They include conditions like congenital anomalies, coronary artery disease, cancers, mental illness, diabetes, obesity, and hypertension. The genetic component of these diseases is variable and does not usually account for more than 10–20% of the attributable risks [1].

Consanguineous marriages continue to be very prevalent in numerous parts of the developing world. In some parts of Africa, the eastern Mediterranean region, Brazil, and southern India, the rate of consanguinity ranges from 25% to 60% [2]. Consanguinity increases the likelihood of homozygosity among the offspring and is associated with higher risks for stillbirths, neonatal or childhood death, congenital malformations, and intellectual disability, and a higher incidence of autosomal recessive conditions. Thus high incidence of consanguinity may contribute to the burden of genetic diseases in some communities in the developing world. The combination of founder effects and geographical isolation has led to clusters of both autosomal recessive and dominant single-gene diseases, as is the case with spinocerebellar atrophy in Cuba, Huntington disease in Venezuela, and albinism in South Africa [2]. The strong cultural practices that favor consanguineous marriages seem to weigh highly against the medical services staffed by medical personal, who are ill-trained to deal with all the issues that confront them in such populations [17].

2. The need for political will and commitment to support the implementation of genetic services

For health planners to weigh fairly the demands of genetic disease control against other health needs, the prevalence and burden of genetic disorders in a defined population needs to be recognized. Prevention programs are often misconceived as expensive and dependent on sophisticated technologies. Such misconceptions can be dispelled only with the availability of sound epidemiological data to convince policy makers of the need for and impact of genetic services. The Cuban genetic program succeeded as a result of the strong political will and support of its government [9]. Similarly, in Malaysia, the first medical genetic service was introduced in 1994 at one of the main teaching hospitals in Kuala Lumpur. This led to the improvement of the medical genetic services in the country with the availability of genetic counseling and diagnostic services for both pediatric and adult-onset inherited conditions at four main centers of medical genetic services in Malaysia [18]. Services for prenatal diagnosis and assisted reproductive technologies were established at tertiary centers and private medical facilities. According to Lee et al., some of the positive outcomes that developed as a result include: governmental recognition of clinical genetics as a subspecialty, increased funding for genetics services, development of medical ethics guidelines, and the establishment of support groups. Policy makers were presented with proposals to develop genetic counseling courses owing to a lack of qualified genetic counselors. Some of the challenges encountered included limited resources and public awareness, ethical dilemmas such as religious and social issues, and inadequate numbers of well-trained genetic health professionals [18].

3. The need to improve epidemiological knowledge about genetic disorders and birth defects

The March of Dimes (MOD) Global Report on Birth Defects stated that worldwide, around 7.9 million births occur each year with serious birth defects and almost 94% of these births occur in the developing countries [19]. According to the joint WHO and MOD meeting report, 7% of all neonatal mortality and 3.3 million deaths of children under age 5 years result from birth defects. It is believed that 70% of birth defects can be prevented through the application of various cost-effective community genetic services [1,20].

The reduction in infant mortality in most developing countries signifies a general improvement in most health indicators as infections and malnutrition are brought under control. This reflects an epidemiological transition with birth defects and genetic diseases accounting for an increasing share of morbidity and mortality. The prevalence of birth defects and genetic disorders in the developing world is underestimated mainly because of a lack of diagnostic capacity and the unreliability of health records and statistics. A registry of birth defects should address not only the collection, analysis, and dissemination of information but also contribute to local interventions like prevention, diagnosis, and treatment. Countries that have maintained birth defect registries such as Mexico, Costa Rica, Cuba, South Africa, and China report that the prevalence of recognizable malformations among newborns is 2–3%, which is similar to that found in the developed countries [2].

4. The need to define the goals of genetic services in terms of both individual and family well-being and public health

As an example, carrier screening programs for recessive conditions particularly prevalent in some populations provides couples and individuals with information necessary for making informed decisions regarding marriage and reproductive options.

5. The need to improve prenatal and perinatal services

Prenatal diagnosis is considered the most practical current application of genetic services, particularly where termination of pregnancy (TOP) is legal. This is the case in China, Southeast Asian countries, India, South Africa, Cuba, and Cyprus, where government-funded prenatal diagnosis is provided. In other countries, although TOP is technically illegal, prenatal diagnosis of fetal abnormalities is an accepted service. However, in countries where the public sector does not fund these services, the growing demand for prenatal diagnosis is channeled to the private sector, creating economic barriers to access [2].

6. The need to organize genetic services in a comprehensive and integrated manner with roots in the primary healthcare level

Public health policies throughout the world have emphasized the need of clearly defining program goals and extending primary healthcare to all segments of the population, while linking the primary care level with secondary and tertiary levels in a regionalized manner. The application of that same strategy is key to the implementation of population-based preventative programs for birth defects and genetic diseases and patient-/family-based genetic services. In developing countries, the main setbacks to increasing access to genetic services have been the shortage of trained specialists, the lack of appropriate technology, the scarcity of funding, and the lack of links with the primary healthcare level. The challenge for developing countries is to define the type of genetic services needed according to the prevalence of genetic diseases. The development and implementation of services with roots in the primary healthcare level that follow sustainable cost-efficient models are crucial [1].

7. The need to select programs and targets according to prevalence, severity, and predicted outcomes
The target population for genetic services may be the entire community (eg, heterozygote detection of carrier status for a prevalent recessive condition like thalassemia), the prenatal population (eg, maternal serum screening or fetal anomaly scan), or special groups identified by the presence of a particular risk factor (maternal age, family history, exposure to teratogens) [21,22].

8. The need to respect ethical principles and cultural diversity
The developing world differs from the west in its cultural diversity. Therefore proper attention must be paid to the ethical, legal, and social issues associated with genetic testing and screening. Genetic testing and screening should be voluntary, offered only in the context of comprehensive genetic service delivery, and accompanied by proper genetic counseling appropriate to the test and clinical context. Such programs should be implemented in a manner that is sensitive to the population's cultural practices and religious views and in accordance with national legislation [1].

In most instances, cultural practices hinder progress of genetics and introduction of genetic services. In Sri Lanka, we found an interesting social marketing initiative aimed at promoting a public health program of carrier screening and matching test results to avoid marriages between carriers of β-thalassemia, where the concept of screening and comparing results was presented as the modern-day equivalent of the traditional cultural practice of matching horoscopes or astrological charts and avoiding mismatched marriages [23].

9. The need to train health professionals in medical genetics
A significant deficiency in most of the developing world is the lack of education of health professionals in genetics. Most medical schools in developing countries do not have formal courses in clinical genetics and physicians have inadequate knowledge of the modern application of genetics in medical practice. In Latin America, clinicians and scientists who received medical genetics training abroad eventually developed successful local postgraduate training programs in medical genetics in their countries, and there are currently more than 500 trained clinical geneticists in the region. The specialty of medical genetics has been recognized in Mexico, Cuba, Brazil, and Argentina [2]. Given the contextual differences in healthcare and social, cultural, religious, and legal factors, the modalities for delivering genetic counseling, including the personnel who provide the service, should be left to each country to determine. Defining the minimum training required for health personnel to provide genetic counseling is the basic priority. Training primary health professionals (general physicians, nurses, social workers) in the core competencies in genetics in developing countries would help address these shortages [1]. Genetics plays a key role in the daily practice of primary care practitioners with regard to detecting and managing the risk of multifactorial disorders, genetic reproductive risks, and targeted drug therapy. The development of appropriate guidelines and web-based information resources and training professionals to utilize these tools will help them make personalized genetic risk assessment a part of holistic, patient-oriented, primary healthcare practice [12].

10. The need to educate the public in genetics
Community-based health educational programs are often valuable to improve community awareness of genetic risk and to reduce the stigma sometimes attached to those identified as carriers of genetic disorders [24]. Public education on genetic disorders and birth defects is of paramount importance, including the appropriate use of the media while respecting ethical principles and cultural diversity. Secondary school curricula should reinforce the role of genetics in human health through the use of common examples [25]. The Community Health Agents Program was established in Brazil in 1991 as part of the Family Health Program. The program is seen as an important facet of primary healthcare strategy, with community health agents trained to identify, refer, guide, and provide follow-up care to families in their community [26].

11. Encourage the formation of parent/patient organizations
The formation of organized groups of individuals and families affected by genetic disorders should be stimulated. Besides providing information to increase the public knowledge about genetics, these organizations help families in many ways to cope with the challenges faced in caring for affected individuals [25].

The challenge for the future, however, lies in incorporating genomics in programs aimed at prevention of common noncommunicable disorders. As predisposing genetic factors to common multifactorial disorders are being recognized, programs should be implemented to identify those factors in the population. Much research is targeted at detecting susceptibility alleles for common diseases in the hope that knowledge may shed light on disease pathogenesis, the causative role of genetic and environmental factors, and the development of newer methods of prevention and treatment. However, for the majority of multifactorial disorders, the underlying genetic contribution remains unknown or ill understood, and most predisposing variants uncovered so far have very low penetrance and hence very low clinical utility. Preventative measures and treatment for conditions such as diabetes, obesity, coronary disease, and some cancers follow the detection of such genetic risk factors [1].

MEDICAL SERVICES INCORPORATING GENETICS AND GENOMICS

Genomics has several distinguishing elements that will have to be fulfilled for the efficient provision of medical genetics:

1. Making genomic-based diagnosis routine
   Developments of genomic-based medical investigations were based on research and novel discovery. However, in the future, technology must evolve to aid clinicians to acquire a complete genomic diagnostic panel as a routine patient management option. For the realization of this goal, advancements must occur in several sectors: strengthening primary and secondary education, conducting public outreach, building healthcare provider genomic competencies, and preparing of the next generation of genomic researchers. Healthcare providers require continuous training in interpretation of results in parallel with the advancement of the complexity of data provided by genetic tests. In addition, training in communication skills for efficient transmission of genetic information to patients and family members is required. The present numbers of clinical geneticists and genetic counselors are insufficient to manage the expanding medical service of genomic medicine. In addition, as genomic investigations begin to be routinely utilized in the clinical setting, the role of interpretation and reporting will expand. Therefore a standard for genomic medicine certification for other subspecialists is urgently needed, as is genomic medicine training for other healthcare workers [27].

2. Defining the genetic component of disease
   All diseases have a genetic component ranging, from single-gene mendelian inheritance patterns to complex diseases with environmental and genetic components. The genetic variations underlying each disease need to be assessed by organizing patient sample collections and analysis of the spectrum of disease conditions. This requires continual, broad, and representative public participation in genomic research. Generating comprehensive data requires organization at an international level with interdisciplinary expertise and high standards of data that ensure broad utility and computational intensity.

   At present a clinically accepted universal database of disease-associated mutations does not exist. Therefore interpretation of variants found is problematic. Use of information in primary literature and general and locus-specific databases is the norm. However, it is disturbing to note that 27% of literature-cited mutations may be incorrect [28]. This issue is being addressed through the establishment of consortia such as those organized through the National Center for Biotechnology Information, the Human Genome Variation Society, and the Human Variome Project [29].

3. Practical systems for clinical genomic information
   Genomic catalogs of data are available at present and are widely used. However, there is a need to further improve existing collections of databases and create novel catalog areas such as functional genomic elements, RNAs, proteins, and other biological molecules. International efforts have created catalogs such as The SNP Consortium, the International HapMap Project, and the 1000 Genomes Project. Another example is the Global Alliance for Genomics and Health (http://genomicsandhealth.org/), which is focused on ensuring effective and responsible sharing of clinical and genomic data around the world [30]. To further improve on the present system of data catalogs, requirements include data analysis, integration, visualization, computational tools and infrastructure advancement, and extensive training of individuals.

   In the practical setting, interpretation and reporting of genetic variants of unknown significance (VUSs) remains a significant challenge. Researchers are prone to report variants that may not be pathogenic for fear of undercalling a pathogenic genotype. However, as genomic data accumulates, the knowledge of variant frequencies and their correlation with disease should improve the interpretation of VUSs [31].

4. Psychosocial and ethical implications
   Genomic research, clinical data, and the utilization of genomic data have numerous societal and ethical implications that demand careful attention. These include psychosocial and ethical issues in genomic medicine, legal public and policy issues, and broader societal issues [32].

   We are living in a time of unparalleled surge in knowledge and rapidly changing technology. Such biotechnology, especially when it involves diverse populations, raises complex ethical, legal, social, and religious issues. Despite this, it is imperative that genetic screening programs are carried out by developing countries. Comprehending the psychosocial impact of a congenital disorder such as Down syndrome on affected individuals and their families necessitates an understanding of the cultural setting in which they are located. A study was carried out to characterize understandings of Down syndrome in Pakistan in a group of health professionals, researchers, and parents of children with Down syndrome. Attitudes toward prenatal testing and TOP demonstrated that a belief in religion was not inevitably associated with a rejection of these technologies. The study reflects the religious, cultural, and economic context of Pakistan, and issues concomitant with raising a child with a learning disability in that country [33].

Over the past decade, significant advances in genetic testing and technologies have altered the clinical management of individuals. Advances in genetic testing are also accompanied by a string of new challenges related to the ethical and social issues for our society. Medical genetics is the field of medicine most centrally involved in providing services to people with genetic disorders and to their families. There are a number of ways in which genetic information is considered to be different from other health information, although the degree and impact of these differences remain under discussion. For example: (1) genetic information may have medical and social implications for an entire family, rather than only the affected individual; (2) genetic discoveries may be predictive of future adverse health events for an individual or family member who is currently unaffected; and (3) genetic information and the choices made in the present may affect future generations. These features of genetic information generate specific ethical considerations, particularly concerning the implementation of medical genetic services [24]. The following four principles, outlined in the WHO 2006 review of ethical issues in medical genetics, are widely accepted as philosophical doctrines of bioethics:

a. *autonomy*: concerning the self-determination of individuals and protecting those individuals with diminished autonomy
b. *beneficence*: giving utmost priority to the well-being of individuals and maximizing benefits to their health
c. *nonmaleficence*: evading and averting harm to individuals or, at least, minimizing harm
d. *justice*: treating individuals and groups impartially, and allocating benefits and burdens of healthcare as impartially as possible in society

The ethical issues arising in relation to genetic testing and screening in developing countries cannot be readily compartmentalized rendering to these principles, because the issues often comprise of questions relating to more than one principle [1]. Thus effectively examining the societal implications and repercussions of genomic advances involves expertise and skills in genomics and clinical medicine and professionals in bioethics, psychology, sociology, anthropology, history, philosophy, law, economics, health services research, and related disciplines [24].

Psychosocial and ethical concerns in genomic medicine include interpreting information; communicating with patients about the ambiguity and evolving nature of expectations based on genomic information; evaluating the usefulness of genomically informed diagnostics and therapeutics; addressing issues related to preimplantation, prenatal, and postnatal genetic diagnoses; and determining how concepts of race and ethnicity relay to the biology of disease and the potential to advance genomic medicine. Psychosocial and ethical issues in genomics research include guaranteeing appropriate well-being of human research participants and addressing the perceptions of risks and benefits of participating in genomic studies; expanding the diversity of research cohorts; integrating biological ancestry markers and ethnicity as variables in genomic studies; and considering vulnerable populations (for example, children and the disabled) and deceased individuals in genomics research [27]. Legal and public policy concerns consist of intellectual property in genomics; insurance compensation for genomic services; regulation of genetic testing; regulating direct-to-consumer genetic testing; the regulation of pharmacogenomics and genomics-based therapeutics; protection against genetic discrimination and stigmatization; and uses of genomics in nonmedical settings. Broader societal issues include the consequences of increasing genomic knowledge for conceptualizing health and disease; for understanding identity at the individual and group levels, including race and ethnicity; for gaining insights about human origins; and for considering genetic determinism, autonomy, and individual responsibility. The patient's cultural background, including beliefs, rituals, and customs, are recognized, valued, respected, and amalgamated into the care plan [27].

Diagnostic genetic testing may be performed on individuals with physical, developmental, or behavioral features associated with a potential genetic syndrome or for pharmacogenetic drug selection and dosing decisions. Developments in genetic research hold particular potential in the diagnosis and treatment of childhood diseases. Predictive genetic testing may be performed on an individual with a positive family history for a particular genetic condition, particularly if early screening or treatment may affect morbidity or mortality. Screening of an individual with symptoms of a genetic disorder, genetic testing, is equivalent to other medical diagnostic evaluations. Parents or guardians should be educated about the potential benefits and potential harms of testing, and their consent should be acquired. Medical benefits of genetic screening comprise of the opportunity of preventative or therapeutic interventions, choices about surveillance, the explanation of diagnosis and prognosis, and recurrence risks. Medical harm occurs if parents or guardians respond to the results by pursuing unproven treatments or precautionary measures, particularly if they are ineffective or have significant adverse effects. Therefore appropriate posttest counseling is important [34].

We are arriving at a fascinating and uncertain era of medical history as existing DNA sequencing technology holds the potential to revolutionize our medical care and predict our future based on information of our own individual inherited and acquired genetics. Potential benefits of carrier screening and detection in childhood include potentially greater acceptance and amalgamation of status into life plans, prevention of the shock and resentment that may accrue when detection is delayed, and greater opportunity for parental guidance in appreciating the nature of the genetic challenge and accessible management options. As stated previously, newborn screening may detect carriers for recessive conditions such as hemoglobinopathies and cystic fibrosis. There is broad consensus that when carriers are identified in newborn screening, carrier status should be revealed to the child's parents or guardians. Furthermore, potential harms of early screening include labeling, stigma, or discrimination, and a potential for misunderstanding and misinterpretation of the distinction between carrier status and affected status [34].

Predictive genetic testing can happen in numerous situations or contexts and can refer to predictive testing of either a childhood-onset or adult-onset condition. Most predictive genetic testing for adult-onset disorders is predispositional. Predictive genetic testing of minors should be considered only if effective medical interventions are available and obtainable to treat, prevent, or delay the course of the disease. Medical harms include misdiagnosis to the degree that a genotype does not correlate with a phenotype, ambiguous results in which a precise phenotype cannot be predicted, and use of ineffective or harmful preventative or therapeutic interventions. Psychosocial benefits include reduction of uncertainty, improbability, and anxiety; the opportunity for psychological modification or adjustment; the capability to make realistic life plans; and sharing the facts with family members. Psychological harms include alteration of self-image, distortion of parental perception of the child, increased anxiety and guilt, altered expectation by self and others, familial stress related to identification of other at-risk family members, difficulty obtaining life and disability insurance, and the detection of misattributed parentage. Reproductive benefits include avoiding the birth of a child with genetic disease or having time to adjust for the birth of a child with genetic disease. Reproductive harms include altering family planning decisions on the basis of social pressures [34].

As public health screening continued to expand, the WHO commissioned a study by Wilson and Jungner, who in 1968 enumerated 10 criteria "to guide the selection of conditions that would be suitable for screening." Although not written specifically for genetic applications, the criteria have served as a policy standard for population-based genetic screening since the 1980s [35]. The criteria are:

a. The condition sought should be an important health problem.
b. There should be an accepted treatment for patients with recognized disease.
c. Facilities for diagnosis and treatment should be available.
d. There should be a recognizable latent or early symptomatic stage.
e. There should be a suitable test or examination.
f. The test should be acceptable to the population.
g. The natural history of the condition, including development from latent to declared disease, should be adequately understood.
h. There should be an agreed policy on whom to treat as patients.
i. The cost of case finding (including diagnosis and treatment of patients diagnosed) should be economically balanced in relation to possible expenditure on medical care as a whole.
j. Case finding should be a continuing process and not a "once and for all" project [34].

In an era of personalized medicine, pharmacogenomics is used to increase therapeutic sensitivity and reduce the rate of adverse drug reactions. The principle of respect for autonomy is upheld in the process of informed consent. When the individual lacks decision-making capability, as all newborn infants do, the standard practice includes surrogate permission for medical interventions. Furthermore, from a global and local economic perspective, it is mandatory to make sure that the personalized medical application of large-scale genomic analysis not be a luxury or a high-cost center, but have the potential to save both lives and healthcare expenses, early disease detection/screening, and more efficient pharmaceutical delivery [34]. To achieve this goal, we need to expand the clinical use of this technology both rapidly and economically while safeguarding the reliability of the techniques and the safety and well-being of patients and research participants [34]. This requires a careful consideration regarding the suitable setting and regulatory structure surrounding genomics, in addition to the development of consensus concerning what constitutes of a genetic test in the time of extensive genomics and informatics.

5. Therapeutics

Genomic information has the potential to improve therapeutics. When genomics focuses on pharmaceuticals, opportunities exist to improve target identification, rational drug design, genomic-based stratification in clinical trials, higher efficacy and fewer adverse effects with genotype-guided drug prescription, and the development of gene therapy strategies.

Developing countries are not only potentially huge drug markets for drug therapeutics but are also depositories of important human genetic diversity. Genotyping studies of various populations in the developing world will become valuable in that understanding this diversity will better define those population subgroups that will benefit more from one particular drug than others, and allow the detection of side effects that might not be seen in populations in the developed world. In spite of limited resources, several developing countries have begun characterizing the genetic variation within their populations in the hopes that this information may be used someday to improve local health (eg, HUGO Pan-Asian Initiative, Indian Genome Variation Consortium, Thailand SNP Discovery Project, Pharmacogenetics for Every Nation Initiative) [36].

SUMMARY

In summary, application of genetics and genomics in healthcare lags behind in the developing world. The examples that we have presented here make it clear that such examples are few and far between and that effective, efficient, equitable, and responsible use of genetics and genomics for the population as a whole, even in "affluent" developing countries, lags far behind that of the West. There is an urgent need for organizations such as the WHO to play a leadership role in promoting genetics and genomics in healthcare and health professional education in developing countries beyond its traditional mandate of prevention and control of birth defects.

REFERENCES

[1] WHO. Medical genetic services in developing countries: the ethical, legal and social implications of genetic testing and screening. 2006.

[2] Penchaszadeh VB. Predictive genetic tests: medical, ethical and social aspects. Medicina 2000;60(5 Pt 2):707–8.

[3] Collins FS, Morgan M, Patrinos A. The Human Genome Project: lessons from large-scale biology. Science 2003;300(5617):286–90.

[4] WHO. Genomics and world health. Geneva: The Advisory Committee on Health Research of the World Health Organization; 2002.

[5] WHO. Guidelines for the control of haemoglobin disorder. Geneva: World Health Organization (WHO/HDP/HB/GL/94.1); 1994.

[6] Verma IC, Saxena R, Lall M, Bijarnia S, Sharma R. Genetic counseling and prenatal diagnosis in India: experience at Sir Ganga Ram hospital. Indian J Pediatr 2003;70(4):293–7.

[7] Paneque HM, Reynaldo AR, Velazquez Perez L, Santos FN, Miranda HE, Real PN, et al. Type 2 spinocerebellar ataxia: an experience in psychological rehabilitation. Rev Neurol 2001;33(11):1001–5.

[8] Angastiniotis M, Kyriakidou S, Hadjiminas M. The Cyprus Thalassemia Control Program. Birth Defects Orig Artic Ser 1988;23(5B):417–32.

[9] Heredero-Baute L. Community-based program for the diagnosis and prevention of genetic disorders in Cuba: twenty years of experience. Community Genet 2004;7(2–3):130–6.

[10] Hilgart JS, Hayward JA, Coles B, Iredale R. Telegenetics: a systematic review of telemedicine in genetics services. Genet Med 2012;14(9):765–76.

[11] Dissanayake VHW, Nisansala D, Sandamal S, Jayasekara RW. Tele-Genetics: Using Low Cost Internet Technology to Provide Genetic Consultations to Rural Areas in Sri Lanka. Sri Lanka J Bio-Medical Inf 2010;1(1):46–8.

[12] Qureshi N, Modell B, Modell M. Timeline: raising the profile of genetics in primary care. Nat Rev Genet 2004;5(10):783–90.

[13] de Abrew A, Dissanayake VHW, Korf BR. Challenges in global genomics education. Appl Transl Genomics 2014;3(4):128–9.

[14] Dissanayake VH, Weerasekera LY, Gammulla CG, Jayasekara RW. Prevalence of genetic thrombophilic polymorphisms in the Sri Lankan population: implications for association study design and clinical genetic testing services. Exp Mol Pathol 2009;87(2):159–62.

[15] Jimenez-Sanchez G, Silva-Zolezzi I, Hidalgo A, March S. Genomic medicine in Mexico: initial steps and the road ahead. Genome Res 2008;18(8):1191–8.

[16] WHO. Control of hereditary diseases. *Technical Report Series*, No. 865. Geneva: World Health Organization; 1996.

[17] Hamamy H. Consanguineous marriages: preconception consultation in primary health care settings. J Community Genet 2012;3(3):185–92.

[18] Lee JM, Thong MK. Genetic counseling services and development of training programs in Malaysia. J Genet Couns 2013;22(6):911–6.

[19] Christianson A, Howson C, Modell B. March of dimes global report on birth defects: The hidden toll of dying and disabled children. 2006.

[20] Sharma R. Birth defects in India: hidden truth, need for urgent attention. Indian J Hum Genet 2013;19(2):125–9.

[21] WHO. Management of birth defects and haemoglobin disorders Report of a Joint WHO-March of Dimes Meeting. Geneva, Switzerland: WHO; 2006.

[22] DiCastro M, Frydman M, Friedman I, Shiri-Sverdlov R, Papa MZ, Goldman B, et al. Genetic counseling in hereditary breast/ovarian cancer in Israel: psychosocial impact and retention of genetic information. Am J Med Genet 2002;111(2):147–51.

[23] Mudiyanse RM. Thalassemia treatment and prevention in Uva Province, Sri Lanka: a public opinion survey. Hemoglobin 2006;30(2):275–89.

[24] WHO. Proposed international guidelines on ethical issues in medical genetics and genetic services (part I). Rev Derecho Genoma Hum 1998;(8):219–23.

[25] Marques-de-Faria AP, Ferraz VE, Acosta AX, Brunoni D. Clinical genetics in developing countries: the case of Brazil. Community Genet 2004;7(2–3):95–105.

[26] Acosta AX, Abe-Sandes K, Giugliani R, Bittles AH. Delivering genetic education and genetic counseling for rare diseases in rural Brazil. J Genet Couns 2013;22(6):830–4.

[27] Green ED, Guyer MS. Charting a course for genomic medicine from base pairs to bedside. Nature 2011;470(7333):204–13.

[28] Bell CJ, Dinwiddie DL, Miller NA, Hateley SL, Ganusova EE, Mudge J, et al. Carrier testing for severe childhood recessive diseases by next-generation sequencing. Sci Transl Med 2011;3(65):65ra4.

[29] Kohonen-Corish MR, Al-Aama JY, Auerbach AD, Axton M, Barash CI, Bernstein I, et al. How to catch all those mutations: the report of the third Human Variome Project Meeting, UNESCO Paris, May 2010. Hum Mutat 2010;31(12):1374–81.

[30] Burton H, Jackson C, Abubakar I. The impact of genomics on public health practice. Br Med Bull 2014;112(1):37–46.

[31] Richards CS, Bale S, Bellissimo DB, Das S, Grody WW, Hegde MR, et al. ACMG recommendations for standards for interpretation and reporting of sequence variations: revisions 2007. Genet Med 2008;10(4):294–300.

[32] Kingsmore SF, Saunders CJ. Deep sequencing of patient genomes for disease diagnosis: when will it become routine? Sci Transl Med 2011;3(87):87ps23.

[33] Bryant LD, Ahmed S, Ahmed M, Jafri H, Raashid Y. 'All is done by Allah': understandings of Down syndrome and prenatal testing in Pakistan. Soc Sci Med 2011;72(8):1393–9.

[34] Ross LF, Saal HM, David KL, Anderson RR, American Academy of Pediatrics, American College of Medical Genetics and Genomics, et al. Technical report: ethical and policy issues in genetic testing and screening of children. Genet Med 2013;15(3):234–45.

[35] Wilson JT, Pettigrew LE, Teasdale GM. Structured interviews for the Glasgow Outcome Scale and the extended Glasgow Outcome Scale: guidelines for their use. J Neurotrauma 1998;15(8):573–85.

[36] WHO. The ethical, legal and social implications of pharmacogenomics in developing countries. 2007.

Chapter 22

Genomic Applications in Forensic Medicine

C. Børsting, N. Morling

Section of Forensic Genetics, Department of Forensic Medicine, Faculty of Health and Medical Sciences, University of Copenhagen, Copenhagen, Denmark

Chapter Outline

Introduction	295
The Basics of Next Generation Sequencing	297
Single Molecule Sequencing	299
Next Generation Sequencing Solutions in Forensic Genetics	300
Short Tandem Repeat Sequencing	302
The First Commercial Next Generation Sequencing Kits for Forensic Genetics	303

New Frontiers in Forensic Genetics	305
Concluding Remarks	306
Acknowledgment	307
References	307

INTRODUCTION

DNA sequencing has a long history in forensic medicine. In the late 1980s and early 1990s, sequencing of mitochondrial DNA (mtDNA) was evaluated and used for case work at a time when restriction fragment length polymorphism (RFLP) analysis was the state of the art for human identification and years before the first short tandem repeat (STR) assays were developed. Successful RFLP analysis required micrograms of preferably intact DNA, and that made the sensitive polymerase chain reaction (PCR)–based mtDNA sequencing method the preferred tool for typing of low amounts of degraded sample materials; eg, hair shafts and old bones [1–3]. Sequencing of the mtDNA control region was used extensively and the European DNA Profiling Group's mtDNA population database project (EMPOP) was initiated in 1999 with the purpose of creating a common forensic standard for mtDNA sequencing and an online mtDNA database with high-quality mtDNA population data [4,5]. Laboratories that qualified by successful participation in EMPOP collaborative exercises submitted mtDNA sequences into EMPOP and with release 11 (October 2013), the EMPOP database contained 34,617 mtDNA sequences from populations all over the world.

Sequencing was conducted with the Sanger dideoxynucleotide triphosphate (ddNTP) chain terminating method [6], where the incorporation of a ddNTP to a growing DNA chain prevented further extension by the DNA polymerase (Fig. 22.1A). Early on, the synthesized DNA fragments were separated by slab gel electrophoresis and detected by either radioactively or fluorescently labeled deoxynucleotide triphosphates (dNTPs) incorporated into the DNA fragments. Subsequent introduction of fluorescently labeled ddNTPs and capillary electrophoresis (CE) platforms [7] increased sensitivity and throughput, and decreased the cost of Sanger sequencing to a level where sequencing of complete genomes became possible. The improvements in CE technology and the development of highly sensitive PCR-based STR assays gradually reduced the need for mtDNA sequencing in forensic genetics during the 1990s. However, the Sanger sequencing method was used continuously for verification and identification of, eg, STR alleles (see references in STRbase, http://www.cstl.nist.gov/strbase/). The ddNTP chain terminating method was also used for the so-called "mini-sequencing" or single base extension (SBE) reaction that was used for typing of single nucleotide polymorphisms (SNPs) [8]. SBE was a post-PCR cyclic reaction where SBE primers hybridized to the PCR products and were extended with a labeled ddNTP complimentary to the nucleotide in the SNP position (Fig. 22.1B). The SBE products were detected by CE, where the length of the extended SBE primer identified the SNP locus and the ddNTP label identified the SNP allele. Panels of SNPs for human identification, pigmentary traits, and ancestry information were identified and SBE assays were validated and used in actual case work [9–14].

Medical and Health Genomics. http://dx.doi.org/10.1016/B978-0-12-420196-5.00022-8

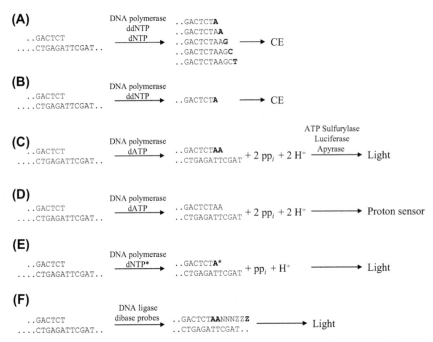

FIGURE 22.1   Sequencing methods. (A) Sanger sequencing. DNA is synthesized in the presence of fluorescently labeled dideoxynucleotide triphosphate *(ddNTPs)*. The differently sized fragments are separated by capillary electrophoresis (CE) and the sequence of fluorescently labeled nucleotides is detected by a camera. (B) Single base extension (SBE). The SBE primers are extended with a fluorescently labeled ddNTP complimentary to the nucleotide in the SNP locus. The extended SBE primers are detected by CE. (C) Pyrosequencing. Nucleotides are added sequentially to the sequencing reaction. Incorporation of one or more nucleotides to the growing strand release one or more pyrophosphates that are used in secondary enzymatic reactions to generate light. The light emission is detected by a camera. (D) Semiconductor sequencing. Nucleotides are added sequentially to the sequencing reaction. Incorporation of one or more nucleotides to the growing strand release one or more hydrogen ions that are detected by an ion sensor. (E) Sequencing by synthesis. DNA synthesis is performed with fluorescently labeled dNTPs with reversible 3′ terminators *(asterisk)*. Each addition of a nucleotide to the growing strand is detected by a camera. The 3′ terminator is chemically removed, allowing for the next nucleotide to be incorporated. (F) Sequencing by ligation. The sequencing primer is hybridized to the target DNA and four sets of four fluorescently labeled dibase probes (all the 16 possible combinations) are added sequentially to the ligase reaction. Successful ligation of a probe to the sequencing primer is detected by a camera. The probes are cleaved (between the N and Z nucleotides) and another cycle of ligations can begin. The bold letters indicate the variant bases of the dibase probes (in this case AA). *ATP*, Adenosine triphosphate; *dATP*, deoxyadenosine triphosphate; *dNTP*, deoxynucleotide triphosphate; H_+, hydrogen ion; pp_i, pyrophosphate.

Pyrosequencing was presented as a real-time sequencing alternative to Sanger sequencing in 1996 [15]. Nucleotides were added sequentially to the DNA synthesis reaction and the released pyrophosphate was used to generate light via a cascade of enzymatic reactions involving the three enzymes: adenosine triphosphate sulfurylase, luciferase, and apyrase (Fig. 22.1C). The light was detected in real-time by a charge-coupled device camera and thus electrophoresis of the sequencing products was not necessary. Pyrosequencing was cheap and fast compared with Sanger sequencing, and the method was applied to mtDNA sequencing [16,17] and later also used for STR sequencing [18]. However, the short sequencing length and especially the limited multiplexing capability of the instruments were not compatible with the low amounts of DNA usually recovered from trace samples and the method was never used in case work.

Even though the first pyrosequencing instruments never found a strong foothold in science, the pyrosequencing technology itself and the idea of real-time sequencing became the foundation on which the ongoing revolution in DNA sequencing was made. The first commercial high-throughput sequencing platform, the Genome Sequencer 20 from 454 Life Sciences, used pyrosequencing [19] and it was possible to sequence the human genome in 5 months at a cost of $1.5 million with this technology [20]. In comparison, the first human genome was sequenced with Sanger sequencing technology over 13 years and a cost of $2700 million [21]. Several high-throughput sequencing methods and platforms have since then been introduced. Most of them have been acquired by larger companies and sometimes the instruments have changed names; eg, Solexa was changed to Illumina. Some have come and gone again; eg, the HeliScope platform from Helicos BioSciences [22,23]; and Roche has recently announced that the production of the highly successful 454 pyrosequencers will be terminated in 2015. Early on, these platforms were usually referred to as *next generation sequencing (NGS)* or massively parallel sequencing platforms. However, with the introduction of single molecule sequencing, some platforms were referred to as *second generation sequencers*

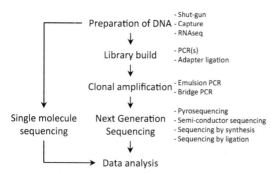

FIGURE 22.2 Work flow for high-throughput sequencing. *PCR,* Polymerase chain reaction; *RNAseq,* RNA sequencing.

and the single molecule sequencer sometimes referred to as *third generation sequencers* or the *next–next generation sequencers*. We will use the general term *NGS* in this review to cover all sequencing methods, except for single molecule sequencing, that have been developed after Sanger sequencing and the early pyrosequencing methods. Single molecule sequencing will be addressed in a separate section. There are many excellent reviews in the literature [21,24–31] that describe the various platforms in this rapidly changing field. In this review, we will focus on the possible applications of NGS in forensic genetics and only give an introduction to high-throughput sequencing (Fig. 22.2).

THE BASICS OF NEXT GENERATION SEQUENCING

The capability of some NGS platforms is so large that the aim of the sequencing assay may simply be to sequence every double-stranded DNA molecule in the sample material. Sequencing without any prior selection of targets is known as *shotgun sequencing* and requires fragmentation of micrograms of DNA into short fragments of 50–500 base pairs either by mechanical force, enzymatic digestion, or random insertion of transposons [21,24,25,31]. If the sample material is complementary DNA (cDNA), shotgun sequencing may generate a gene expression profile of the sample that is known as *RNA sequencing* [32,33].

The alternative to shotgun sequencing is usually called *targeted (re)sequencing,* and involves an initial enrichment step that either amplifies the selected regions by PCR, uses probes to capture the regions, or uses a combination of probes and enzymatic reactions [34–36]. The probes may be attached to a solid surface (eg, a slide or a bead) or they may be biotinylated and hybridize to their targets in solution. Either way, the purpose is to capture the selected genomic regions and eliminate the unwanted fragments of DNA. The DNA is subsequently eluted and used for sequencing. If the probes are used in combination with a DNA polymerase, the captured fragments are used as templates for DNA synthesis. The probes may be

biotinylated and the primer-extended probes captured by streptavidin beads, or the DNA synthesis is followed by a DNA ligase reaction that generates a circular product resistant to subsequent exonuclease treatment. The newly synthesized DNA is isolated from the genomic DNA and used in the downstream NGS reaction. Large portions of the genome may be captured with these technologies; eg, all the known coding regions (known as *exome sequencing*) or panels of relevant genes related to particular diseases. There are countless numbers of commercial capture assays available from different companies; eg, the Sure-Select human all exon kit (Agilent), the HaloPlex exome kit (Agilent), the Ion AmpliSeq Exome RDY kit (Life Technologies), the Nextera Rapid Capture Exome kit (Illumina), the SeqCap EZ human exome library (Roche NimbleGen), the TruSight One sequencing panel (Illumina) that targets 4800 genes or the Ion AmpliSeq Cancer panel (Life Technologies) that targets 400 genes by PCR. These companies also provide services for generation of customized panels defined by the user for specific projects or purposes. The major advantage of capture methods is that the majority of the sequencing capacity is focused on the regions of interest. This allows for more efficient sequencing experiments either by allowing more individuals to be sequenced at the same time or making the number of sequences for each nucleotide position (known as the *coverage* or *sequencing depth*) higher. The PCR-based capture method is by far the most sensitive and requires <10 ng DNA per multiplex reaction, whereas the probe-based methods typically requires 50–500 ng DNA. In contrast, the PCR-based captures are limited by the level of multiplexing capability (up to 6144 amplicons with the Ion AmpliSeq technology).

Once the DNA has been prepared for either shotgun or capture sequencing, the fragments are used to generate a library. The library is constructed by ligating adapters to the fragments or by one or two PCR reactions where the PCR primers are tagged with sequences needed for the downstream reactions. The adapters or the PCR primer tags may include specific sequences for clonal amplification of the library (see below), target sequences for the NGS reaction, a key sequence with four to eight nucleotides used for quality control of the NGS reaction, and a 6- to 10-nucleotide barcode for identification of the sample. The various sequence elements are combined in various ways depending on the assay and the NGS platform; eg, the barcode can be left out entirely if only one sample is sequenced (typical for shotgun sequencing), barcodes may be placed in both ends or in only one end of the library, or different barcodes may be used in either end to have a high number of combinations when many samples are sequenced in the same experiment. The construction of the library is the critical step of the experimental design. The choice of barcodes dictates how many samples can be sequenced. The choice of key and tag

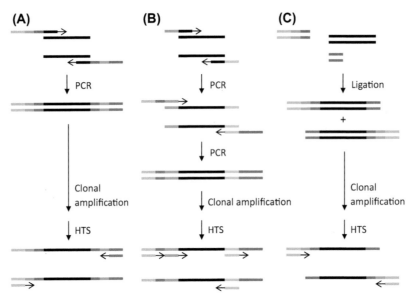

FIGURE 22.3 Examples of library building and sequencing strategies. (A) The library is generated by one polymerase chain reaction (PCR). The PCR primers include five elements: the target sequence (*black*), the barcode for sample identification (*red*), the key sequence for sequence quality control (*blue*), and sequencing targets (*orange* and *purple*). One of the sequencing targets is also used to hybridize the library to the solid surface during the clonal amplification step. With two sequencing targets, it is possible to perform directional sequencing of only one strand by choosing a sequencing primer complementary to either the *orange* or the *purple* sequencing target. With only one sequencing target (when the *orange* and the *purple* sequences are the same), both strands would be sequenced in the next generation sequencing (NGS) reaction. (B) The library is generated by two PCRs. In the first PCR, the primers include the target sequence (*black*) and the sequencing targets (*light green* and *dark green*). In the second PCR, the primers hybridize to the sequencing targets and include tags with the barcode (*red*) and sequences for hybridization to the solid surface used for the clonal amplification. The target sequence (*black*) is sequenced via the two sequencing targets (*green*), whereas the barcodes are sequenced in separate reactions. (C) The library is generated by ligation of adapters to the fragmented genomic DNA. One adapter includes the barcode for sample identification (*red*), the key sequence for sequence quality control (*blue*), and the sequencing target (*orange*). The second adapter includes the sequence for hybridization to the solid surface used for the clonal amplification. Four different products will be generated by the ligation: the two products shown in (C), where two different adapters are ligated to the DNA fragment; and two products where the same adapter ligates to both ends. The latter products cannot be used in the downstream reactions. Sequencing is conducted from hybridization of a sequencing primer complimentary to the sequencing target (*orange*). Both strands will be sequenced because the adapter with the sequencing target ligates to either the forward or the reverse strand in equal numbers. *HTS,* High-throughput sequencing.

sequences for the clonal amplification and NGS sequencing partly dictates the choice of NGS platform. Three examples of library constructions are described in Fig. 22.3.

Multiple libraries may be pooled in equal amounts before the clonal amplification. The number of samples that can be analyzed in the same experiment depends on a number of factors: (1) the number of available barcodes, (2) the sequencing capacity of the NGS platform, (3) the numbers and sizes of targeted regions, and (4) the desired sequencing depth.

The library pool is used as a target for the clonal amplification step. Individual DNA molecules are hybridized to a primer on a solid surface and each molecule is amplified by PCR in a reaction that is isolated from the other DNA molecules in the library pool (thus the name *clonal amplification*). The physical separation of the molecules is secured by hybridization of one DNA molecule to one bead and generation of an oil–water emulsion with one bead per droplet [emulsion PCR (emPCR)] [19,37] or by hybridizing the DNA molecules to a slide (bridge PCR) [38]. Millions of individual DNA molecules (clones) may be amplified

simultaneously and, after amplification, thousands of copies of each original DNA molecule form an immobilized "cluster of DNA" on the bead or the slide. Each DNA cluster forms an ideal target for sequencing and all the clusters may be efficiently sequenced in parallel on an NGS platform (thus the name *massively parallel sequencing*). The beads from the emPCR are placed in picoliter-sized wells with one bead per well, whereas the slide from the bridge PCR is used directly for sequencing.

The DNA sequence of each cluster is determined in real time by one of four methods (see Fig. 22.1C–F): pyrosequencing (Roche 454 sequencing), semiconductor sequencing (Thermo Fisher Scientific Ion Torrent), sequencing by synthesis (Illumina), or sequencing by ligation (Thermo Fisher Scientific SOLiD and BGI-Shenzhen Complete Genomics). In pyrosequencing and semiconductor sequencing, the nucleotides are added sequentially to the reaction. In some clusters, no DNA synthesis will take place because the added nucleotide cannot extend the growing strand. In others, one or more nucleotides will be added and the generated light signal or the number of protons released will be detected and interpreted. With the sequencing by synthesis method, all four fluorescently labeled nucleotides are present in the reaction and one nucleotide is added to the growing DNA strand in all clusters. The nucleotides are reversibly blocked in the 3′ position, which prevents incorporation of more than one nucleotide at the time. This is an advantage when the sequence contains stretches of the same nucleotide (homopolymer stretches). In pyrosequencing and semiconductor sequencing, several nucleotides will be incorporated at homopolymer stretches and more light and protons will be detected, respectively. However, if the stretch is longer than five nucleotides, it may be difficult to deduce the correct number of nucleotides in the homopolymer. Sanger sequencing suffered from a similar problem when similar sized fragments with the same ddNTP were difficult to separate by electrophoresis. Sequencing by ligation is different from the other methods because it involves ligation of fluorescently labeled probes to a primer. The probes may vary in one (BGI-Shenzhen Complete Genomics) or two (Thermo Fisher Scientific SOLiD) positions. Only probes that are perfectly complementary to the target sequence will ligate to the growing chain of primer and probes. Consecutive rounds of ligations and cleavage reactions generate a patchwork of fluorescent signals from the same cluster that may be combined into a complete sequence [21,24,25,31].

The maximum number of bases that may be sequenced in each cluster (known as the *read length*) varies between the NGS methods. The read length has been an important focus point for commercial competition and considerable improvements in read length has been achieved for pyrosequencing and sequencing by synthesis platforms. Pyrosequencing generates read lengths that are comparable to the read lengths of the Sanger sequencing method (600–1000 bps) and the sequencing by synthesis platforms have reached a read length of 300 bps. The first semiconductor platform was launched in 2011 and the read lengths are now up to 400 bps. In contrast, sequencing by ligation generates very short read lengths (<75 bp) and it has not changed much since the platforms were released in 2006.

Another major focus point for commercial competition between NGS platforms has been the overall sequencing capacity or the total number of clusters (or reads) that are sequenced per run. The sequencing capacity is sometimes calculated as the number of sequenced bases per run; however, this may be misleading because the read lengths vary between platforms. The HiSeq 2500 system (Illumina) has the largest capacity and can sequence four billion clusters in one experiment. It takes 5–11 days to complete a run; however, the experiment may generate enough sequences to cover 5–10 (almost) complete human genomes with an average coverage of more than 30. The smallest of the commercial high-throughput sequencing platforms is the GS Junior System (Roche). It can sequence 100,000 clusters in

10 hours and belongs to one of the so-called "bench-top" sequencers that also include the MiSeq (Illumina), the NextSeq 500 (Illumina), the Ion Proton system (Thermo Fisher Scientific Ion Torrent), and the Ion PGM system (Thermo Fisher Scientific Ion Torrent). The capacity of these instruments is in the range from 5 to 400 million clusters and the run times range from 2 to 55 hours.

A high degree of flexibility in the experimental design is possible on the sequencing by synthesis and semiconductor platforms. There are different sizes of flow cells (Illumina) or chips (Ion Torrent) available that vary in the number of clusters that may be sequenced and there are different reagent kits available that vary in the number of nucleotide cycles. The number of nucleotide cycles regulates the read lengths and also affects the run time of the instrument (more cycles generate longer reads and take longer time). In general, the run time on the pyrosequencers and the semiconductor platforms are relatively short because signal detection is performed in real time, whereas signal detection on the sequencing by synthesis and sequencing by ligation platforms is done by imaging, which makes the run times longer. However, the manual preparation of the samples for the sequencing by synthesis platform is short compared with the other platforms, because the cluster generation by bridge PCR and the sequencing reaction are an automated protocol on the flow cell performed by the NGS instrument. The other three sequencing methods use libraries that are clonally amplified by emPCR, which involves many pipetting steps and considerable hands-on time. The large flexibility of the bench-top instruments makes them highly suitable for capture-based sequencing experiments where the numbers of samples or the sizes of the captured regions may vary from project to project or from experiment to experiment. This makes the platforms ideal for research. However, the scalability also makes the platforms interesting for diagnostic laboratories that perform routine genetic investigations.

SINGLE MOLECULE SEQUENCING

Detection of the sequence of a single DNA molecule instead of a cluster of clonally amplified DNA is often referred to as *third generation sequencing* [26,29,30]. With these methods, the original DNA or RNA molecules may be analyzed and any biases generated by the capture and clonal amplification steps are eliminated (see Fig. 22.2).

The HeliScope platform (Helicos Biosciences) was the first commercial single molecule sequencing platform to be launched [22,23]. However, the company had a very short life time and the HeliScope platform is no longer produced. Soon after, the PacBio platform (Pacific Biosciences) was launched [39,40] and recently, a beta version of the MinION sequencer (Oxford Nanopore) was released for testing by members of the MinION access program. The HeliScope and PacBio platforms use variations of the sequencing by synthesis technology (see Fig. 22.1E). For the HeliScope platform, the original DNA is fragmented and a poly(A) tail is added to the fragments. This library is subsequently hybridized to anchored poly(T) probes on a slide and the sequences are determined by primer extension using cycles of sequential addition of Cy5-labelled nucleotides and fluorescence imaging. The nucleotides are reversibly blocked in the 3′ end to ensure that only one nucleotide is incorporated per cycle. The run time on the HeliScope platform is 2–9 days and the read lengths are only 50 nucleotides. However, up to 1.5 billion reads may be generated. On the PacBio platform, DNA polymerase/DNA template complexes are immobilized at the bottom of zero-mode waveguide (ZMW) wells that are zeptoliter (10^{-21} L)–sized wells only nanometers in diameter. One DNA polymerase/DNA template complex fits into one ZMW well and the template is sequenced using four different fluorescently labeled nucleotides. The fluorophore is linked to the terminal phosphate and it is released when the nucleotide is incorporated into the growing strand. The fluorophores are excited by multiple lasers and the pulse of fluorescence is monitored by a camera. The duration and intensity of the pulse determines the identity of the incorporated nucleotide. All four nucleotides are added to the reaction and the generated signals are detected in real-time with a speed of approximately five nucleotides per second. That makes the run time on the PacBio platform short, and it usually last less than 1 hour. The PacBio may generate read lengths of more than 15,000 nucleotides and the maximum capacity is currently 150,000 ZMWs. Furthermore, the PacBio platform uses the ɸ29 DNA polymerase that is capable of multiple displacement amplification. Thus a circular DNA template may be sequenced several times in one experiment and a single ZMW well may generate a sequencing depth of 10 or more, depending of the size of the template DNA and the read length.

The MinION platform uses a completely different technology based on the transport of DNA molecules through a nanopore embedded in a lipid bilayer or a synthetic polymer [30,41,42]. An electric field across the membrane will drive the DNA molecules through the pore and the current of (other) ions through the pore will be partly blocked as the DNA passes through. The decrease in current amplitude is detected and used to determine the nucleotide sequence of the DNA passing through the pore. The pore used by the MinION is not known; however, many channel proteins, synthetic solid-state nanopores, and even scaffold structures of DNA or proteins have been used for nanopore sequencing. On the MinION, a protein/DNA complex binds reversibly to the pore. The undisclosed protein unfolds the double-stranded DNA and single-stranded DNA pass through the pore. A hairpin is attached to the DNA molecule during sample preparation and when the hairpin sequence has passed through the pore, the reverse strand follows and

the sequence of the complementary strand may be determined. The MinION access program was initiated in the spring of 2014 and peer-reviewed results are limited [43]. However, in our own experience, the MinION may generate read lengths that are longer than 70,000 nucleotides and the total numbers of reads are 2000–10,000 [Jill Olofsson, unpublished results].

The long read lengths of the PacBio and MinION platforms will make it possible to determine the haplotype of an individual in long stretches of DNA and will simplify the assembly of genome regions with multiple duplications and repeats [44,45]. Furthermore it may be possible to determine epigenetic modifications of the DNA in real time, because modified nucleotides, eg., 5-methylcytosine, 5-hydroxymethylcytosine, and N6-methyladenine, give off another fluorescent pulse on the PacBio and decrease the current amplitude differently in nanopores than unmodified nucleotides [46–49].

Both platforms suffer from very high error rates because current methods of signal detection are inadequate. The base call error rate is estimated to be >15% on the PacBio [40] and >30% on the MinION [43] [Jill Olofsson, unpublished results]. The errors on the PacBio seem random and the quality of the consensus sequence generated in each ZMW may therefore be improved by increasing the sequencing depth using circular DNA templates and ɸ29 DNA polymerase (see previous discussion). In contrast, insertion/deletion (indel) errors are common on the MinION, which makes it very difficult to align the generated sequences accurately and exploit the sequence information in the long reads. Another disadvantage is the relatively large amount of input DNA required. Even though individual DNA molecules are sequenced on these platforms, the amount of input DNA is 250–5000 ng on the PacBio and >1000 ng on the MinION.

NEXT GENERATION SEQUENCING SOLUTIONS IN FORENSIC GENETICS

The idea of sequencing every DNA (and/or RNA) molecule in the sample is very intriguing to a forensic geneticist, who is used to dealing with the challenge of obtaining sufficient information from trace samples that often contain DNA from more than one contributor. However, shotgun sequencing requires micrograms of DNA and is not applicable for many of the forensic samples. Also, reproducibility may be impossible, because the shotgun experiment is not directed toward specific targets but generates sequences from random positions in the genome. Thus two shotgun sequencing experiments of the same sample, or, eg, a trace sample and a reference sample from a suspect, will generate different results. Furthermore shotgun sequencing requires exhaustive data analyses that are both time-consuming and may generate different DNA profiles, depending on the choice of NGS platform and alignment software. Concordance studies between platforms have demonstrated that as many as 20% of the SNPs and 80% of the indels called by one platform were not reproduced by typing the same sample on another platform [50–53]. Large portions of the inconsistencies were seen in regions of low (or no) sequencing depth for one or both platforms or caused by systematic errors introduced by the different methods of alignment and variant calling. Even though the capacity of some NGS platforms is huge, they are only just able to sequence genomes the size of the human genome, and large portions of the genome are only covered by a low number of reads in typical shotgun sequencing experiments. This leads to a high risk of misinterpretation of the sequencing data and to the lack of reproducibility observed in concordance studies. Finally, full genome sequencing seems excessive in most forensic genetic cases where the purpose is primarily to establish the identity of the individual(s) contributing to the sample and possibly estimate any phenotypical characteristics of the individual(s) or identify the specific tissue type(s) in the sample. This requires relatively few markers and a capture-based approach will be much more economical and require less sample material.

Today, the core forensic markers are typed with PCR-CE and there are individual assays for autosomal STRs, Y-chromosome STRs, X-chromosome STRs, indels, mtDNA SNPs, autosomal SNPs, Y-chromosome SNPs, ancestry informative markers (AIMs), phenotypical markers, messenger RNA (mRNA), and so on. PCR-CE may be performed in 1 work day, whereas NGS takes minimally 2–3 days. However, one of the major advantages of NGS is that all (or most) of the PCR-CE assays may be combined into a single NGS assay if it is possible to develop a capture for the relevant loci. One NGS assay with many different markers will save time in cases where supplementary investigations are needed and reduce the overall time a sample is processed in the laboratory. Among the various capture methods, PCR continues to be the most sensitive and it is the only method that approaches the level of sensitivity required for forensic genetic case work. Another important advantage is that the fragments do not need to be separated by lengths in CE and thus all the analyzed fragments can be designed to be as short as possible, which will improve the chance of typing degraded DNA/RNA.

Combining nuclear markers with mtDNA or mRNA markers in a sequencing assay may prove to be difficult. DNA, mtDNA, and RNA may be coextracted and the RNA converted to cDNA in a separate reverse transcriptase reaction. However, the large variation in target copy numbers will make construction of a combined multiplex PCR very difficult. Also, it is important to keep in mind that mRNA sequencing needs to be semiquantitative to allow tissue identification of the sample and that genomic DNA is unwanted in cDNA analyses. Nevertheless, it may be possible to pool PCR products from separate PCR captures of nuclear DNA, mtDNA, and cDNA before the library build (adapter ligation or the second PCR) or, more likely, to pool DNA, mtDNA, and cDNA libraries before the clonal amplification step. This way it should be possible to sequence all relevant markers in a single sequencing reaction. However, it is uncertain whether this is a practical solution for case work, because the information obtained from mRNA and mtDNA is not needed in all cases and the case officer needs to process all the sequencing data if the information is generated, even though the information is irrelevant to the case. Also, sequencing of mtDNA and cDNA will take up a large portion of the sequencing capacity, which eventually will result in fewer samples per sequencing run and a higher cost of the investigation. Similarly, it may be argued that ancestry information, phenotypical traits, or certain human identification markers are irrelevant in other case-work scenarios. Therefore flexible NGS solutions for various types of cases, including assays with large number of markers, will be preferred.

It is generally accepted that sequencing by ligations has the lowest error rate among the NGS methods, followed by sequencing by synthesis, semiconductor sequencing, and pyrosequencing in that order [29,50,52,53]. However, these error rates are from genome sequencing studies and can be misleading, because the errors are unevenly distributed and typically related to specific sequence elements; eg, sequencing of homopolymer regions (see previous discussion), and it would be much too simple to state that sequencing by ligations is the best platform for forensic genetic applications (it is not because of the short read lengths). In order to evaluate the quality of a given NGS platform/assay, it is necessary to properly validate the genotypes against existing methods. Recently there have been numerous reports in the literature from molecular diagnostics laboratories where NGS results have been compared to mainly Sanger sequencing [54–60]. The conclusion drawn from these studies is that probe or PCR capture-based NGS analyses have matured sufficiently to be used in clinical diagnostics and will gradually replace Sanger sequencing as the gold standard. The main concerns are the data analyses and the overwhelming number of (new) variants that are detected. Each variant must be evaluated and classified as benign or disease-related, and this may be difficult for the local clinician. Another concern has been the detection of variants that are not related to the disease under investigation. Large capture-based investigations, eg, exome sequencing, may reveal variants in other genes that may be disease-related. How to handle this information raises ethical considerations for the clinician. These discussions are interesting because forensic genetics face many of the same challenges and will have to consider many of the same questions on how much and which loci to sequence, what to report, and whether ignoring sequence information is prudent, and certainly, there will be a number of ethical and legislative considerations by introducing NGS in forensic genetics.

Short Tandem Repeat Sequencing

STRs are essential to crime case work and will continue to be so because of the large national DNA databases with STR profiles from criminal offenders and irreplaceable trace samples from old cases. Consequently any NGS assay designed for forensic genetics must be able to sequence the core STR loci. However, most NGS studies focus on SNPs, small indels, and copy number variations, whereas repeats have not attracted much attention even though repeats cover almost half, and STRs alone 15%, of the human genome [61]. In the years after the first NGS platforms were launched, the read lengths of most instruments were too short to span many repeat structures, which made it difficult to align reads with repetitive sequences, and often these reads were simply ignored. The pyrosequencers were the only platforms with sufficient read length to sequence the core STR loci used in forensic genetics and most of the forensic literature with NGS STR data was produced with pyrosequencing technology [62–70]. In a few reports, sequencing by synthesis [71–73] and recently also semiconductor sequencing [70,74] were used. The libraries were constructed directly by PCR or by adapter ligation as described in Fig. 22.3A and C, respectively.

In contrast to fragment length analysis by PCR-CE, sequencing reveals the true variation of STR loci. Previously unknown STR alleles and more overall variability have been found by NGS of mainly complex and compound STRs [65,66,69], whereas few new alleles have been detected by sequencing of simple STRs [62,69,74]. Complex and compound STRs consist of different subrepeats and, if the individual subrepeats are polymorphic, the number of possible alleles will be much higher than in simple repeats (it is equivalent to having two STRs in a haplotype). In one study of 197 Danes, 53 different alleles in *D12S391* were detected by NGS, whereas only 15 different alleles were detected by PCR-CE. Many alleles in *D12S391* have the same length but different sequence compositions in the repeat region. So far, the top score is eight different alleles with a length equivalent to 21 repeats (Table 22.1) [66]. Sequence variations may also be found in the flanking regions of STRs [62,64,75]; eg, the rs6736691 SNP near *D2S1338* [75]. One SNP with a minor allele frequency of a reasonably size may generate two families of SNP–STR haplotypes and increase the number of SNP–STR alleles by a factor of two.

Discovery of many new STR and SNP–STR alleles with the same sizes makes the old PCR-CE–based nomenclature for STR alleles inadequate. A new transparent description of STR sequences is much in demand and the International Society of Forensic Genetics (ISFG) has initiated a working

TABLE 22.1 Examples of *D12S391* Alleles With the Same Length

D12S391[21]AGAT[11]AGAC[9]AGAT[1]
D12S391[21]AGAT[11]AGAC[10]
D12S391[21]AGAT[12]AGAC[8]AGAT[1]
D12S391[21]AGAT[12]AGAC[9]
D12S391[21]AGAT[13]AGAC[7]AGAT[1]
D12S391[21]AGAT[13]AGAC[8]
D12S391[21]AGAT[13]GGAC[1]AGAC[7]
D12S391[21]AGAT[14]AGAC[6]AGAT[1]

group with the purpose of finding a common definition for naming sequenced STR alleles. In this review (see Table 22.1; Figs. 22.4 and 22.5), we have used the nomenclature of Gelardi et al. [66], where the name is divided into four elements: (1) the locus name, (2) the length of the repeat region divided by the length of the repeat unit, (3) the sequence(s) of the repeat unit(s) followed by the number of repeats, and (4) variations in the flanking regions.

More variable loci also mean more statistical power of the investigations and will reduce the number of loci that needs to be typed to solve a case. In the largest population study to date, the match probability decreased from 0.0001 to 0.000005 and the typical paternity index increased from 59 to 415 when three STRs *D3S1358, D12S391,* and *D21S11* were typed by NGS in 197 Danes and the results were compared with PCR-CE results [66]. These three loci are highly polymorphic STRs with more than one subrepeat, and the number of detected alleles was almost tripled by sequencing compared with PCR-CE. For simple STRs, the number of detected alleles is not expected to increase by a factor of two or three and the difference in statistical power between PCR-CE and NGS results will be smaller.

Another interesting observation from the same study was that approximately 30% of the homozygous genotype calls by PCR-CE turned out to be heterozygous when the individuals were sequenced. This demonstrates another important advantage of NGS of STRs. Sequencing of complex and compound STRs with many alleles of the same size may simplify mixture interpretation if the contributors have alleles of the same size with different sequence compositions or if the true allele of the minor contributor has a different sequence than the stutter artifact of the major contributor. It was recently demonstrated that sequences from the minor contributor in 1:100 and 1:50 mixtures were detectable by NGS [74,76], something that is not possible with the PCR-CE technology. In these types of mixtures, the reads from the minor contributor will be difficult to separate from stutters and noise sequences; however, the mere

(A)

BarcodeId	AMELX	AMELY	CSF1PO	D16S539	D3S1358	D5S818	D7S820	D8S1179	TH01	TPOX	vWA
IonXpress_049	1	-	11	12	15, 16	12	10, 11	8, 11	6, 9.3	8, 11	14, 16
IonXpress_050	1	-	11, 13	10, 11	15, 17	11	9, 10	10, 14	9, 9.3	8	15
IonXpress_051	1	-	11, 12	11	16, 18	12	10, 11	9, 14	9	8, 10	14, 19
IonXpress_052	1	-	10, 11	11	16, 17	10, 11	9, 12	10, 13	6, 9.3	8	16
IonXpress_053	1	-	10, 11	12, 14	17	11, 12	11	10, 14	6, 8	8	16, 20
IonXpress_054	1	-	9, 13	11, 12	16, 17	12	10	*14*	9.3	8, 9	15, 18
IonXpress_055	1	1	12	12	15, 17	11, 12	8, 10	10, 13	6, 7	8, 11	17, 18
IonXpress_056	1	1	10, 11	12, 13	15, 18	11, 12	9, 11	12, 13	7, 9	8	17, 18
IonXpress_057	1	-	12, 13	9, 12	14, 15	12	10, 12	10, 13	6, 9.3	11, 12	14, 16
IonXpress_058	1	1	11	11	15, 16	11, 12	8	13	6	8, 11	14, 18
IonXpress_059	1	1	10, 11	11, 12	16, 17	11, 12	11, 13	13, 14	6, 7	8, 9	15, 18
IonXpress_060	1	1	10, 12	9, 12	15, 18	11, 13	9, 11	13	9	8	17, 18
IonXpress_061	1	1	11	9, 11	17, 18	11	9	13, 15	6	8	16, 17
IonXpress_062	1	1	12, 14	10, 13	17, 18	11	10, 10.1	11, 12	6, 9.3	8	18
IonXpress_063	1	-	12	10, 11	15, 16	13	8, 12	14	7	8, 11	15, 17
IonXpress_064	1	1	10, 12	11, 12	16, 19	12	7, 12	12, 18	7, 9.3	11	17

Locus	Allele	Counts	Total cov	Sequence
D8S1179	14	3995	8655	TATCTATCTATCTATCTATCTATCTATCTATCTATCTATCTATCTATCTATCTATCTATCTATC
D8S1179	14	3938	8655	TATCTATCT**G**TCTATCTATCTATCTATCTATCTATCTATCTATCTATCTATCTATCTATC

(B)

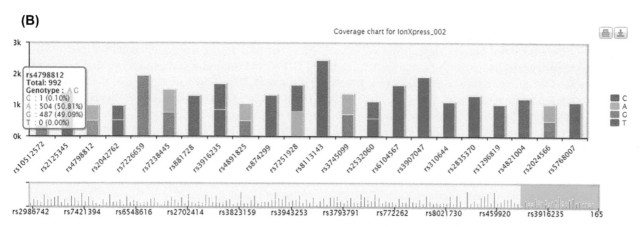

FIGURE 22.4 Analysis of short tandem repeats (STRs) and single nucleotide polymorphisms (SNPs) using the Torrent Suite Server. (A) Profile summaries of 15 samples typed with the Ion Torrent HID STR 10-plex. The *D8S1179* locus in the individual labeled with the IonXpress 054 barcode is highlighted and some details of the next generation sequencing (NGS) reaction for this locus are shown in the smaller graph (*below*). The individual has two *D8S1179* alleles of the same length and the single nucleotide difference is highlighted. The allele balance of the two alleles was almost 1:1 (3995 reads of *D8S1179*[14]TCTA[14] versus 3938 reads of *D8S1179*[14] TCTA[1]TCTG[1]TCTA[12]). Please note that the Torrent Suite Server displays the repeat as TATC repeats, whereas *D8S1179* is defined as TCTA repeats in the forensic genetic literature. (B) Coverage chart of an individual typed with the HID-Ion AmpliSeq Ancestry Panel. Only 22 of the 173 SNPs in the panel are shown in the chart, as indicated by the *green (light gray in print versions) bar* below the chart. Details of the NGS reaction for the rs4798812 locus are shown in a box on the *left*. The allele balance of the two alleles was almost 1:1 (504 reads with an A and 487 reads with a G). One read was called as a C. This base call was most likely an error caused by the PCR.

fact that they could be identified opens up for new possibilities in mixture interpretation and it is certainly something that should be explored further.

The First Commercial Next Generation Sequencing Kits for Forensic Genetics

Thermo Fisher Scientific launched two SNP typing assays in 2014 designed for the Ion PGM System: (1) the HID-Ion AmpliSeq Identity Panel for human identification [76,77]

that amplifies 124 autosomal SNPs, including most of the SNP*for*ID, [78] individual identification SNPs (IISNPs) [79], and 34 Y-chromosome SNPs; and (2) the HID-Ion AmpliSeq Ancestry Panel for ancestry estimation that include most of the AIMs in the Seldin [80] and Kidd laboratory selection panels [81]. Furthermore, Thermo Fisher Scientific is working on a panel of core forensic STRs and an early version of this panel, the Ion Torrent HID STR 10-plex, has been tested by forensic laboratories [74]. The assays use the Ion AmpliSeq technology, in which the PCR

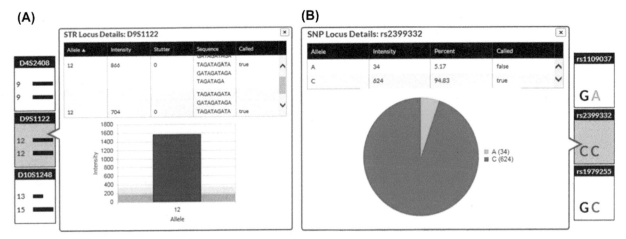

FIGURE 22.5 Analysis of short tandem repeats (STRs) and single nucleotide polymorphisms (SNPs) using the ForenSeq Universal Analysis Software. Results from two different samples typed with the ForenSeq DNA Signature Prep Kit are shown in (A) and (B). Genotype calls for each locus are shown in *small boxes*. (A) The next generation sequencing (NGS) result for *D9S1122* is highlighted and some details are shown in the *large box*. The individual has two *D9S1122* alleles of the same length. The allele balance was approximately 1.2:1 (an intensity of 866 for the *D9S1122*[12]TAGA[1]TCGA[1] TAGA[10] allele versus an intensity of 704 for the *D9S1122*[12]TAGA[12] allele). (B) The NGS result for rs2399332 is highlighted and some details are shown in the *large box*. This sample was a 1:10 mixture and the ForenSeq Universal Analysis Software correctly designated the sample as a possible mixture (not shown). For rs2399332, the genotypes of the major and minor contributor are C and AC, respectively. The allele balance was approximately 1:20 (A versus C) as expected. Please note that the A allele was not called by the ForenSeq Universal Analysis Software.

primers are partly degraded before adapter ligation. This reduces the lengths of the fragments to be sequenced and it removes most of the primer-dimers from the library, which makes the sequencing more efficient. The PCR fragments are clonally amplified by emPCR and sequenced in both directions using semiconductor sequencing technology (see Fig. 22.1D). The library construction and sequencing strategy are described in Figure 22.3C. The sequence data are analyzed using the analysis software on the Torrent Suite Server (see Fig. 22.4). The sensitivity of the two SNP typing assays, that both amplify >120 SNPs in one PCR, was 0.5–1 ng ([76]; unpublished results), whereas full STR profiles were obtained from only 50 pg with the Ion Torrent HID STR 10-plex [74]. It was also noteworthy that the STR assay generated full STR profiles from degraded samples, whereas PCR-CE typing of the same samples resulted in partial profiles. This was most likely because of the short (<170 bp) PCR products generated by the Ion Torrent HID STR 10-plex.

So far, the strategy of Thermo Fisher Scientific has been to develop assays that may be used as supplement to PCR-CE typing. In contrast, Illumina has announced that their strategy is to replace PCR-CE with PCR-NGS. At the time of writing of this review, Illumina is conducting beta tests of their new ForenSeq DNA Signature Prep Kit together with selected forensic laboratories and Illumina plans to launch the kit in the fall of 2014. The ForenSeq DNA Signature Prep Kit amplifies 27 autosomal STRs, eight X-STRs, 25 Y-STRs, 95 autosomal human identification SNPs, 56 autosomal AIMs, and 24 autosomal SNPs associated with pigmentary traits, all in one multiplex PCR. The multiplex includes, among others,

all of the STR loci in the Combined DNA indexing system (CODIS) and European standard set, most of the SNP*for*ID [78] and IISNPs, [79] and all of the HIrisPlex loci [82]. The ForenSeq DNA Signature Prep Kit will be introduced together with the MiSeq FGx platform, a MiSeq developed specifically for forensic genomics. The MiSeq FGx platform is supported by the ForenSeq Universal Analysis Software that manages both the experimental set-up and data analysis (see Fig. 22.5). The PCR fragments are clonally amplified by bridge PCR and sequenced with the sequencing by synthesis technology (see Fig. 22.1E). The library construction and sequencing strategy are described in Fig. 22.3B. The maximum read length of the standard MiSeq reagent kits is 300 nucleotides. However, for the ForenSeq DNA Signature Prep Kit, 350 cycles of sequential addition of labeled nucleotides and fluorescence imaging is performed on the forward strand to allow sequencing of the longest STRs. The reverse strand and the barcodes are sequenced in short separate reactions. Thus the reverse strand is only sequenced to confirm the sequence of the reverse PCR primer and the first few nucleotides of the amplicon. In our hands, the ForenSeq DNA Signature Prep Kit generated results from 50 pg to 10 ng of input DNA. However, detailed evaluation of the sequence data is required and has not been completed at the time of writing. Proper analyses of genotype concordance, degraded samples, and mixtures are pending.

One of the major challenges will be to develop a forensic NGS tool for analysis and reporting of the sequence data. With NGS data, it is not possible to analyze the sequences manually or even to analyze every genotype call manually. Therefore the software solution must be completely trustworthy and thoroughly validated before it can be used in real case work. Thermo Fisher Scientific and Illumina have developed software solutions for analysis of their kits; however, they are not sufficiently sophisticated for forensic genetics. There is a tendency to analyze all STRs or all SNPs with the same criteria and there are very few or no options for the user to alter the parameters for analysis. It is well known that different loci must be analyzed with different criteria and historically, individual laboratories have defined different analysis parameters based on in-house validation studies and according to different standards of accreditation [83,84]. This will also be necessary with NGS kits and the software must be able to accommodate this demand from the forensic community. A "black box" for analysis is not acceptable and will probably not be used by many forensic laboratories [85]. For each STR locus, there should be user-defined options for: (1) the minimum number of reads used to call the genotype, (2) acceptable stutter ratios (reads with the same sequence as the genotype except for the number of repeats), (3) acceptable noise ratios (reads that are not identical to the genotype), and (4) acceptable allele balances. The latter may depend on the lengths of the two alleles because of the tendency to generate more reads of the shortest allele [64,74]. Thus the acceptable allele balance may vary depending on the length difference between the alleles. The software should also be able to identify two STR alleles of the same size but with different sequences, and it should be able to name the alleles according the nomenclature suggested by the ISFG. Figs. 22.4A and 22.5A show two examples where the heterozygous individual is called as homozygous because the primary report is too simplified, and the consequence is that manual intervention is required to analyze the results properly. The software should be able to identify SNP–STR variants by analyzing the flanking sequences for variants, which is not possible with the current software solutions from Thermo Fisher Scientific and Illumina. For each SNP locus, the same user-defined options should be available except for the acceptable stutter ratio and the length-dependent allele balance, which is not relevant for SNPs. In addition, there should be an option to define a maximum threshold of reads with base calls of the second known SNP allele in the case of a homozygous genotype call. For biallelic SNPs, the only way to identify mixtures is to look for allelic imbalances, and this is an important quality assurance for SNP typing assays [9,10,76]. Fig. 22.5B shows an example in which a mixture sample was called as homozygous even though 5% of the reads had sequences with the second allele and clearly indicated that the sample was heterozygous.

Future software should have a specific module for mixture interpretation that may be used once a sample has been identified as a mixture. NGS offers new possibilities for analysis of mixtures, and because the forensic community is only beginning to explore the use of NGS, such tools have not been developed. Future software modules should also be used to estimate biogeographical ancestry, mtDNA haplogroups, Y-chromosome haplogroups, tissue identification, and phenotypes. It will be important for the development of these modules that the forensic genetic community and the manufacturers of commercial kits engage in a close collaboration and that the software algorithms are well described in the user manual or scientific papers to simplify future accreditation attempts of NGS in forensic laboratories. Preferably, the analyses should apply to recommendations from ISFG and similar forensic standardization bodies.

New Frontiers in Forensic Genetics

Besides analysis of classical forensic markers, NGS makes it possible to expand forensic genetic investigations to new areas related to forensic medicine. When a person dies unexpectedly and for no apparent reason, shotgun or exome sequencing may identify genetic variants associated with known diseases and assist the pathologist in finding the cause of death. In Denmark, it is estimated that approximately 20% of all deaths are caused by sudden unexpected cardiac arrest and one-third of these deaths remain unexplained even after autopsy [86]. It is assumed that many of these individuals have a genetic disorder and that sequencing of selected genes may identify disease-related variants. Genetic testing will not only improve the diagnostic rate and add important information to research in cardiac diseases, it will also allow for identification of relatives with the same genetic disorder and initiation of treatment of these relatives. In a similar way, NGS may be used to screen for variants in genes that are involved in the metabolism of particular drugs and supplement toxicology investigations of a deceased individual in order to assess whether an unexpected death was accidental or premeditated [87,88]. It will also be possible to investigate DNA from bacteria, viruses, phages, and fungi from the deceased person either to identify disease-causing microorganisms or to look for imbalances in the microbial communities that may give clues to the cause of death [89–91].

Sequencing of the microbiome in swabs or soil samples has demonstrated large differences in the different taxa found at different locations [92–94]. This may be used to find similarities between trace and reference samples. However, it should be emphasized that perfect matches or even exclusions are unlikely because the microbiome is constantly changing under the influence of environmental factors such as temperature, humidity, and time of sampling. Also, it was found that samples taken a few meters apart at the same time only shared 50% of the microbiome diversity. Nevertheless, the variation between sampling sites was much higher [94].

Box 22.1 Forensic Genetics and Genomics: Key Points

1. The utilization of polymorphic DNA markers, minisatellites (variable number tandem repeats), and microsatellites [short tandem repeats (STRs)] for human identification in forensic genetics was originally proposed by Sir Alec Jeffreys, University of Leicester, United Kingdom.

2. Many STRs' loci are used. Unlike single nucleotide polymorphisms (SNPs), which most commonly have two alleles, STRs have numerous alleles that are defined by the number of repeats at each locus. Because of the polymorphic nature, STRs offer very high weights of the evidence when a "match" between the DNA profiles of, eg, a crime scene sample and a suspect, is observed. In practice, about 10–15 carefully chosen STR loci are sufficient to identify an individual in most cases. In the United States, the Federal Bureau of Investigation (FBI) uses a panel of 13 STRs (plus the *AMELX/AMELY* loci for sex determination) for its CODIS database. In the European Union, 12 STRs are recommended. These STR markers, together with further markers, are implemented into a number of commercial kits.

3. The current applications for DNA in forensic laboratories are focused on investigations in crime, relationship, immigration, and identification cases:
   a. matching a crime scene DNA sample to a reference sample (ie, a cheek swab) to implicate a suspect, either directly or indirectly (ie, using a close relative)
   b. searching for a match among the profiles of criminals in crime DNA databases, like the CODIS database of the FBI
   c. comparing DNA profiles from samples from, eg, two crime scenes to learn whether the same individual(s) may have been involved
   d. identification of human remains or dead bodies.
   e. notably, the STR profiles in CODIS are designed for matching DNA samples; however, STRs are not suited for distinguishing an individual's ancestry or other physical or medical traits.

4. There is an ongoing dispute between different legal systems whether searches in crime DNA databases can be used to identify relatives that may match a crime-related DNA profile. This approach, commonly referred to as *familial searching,* is only used in a few countries and US states because of concerns about legal issues and human rights.

5. In addition to the commonly used autosomal STR panels, Y-STRs (microsatellites on the Y chromosome) are sometimes used in genealogical DNA testing. This is particularly useful for matching males of the same paternal lineage because these individuals share the same Y chromosome.

6. The use of mitochondrial DNA (mtDNA) sequence profiling is used by many forensic laboratories for analyzing hair, teeth, bones, and other material obtained from scenes of the crime, major accidents, and disasters including acts of terrorism and war. Analysis of mtDNA can be performed in some cases when STR typing is impossible because of degradation of the DNA. Mitochondrial DNA is inherited maternally and lacks recombination. Databases of mtDNA control-region nucleotide sequences are available to the forensic and research communities. The mtDNA Population Database program, managed by the FBI, has two components: population data, and specialized software designed to search these data. Similarly, the European DNA Profiling Mitochondrial DNA Population Database (EMPOP) holds mtDNA sequences from all over the world. Each mtDNA haplotype in the database is quality controlled using logical and phylogenetic tools.

7. Ethical concerns relating to the inappropriate use of forensic DNA profiles have been raised. The legal systems in the various jurisdiction are expected to address personal interests and human rights of individuals that are investigated.

CONCLUDING REMARKS

High-throughput sequencing has accelerated research in many areas of biology and applied science. In the last few years, the use of NGS in forensic genetics has been debated and now, we are beginning to see applications directed specifically for human identification and determination of phenotypical traits (Box 22.1). The advantages of NGS compared with the traditional PCR-CE methods are many and there is little doubt that NGS will be implemented and used in forensic laboratories in the future. Prices of instruments and kits will determine how fast the transition from CE to NGS will be and how large a fraction of cases will be investigated by NGS. Development and validation of software solutions will be other critical aspects of the introduction of NGS into forensic genetics. Both commercial companies and forensic laboratories [63,70,72,73] have initiated the process;

however, the current software solutions are not sufficiently advanced and more work and collaboration between the companies and the forensic community is necessary.

Among the various platforms, the bench-top sequencers seem to be applicable for case work in terms of daily throughput, flexibility, run time, and instrument cost. Currently, PCR-based capture methods combined with sequencing by synthesis and semiconductor sequencing are the most promising technologies. However, high-throughput sequencing has evolved dramatically in the last decade and there is reason to believe that the development will continue. PCR may be replaced with probe capture methods if the sensitivity can be improved, and single molecule sequencers may render NGS platforms obsolete in the coming years if new landmarks in signal detection can be developed and the base call error rates can be reduced to an acceptable level.

ACKNOWLEDGMENT

We thank Vania Pereira, Jill Olofsson, Jeppe D. Andersen, and Marie-Louise Kampmann for helpful discussions.

REFERENCES

[1] Higuchi R, von Beroldingen CH, Sensabaugh GF, Erlich HA. DNA typing from single hairs. Nature 1988;332:543–6.

[2] Ginther C, Issel-Tarver L, King MC. Identifying individuals by sequencing mitochondrial DNA from teeth. Nat Genet 1992;2:135–8.

[3] Sullivan KM, Hopgood R, Gill P. Identification of human remains by amplification and automated sequencing of mitochondrial DNA. Int J Leg Med 1992;105:83–6.

[4] Parson W, Brandstatter A, Alonso A, Brandt N, Brinkmann B, Carracedo A, et al. The EDNAP mitochondrial DNA population database (EMPOP) collaborative exercises: organisation, results and perspectives. Forensic Sci Int 2004;139:215–26.

[5] Parson W, Bandelt HJ. Extended guidelines for mtDNA typing of population data in forensic science. Forensic Sci Int Genet 2007;1:13–9.

[6] Sanger F, Nicklen S, Coulson AR. DNA sequencing with chain-terminating inhibitors. Proc Natl Acad Sci USA 1977;74:5463–7.

[7] Mitchelson KR. The use of capillary electrophoresis for DNA polymorphism analysis. Mol Biotechnol 2003;24:41–68.

[8] Børsting C, Morling N. Single nucleotide polymorphisms. In: Siegel JA, Pekka S, editors. Encyclopedia of forensic sciences. Maryland, USA: Elsevier; 2013. p. 233–9.

[9] Børsting C, Rockenbauer E, Morling N. Validation of a single nucleotide polymorphism (SNP) typing assay with 49 SNPs for forensic genetic testing in a laboratory accredited according to the ISO 17025 standard. Forensic Sci Int Genet 2009;4:34–42.

[10] Børsting C, Mikkelsen M, Morling N. Kinship analysis with diallelic SNPs: experiences with the SNPforID multiplex in an ISO17025-accreditated laboratory. Transfus Med Hemother 2012;39:195–201.

[11] Walsh S, Lindenbergh A, Zuniga SB, Sijen T, de Knijff P, Kayser M, et al. Developmental validation of the IrisPlex system: determination of blue and brown iris colour for forensic intelligence. Forensic Sci Int Genet 2011;5:464–71.

[12] Walsh S, Chaitanya L, Clarisse L, Wirken L, Draus-Barini J, Kovatsi L, et al. Developmental validation of the HIrisPlex system: DNA-based eye and hair colour prediction for forensic and anthropological usage. Forensic Sci Int Genet 2014;9:150–61.

[13] Phillips C, Salas A, Sanchez JJ, Fondevila M, Gomez-Tato A, Alvarez-Dios J, et al. Consortium: inferring ancestral origin using a single multiplex assay of ancestry-informative marker SNPs. Forensic Sci Int Genet 2007;1:273–80.

[14] Phillips C, Prieto L, Fondevila M, Salas A, Gomez-Tato A, Alvarez-Dios J, et al. Ancestry analysis in the 11-M Madrid bomb attack investigation. PLoS One 2009;4:e6583.

[15] Ronaghi M, Karamohamed S, Pettersson B, Uhlen M, Nyren P. Real-time DNA sequencing using detection of pyrophosphate release. Anal Biochem 1996;242:84–9.

[16] Andreasson H, Asp A, Alderborn A, Gyllensten U, Allen M. Mitochondrial sequence analysis for forensic identification using pyrosequencing technology. Biotechniques 2002;32:124–6; 128, 130–3.

[17] Andreasson H, Nilsson M, Budowle B, Frisk S, Allen M. Quantification of mtDNA mixtures in forensic evidence material using pyrosequencing. Int J Leg Med 2006;120:383–90.

[18] Divne AM, Edlund H, Allen M. Forensic analysis of autosomal STR markers using pyrosequencing. Forensic Sci Int Genet 2010;4:122–9.

[19] Margulies M, Egholm M, Altman WE, Attiya S, Bader JS, Bemben LA, et al. Genome sequencing in microfabricated high-density picolitre reactors. Nature 2005;437:376–80.

[20] Wheeler DA, Srinivasan M, Egholm M, Shen Y, Chen L, McGuire A, et al. The complete genome of an individual by massively parallel DNA sequencing. Nature 2008;452:872–6.

[21] Voelkerding KV, Dames SA, Durtschi JD. Next-generation sequencing: from basic research to diagnostics. Clin Chem 2009;55:641–58.

[22] Braslavsky I, Hebert B, Kartalov E, Quake SR. Sequence information can be obtained from single DNA molecules. Proc Natl Acad Sci USA 2003;100:3960–4.

[23] Harris TD, Buzby PR, Babcock H, Beer E, Bowers J, Braslavsky I, et al. Single-molecule DNA sequencing of a viral genome. Science 2008;320:106–9.

[24] Mardis ER. Next-generation DNA sequencing methods. Annu Rev Genomics Hum Genet 2008;9:387–402.

[25] Metzker ML. Sequencing technologies: the next generation. Nat Rev Genet 2010;11:31–46.

[26] Korlach J, Bjornson KP, Chaudhuri BP, Cicero RL, Flusberg BA, Gray JJ, et al. Real-time DNA sequencing from single polymerase molecules. Methods Enzymol 2010;472:431–55.

[27] Nielsen R, Paul JS, Albrechtsen A, Song YS. Genotype and SNP calling from next-generation sequencing data. Nat Rev Genet 2011;12:443–51.

[28] Altmann A, Weber P, Bader D, Preuss M, Binder EB, Muller-Myhsok B. A beginners guide to SNP calling from high-throughput DNA-sequencing data. Hum Genet 2012;131:1541–54.

[29] Ozsolak F. Third-generation sequencing techniques and applications to drug discovery. Expert Opin Drug Discov 2012;7:231–43.

[30] Haque F, Li J, Wu HC, Liang XJ, Guo P. Solid-state and biological nanopore for real-time sensing of single chemical and sequencing of DNA. Nano Today 2013;8:56–74.

[31] Goldman D, Domschke K. Making sense of deep sequencing. Int J Neuropsychopharmacol 2014:1–9.

[32] Cloonan N, Grimmond SM. Transcriptome content and dynamics at single-nucleotide resolution. Genome Biol 2008;9:234.

[33] Yano M, Ohtsuka T, Okano H. RNA-binding protein research with transcriptome-wide technologies in neural development. Cell Tissue Res 2014;359:135–44.

[34] Mamanova L, Coffey AJ, Scott CE, Kozarewa I, Turner EH, Kumar A, et al. Target-enrichment strategies for next-generation sequencing. Nat Methods 2010;7:111–8.

[35] Liu X, Wang J, Chen L. Whole-exome sequencing reveals recurrent somatic mutation networks in cancer. Cancer Lett 2013;340:270–6.

[36] Hagemann IS, Cottrell CE, Lockwood CM. Design of targeted, capture-based, next generation sequencing tests for precision cancer therapy. Cancer Genet 2013;206:420–31.

[37] Dressman D, Yan H, Traverso G, Kinzler KW, Vogelstein B. Transforming single DNA molecules into fluorescent magnetic particles for detection and enumeration of genetic variations. Proc Natl Acad Sci USA 2003;100:8817–22.

[38] Fedurco M, Romieu A, Williams S, Lawrence I, Turcatti G. BTA, a novel reagent for DNA attachment on glass and efficient generation of solid-phase amplified DNA colonies. Nucleic Acids Res 2006;34:e22.

[39] Lundquist PM, Zhong CF, Zhao P, Tomaney AB, Peluso PS, Dixon J, et al. Parallel confocal detection of single molecules in real time. Opt Lett 2008;33:1026–8.

[40] Eid J, Fehr A, Gray J, Luong K, Lyle J, Otto G, et al. Turner: real-time DNA sequencing from single polymerase molecules. Science 2009;323:133–8.

[41] Stoloff DH, Wanunu M. Recent trends in nanopores for biotechnology. Curr Opin Biotechnol 2013;24:699–704.

[42] Liang L, Wang Q, Agren H, Tu Y. Computational studies of DNA sequencing with solid-state nanopores: key issues and future prospects. Front Chem 2014;2:1–4.

[43] Mikheyev AS, Tin MM. A first look at the Oxford Nanopore MinION sequencer. Mol Ecol Resour 2014;14:1097–102.

[44] Chin CS, Alexander DH, Marks P, Klammer AA, Drake J, Heiner C, et al. Nonhybrid, finished microbial genome assemblies from long-read SMRT sequencing data. Nat Methods 2013;10:563–9.

[45] Huddleston J, Ranade S, Malig M, Antonacci F, Chaisson M, Hon L, et al. Reconstructing complex regions of genomes using long-read sequencing technology. Genome Res 2014;24:688–96.

[46] Flusberg BA, Webster DR, Lee JH, Travers KJ, Olivares EC, Clark TA, et al. Direct detection of DNA methylation during single-molecule, real-time sequencing. Nat Methods 2010;7:461–5.

[47] Wallace EV, Stoddart D, Heron AJ, Mikhailova E, Maglia G, Donohoe TJ, et al. Identification of epigenetic DNA modifications with a protein nanopore. Chem Commun (Camb) 2010;46:8195–7.

[48] Clark TA, Murray IA, Morgan RD, Kislyuk AO, Spittle KE, Boitano M, et al. Characterization of DNA methyltransferase specificities using single-molecule, real-time DNA sequencing. Nucleic Acids Res 2012;40:e29.

[49] Schreiber J, Wescoe ZL, Abu-Shumays R, Vivian JT, Baatar B, Karplus K, et al. Error rates for nanopore discrimination among cytosine, methylcytosine, and hydroxymethylcytosine along individual DNA strands. Proc Natl Acad Sci USA 2013;110:18910–5.

[50] Nothnagel M, Herrmann A, Wolf A, Schreiber S, Platzer M, Siebert R, et al. Technology-specific error signatures in the 1000 Genomes Project data. Hum Genet 2011;130:505–16.

[51] Lam HY, Clark MJ, Chen R, Chen R, Natsoulis G, O'Huallachain M, et al. Performance comparison of whole-genome sequencing platforms. Nat Biotechnol 2012;30:78–82.

[52] Ratan A, Miller W, Guillory J, Stinson J, Seshagiri S, Schuster SC. Comparison of sequencing platforms for single nucleotide variant calls in a human sample. PLoS One 2013;8:e55089.

[53] Rieber N, Zapatka M, Lasitschka B, Jones D, Northcott P, Hutter B, et al. Coverage bias and sensitivity of variant calling for four whole-genome sequencing technologies. PLoS One 2013;8:e66621.

[54] Rehm HL. Disease-targeted sequencing: a cornerstone in the clinic. Nat Rev Genet 2013;14:295–300.

[55] Beadling C, Neff TL, Heinrich MC, Rhodes K, Thornton M, Leamon J, et al. Combining highly multiplexed PCR with semiconductor-based sequencing for rapid cancer genotyping. J Mol Diagn 2013;15:171–6.

[56] Wong SQ, Li J, Tan AY, Vedururu R, Pang JM, Do H, et al. Sequence artefacts in a prospective series of formalin-fixed tumours tested for mutations in hotspot regions by massively parallel sequencing. BMC Med Genomics 2014;7:23.

[57] Castera L, Krieger S, Rousselin A, Legros A, Baumann JJ, Bruet O, et al. Next-generation sequencing for the diagnosis of hereditary breast and ovarian cancer using genomic capture targeting multiple candidate genes. Eur J Hum Genet 2014;22:1305–11.

[58] Millat G, Chanavat V, Rousson R. Evaluation of a new high-throughput next-generation sequencing method based on a custom AmpliSeq library and ion torrent PGM sequencing for the rapid detection of genetic variations in long QT syndrome. Mol Diagn Ther 2014;18:533–9.

[59] Tarabeux J, Zeitouni B, Moncoutier V, Tenreiro H, Abidallah K, Lair S, et al. Streamlined ion torrent PGM-based diagnostics: *BRCA1* and *BRCA2* genes as a model. Eur J Hum Genet 2014;22:535–41.

[60] Tsongalis GJ, Peterson JD, de Abreu FB, Tunkey CD, Gallagher TL, Strausbaugh LD, et al. Routine use of the Ion Torrent AmpliSeq Cancer Hotspot Panel for identification of clinically actionable somatic mutations. Clin Chem Lab Med 2014;52:707–14.

[61] Treangen TJ, Salzberg SL. Repetitive DNA and next-generation sequencing: computational challenges and solutions. Nat Rev Genet 2012;13:36–46.

[62] Fordyce SL, Avila-Arcos MC, Rockenbauer E, Børsting C, Frank-Hansen R, Petersen FT, et al. High-throughput sequencing of core STR loci for forensic genetic investigations using the Roche Genome Sequencer FLX platform. Biotechniques 2011;51:127–33.

[63] Van Neste C, Van Nieuwerburgh F, Van Hoofstat D, Deforce D. Forensic STR analysis using massive parallel sequencing. Forensic Sci Int Genet 2012;6:810–8.

[64] Rockenbauer E, Hansen S, Mikkelsen M, Børsting C, Morling N. Characterization of mutations and sequence variants in the *D21S11* locus by next generation sequencing. Forensic Sci Int Genet 2014;8:68–72.

[65] Dalsgaard S, Rockenbauer E, Buchard A, Mogensen HS, Frank-Hansen R, Børsting C, et al. Non-uniform phenotyping of *D12S391* resolved by second generation sequencing. Forensic Sci Int Genet 2014;8:195–9.

[66] Gelardi C, Rockenbauer E, Dalsgaard S, Børsting C, Morling N. Second generation sequencing of three STRs *D3S1358*, *D12S391* and *D21S11* in Danes and a new nomenclature for sequenced STR alleles. Forensic Sci Int Genet 2014;12:38–41.

[67] Tomas C, Mogensen HS, Friis SL, Hallenberg C, Stene MC, Morling N. Concordance study and population frequencies for 16 autosomal STRs analyzed with PowerPlex(R) ESI 17 and AmpFlSTR(R) NGM SElect in Somalis, Danes and Greenlanders. Forensic Sci Int Genet 2014;11:e18–21.

[68] Westen AA, Kraaijenbrink T, Robles de Medina EA, Harteveld J, Willemse P, et al. Comparing six commercial autosomal STR kits in a large Dutch population sample. Forensic Sci Int Genet 2014;10:55–63.

[69] Scheible M, Loreille O, Just R, Irwin J. Short tandem repeat typing on the 454 platform: strategies and considerations for targeted sequencing of common forensic markers. Forensic Sci Int Genet 2014;12:107–19.

[70] Anvar SY, van der Gaag KJ, van der Heijden JW, Veltrop MH, Vossen RH, de Leeuw RH, et al. TSSV: a tool for characterization of complex allelic variants in pure and mixed genomes. Bioinformatics 2014;30:1651–9.

[71] Bornman DM, Hester ME, Schuetter JM, Kasoji MD, Minard-Smith A, Barden CA, et al. Short-read, high-throughput sequencing technology for STR genotyping. Biotechniques 2012:1–6. http://dx.doi.org/10.2144/000113857.

[72] Warshauer DH, Lin D, Hari K, Jain R, Davis C, Larue B, et al. STRait Razor: a length-based forensic STR allele-calling tool for use with second generation sequencing data. Forensic Sci Int Genet 2013;7:409–17.

[73] Van Neste C, Vandewoestyne M, Van Criekinge W, Deforce D, Van Nieuwerburgh F. My-Forensic-Loci-queries (MyFLq) framework for analysis of forensic STR data generated by massive parallel sequencing. Forensic Sci Int Genet 2014;9:1–8.

[74] Fordyce SL, Mogensen HS, Borsting C, Lagace RE, Chang CW, Rajagopalan N, et al. Second-generation sequencing of forensic STRs using the ion torrent HID STR 10-plex and the ion PGM. Forensic Sci Int Genet 2015;14:132–40.

[75] Dalsgaard S, Rockenbauer E, Gelardi C, Børsting C, Fordyce SL, Morling N. Characterization of mutations and sequence variations in complex STR loci by second generation sequencing. Forensic Sci Int Genet 2013;4(Suppl.):e218–9.

[76] Børsting C, Fordyce SL, Olofsson J, Mogensen HS, Morling N. Evaluation of the Ion Torrent HID SNP 169-plex: a SNP typing assay developed for human identification by second generation sequencing. Forensic Sci Int Genet 2014;12:144–54.

[77] Seo SB, King JL, Warshauer DH, Davis CP, Ge J, Budowle B. Single nucleotide polymorphism typing with massively parallel sequencing for human identification. Int J Leg Med 2013;127:1079–86.

[78] Sanchez JJ, Phillips C, Børsting C, Balogh K, Bogus M, Fondevila M, et al. A multiplex assay with 52 single nucleotide polymorphisms for human identification. Electrophoresis 2006;27:1713–24.

[79] Pakstis AJ, Speed WC, Fang R, Hyland FC, Furtado MR, Kidd JR, et al. SNPs for a universal individual identification panel. Hum Genet 2010;127:315–24.

[80] Nassir R, Kosoy R, Tian C, White PA, Butler LM, Silva G, et al. An ancestry informative marker set for determining continental origin: validation and extension using human genome diversity panels. BMC Genet 2009;10:39.

[81] Nievergelt CM, Maihofer AX, Shekhtman T, Libiger O, Wang X, Kidd KK, et al. Inference of human continental origin and admixture proportions using a highly discriminative ancestry informative 41-SNP panel. Investig Genet 2013;4:13.

[82] Walsh S, Liu F, Wollstein A, Kovatsi L, Ralf A, Kosiniak-Kamysz A, et al. The HIrisPlex system for simultaneous prediction of hair and eye colour from DNA. Forensic Sci Int Genet 2013;7:98–115.

[83] Poulsen L, Friis SL, Hallenberg C, Simonsen BT, Morling N. A report of the 2009-2011 paternity and relationship testing workshops of the English speaking working group of the International Society for Forensic Genetics. Forensic Sci Int Genet 2014;9:e1–2.

[84] Thomsen AR, Hallenberg C, Simonsen BT, Langkjaer RB, Morling N. A report of the 2002–2008 paternity testing workshops of the English speaking working group of the International Society for Forensic Genetics. Forensic Sci Int Genet 2009;3:214–21.

[85] Gjertson DW, Brenner CH, Baur MP, Carracedo A, Guidet F, Luque JA, et al. ISFG: recommendations on biostatistics in paternity testing. Forensic Sci Int Genet 2007;1:223–31.

[86] Hertz CL, Ferrero-Miliani L, Frank-Hansen R, Morling N, Bundgaard H. A comparison of genetic findings in sudden cardiac death victims and cardiac patients: the importance of phenotypic classification. Eur Eur Pacing, Arrhythm Cardiac Electrophysiol 2014. http://dx.doi.org/10.1093/europace/euu210.

[87] Visser LE, van Schaik RH, van Vliet M, Trienekens PH, De Smet PA, Vulto AG, et al. Allelic variants of cytochrome P450 2C9 modify the interaction between nonsteroidal anti-inflammatory drugs and coumarin anticoagulants. Clin Pharmacol Ther 2005;77:479–85.

[88] Sullivan D, Pinsonneault JK, Papp AC, Zhu H, Lemeshow S, Mash DC, et al. Dopamine transporter *DAT* and receptor *DRD2* variants affect risk of lethal cocaine abuse: a gene–gene–environment interaction. Transl Psychiatry 2013;3:e222.

[89] Cox MJ, Cookson WO, Moffatt MF. Sequencing the human microbiome in health and disease. Hum Mol Genet 2013;22:R88–94.

[90] Quinones-Mateu ME, Avila S, Reyes-Teran G, Martinez MA. Deep sequencing: becoming a critical tool in clinical virology. J Clin Virol 2014;61:9–19.

[91] Frey KG, Herrera-Galeano JE, Redden CL, Luu TV, Servetas SL, Mateczun AJ, et al. Comparison of three next-generation sequencing platforms for metagenomic sequencing and identification of pathogens in blood. BMC Genomics 2014;15:96.

[92] Tringe SG, Rubin EM. Metagenomics: DNA sequencing of environmental samples. Nat Rev Genet 2005;6:805–14.

[93] Giampaoli S, Berti A, Di Maggio RM, Pilli E, Valentini A, Valeriani F, et al. The environmental biological signature: NGS profiling for forensic comparison of soils. Forensic Sci Int 2014;240:41–7.

[94] Young JM, Weyrich LS, Cooper A. Forensic soil DNA analysis using high-throughput sequencing: a comparison of four molecular markers. Forensic Sci Int Genet 2014;13:176–84.

Chapter 23

Public and Population Health Genomics

A.L. Wise, T.A. Manolio

National Institutes of Health (NIH), Bethesda, MD, United States

Chapter Outline

Introduction	311	Alzheimer Dementia	313
Breast Cancer	311	Cystic Fibrosis	315
Colorectal Cancer	312	Cross-Cutting Issues of Population Genomics	315
Bronchial Asthma	313	Summary	316
Crohn Disease	313	References	316

INTRODUCTION

Population sciences such as epidemiology focus on studying whole populations rather than individuals. Through studying environmental, genomic, and social factors that affect human health, population level interventions can be identified. The social sciences focus on studying society and human behavior through fields such as anthropology, economics, law, psychology, and sociology. The study of the ethical, legal, and social implications (ELSI) of genomics plays an important role in applying genomics to population health.

Although the fields of population genomics and genomic medicine look to prevent or treat disease through different perspectives, they can act complementarily to enhance overall health outcomes for both individuals and populations at large. Population genomics seeks to integrate knowledge from genomic, population, and social sciences to improve population health.

The genomic sciences focus on studying whole genomes, such as the entire DNA sequence making up the human genome. Through studying genomics, genetic variants influencing human health can be identified. Studies of particular genes can then further elucidate the function of genetic variants. The three major disciplines contributing to population genomics (genomics, population science, and social sciences) explore two cross-cutting issues: global health, and population versus individual health. This aspect of the emerging field of population genomics is discussed in this chapter using specific examples from diverse diseases such as breast cancer, colorectal cancer, bronchial asthma, Crohn disease, Alzheimer dementia, and cystic fibrosis.

BREAST CANCER

Around one in nine women worldwide develop breast cancer during their lifetime. A family history of breast cancer along with ovarian cancer may be encountered in approximately 5% of the affected women. A number of disease-causing mutations in the two major genes, *BRCA1* and *BRCA2*, are now recorded worldwide. In addition, several variants may occur in 5–10% of breast cancer cases; these variants are found in less than 1% of the general population [1]. Specific populations, such as those of Ashkenazi Jewish descent, have an increased incidence of *BRCA1* or *BRCA2* variants. Two variants in *BRCA1* and one in *BRCA2* are found at a rate 5 times higher in Ashkenazi Jews than in the general population [2,3].

For these reasons, family members of those with known *BRCA1/2* variants or those with a family history of breast cancer may be offered genetic testing. Men with *BRCA1/2* variants are also at an increased risk of developing breast cancer [4,5]. Thousands of variants have been discovered in *BRCA1* and *BRCA2*, yet only a minority have a known deleterious effect [6]. Genetic testing therefore has the possibility of finding a variant of unknown effect, for which the functional significance is unclear. One recent study found that 10% of women undergoing *BRCA1* and *BRCA2* testing receive an ambiguous test result because of the detection of a variant of unknown significance [7]. Thus although deleterious variants are known to increase the risk of developing breast cancer approximately five-fold, deciding how to react to variants of unknown effect can be challenging for all involved including both clinicians and patients [8].

COLORECTAL CANCER

Genomic information can be used clinically to inform disease risk, diagnosis, drug selection, and drug dosing. Colorectal cancer provides a good example of an area where population level screening along with genomic medicine approaches are coming together to improve overall population health.

Over 1 million individuals are diagnosed with colorectal cancer each year worldwide, accounting for approximately 9–10% of cancer diagnoses in 2008 [9]. It is the third leading cause of cancer-related death in the United States and the fourth worldwide [9,10]. In colorectal cancer, a patient's genomic information can be used to determine risk of inherited colorectal cancer syndromes, whether certain biological agents will work in specific patients, and what starting dose to use on specific chemotherapeutics.

As many as 20–25% of colorectal cancer cases have a family history of colorectal cancer (two or more first-degree relatives with colorectal cancer), yet only 5–6% have an established familial genetic syndrome with a known genetic variant [11,12]. Of those with established familial genetic syndromes, approximately 3% will be diagnosed with Lynch syndrome (including variants in the genes *MLH1, MSH2, MSH6, PMS2*, and *EPCAM*) and 1% with familial adenomatous polyposis (including variants in *APC* and *MUTYH*) [11]. Individuals with a family history of colorectal cancer have a two- to threefold greater risk of developing colorectal cancer than the general population and thus genetic testing for an individual with a known family history has substantial public health benefit [11]. Those with a known family history of colorectal cancer are also recommended for screening at younger ages, typically 10 years younger than the onset of the youngest case in their family.

Genetic testing is also used to determine treatment options in colorectal cancer [11,12] (Table 23.1). For example, genetic variants that make *KRAS* constitutively active have been shown to provide resistance to monoclonal antibodies directed against the upstream epidermal growth factor receptor (EGFR), because both are components of a cellular pathway leading to abnormal cell growth and cancer. Thus cetuximab and panitumumab (anti-EGFR antibodies) are given only to individuals with normally functioning *KRAS*, where blocking EGFR can have an effect [11].

Pharmacogenomics can also be useful in determining drug dosage for colorectal cancer. For example, the FDA recommends testing for *UGT1A1* variants when administering irinotecan, because individuals homozygous or heterozygous for the *UGT1A1*28* allele are at increased risk of developing neutropenia and severe infections [13]. Individuals with inactivating *UGT1A1* variants are therefore recommended to be started at a lower dosage of irinotecan to reduce the risk of neutropenia [13].

In addition to modifying drug dosing, pharmacogenomic information can also be used in drug selection to choose agents more likely to give a beneficial response based on a patient's genetically driven ability to metabolize them. For example, in patients of Asian ancestry given carbamazepine (used to treat epilepsy and bipolar disorder) the *HLA-B*1502* allele has been associated with Stevens-Johnson syndrome/toxic epidermal necrolysis, a life-threatening skin condition. This allele can be found in over 15% of the population in some regions in Asia including Hong Kong, Thailand, Malaysia, and parts of the Philippines, and is very rare in other populations outside Asia [14] (Table 23.2).

Within populations of Asian ancestry, there can also be great variation, such as is seen within China where the *HLA-B*1502* allele prevalence varies from 0 to 36%, depending upon ethnicity. Thus the US Food and Drug

TABLE 23.1 Pharmacogenomic Variants in Colorectal Cancer [11].

	Function	Consequences
BRAF	Downstream pathways constitutively active	Resistance to anti–epidermal growth factor receptor monoclonal antibodies
ERCC-1	DNA excision repair	Resistance to platinum-based chemotherapy drugs
Interleukin 8	Increased *VEGF* expression	Increased cancer recurrence
KRAS	Downstream pathways constitutively active	Resistance to anti–epidermal growth factor receptor monoclonal antibodies
Microsatellite instability	Reduced DNA repair	Improved prognosis
TSER	Increased or decreased thymidylate synthase, depending on variant	Response to fluorouracil reduced/increased (negative relationship)
UTGA1	Responsible for metabolism of irinotecan	Dosing for irinotecan
VEGF	Increased *VEGF* expression	Increased cancer recurrence

TABLE 23.2 *HLA-B*1502* Allele Prevalence in Worldwide Populations [46].

Population	*HLA-B*1502* Allele Prevalence
China	0–36%
Indonesia	11–17%
Malaysia	21–6%
Vietnam	14%
Thailand	8–9%
India	0–6%
Singapore	6%
Taiwan	4–6%
USA	0–4%
South Korea	0.2–2%
Australia	0–1%
Japan	0.10%
Germany	0%
Brazil	0%
Bulgaria	0%
Burkina Faso	0%
Cuba	0%
Ireland	0%
Italy	0%
Mexico	0%
Morocco	0%
Oman	0%
South Africa	0%

Administration (FDA) recommends genetic testing for the *HLA-B*1502* variant before prescribing carbamazepine for patients of Asian ancestry and for those with one or two copies of the variant only treating with carbamazepine when the benefits outweigh the risks of the drug [14].

Bronchial Asthma

Population variation is an important consideration when studying common complex conditions that are influenced by multiple genetic, environmental, and social risk factors, such as bronchial asthma. Over 300 million individuals of all ages have asthma worldwide [15]. Prevalence estimates can vary greatly by ethnicity; however, from 2% to 33% [16]. In the United States, prevalence ranges from approximately 8% in European Americans to 12% in African

Americans and 7% in Hispanic Americans [17]. Within admixed populations, such as Hispanic Americans, even greater variation can be seen when populations are further substratified, with Mexican American populations around 6%, whereas Puerto Rican populations are closer to 19% [17]. Genetic studies have shown that at least some of this variation is caused by differences in genetic variants, with 35–80% of the variation in asthma heritability explained by genetic factors [18,19]. For example, variants in *ADAM33* have been seen in European, African American, and some Hispanic populations, but not in other European American, Mexican, Puerto Rican, and Korean populations, all of which found different variants in *ADAM33* associated with asthma [20] (Fig. 23.1).

Studying the interplay between environmental, genetic, and social risk factors is also critical to understanding the etiology of this complex disease. For example, the effect of air pollution on asthma case reports is modified by genetic factors as well, showing potential gene–environment interactions. A key measure of air pollution is PM10, the concentration in parts per million of particulate matter 10 μm in diameter or less, which can penetrate and irritate small airways. PM10 has been shown in multiple epidemiological studies to be an independent risk factor for increased respiratory symptoms including asthma [21–25]. Similarly variants in over 100 genes have been associated with asthma in genome-wide association studies (GWASs) [20,26,27]. Looking at the two risk factors together, however, reveals a potential gene–environment interaction where variants in *GSTP1*, *SOD2*, and *NFE2L2*, all related to oxidative stress pathways, were also associated with increased hospital admissions for asthma-related symptoms during days with high PM10 levels [21].

CROHN DISEASE

Genomics can also be used to help identify and better define environmental risk factors in population studies. For example, genomic data profiling the bacteria inhabiting the human gut, or gut microbiome, has revealed differences in the bacterial populations present in individuals with Crohn disease (a form of inflammatory bowel disease) [28]. The genomic signatures of the gut microbiome in patients with Crohn disease shows some bacterial populations to be decreased, whereas others are more abundant [28–31].

ALZHEIMER DEMENTIA

Many genetic loci have also been associated with multiple phenotypes, as evidenced in the National Human Genome Research Institute Catalog of Published GWASs [27,32]. (Fig. 23.2). Such pleiotropic genes (genes associated with multiple phenotypes) can present additional challenges when considering the ELSI of returning genetic testing

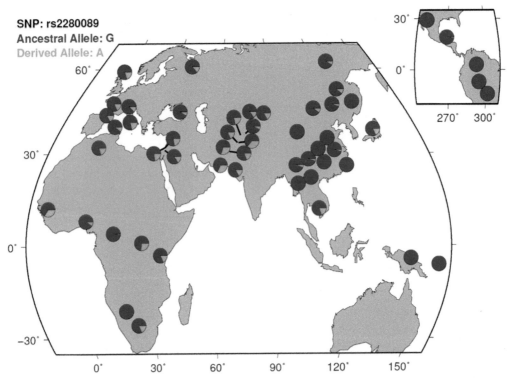

FIGURE 23.1 Example of the variation in allele incidence by population for rs2280089 in *ADAM33*. The *A* allele has been previously associated with predisposition to asthma and bronchial hyperresponsiveness in populations from the United States, United Kingdom, and China [48–51].

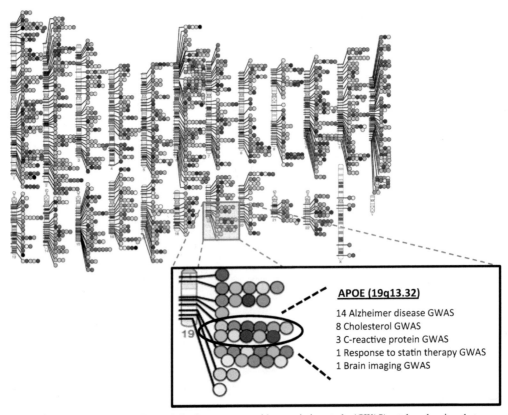

FIGURE 23.2 The National Human Genome Research Institute genome-wide association study *(GWAS)* catalog showing that many genetic loci are associated with multiple phenotypes. *APOE* on 19q13.32 is highlighted along with examples of the disease phenotypes associated with the gene [27].

results. For example, variants in *APOE* are associated with multiple phenotypes, including Alzheimer dementia, cholesterol level, coronary disease, C-reactive protein, hyperlipoproteinemia type III, low-density lipoprotein level, macular degeneration, and response to statin therapy [33]. The *APOE*e4* variant in particular has been associated with increased risk for developing both Alzheimer dementia and atherosclerosis along with a protective effect against developing macular degeneration [33].

CYSTIC FIBROSIS

In many ways the family serves as an intermediary between individual- and population-level views of health. It is an important viewpoint that should be considered in population genomics, because genomic information is inherently relevant not only to the individual tested, but also to their family members with whom they share a large proportion of their genetic variants. How and with whom such family-related health information can or should be shared is an important consideration for advancing both individual and family health.

The availability of genomic information is also blurring the line between population and individual level views of health. For example, genetic testing for cystic fibrosis spans population screening–based carrier, prenatal, and newborn tests to individualized genomic medicine–based diagnostic and pharmacogenomic testing for treatment selection. From the population screening perspective, genetic testing is offered to prospective parents of European decent and others who may be at increased risk of having a child affected by cystic fibrosis, because the prevalence of cystic fibrosis is highest in Northern Europe [34].

Over 1500 variants have been found in the *CFTR* gene, but the functional significance of many is unknown, with the most common variant associated with cystic fibrosis being ΔF508 [34]. In 2012, the FDA approved ivacaftor, the first drug to treat a specific cystic fibrosis variant, G551D in *CFTR* [35,36] (Table 23.3). The G551D variant impairs the ability of the *CFTR* channel to open [34,35,37]. Ivacaftor functions by increasing the likelihood of the *CFTR* channel being open, improving chloride transport and restoring the function of the *CFTR* gene [35–37]. As the cost of genomic

sequencing continues to drop and electronic health records improve, the cost of collecting and interpreting genomic data may fall below the cost of conducting individual genetic tests, further blurring the line between clinical and public health data.

CROSS-CUTTING ISSUES OF POPULATION GENOMICS

Although all three of the population sciences contributing to population genomics work together, there are also some issues that more broadly span the field of population genomics and its relationship to medicine and public health. Touched upon in many of the examples discussed, it is important to consider the broader implications of population genomics to global health and how population and individual level views of health can work together to improve health worldwide.

Cardiovascular disease is a leading cause of death worldwide, with over 13.5 million deaths from ischemic heart disease, stroke, or another form of cerebrovascular disease in 2008, and is highly amenable to study using population genomics techniques [38]. For example, adding rs10757274 genotyping to the Framingham risk score improved its ability to determine individuals who would suffer later cardiovascular events independent of family history [39]. Such models can be used to screen populations to determine individuals at increased risk of disease and recommend further testing and individualized genomic medicine.

Whereas chronic conditions such as cardiovascular disease make up the majority of deaths in the developed world, infections are still a major health concern within developing countries and are equally amenable to study using population genomics. Genomics has made possible the rapid identification of the organisms causing recent pandemic outbreaks including H1N1 and severe acute respiratory syndrome, as well as identifying the source of foodborne illness. The availability of genomic sequence information on malaria parasites, mosquito vectors, and their human hosts are all being leveraged to produce more rapid diagnosis and better drugs, vaccines, and intervention strategies to fight malaria [40,41].

To maximize the benefit of population genomics advances to global health, it is also important to include multiple populations of diverse age, ethnicity, and gender in disease research. As evidenced by the example of asthma genomics, the prevalence of disease can be highly variable across ancestral groups and genetic variants often vary in incidence as well. Thus although a single pathway may be implicated in disease across many populations, the most common variant in each population may lie in different genes or gene regions.

Local environmental and social factors that impact disease and population health should also be incorporated into

TABLE 23.3 National Heart, Lung, and Blood Institute Exome Sequencing Project Results for African-American and European-American Participants for the G551D Variant (rs75527207) Associated with Cystic Fibrosis [47].

	Allele Count A	Allele Count G
African-American	0	4406
European-American	18	8582

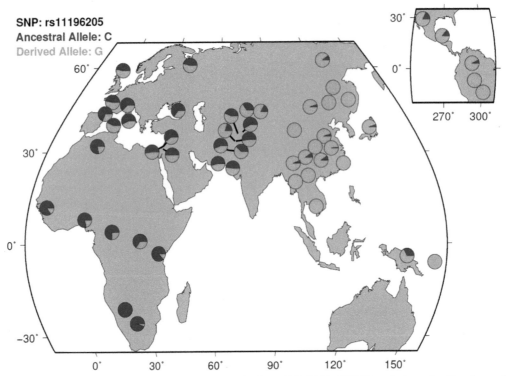

FIGURE 23.3 Example of the variation in allele frequency by population for rs11196205 in *TCF7L2*, a single nucleotide polymorphism previously associated with type 2 diabetes mellitus risk [44,48].

studies of population genomics to produce the most complete picture of disease etiology. For example, the prevalence of type 2 diabetes mellitus is increasing globally and has been associated with multiple genetic (more than 60 genes to date), epigenetic (such as methylation or histone modification), environmental (such as diet), and social factors (such as exercise), all of which contribute to this complex disease [42]. The prevalence of type 2 diabetes mellitus varies by country from approximately 5–29%, with risk alleles such as the C allele in rs11196205 decreasing in incidence from sub-Saharan Africa to Asia (Fig. 23.3) [43,44]. Effects of other risk factors also vary across different populations, with the relative risk of type 2 diabetes mellitus for each 5 kg/m^2 increase in body mass index; for example, being 2.4 in Asian Americans, 2.2 in Hispanic Americans, 2.0 in European Americans, and 1.6 in African Americans [45].

SUMMARY

In this chapter we have explored how the integration of genomic, population, and social sciences in population genomics can improve health through examples in pharmacogenomics, population variation, and genetic pleiotropy. We have also investigated cross cutting issues in global health and population versus individual health where population genomics can play a crucial role in the translation of genomic health discoveries worldwide and population

screening can work together with genomic medicine to provide the greatest health benefit to both individuals and populations at large. Thus, multidisciplinary research in population genomics, can improve clinical care through understanding of the genetic variation in populations that contributes to complex disease.

REFERENCES

[1] Schwartz GF, Hughes KS, Lynch HT, et al. Proceedings of the International Consensus Conference on Breast Cancer Risk, Genetics, & Risk Management, April, 2007. Cancer November 15, 2008;113(10):2627–37.

[2] Struewing JP, Hartge P, Wacholder S, et al. The risk of cancer associated with specific mutations of *BRCA1* and *BRCA2* among Ashkenazi Jews. N Engl J Med May 15, 1997;336(20):1401–8.

[3] Warner E, Foulkes W, Goodwin P, et al. Prevalence and penetrance of *BRCA1* and *BRCA2* gene mutations in unselected Ashkenazi Jewish women with breast cancer. J Natl Cancer Inst July 21, 1999;91(14):1241–7.

[4] Thompson D, Easton DF. Cancer incidence in *BRCA1* mutation carriers. J Natl Cancer Inst September 18, 2002;94(18):1358–65.

[5] Cancer risks in *BRCA2* mutation carriers: the Breast Cancer Linkage Consortium. J Natl Cancer Inst August 4, 1999;91(15):1310–6.

[6] An open access on-line breast cancer mutation data base. http://research.nhgri.nih.gov/bic/?CFID=313172&CFTOKEN=38988484.

[7] Peshkin BN, DeMarco TA, Brogan BM, Lerman C, Isaacs C. *BRCA1/2* testing: complex themes in result interpretation. J Clin Oncol May 1, 2001;19(9):2555–65.

[8] Howlader N, Noone A, Krapcho M, et al, editors. Seer cancer statistics review, 1975–2009 (Vintage 2009 populations). Bethesda, MD: National Cancer Institute; 2012. http://seer.cancer.gov/csr/1975_2009_pops09/. [accessed 27.06.12].

[9] Bray F, Ren JS, Masuyer E, Ferlay J. Global estimates of cancer prevalence for 27 sites in the adult population in 2008. Int J Cancer March 1, 2013;132(5):1133–45. http://dx.doi.org/10.1002/ijc.27711. Epub 2012 Jul 26.

[10] Society AC. Cancer facts & figures 2012. 2012. http://www.cancer.org/acs/groups/content/@epidemiologysurveilance/documents/document/acspc-031941.pdf. [accessed 27.06.12].

[11] Cunningham D, Atkin W, Lenz HJ, et al. Colorectal cancer. Lancet March 20,2010;375(9719):1030–47.

[12] Gala M, Chung DC. Hereditary colon cancer syndromes. Semin Oncol August 2011;38(4):490–9.

[13] Administration UFaD. Dugs@FDA: irinotecan label and approval history. http://www.accessdata.fda.gov/scripts/cder/drugsatfda/index.cfm?fuseaction=Search.Label_ApprovalHistory - labelinfo.

[14] Administration UFaD. Drugs@FDA: carbamazepine label and approval history. http://www.accessdata.fda.gov/scripts/cder/drugsatfda/.

[15] Masoli M, Fabian D, Holt S, Beasley R. The global burden of asthma: executive summary of the GINA Dissemination Committee report. Allergy May 2004;59(5):469–78.

[16] Sembajwe G, Cifuentes M, Tak SW, Kriebel D, Gore R, Punnett L. National income, self-reported wheezing and asthma diagnosis from the world health survey. Eur Respir J February 2010;35(2):279–86.

[17] Current asthma prevalence percents by age. United States: National Health Interview Survey (NHIS) Data; 2010. http://www.cdc.gov/asthma/nhis/2010/table4-1.htm.

[18] Nieminen MM, Kaprio J, Koskenvuo M. A population-based study of bronchial asthma in adult twin pairs. Chest July 1991;100(1):70–5.

[19] Duffy DL, Martin NG, Battistutta D, Hopper JL, Mathews JD. Genetics of asthma and hay fever in Australian twins. Am Rev Respir Dis December 1990;142(6 Pt 1):1351–8.

[20] Drake KA, Galanter JM, Burchard EG. Race, ethnicity and social class and the complex etiologies of asthma. Pharmacogenomics April 2008;9(4):453–62.

[21] Canova C, Dunster C, Kelly FJ, et al. PM10-induced hospital admissions for asthma and chronic obstructive pulmonary disease: the modifying effect of individual characteristics. Epidemiology July 2012;23(4):607–15.

[22] Weinmayr G, Romeo E, De Sario M, Weiland SK, Forastiere F. Short-term effects of PM10 and NO$_2$ on respiratory health among children with asthma or asthma-like symptoms: a systematic review and meta-analysis. Environ Health Perspect April 2010;118(4):449–57.

[23] Preutthipan A, Udomsubpayakul U, Chaisupamongkollarp T, Pentamwa P. Effect of PM10 pollution in Bangkok on children with and without asthma. Pediatr Pulmonol March 2004;37(3):187–92.

[24] Gordian ME, Choudhury AH. PM10 and asthma medication in school children. Arch Environ Health January 2003;58(1):42–7.

[25] Donaldson K, Gilmour MI, MacNee W. Asthma and PM10. Respir Res 2000;1(1):12–5.

[26] Ober C, Hoffjan S. Asthma genetics 2006: the long and winding road to gene discovery. Genes Immun March 2006;7(2):95–100.

[27] Hindorff LA, Sethupathy P, Junkins HA, et al. Potential etiologic and functional implications of genome-wide association loci for human diseases and traits. Proc Natl Acad Sci USA June 9, 2009;106(23):9362–7.

[28] Nagalingam NA, Lynch SV. Role of the microbiota in inflammatory bowel diseases. Inflamm Bowel Dis May 2012;18(5):968–84.

[29] Frank DN, Robertson CE, Hamm CM, et al. Disease phenotype and genotype are associated with shifts in intestinal-associated microbiota in inflammatory bowel diseases. Inflamm Bowel Dis January 2011;17(1):179–84.

[30] Walker AW, Sanderson JD, Churcher C, et al. High-throughput clone library analysis of the mucosa-associated microbiota reveals dysbiosis and differences between inflamed and non-inflamed regions of the intestine in inflammatory bowel disease. BMC Microbiol 2011;11:7.

[31] Gophna U, Sommerfeld K, Gophna S, Doolittle WF, Veldhuyzen van Zanten SJ. Differences between tissue-associated intestinal microfloras of patients with Crohn's disease and ulcerative colitis. J Clin Microbiol November 2006;44(11):4136–41.

[32] Sivakumaran S, Agakov F, Theodoratou E, et al. Abundant pleiotropy in human complex diseases and traits. Am J Hum Genet November 11, 2011;89(5):607–18.

[33] Apolipoprotein E. Online Mendelian Inheritance in Man. OMIM; 2012. http://omim.org/entry/107741?search=APOE&highlight=apoe - contributors-shutter. [accessed 24.06.12].

[34] O'Sullivan BP, Freedman SD. Cystic fibrosis. Lancet May 30, 2009;373(9678):1891–904.

[35] Davis PB, Yasothan U, Kirkpatrick P. Ivacaftor. Nat Rev Drug Discov May 2012;11(5):349–50.

[36] Administration UFaD. Drugs@FDA: ivacaftor label and approval history. http://www.accessdata.fda.gov/scripts/cder/drugsatfda/index.cfm?fuseaction=Search.Label_ApprovalHistory - labelinfo.

[37] Van Goor F, Hadida S, Grootenhuis PD, et al. Rescue of CF airway epithelial cell function in vitro by a *CFTR* potentiator, VX-770. Proc Natl Acad Sci U. S. A November 3, 2009;106(44):18825–30.

[38] Death: Top 10 Causes. WHO Factsheet. http://www.who.int/mediacentre/factsheets/fs310/en/index.html.

[39] Talmud PJ, Cooper JA, Palmen J, et al. Chromosome 9p21.3 coronary heart disease locus genotype and prospective risk of CHD in healthy middle-aged men. Clin Chem March 2008;54(3):467–74.

[40] Volkman SK, Neafsey DE, Schaffner SF, Park DJ, Wirth DF. Harnessing genomics and genome biology to understand malaria biology. Nat Rev Genet May 2012;13(5):315–28.

[41] Agnandji ST, Lell B, Soulanoudjingar SS, et al. First results of phase 3 trial of RTS,S/AS01 malaria vaccine in African children. N Engl J Med November 17, 2011;365(20):1863–75.

[42] McCarthy MI. Genomics, type 2 diabetes, and obesity. N Engl J Med December 9, 2010;363(24):2339–50.

[43] Danaei G, Finucane MM, Lu Y, et al. National, regional, and global trends in fasting plasma glucose and diabetes prevalence since 1980: systematic analysis of health examination surveys and epidemiological studies with 370 country-years and 2.7 million participants. Lancet July 2, 2011;378(9785):31–40.

[44] Chen R, Corona E, Sikora M, et al. Type 2 diabetes risk alleles demonstrate extreme directional differentiation among human populations, compared to other diseases. PLoS Genet April 2012;8(4):e1002621.

[45] Shai I, Jiang R, Manson JE, et al. Ethnicity, obesity, and risk of type 2 diabetes in women: a 20-year follow-up study. Diabetes Care July 2006;29(7):1585–90.

[46] Gonzalez-Galarza FF, Christmas S, Middleton D, Jones AR. Allele frequency net: a database and online repository for immune gene frequencies in worldwide populations. Nucleic Acids Res January 2011;39(Database issue):D913–9.

[47] Exome Variant Server. http://evs.gs.washington.edu/EVS/.

[48] Li JZ, Absher DM, Tang H, et al. Worldwide human relationships inferred from genome-wide patterns of variation. Science February 22, 2008;319(5866):1100–4.

[49] Raby BA, Silverman EK, Kwiatkowski DJ, Lange C, Lazarus R, Weiss ST. *ADAM33* polymorphisms and phenotype associations in childhood asthma. J Allergy Clin Immunol June 2004;113(6):1071–8.

[50] Qu S, Sun D, Wang Y, Zhang C, Lv Y, Yao L. Association of *ADAM33* polymorphisms with childhood asthma in a northern Chinese population. Exp Mol Pathol December 2011;91(3):775–9.

[51] Van Eerdewegh P, Little RD, Dupuis J, et al. Association of the *ADAM33* gene with asthma and bronchial hyperresponsiveness. Nature July 25, 2002;418(6896):426–30.

Glossary: Medical and Health Genomics[1]

ACRONYMS AND ABBREVIATIONS

AAMC Association of American Medical Colleges
ABGC American Board of Genetic Counseling
ABMGG American Board of Medical Genetics and Genomics
ACCME Accreditation Council for Continuing Medical Education
aCGH Array comparative genome hybridization
ACGME Accreditation Council for Graduate Medical Education
ACMG American College of Medical Genetics and Genomics
APHMG Association of Professors of Human and Medical Genetics
API Application program interface
CDAC Consortium Data Access Committee
CMA Chromosomal microarray
CME Continuing medical education
CNV Copy number variant
CVD Cardiovascular disease
DAC Data access committee
DACO Data Access Compliance Office
dbGaP Database of Genotypes and Phenotypes
DDD Deciphering Developmental Disorders
DDG2P Developmental Disorder Genotype to Phenotype Database
EAGDA Expert Advisory Group on Data Access
EGA European Genome–Phenome Archive
GA4GH Global Alliance for Genomics and Health
GAIN Genetic Association Information Network
GoNL Genome of the Netherlands Project
HCM Hypertrophic cardiomyopathy
HGP Human Genome Project
HGVS Human Genome Variation Society
HHMI Howard Hughes Medical Institute
HPO Human Phenotype Ontology
HUGO Human Genome Organization
IBD Inflammatory bowel disease
ICC Inherited cardiac condition
ICGC International Cancer Genome Consortium
ID Intellectual disability
IDAC International Data Access Committee
Indel Insertion and deletion

ISCC Intersociety Coordinating Committee (for coordination of initiatives in genomic medicine)
Kb Kilobase pairs (1000 base pairs DNA)
LCME Liaison Committee on Medical Education
Mb Megabase pairs (1,000,000 base pairs DNA)
MCAT Medical College Admission Test
MDD Major depressive disorder
MLPA Multiplex ligation-dependent probe amplification
MOC Maintenance of certification
NCHPEG National Coalition for Health Professional Education in Genetics
NCI National Cancer Institute
NGS Next generation (DNA and RNA) sequencing
NIH National Institutes of Health
OECD Organization for Economic Cooperation and Development
OSCE Objective structured clinical examination
PI Principle investigator
RRC Residency review committee
SCFA Short chain fatty acid
SNP Single nucleotide polymorphism
SNV Single nucleotide variant
SV Structural variant
UKBN UK DNA Banking Network
UPD Uniparental disomy
USMLE US Medical Licensing Examination
WES Whole exome sequencing
WGS Whole genome sequencing
WTCCC Wellcome Trust Case Control Consortium

TERMS AND PHRASES

Acrocentric A chromosome having the centromere close to one end.
Algorithm A step-by-step method for solving a computational problem.
Allele An alternative form of a gene at the same chromosomal locus.
Allelic heterogeneity Different alleles for one gene.
Alternative splicing A regulatory mechanism by which variations in the incorporation of coding regions (see *Exon*) of the gene into messenger RNA (mRNA) lead to the production of more than one related protein or isoform.
Alu repeat (or sequence) One of a family of about 750,000 interspersed sequences in the human genome that are thought to have originated from *7SL* RNA gene.
Alu-PCR A polymerase chain reaction (PCR) reaction that uses an oligonucleotide primer with a sequence derived from the Alu repeat.

---

1. Compiled by the author from various sources, including his personal collection of genetic/genomic terms, definitions, phrases, annotations, and quotations. Several of these appear in many books, are accessible in the public domain, and are commonly used in all aspects of clinical and nonclinical practice. The author does not claim any form of ownership, and holds the opinion that no permission is required for anyone else using any information or material from this glossary.

Amino acid A chemical subunit of a protein. Amino acids polymerize to form linear chains linked by peptide bonds called *polypeptides*. All proteins are made from 20 naturally occurring amino acids.

Amplification refractory mutation system (ARMS) An allele-specific PCR amplification reaction; commonly used for carrier detection, for example cystic fibrosis.

Annotation The descriptive text that accompanies a sequence in a database method.

Anticipation A phenomenon in which the age of onset of a disorder is reduced and/or the severity of the phenotype is increased in successive generations.

Antigen A molecule that is perceived by the immune system to be foreign.

Apoptosis Programmed cell death.

Array backbone Oligonucleotide probes that collectively provide coverage across the whole genome. Probes are usually evenly spaced throughout the genome; the higher the density of probes, the smaller the gap between probes. Additional probes can then be layered onto the backbone in specific regions of interest. These pertain to usually known microdeletion syndromes such as in the DiGeorge syndrome critical region and the Prader-Willi/Angelman syndrome region.

(Array) comparative genomic hybridization (CGH, aCGH) A molecular cytogenetic method for analyzing copy number variants (CNVs) relative to ploidy level in the DNA of a test sample compared with a reference sample, without the need for culturing cells. This technique was originally developed for the evaluation of the differences between the chromosomal complements of solid tumor and normal tissue. This technique is now the first choice to detect unbalanced chromosomal abnormalities in unexplained developmental delay; however this method is not applicable to detect balanced chromosomal abnormalities such as reciprocal translocations, inversions or ring chromosomes because these do not affect copy numbers.

Array content A collective term for the oligonucleotide probes contained on a microarray.

Array density The number of oligonucleotide probes present on a microarray. This can typically range from 60,000–4,000,000 unique probes.

Autosome Any chromosome other than a sex chromosome (X or Y) and the mitochondrial chromosome.

Autozygosity mapping A form of genetic mapping for autosomal recessive disorders in which affected individuals are expected to have two identical disease alleles by descent.

Autozygosity In an inbred person, homozygosity for alleles identical by descent.

Bacterial artificial chromosome (BAC) DNA vectors into which large DNA fragments can be inserted and cloned in a bacterial host.

Base See *Nucleotide*.

Bioinformatics An applied computational system that includes development and utilization of facilities to store, analyze, and interpret biological data.

Biotechnology The industrial application of biological processes, particularly recombinant-DNA technology and genetic engineering.

Blastocyst The mammalian embryo at the stage at which it is implanted into the wall of the uterus.

Candidate gene Any gene that by virtue of a known property (function, expression pattern, chromosomal location, or structural motif) is considered as a possible locus for a given disease.

Carrier testing Carried out to determine whether an individual carries one copy of an altered gene for a particular recessive disease.

Carrier A person who carries an allele for a recessive disease (see *Heterozygote*) without the disease phenotype but who can pass it on to the next generation.

cDNA (complementary DNA) A piece of DNA copied in vitro from mRNA by a reverse transcription enzyme.

Cell cycle A series of tightly regulated steps that a cell goes through from its creation to division to form two daughter cells.

Central dogma A term proposed by Francis Crick in 1957: "DNA is transcribed into RNA which is translated into protein."

Centimorgan (cM) A unit of genetic distance equivalent to 1% probability of recombination during meiosis. One centimorgan is equivalent, on average, to a physical distance of approximately 1 megabase in the human genome.

Centromere The constricted region near the center of a chromosome that has a critical role in cell division.

Chromosome painting Fluorescent labeling of whole chromosomes by a fluorescence in situ hybridization (FISH) procedure, in which labeled probes each consist of a complex mixture of different DNA sequences from a single chromosome.

Chromosome Subcellular structures that contain and convey the genetic material of an organism.

Clinical genomics The application of large-scale, high-throughput genomics technologies in clinical settings, such as clinical trials or primary care of patients.

Clinical ontology Defined as hierarchies of concepts that apply to controlled syntax, database schema, semantic networks, or thesauruses. An ontological approach may be used to extract knowledge about disease progression, disease presentation, and comorbidities (see *Gene annotation*).

Clinical proteomics Translational application of new protein-based technologies in clinical medicine. For example, serum proteomic pattern diagnostics is a new concept in multiplexed biomarker analysis that may offer a new and exciting method for earlier and more accurate disease prediction.

Clinical sensitivity The proportion of persons with a disease phenotype who test positive (see *Positive predictive probability*)

Clinical specificity The proportion of persons without a disease phenotype who test negative; referred to in terms of *positive predictive probability* calculated with sensitivity data.

Clone A line of cells derived from a single cell and therefore carrying identical genetic material.

Cloning vector A DNA construct such as a plasmid, modified viral genome (bacteriophage or phage), or artificial chromosome that can be used to carry a gene or fragment of DNA for purposes of cloning (for example, a bacterial, yeast, or mammalian cell).

Coding DNA (sequence) The portion of a gene that is transcribed into mRNA.

Codon A three-base sequence of DNA or RNA that specifies a single amino acid.

Companion diagnostics The increasing trend for genetic diagnostics and therapeutics to become intertwined. These diagnostics identify the subset of patients who would benefit from a drug.

Comparative genomics The comparison of genome structure and function across different species in order to further our understanding of biological mechanisms and evolutionary processes.

Complementary DNA (cDNA) DNA generated from an expressed mRNA through a process known as *reverse transcription*.

Complex diseases Diseases characterized by risk to relatives of an affected individual that is greater than the incidence of the disorder in the population.

Complex trait One which is not strictly mendelian (ie, dominant, recessive, or sex-linked) and may involve the interaction of two or more genes to produce a phenotype, or may involve gene–environment interactions.

Congenital Any trait, condition, or disorder that exists from birth.

Consanguinity Marriage between two individuals having common ancestral parents, commonly between first cousins; an approved practice in some communities who share social, cultural, and religious beliefs. In genetic terms, two such individuals could be heterozygous by descent for an allele expressed as *coefficient of relationship*, and any offspring could be therefore homozygous by descent for the same allele expressed as *coefficient of inbreeding*.

Conserved sequence A base sequence in a DNA molecule (or an amino acid sequence in a protein) that has remained essentially unchanged throughout evolution.

Constitutional mutation A mutation that is inherited and therefore present in all cells containing the relevant nucleic acid.

Contig A consensus sequence generated from a set of overlapping sequence fragments that represent a large piece of DNA, usually a genomic region from a particular chromosome.

Copy number The number of different copies of a particular DNA sequence in a genome.

Copy number variant (CNV) Sequences typically 1 Kb–3 Mb in size, deleted or inserted into the genome; CNVs can be both benign or pathogenic.

CpG island A short stretch of DNA, often less than 1 Kb, containing CpG dinucleotides that are unmethylated and present at the expected frequency. CpG islands often occur at transcriptionally active DNA.

Cytogenomic microarray Genomic approach to detection of copy number variants

Cytoplasm The internal matrix of a cell. The cytoplasm is the area between the outer periphery of a cell (the cell membrane) and the nucleus (in a eukaryotic cell).

De novo mutation A mutation that arises in the proband but is not present in the mother's or father's germline.

Demographic transition The change in the society from extreme poverty to a stronger economy, often associated by a transition in the pattern of diseases from malnutrition and infection to the intractable conditions of middle and old age; for example, cardiovascular disease, diabetes, and cancer.

Denaturation Dissociation of complementary strands to produce single-stranded DNA and/or RNA.

Determinism (genetic) The philosophical doctrine that human action is not free but determined by genetic factors.

Diploid A genome (the total DNA content contained in each cell) that consists of two homologous copies of each chromosome.

Direct-to-consumer testing Genomic testing provided without intervention of a health professional.

Disease etiology Any factor or series of related events directly or indirectly causing a disease. For example, the genomics revolution has improved our understanding of disease determinants and provided a deeper understanding of molecular mechanisms and biological processes (see *Systems biology*).

Disease expression When a disease genotype is manifested in the phenotype.

Disease interventions A term used in genomics that refers to development of a new generation of therapeutics based on genes.

Disease management A continuous, coordinated healthcare process that seeks to manage and improve the health status of a patient over the entire course of a disease. The term may also apply to a patient population. Disease management services include disease-prevention efforts and as well as patient management.

Disease phenotype Includes disease-related changes in tissues as judged by gross anatomical, histological, and molecular pathological changes. Gene and protein expression analysis and interpretation studies, particularly at the whole-genome level, are able to distinguish among apparently similar phenotypes.

Disease A fluid concept influenced by societal and cultural attitudes that change with time and in response to new scientific and medical discoveries. The human genome sequence will dramatically alter how we define, prevent, and treat disease. Similar collections of symptoms and signs (phenotypes) may have very different underlying genetic constitutions (genotypes). As genetic capabilities increase, additional tools will become available to subdivide disease designations that are clinically identical (see *Taxonomy of disease*).

Diversity, genomic The number of base differences between two genomes divided by the genome size.

Deoxyribonucleic acid (DNA) The chemical that comprises the genetic material of all cellular organisms.

DNA fingerprinting Use of a hypervariable minisatellite probe (usually those developed by Alec Jeffreys) on a Southern blot to produce an individual-specific series of bands for identification of individuals or relationships.

DNA library A collection of cell clones containing different recombinant DNA clones.

DNA sequencing Technologies through which the order of base pairs in a DNA molecule can be determined.

Domain A discrete portion of a protein with its own function. The combination of domains in a single protein determines its overall function.

Dominant negative mutation A mutation that results in a mutant gene product that can inhibit the function of the wild-type gene product in heterozygotes.

Dominant An allele (or the trait encoded by that allele) that produces its characteristic phenotype when present in the heterozygous form.

Dosage effect The number of copies of a gene; variation in the number of copies can result in aberrant gene expression or be associated with disease phenotype; also used in the context of haploinsufficiency.

Drug design Development of new classes of medicines based on a reasoned approach using gene sequence and protein structure function information rather than the traditional trial-and-error method.

Drug interaction Adverse drug interaction, drug–drug interaction, drug–laboratory interaction, or drug–food interaction. It is defined as an action of a drug on the effectiveness or toxicity of another drug.

Electronic health record (EHR) A real-time patient health record with access to evidence-based decision-support tools that can be used to aid clinicians in decision making, automating and streamlining the clinician's workflow, thereby ensuring that all clinical information is communicated. It can also support the collection of data for uses other than clinical care, such as billing, quality management, outcome reporting, and public health disease surveillance and reporting.

Embryonic stem (ES) cells A cell line derived from undifferentiated, pluripotent cells from the embryo compared with those harnessed from induced adult somatic cells, also referred to as *induced adult pluripotent cells (IAPCs).*

Enhancer A regulatory DNA sequence that increases transcription of a gene. An enhancer can function in either orientation, and it may be located up to several thousand base pairs upstream or downstream from the gene it regulates.

Entrustable professional activities Activities that a physician would be expected to be able to carry out without supervision.

Environmental factors May include chemical factors, dietary factors, infectious agents, and physical and social factors.

Enzyme A protein that acts as a biological catalyst that controls the rate of a biochemical reaction within a cell.

Epigenetic A term describing nonmutational phenomena, such as methylation and histone modification, that modify the expression of a gene.

Euchromatin The fraction of the nuclear genome that contains transcriptionally active DNA and that, unlike heterochromatin, adopts a relatively extended conformation.

Eukaryote An organism whose cells show internal compartmentalization in the form of membrane-bound organelles (includes animals, plants, fungi, and algae).

Exome The portion of the genome (ie, genes) that encode proteins.

Exome sequencing DNA sequencing of the protein-encoding components of the genome

Exon The sections of a gene that code for all of its functional product. Eukaryotic genes may contain many exons interspersed with noncoding introns. An exon is represented in the mature mRNA product—the portions of an mRNA molecule that are left after all introns are spliced out, which serves as a template for protein synthesis.

Family history An essential tool in clinical genetics. Interpreting the family history can be complicated by many factors, including small families, incomplete or erroneous family histories, consanguinity, variable penetrance, and the current lack of real understanding of the multiple genes involved in polygenic (complex) diseases.

Fluorescence in situ hybridization (FISH) A form of chromosome in situ hybridization in which a nucleic acid probe is labeled by incorporation of a fluorophore, a chemical group that fluoresces when exposed to ultraviolet irradiation.

Founder effect Changes in allelic frequencies that occur when a small group is separated from a large population and is established in a new location.

Frameshift mutation The addition or deletion of a number of DNA bases that is not a multiple of three, thus causing a shift in the reading frame of the gene. This shift leads to a change in the reading frame of all parts of a gene that are downstream from the mutation, leading to a premature stop codon, and thus to a truncated protein product.

Functional genomics The development and implementation of technologies to characterize the mechanisms through which genes and their products function and interact with each other and with the environment (see *Transcriptomics*)

Gain-of-function mutation A mutation that produces a protein that takes on a new or enhanced function.

Gene expression The process through which a gene is activated at a particular time and place so that its functional product is produced; eg, transcription into mRNA followed by translation into protein.

Gene expression profile The pattern of changes in the expression of a specific set of genes that is relevant to a disease or treatment. The detection of this pattern depends upon the use of specific gene-expression measurement techniques.

Gene family A group of closely related genes that make similar protein products.

Gene knockouts A commonly used technique to demonstrate the phenotypical effects and/or variation related to a particular gene in a model organism; for example, in the mouse (see *Knockout*); the absence of many genes may have no apparent effect upon phenotypes (though stress situations may reveal specific susceptibilities). Other single knockouts may have a catastrophic effect upon the organism, or be lethal so that the organism cannot develop at all.

Gene regulatory network A functional map of the relationships between a number of different genes and gene products (proteins), regulatory molecules, and so forth that define the regulatory response of a cell with respect to a particular physiological function.

Gene therapy A therapeutic medical procedure that involves either replacing/manipulating or supplementing nonfunctional genes with healthy genes. Gene therapy can be targeted to somatic (body) or germ (egg and sperm) cells. In somatic gene therapy, the recipient's genome is changed, but the change is not passed along to the next generation. In germline gene therapy, the parent's egg or sperm cells are changed with the goal of passing on the changes to their offspring.

Gene-based therapy Refers to all treatment regimens that employ or target genetic material; examples include: (1) transfection (introducing cells whose genetic makeup is modified), (2) antisense therapy, and (3) naked DNA vaccination.

Gene The fundamental unit of heredity; in molecular terms, a gene comprises a length of DNA that encodes a functional product, which may be a polypeptide (a whole or constituent part of a protein or an enzyme) or a ribonucleic acid. It includes regions that precede and follow the coding region, as well as introns and exons. The exact boundaries of a gene are often poorly defined, because many promoter and enhancer regions dispersed over many kilobases may influence transcription.

Genetic (genomic) epidemiology A field of research in which correlations are sought between phenotypical trends and genetic or genomic variation across population groups.

Genetic architecture The full range of genetic effects on a trait. Genetic architecture is a moving target that changes according to gene and genotype frequencies, distributions of environmental factors, and such biological properties as age and sex

Genetic code The relationship between the order of nucleotide bases in the coding region of a gene and the order of amino acids in the polypeptide product. It is a universal, triplet, nonoverlapping code, such that each set of three bases (termed a *codon*) specifies which of the 20 amino acids is present in the polypeptide chain product of a particular position.

Genetic counseling An important process for individuals and families who have a genetic disease or who are at risk for such a disease. Genetic counseling provides patients and other family members information about their condition and helps them make informed decisions.

Genetic determinism The unsubstantiated theory that genetic factors determine a person's health, behavior, intelligence, or other complex attributes.

Genetic discrimination Unfavorable discrimination against an individual, a family, a community, or an ethnic group on the basis of genetic information. Discrimination may include societal segregation, political persecution, restriction of opportunities for education and training, lack of or restricted employment prospects, and inadequate personal financial planning; for example, life insurance and mortgages.

Genetic engineering The use of molecular biology techniques such as restriction enzymes, ligation, and cloning to transfer genes among organisms (also known as *recombinant DNA cloning*).

Genetic map A map showing the positions of genetic markers along the length of a chromosome relative to each other (genetic map) or in absolute distances from each other.

Genetic screening Testing a population group to identify a subset of individuals at high risk of having or transmitting a specific genetic disorder.

Genetic susceptibility Predisposition to a particular disease attributed to the presence of a specific allele or combination of alleles in an individual's genome.

Genetic test An analysis performed on human DNA, RNA, genes, and/or chromosomes to detect heritable or acquired genotypes. A genetic test also is the analysis of human proteins and certain metabolites, which are predominantly used to detect heritable or acquired genotypes, mutations, or phenotypes.

Genetic testing Strictly refers to testing for a specific chromosomal abnormality or a DNA (nuclear or mitochondrial) mutation already known to exist in a family member. This includes diagnostic testing (postnatal or prenatal), presymptomatic or predictive genetic testing, or testing for establishing the carrier status. The individual concerned should have been offered full information on all aspects of the genetic test through the process of nonjudgmental and nondirective genetic counseling. Most laboratories require a formal, fully informed, signed consent before carrying out the test. Genetic testing commonly involves DNA/RNA-based tests for single-gene variants, complex genotypes, acquired mutations, and measures of gene expression. Epidemiological studies are needed to establish clinical validity of each method to establish its sensitivity, specificity, and predictive value.

Genetics Refers to the study of heredity, genes, and genetic material. In contrast to genomics, genetics is traditionally related to lower-throughput, smaller-scale emphasis on single genes, rather than on studying the structure, organization, and function of many genes.

Genome annotation The process through which landmarks in a genomic sequence are characterized using computational and other means; for example, genes are identified, predictions made as to the function of their products, their regulatory regions defined, and intergenic regions characterized (see *Annotation*).

Genome ontology A standard set of consistent nomenclature systems that can be used to describe gene and protein functions in all organisms based on molecular function, biological process, and cellular location.

Genome The complete set of chromosomal and extrachromosomal DNA/RNA of an organism, a cell, an organelle, or a virus.

Genomic drugs Drugs based on molecular targets; genomic knowledge of the genes involved in diseases, disease pathways, and drug-response (see *Pharmacogenomics*).

Genomic instability An increased tendency of the genome to acquire mutations when various processes involved in maintaining and replicating the genome are dysfunctional.

Genomic profiling Complete genomic sequence of an individual, including the expression profile. This would be targeted to specific requirements; for example, most common complex diseases (diabetes, hypertension, and coronary heart disease).

Genomics The study of the genome and its action. The term is commonly used to refer large-scale, high-throughput molecular analyses of multiple genes, gene products, or regions of genetic material (DNA and RNA). The term also includes the comparative aspect of genomes of various species, their evolution, and how they relate to each other (see *Comparative genomics*).

Genotype The genetic constitution of an organism; commonly used in reference to a specific disease or trait.

Germline cell A cell with a haploid chromosome content (also referred to as a *gamete*); in animals, sperm or egg; and in plants, pollen or ovum.

Germline mosaic (germinal mosaic, gonadal mosaic, gonosomal mosaic) An individual who has a subset of germline cells carrying a mutation that is not found in other germline cells.

Germline mutation A gene change in the body's reproductive cells (egg or sperm) that becomes incorporated into the DNA of every cell in the body of offspring; germline mutations are passed on from parents to offspring, also called *hereditary mutation*.

Haploid Describing a cell (typically a gamete) that has only a single copy of each chromosome (ie, 23 chromosomes in humans).

Haplotype A series of closely linked loci on a particular chromosome that tend to be inherited together as a block.

Heterozygosity The presence of different alleles of a gene in one individual or in a population; a measure of genetic diversity.

Heterozygote Refers to a particular allele of a gene at a defined chromosome locus. A heterozygote has a different allelic form of the gene at each of the two homologous chromosomes

Homology Similarity between two sequences because of their evolution from a common ancestor, often referred to as *homologs*.

Homozygote The same allelic form of a gene on each of the two homologous chromosomes.

Human Genome Project A program to determine the sequence of the entire three billion bases of the human genome.

Identity by descent Alleles in an individual or in two people that are identical because they have been inherited from the same common ancestor, as opposed to "identity by state"; that is, coincidental possession of similar alleles in unrelated individuals (see *Consanguinity*).

In situ hybridization Hybridization of a labeled nucleic acid to a target nucleic acid that is typically immobilized on a microscopic slide, such as DNA of denatured metaphase chromosomes (as in FISH) or the RNA in a section of tissue [as in tissue in situ hybridization (TISH)].

In vitro (Latin) Literally "in glass," meaning outside of the organism in the laboratory, usually is a tissue culture.

In vivo (Latin) Literally "in life," meaning within a living organism.

Incidental findings (IFs) DNA sequence variants that cause a deleterious phenotype in a subject that are not linked to the primary clinical presentation. For example, a coincidental finding may be that the subject has a breast cancer *(BRCA1)* mutation.

Indel A variation (benign or pathogenic) caused by the insertion or deletion of 2 or more bases of DNA.

Intron A noncoding sequence within eukaryotic genes that separates the exons (coding regions). Introns are spliced out of the messenger RNA molecule created from a gene after transcription and before protein translation (protein synthesis).

Isoforms/isozymes Alternative forms of protein/enzyme resulting from posttranslational modifications of the gene product.

Knockout A technique used primarily in mouse genetics to inactivate a particular gene in order to define its function.

Ligase An enzyme that can use adenosine triphosphate to create phosphate bonds between the ends of two DNA fragments, effectively joining two DNA molecules into one.

Linkage analysis A process of locating genes on the chromosome by measuring recombination rates between phenotypical and genetic markers (see *Lod score*)

Linkage disequilibrium The nonrandom association in a population of alleles at nearby loci.

Linkage The phenomenon whereby pairs of genes that are located in close proximity on the same chromosome tend to be coinherited.

Locus The specific site on a chromosome at which a particular gene or other DNA landmark is located.

Lod score A measure of the likelihood of genetic linkage between loci. A lod score greater than +3 is often taken as evidence of linkage; one that is less than −2 is often taken as evidence against linkage.

Loss of heterozygosity (LOH) Loss of alleles on one chromosome detected by assaying for markers for which an individual is constitutionally heterozygous.

Loss-of-function mutation A mutation that decreases the production or function (or both) of the gene product.

Lyonization The process of random X chromosome inactivation in mammals.

Meiosis Reductive cell division occurring exclusively in testes and ovaries and resulting in the production of haploid cells, including sperm cells and egg cells.

Mendelian genetics Classical genetics, focuses on monogenic genes with high penetrance. Mendelian genetics is a true paradigm and is used in discussing the mode of inheritance (see *Monogenic disease*).

Mendelian segregation The process whereby individuals inherit and transmit to their offspring one of the two alleles present in homologous chromosomes.

Messenger RNA (mRNA) RNA molecules that are synthesized from a DNA template in the nucleus (a gene) and transported to ribosomes in the cytoplasm where they serve as a template for the synthesis of protein (translation).

Microarray diagnostics A rapidly developing tool increasingly used in pharmaceutical and genomics research that has the potential for applications in high-throughput diagnostic devices. Microarrays can be made of DNA sequences with known gene mutations and polymorphisms, as well as selected protein molecules.

Microsatellite DNA Small array (often less than 0.1 kb) of short, tandemly repeated DNA sequences.

Minisatellite DNA An intermediate-size array (often 0.1–20 kb long) of short, tandemly repeated DNA sequences. Hypervariable minisatellite DNA is the basis of DNA "fingerprinting" and many variable number tandem repeat markers.

Missense mutation Substitution of a single DNA base that results in a codon that specifies an alternative amino acid.

Mitochondria Cellular organelles present in eukaryotic organisms that enable aerobic respiration and generate the energy to drive cellular processes. Each mitochondrion contains a small amount of circular DNA encoding a small number of genes (approximately 50).

Mitosis Cell division in somatic cells.

Modifier gene A gene whose expression can influence a phenotype resulting from mutation at another locus.

Molecular genetic testing Molecular genetic testing for use in patient diagnosis, management, and genetic counseling; this is increasingly used in presymptomatic (predictive) genetic testing of at-risk family members using a previously known disease-causing mutation in the family.

Mosaic A genetic mosaic is an individual who has two or more genetically different cell lines derived from a single zygote.

Motif A DNA sequence pattern within a gene that, because of its similarity to sequences in other known genes, suggests a possible function of the gene, its protein products, or both.

Multifactorial disease Any disease or disorder caused by the interaction of multiple genetic (polygenic) and environmental factors.

Multigene family A set of evolutionarily related loci within a genome, at least one of which can encode a functional product.

Mutation A heritable alteration in the DNA sequence.

Natural selection The process whereby some of the inherited genetic variation within a population will affect the ability of individuals to survive to reproduce ("fitness").

Newborn screening Testing all newborns for findings indicative of disorders, such as inborn errors of metabolism.

Noncoding sequence A region of DNA that is not translated into protein. Some noncoding sequences are regulatory portions of genes and others may serve structural purposes (telomeres, centromeres), whereas still others may not have any function.

Nonsense mutation Substitution of a single DNA base that leads in a stop codon, thus leading to the truncation of a protein.

Nucleotide A subunit of the DNA or RNA molecule. A nucleotide is a base molecule [adenine, cytosine, guanine, and thymine (A, C, G, and T) in the case of DNA], linked to a sugar molecule (deoxyribose or ribose) and phosphate groups.

Online Mendelian Inheritance in Man (OMIM) A regularly updated electronic catalog of inherited human disorders and phenotypical traits accessible on the National Center for Biotechnology Information network. Each entry is designated by a number (MIM number).

Oncogene An acquired mutant form of a gene that acts to transform a normal cell into a cancerous one.

Ortholog One of set of homologous genes or proteins that perform similar functions in different species; that is, identical genes from different species, such as *SRY* in humans and *Sry* in mice.

Paralog Similar genes (members of a gene family) or proteins (homologous) in a single species or different species that perform different functions.

Penetrance The likelihood that a person carrying a particular mutant gene will have an altered phenotype (see *Phenotype*).

Pharmacogenetics Specific genes and variants that influence drug metabolism and response of an individual to a drug, often inherited in the mendelian manner

Pharmacogenomics The identification of the genes and genomic variants that influence individual variation in the efficacy or toxicity of therapeutic agents, and the application of this information in new drug development and clinical practice.

Phenotype The clinical and/or any other manifestation or expression of a condition, such as a biochemical immunological alteration of a specific gene or genes, environmental factors, or both.

Plasmid Circular extrachromosomal DNA molecules present in bacteria and yeast. Plasmids replicate autonomously each time a bacterium divides and are transmitted to the daughter cells. DNA segments are commonly cloned using plasmid vectors.

Point mutation The substitution of a single DNA base in the normal DNA sequence.

Polygenic trait or character A character or trait determined by the combined action of a number of loci, each with a small effect.

Polymerase chain reaction (PCR) A molecular biology technique developed in the mid-1980s through which specific DNA segments may be amplified selectively.

Polymorphism The stable existence of two or more variant allelic forms of a gene within a particular population or among different populations.

Positional cloning The technique through which candidate genes are located in the genome through their coinheritance with linked markers. It allows genes to be identified that lack information regarding the biochemical actions of their functional product.

Posttranscriptional modification A series of steps through which protein molecules are biochemically modified within a cell after synthesis by translation of mRNA. A protein may undergo a complex series of modifications in different cellular compartments before its final functional form is produced.

Preimplantation genetic diagnosis (PIGD) Used following in vitro fertilization to diagnose a genetic disease or condition in a preimplantation embryo.

Preconception screening Testing couples for carrier status of autosomal recessive traits to provide counseling regarding risks to a future child.

Predictive testing Determines the probability that a healthy individual with or without a family history of a certain disease might develop that disease.

Predisposition, genetic Increased susceptibility to a particular disease attributed to the presence of one or more gene mutations, and/or a combination of alleles (haplotype), not necessarily abnormal, that is associated with an increased risk for the disease, and/or a family history that indicates an increased risk for the disease.

Prenatal diagnosis Used to diagnose a genetic disease or condition in a developing fetus.

Presymptomatic test See *Predictive testing*.

Primer A short nucleic acid sequence, often a synthetic oligonucleotide, that binds specifically to a single strand of a target nucleic acid sequence and initiates synthesis, using a suitable polymerase, of a complementary strand.

Probe A DNA or RNA fragment that has been labeled in some way and used in a molecular hybridization assay to identify closely related DNA or RNA sequences.

Prokaryote An organism or cell lacking a nucleus, and other membrane-bound organelles. Bacteria are prokaryotic organisms.

Promoter A combination of short-sequence elements to which RNA polymerase binds in order to initiate transcription of a gene.

Protein truncation test A method of screening for chain-terminating mutations by artificially expressing a mutant allele in a coupled transcription-translation system.

Protein The biological effector molecule encoded by sequences of a gene. A protein molecule consists of one or more polypeptide chains of amino acid subunits. The functional action of a protein depends on its three-dimensional structure, which is determined by its amino acid composition.

Proteome All of the proteins present in a cell or organism

Proteomics The development and application of techniques to investigate the protein products of the genome and how they interact to determine biological functions.

Protooncogene A cellular gene that, when mutated, is inappropriately expressed and becomes an oncogene.

Pseudogene A DNA sequence that shows a high degree of sequence homology to a nonallelic functional gene, but is itself nonfunctional.

Recessive An allele that has no phenotypical effect in the heterozygous state.

Recombinant DNA technology The use of molecular biology techniques such as restriction enzymes, ligation, and cloning to transfer genes among organisms (see *Genetic engineering*).

Regulatory mutation A mutation in a region of the genome that does not encode a protein but affects the expression of a gene

Regulatory sequence A DNA sequence to which specific proteins bind to activate or repress the expression of a gene.

Replication A process by which a new DNA strand is synthesized by copying an existing strand, using it as a template for the addition of a complementary bases, catalyzed by a DNA polymerase enzyme.

Restriction enzymes A family of enzymes derived from bacteria that cut DNA at specific sequences of bases

Restriction fragment length polymorphism (RFLP) A polymorphism due to difference in size of allelic restriction fragments as a result of restriction site polymorphism.

Ribonucleic acid (RNA) A single-stranded nucleic acid molecule comprising a linear chain made up of four nucleotide subunits (A, C, G, and U). There are three types of RNA: messenger, transfer, and ribosomal.

Risk communication An important aspect of genetic counseling that involves pedigree analysis, interpretation of the inheritance pattern, genetic risk assessment, and explanation to the family member (or the family).

Reverse transcriptase–PCR (RT-PCR) A PCR reaction in which the target DNA is a cDNA copied by reverse transcriptase from an mRNA source.

Screening Carrying out of a test or tests, examination(s), or procedure(s) in order to expose undetected abnormalities, unrecognized (incipient) diseases, or defects; examples are the early diagnosis of cancer using mass X-ray mammography for breast cancer, and cervical smears for cancer of the cervix.

Segregation The separation of chromosomes (and the alleles they carry) during meiosis; alleles on different chromosomes segregate randomly among the gametes (and the progeny).

Sensitivity (of a screening test) The extent (usually expressed as a percentage) to which a method gives results that are free from false-negatives; the fewer the false-negatives, the greater the test's sensitivity. Quantitatively, sensitivity is the proportion of truly diseased persons in the screened population who are identified as diseased by the screening test.

Sex chromosome The pair of chromosomes that determines the sex (gender) of an organism. In man, one X and one Y chromosome constitute a male, whereas two X chromosomes constitute a female.

Signal transduction The molecular pathways through which a cell senses changes in its external environment and changes its gene-expression patterns in response.

Single nucleotide polymorphism (SNP) A common variant in the genome sequence; the human genome contains about 10 million SNPs

Single Nucleotide Variant (SNV) A DNA sequence variation within an individual.

Somatic All of the cells in the body that are not gametes (germline).

Southern blot hybridization A form of molecular hybridization in which the target nucleic acid consists of DNA molecules that have been size-fractioned by gel electrophoresis and subsequently transferred to a nitrocellulose or nylon membrane.

Splicing A process by which introns are removed from a mRNA before translation, and the exons adjoined.

Stem cell A cell that has the potential to differentiate into a variety of different cell types depending on the environmental stimuli it receives.

Stop codon A codon that leads to the termination of a protein rather than to the addition of an amino acid. The three stop codons are TGA, TAA, and TAG.

Substitution mutation A mutation caused by the substitution of one nucleotide for another. The subsequent change in the triplet code may or may not result in a change in amino acid (benign or neutral).

Systems biology The simultaneous measurement of thousands of molecular components (such as transcripts, proteins, and metabolites) and integration of these disparate data sets with clinical endpoints, in a biologically relevant manner; this model can be applied in understanding the etiology of disease.

Targeted capture A collection of oligonucleotide probes designed to selectively target specific regions of the genome. For example, a targeted capture for HCM would isolate the sarcomeric protein genes known to be associated with HCM.

Telomere The natural end of the chromosome.

Transcription The process through which a gene is expressed to generate a complementary RNA molecule on a DNA template using RNA polymerase.

Transcription factor A protein that binds DNA at specific sequences and regulates the transcription of specific genes.

Transcriptome The total mRNA expressed in a cell or tissue at a given point in time; this field of genomics is referred as *transcriptomics*.

Transfection A process by which new DNA is inserted in a eukaryotic cell, allowing stable integration into the cell's genome.

Transgene A gene from one source that has been incorporated into the genome of another organism (transgenic animal or organism).

Translation A process through which a polypeptide chain of amino acid molecules is generated as directed by the sequence of a particular messenger RNA sequence.

Tumor suppressor gene A gene that serves to protect cells from entering a cancerous state; according to Knudson's "two-hit" hypothesis, both alleles of a particular tumor suppressor gene must acquire a mutation before the cell will enter a transformed, cancerous state.

Whole exome sequencing (WES) DNA sequencing of all exomes (coding regions) spread across the entire genome.

Whole genome sequencing (WGS) DNA sequencing (coding, noncoding, and other components) of the entire genome.

X-chromosome inactivation Random inactivation of one of the two X chromosomes in mammals by a specialized form of genetic imprinting (see *Lyonization*).

Zinc finger A polypeptide motif that is stabilized by binding a zinc atom and confers on proteins an ability to bind specifically to DNA sequences; commonly found in transcription factors.

Index

'*Note*: Page numbers followed by "f" indicate figures, "t" indicate tables and "b" indicate boxes.'

A

AAMC. *See* Association of American Medical Colleges (AAMC)
Abacavir, 255
ABC. *See* ATP-binding cassette (ABC)
ABCB1 efflux transporter. *See* P-glycoprotein efflux transporter (ABCB1 efflux transporter)
Aberrant messenger RNA splicing, variants causing, 45
ABGC. *See* American Board of Genetic Counseling (ABGC)
ABMGG. *See* American Board of Medical Genetics and Genomics (ABMGG)
Abnormal chromosomes, 261
Abnormal ovulation disorders, 214–215
POF, 215–217
ACE inhibitors. See Angiotensin-converting enzyme inhibitors (*ACE inhibitors*)
Acenocoumarol, 126–127
aCGH. *See* array comparative genome hybridization (aCGH)
ACMG. *See* American College of Medical Genetics and Genomics (ACMGG)
ACMGG. *See* American College of Medical Genetics and Genomics (ACMGG)
ACOG. *See* American College of Obstetricians and Gynecologists (ACOG)
"ACT sheets", 238
Actinomyces, 110–111
Actins, 40
Actionable pharmacogenetic information, 122
Active genes, 51
Acute dissection management, 183
ADD. *See* ATRX-DNMT3-DNMT3L (ADD)
Adenosine triphosphatase (ATPase), 250–251
Adenosine triphosphate (ATP), 130
β-Adrenoreceptor antagonists, 127–128
ADRs. *See* Adverse drug reactions (ADRs)
Adverse drug reactions (ADRs), 121, 255–257
Adverse events, 248–249
AGT genes. *See* Angiotensinogen genes (AGT genes)
AIMs. *See* Ancestry informative markers (AIMs)
Akkermansia muciniphila (*A. muciniphila*), 113
ALK. *See* Anaplastic lymphoma kinase (ALK)
Allele, 7
Alpha thalassemia phenotype (*ATRX*), 148
Alzheimer dementia, 313–315

American Academy of Pediatrics, 244
American Board of Genetic Counseling (ABGC), 240
American Board of Medical Genetics and Genomics (ABMGG), 240
American College of Medical Genetics and Genomics (ACMGG), 19, 37, 195–196, 204–205, 238
American College of Obstetricians and Gynecologists (ACOG), 19
Amino acid
disorders, 94
variant, 229–230
AMR. *See* Antimicrobial resistance (AMR)
Amsterdam criteria, 280b
Anaplastic lymphoma kinase (ALK), 130
Ancestry informative markers (AIMs), 301
Aneuploidy pregnancy, 223
Angiotensin II type 1 receptor blockers (ARBs), 183
Angiotensin II type I receptor (AT1), 127
Angiotensin-converting enzyme inhibitors (*ACE inhibitors*), 127–128
Angiotensinogen genes (AGT genes), 127–128
Antiarrhythmics, 128
Anticipation, 155
Antihypertensive agents, 127–128
Antimicrobial resistance (AMR), 101
Antiplatelet agents, 127
aspirin, 127
clinical aspirin resistance, 127
Antiplatelet effect, 127
Aortopathies, 40–41
APHMG. *See* Association of Professors of Human and Medical Genetics (APHMG)
Application program interfaces (APIs), 48
ARBs. *See* Angiotensin II type 1 receptor blockers (ARBs)
array comparative genome hybridization (aCGH), 20, 213, 222
Array density, 37–38
ART. *See* Assisted reproductive technology (ART)
"As-needed" basis, 39–40
Ascertainment bias, 280b
ASDs. *See* Autism spectrum disorders (ASDs)
Aspirin, 127
resistance, 127
response signature, 127

Assisted reproductive technology (ART), 219
Association of American Medical Colleges (AAMC), 242
Association of Professors of Human and Medical Genetics (APHMG), 241–242
AT1. *See* Angiotensin II type I receptor (AT1)
ATP. *See* Adenosine triphosphate (ATP)
ATP-binding cassette (ABC), 130
ATPase. *See* Adenosine triphosphatase (ATPase)
ATR-X syndrome. *See* α-Thalassemia X-linked mental retardation syndrome (ATR-X syndrome)
ATRX. See Alpha thalassemia phenotype (*ATRX*)
ATRX-DNMT3-DNMT3L (ADD), 61–62
Autism spectrum disorders (ASDs), 20
Autoimmune disorders, 139
Autosomal abnormalities in POF, 217
Autosomes, 2
5-Aza-2-deoxycitidine (5-aza DC), 56
Azathioprine, 131, 254
Azoospermia factor region (*AZF region*), 219

B

Backbone design, 37–38
Bacteroides fragilis (*B fragilis*), 112
*BAP*1 gene, 281b
Basal cell carcinomas (BCCs), 266b
Basal cell nevus syndrome. *See* Gorlin syndrome
Bayesian Evolutionary Analysis Sampling Trees (BEAST), 102–103
BCCs. *See* Basal cell carcinomas (BCCs)
BCR-ABL tyrosine kinase, 130
BD. *See* Bipolar disorder (BD)
BDNF. *See* Brain-derived neurotrophic factor (BDNF)
BEAST. *See* Bayesian Evolutionary Analysis Sampling Trees (BEAST)
"Bench-top" sequencers, 299
Benefit-risk ratio, 121, 126
Berlin criteria, 181
Bermuda Principles, 30
Bifidobacteria, 115
Bifidobacterium, 109–110, 113, 115–116
B. breve, 115–116
B. infantis, 114
B. longum, 115–116
Binary fingerprint-based method, 84

Bioinformatics, 10, 17
  pipelines, 43
Biomarker
  detection, 93
  discovery, 93
  for psychiatric diseases, 94–95
Bipolar disorder (BD), 95
BK1. *See* Bradykinin type I (BK1)
Blautia, 115
Blepharophimosis epicanthus inversus
      syndrome (BPES), 217
Blockbuster
  business model, 247
  drug, 247
β blockers, 128
  genes, 128
*BMP*15. *See* Bone morphogenetic protein 15
      gene (*BMP*15)
Bolam test, 193–194
Bone morphogenetic protein 15 gene (*BMP*15),
      217
BPES. *See* Blepharophimosis epicanthus
      inversus syndrome (BPES)
Bradykinin type I (BK1), 127
BRAF mutations, 23
Brain-derived neurotrophic factor (BDNF), 61
*BRCA*2 mutation, 197
Breast cancer, 311
Breast Cancer Information Core, 47
Bronchial asthma, 313
Buffering system, 62
Burrows-Wheeler Aligner (BWA), 43
Butyrate, 112–113
BWA. *See* Burrows-Wheeler Aligner (BWA)

C

Caenorhabditis elegans (*C elegans*), 114
Cancer
  diagnosis, 22–23
  epigenetic mechanisms in, 62–63
  genome atlas, 22
  genomics, 22
  metabolomics, 94
  methylome, 63
  surveillance, 265
  targeted therapeutics development for, 24
Cancer genetics and genomics
  genetic testing
    in cancer predisposition syndromes,
        264–266
    and mainstreaming, 281–283
  inherited cancers, 261
    cancer predisposition genes identification,
        261–263
    genes, 266–281
    high, medium, and smaller risks,
        263–264
    oncogenes, 261–262
    TSGs, 261–262
  microscopic analysis, 261
Cancer predisposition genes
  identification, 261–263
  by phenotypic association, 267t–279t
  phenotypic effects of mutations in, 266–280

Cancer predisposition syndromes, 266
  genetic testing in, 264
    cancer surveillance, 265
    pharmacological management, 266
    prophylactic surgery, 265–266
    risk information as therapy, 264–265
Capillary electrophoresis (CE), 94, 295
Capillary-based Sanger sequencing technology, 16
Capnocytophaga, 110–111
3C approaches. *See*
      Chromosome-conformation-capture
      approaches (3C approaches)
Carbamazepine (CBZ), 131–133
CARD15. *See* Caspase recruitment domain-
      containing protein 15 (CARD15)
Cardiofaciocutaneous syndrome (CFC), 184
Cardiology, 126. *See also* Oncology;
      Rheumatology
  β blockers, 128
    genes, 128
  antiarrhythmics, 128
  antihypertensive agents, 127–128
  antiplatelet agents, 127
    aspirin, 127
    clinical aspirin resistance, 127
  lipid-lowering drugs, 127
    statins, 127
  oral anticoagulants, 126–127
    multicenter randomized controlled clinical
        trials, 126
Cardiovascular diseases (CVDs), 114, 126, 315.
      See Inflammatory bowel disease (IBD)
Carnegie Foundation for Advancement of
      Teaching, 242
Carr-Purcell-Meiboom-Gill pulse sequence, 90
Case-Control Consortium Data Access
      Committee (CDAC), 30–31
Caspase recruitment domain-containing protein
      15 (CARD15), 112
Castration-resistant prostate cancer (CRPC), 23
Castration-sensitive prostate cancer (CSPC), 23
"Cause and effect" relationship, 139
CBZ. *See* Carbamazepine (CBZ)
*CCR*5. *See* Chemokine receptor 5 (*CCR*5)
CD. *See* Crohn disease (CD)
CDAC. *See* Case-Control Consortium Data
      Access Committee (CDAC)
*CDKL*5. *See* Cyclin-dependent kinase 5
      (*CDKL*5)
cDNA. *See* complementary DNA (cDNA)
CE. *See* Capillary electrophoresis (CE)
CELF1. *See* CUGBP/Elav-like family member
      1 (CELF1)
Celiac disease, 97
Cell division, 4, 5f
cell-free DNA (cfDNA), 19
Cell-free fetal DNA, 223–224
Cellular pathways, 180
Central nervous system (CNS), 114
Centromeric clusters, 51–52
CES1. *See* Esterases (CES1)
Cetuximab, 130
CF. *See* Cystic fibrosis (CF)
CFC. *See* Cardiofaciocutaneous syndrome (CFC)
cfDNA. *See* cell-free DNA (cfDNA)

CFTR. *See* Cystic fibrosis transmembrane
      conductance regulator (CFTR)
CHARGE syndrome. *See* Coloboma of
      eye, heart anomaly, choanal atresia,
      retardation, and genital and ear
      anomalies syndrome (CHARGE
      syndrome)
*CHD*7 gene. *See* Chromodomain helicase DNA
      binding protein 7 gene (*CHD*7 gene)
Chemokine receptor 5 (*CCR*5), 253
Children's Mitochondrial Disease Network, 82
ChIP assay. *See* Chromatin immunoprecipi-
      tation assay (ChIP assay)
Chlamydia pneumonia (*C. pneumonia*), 114
Chromatin, 10, 51
Chromatin immunoprecipitation assay (ChIP
      assay), 52
Chromodomain helicase DNA binding protein 7
      gene (*CHD*7 gene), 19–20
Chromosomal disorders, 141–142
Chromosomal microarray (CMA), 37–38
Chromosome-conformation-capture approaches
      (3C approaches), 52–53
Chromosomes, 2, 3f
Chronic myeloid leukemia (CML), 130, 253
CIC. *See* Confidence in chemistry (CIC)
Ciliopathies, 180
CIR. *See* Confidence in rationale (CIR)
CIS. *See* Confidence in safety (CIS)
cis variant, 45
Clarification of Optimal Coagulation through
      Genetics (COAG), 126
ClinGen, 48
ClinGen. *See* Clinical Genome Resource
      (ClinGen)
Clinical aspirin resistance, 127
Clinical bioinformatics, 42
Clinical cancer genetics, 281
Clinical development, 249
  human studies, 249b
  phase I trials, 249
  phase II trials, 249
  phase III trials, 250
  phase IV trials, 250
Clinical exome, 38–39
Clinical Genome Resource (ClinGen), 204
Clinical phenotype-based gene prioritization
      method, 84–85
Clinical phenotyping, 42–43
Clinical surveillance, 265
Clinical testing, nondetection of causative
      genetic changes through, 282–283
Clonal amplification, 298
Clopidogrel, 127
Clostridia, 115
Clostridium difficile (*C. difficile*), 18, 115–116
CMA. *See* Chromosomal microarray (CMA)
"CMA-negative" subjects, 41
CME. *See* Continuing medical education
      (CME)
CML. *See* Chronic myeloid leukemia (CML)
CMT1. *See* Type 1 CMTD (CMT1)
CNCs. *See* Conserved noncoding DNA
      sequences (CNCs)
CNS. *See* Central nervous system (CNS)

CNVs. *See* Copy-number variants (CNVs)
CNVs. *See* Copy-number variants (CNVs)
COAG. *See* Clarification of Optimal
      Coagulation through Genetics (COAG)
coding RNA (cRNAs), 10
coding SNP (cSNP), 252–253
Codons, 2–4
"Collaboration" requirement, 33
Coloboma of eye, heart anomaly, choanal
      atresia, retardation, and genital and
      ear anomalies syndrome (CHARGE
      syndrome), 19–20
Colonization resistance, 18
Colorectal cancer, 280b, 312
   Asian ancestry, 312–313
   bronchial asthma, 313
   HLA-B
      1502 allele prevalence, 313t
   pharmacogenomic information, 312
   pharmacogenomic variants, 312t
   variation in allele incidence, 314f
Community Health Agents Program, 289
Community-based health educational
      programs, 289
Comparative genomics, 4
   analysis, 168
complementary DNA (cDNA), 297
Complex genomic diseases, 156–157
Complex trait genetics, 251
Compound screening, genetic variation effect
      on, 252–253
Compound-specific databases, 90–93
Computational technologies, 16–17
Confidence in chemistry (CIC), 253
Confidence in rationale (CIR), 250
Confidence in safety (CIS), 252
Confidentiality, 192b–193b
Confirm phase, 250
Congenital anomalies of reproductive tract,
      213–214
Consent, 32–33, 192b–193b
   for genome testing in clinical practice,
      191–195
Conserved noncoding DNA sequences
      (CNCs), 168
Contiguous gene syndromes as genomic
      disorders, 151t
Continuing medical education (CME), 243–244
Controlled-access databases, 30–32
Conventional karyotyping, 37–38
Copy-number variants (CNVs), 7, 17, 37,
      174–175, 215
Corynebacterium, 110–111
Costello syndrome (CS), 184
Counseling, 24
Coverage, 297
COX-1. *See* Cyclooxygenase (COX-1)
COX2-inhibitor lumiracoxib, 133
Crizotinib, 24, 130
cRNAs. *See* coding RNA (cRNAs)
Crohn disease (CD), 96, 313
Crowd sourcing and funding strategy, 82
CRPC. *See* Castration-resistant prostate
      cancer (CRPC)
CS. *See* Costello syndrome (CS)

cSNP. *See* coding SNP (cSNP)
CSPC. *See* Castration-sensitive prostate cancer
      (CSPC)
CTCF binding sites, 55, 63
CTD. *See* C-terminal domain (CTD)
C-terminal domain (CTD), 61
Cuban approach, 286
CUGBP/Elav-like family member 1
      (CELF1), 55–56
CVDs. *See* Cardiovascular diseases (CVDs)
Cyclin-dependent kinase 5 (*CDKL5*), 61
Cyclooxygenase (COX-1), 127
CYP enzyme. *See* Cytochrome P450 enzyme
      (CYP enzyme)
CYP2C19 gene, 126–127
CYP2D6. *See* Cytochrome P450 2D6
      (CYP2D6)
Cystic fibrosis (CF), 24, 315
   heterozygotes, 8
Cystic fibrosis transmembrane conductance
      regulator (CFTR), 218–219
Cytochrome P450 2D6 (CYP2D6), 254
Cytochrome P450 enzyme (CYP enzyme), 254

D

D-loop. *See* Displacement loop (D-loop)
Dabigatran etexilate, 127
DACO. *See* Data Access Committee Office
      (DACO)
DACs. *See* Data access committees (DACs)
Data
   access agreement, 31
   interpretation problem, 197–198
      diagnostic *vs.* population screening, 198
      difficulties in clinical interpretation, 198b
      genomic medicine, 197–198
   management, 192b–193b
   producers, 33–34
   sharing, 29–30, 48
   users interests and benefits, 33–34
Data Access Committee Office (DACO), 30–31
Data access committees (DACs), 29–30
   governance approach to data sharing, 30
Database of Genotypes and Phenotypes
      (dbGaP), 29, 31–32
dbGaP. *See* Database of Genotypes and
      Phenotypes (dbGaP)
DDD. *See* Deciphering Developmental
      Disorders (DDD)
ddNTP. *See* dideoxynucleotide triphosphate
      (ddNTP)
De Novo mutations, 46
Debrisoquine, 254
Deciphering Developmental Disorders (DDD),
      41–42
Deep phenotype, 42
Degenerative disorders, 139
Deoxynucleotide triphosphates (dNTPs), 295
Deoxyribonucleic acid (DNA), 2, 297
   methylation, 56, 147
   replication, 60
   sequence analysis, 223
   sequence variants interpretation, 44–45
   synthesis, 297

testing, 79–80
   transposon fossils, 169
Depression, 114
DHFR. *See* Dihydrofolate
      reductase (DHFR)
Diagnosis, 37, 239
Diagnostic(s), 285
   genetic testing, 291
   odyssey, 239
   screening, 198
   techniques, 101
   testing strategy, 39–40
dideoxynucleotide triphosphate
      (ddNTP), 295
Differentially methylated regions (DMRs),
      219–221
Dihydrofolate reductase (DHFR), 130–131
Dihydropyrimidine dehydrogenase gene (*DPYD*
      gene), 129
2,4-Dihydroxypyrimidine, 95
DILI. *See* Drug induced skin, muscle, and liver
      injury (DILI)
Diploid, 4
Direct-to-consumer testing, 205
Disease
   associated with variation, 112–115
      correlation in human-associated
         microbiota, 116t
      CVD, 114–115
      IBD, 112
      major depressive disorder, 114
      obesity, 112–113
      other diseases, 115
      T2D, 113–114
   classification improvement, 256–257
   disease-specific databases, 47
   nosology, 179–180
   ontology, 84
Disparate resources, independent interrogation
      of, 44–45
Displacement loop (D-loop), 9
Disruption of normal splicing, 45
DITdP. *See* Drug-induced torsade de pointes
      (DITdP)
Dlg5. *See* Large homologue 5 (*Drosophila*) (*Dlg5*)
Dlx5 encode proteins, 61
Dlx6 encode proteins, 61
DM. *See* Myotonic dystrophy (DM)
DMD. *See* Duchenne muscular dystrophy
      (DMD); Dystrophin (*DMD*)
DMRs. *See* Differentially methylated regions
      (DMRs)
DNA. *See* Deoxyribonucleic acid (DNA)
DNA methyltransferases (DNMTs), 147
DNMT3B3 isoform, 59
DNMTs. *See* DNA methyltransferases
      (DNMTs)
dNTPs. *See* Deoxynucleotide triphosphates
      (dNTPs)
Double Helix curriculum, 243
Down syndrome, 223, 239
   Karyotype of female with, 141f
Downstream uses, 33
DPYD gene. *See* Dihydropyrimidine dehydro-
      genase gene (*DPYD* gene)

Dravet syndrome, 46
Drug
  drug-metabolizing enzymes, 254
  drug-specific targets approach, 252
    genetic variation effect on compound
      screening, 252–253
  efficacy, 131
  labels, 122
  targets, choosing best, 250–251
  toxicity, 257
Drug discovery
  and development process, 248
  genomics
    choosing best drug targets, 250–251
    complex trait genetics, 251
    drug-specific targets approach, 252
    drugs with targets by human genetics, 251t
    genetic variation effect on compound
      screening, 252–253
    single gene disorders/traits, 251–252
Drug induced skin, muscle, and liver injury
  (DILI), 125, 133
Drug-induced torsade de pointes (DITdP), 128
Duchenne muscular dystrophy (DMD), 38
*DUX*4, 58
Dystrophin (*DMD*), 166

E
EBV. *See* Epstein-Barr virus (EBV)
Ectopia lentis (EL), 181–182
Education, 285
EGA. *See* European Genome-Phenome Archive
  (EGA)
EGFR. *See* Epidermal growth factor receptor
  (EGFR)
EL. *See* Ectopia lentis (EL)
ELSI. *See* Ethical, legal, and social implications
  (ELSI)
EMA. *See* European Medicines Agency (EMA)
eMERGE-PGx project, 126
emPCR. *See* emulsion PCR (emPCR)
EMPOP. *See* European DNA Profiling
  Mitochondrial DNA Population
  Database (EMPOP)
emulsion PCR (emPCR), 298
Encyclopedia of DNA Elements project
  (ENCODE project), 52, 163
Endogenous processes, 95
Endometriosis, 218
  endometriosis-related infertility, 218
Endometrium disorders, 217–218
Enforceability, 33
Enterococcus faecium (*E. faecium*), 112
Entrustable professional activities, 241, 241t
Epidermal growth factor receptor (EGFR), 130,
  231–232, 247, 257, 312
Epigenetic diseases, 51, 145–147. *See also*
  Genome architecture disorders
  with chromosomal alterations, 148
  cluster of genes on 11p15. 5, 149f
  epigenetic changes, 149
  epigenetic initiation and silencing, 147
  epigenetic mutations, 147–148
  origin of uniparental disomy 15 in
    Prader-Willi syndrome, 148f

pedigree of family with MERRF, 147f
pedigree showing paternal transmission of
  paraganglioma, 149f
recognizable epigenetic dysmorphic
  syndromes, 148t
Epigenetic(s), 192b–193b
  code, 52
  in male factor infertility, 219–221
  mechanisms in cancer, 62–63
  molecular pathogenic mechanisms, 56
  silencing, 147–149, 157
Epigenome, 9–10
Epigenomics in male factor infertility, 219–221
Epigenotype, 51
  FXN, 53f
  and gene expression regulation, 51–53
  and human disease, 53
    ATR-X syndrome, 61–62
    FRDA, 54–55
    FSHD, 57–59
    FXS, 56–57
    ICF syndrome, 59–60
    myotonic dystrophy, 55–56
    Rett syndrome, 60–61
    trinucleotide repeats, 54
Epimutation, 282–283
Epistasis, 192b–193b
Epithelial surfaces, 109
Epstein-Barr virus (EBV), 23
Erlotinib (Tarceva), 24
Escherichia coli (*E. coli*), 114
Esterases (CES1), 127
Estrogen receptor, 94
  antagonists, 129
Ethical, legal, and social implications (ELSI), 311
EU-PACT. *See* European Pharmacogenetics of
  Anticoagulant Therapy (EU-PACT)
Eubacterium, 114–115
Euchromatin, 51
Eukaryotes, 2
European DNA Profiling Mitochondrial DNA
  Population Database (EMPOP), 295,
  306b
European Genome-Phenome Archive
  (EGA), 29–31
European Medicines Agency (EMA), 122
European Pharmacogenetics of Anticoagulant
  Therapy (EU-PACT), 126
ExAC. *See* Exome Aggregation Consortium
  (ExAC)
Exome, 39
  sequencing, 39, 297
Exome Aggregation Consortium (ExAC), 204
Exons, 2–4
Expanded GAA repeat, 54
Exposome, 95–96
*EZH*2. *See* Histone methyltransferase enhancer
  of zeste homologue 2 (*EZH*2)

F
Facioscapulohumeral dystrophy (FSHD), 57
  ChIP analysis, 58
  D4Z4 repeats, 57
  *DUX*4, 58
  epigenetic mechanisms, 58–59

genetic basis, 57–58
  ICF syndrome, 58
Facultative anaerobes, 111
Faecalibacterium prausnitzii
  (*F. prausnitzii*), 113
"Fair access" approach, 30
Familial implications, 192b–193b
Familial POF, 215
Familial searching, 306b
Family communication, 205
Fatty acid oxidation defects, 94
FBI. *See* Federal Bureau of Investigation (FBI)
FBN. *See* Fibrillin gene (*FBN*)
FDA. *See* US Food and Drug
  Administration (FDA)
Fecal therapy, 103
Federal Bureau of Investigation (FBI), 306b
Female infertility, 213. *See also* Male factor
  infertility
  abnormal ovulation disorders, 214–215
    POF, 215–217
  congenital anomalies of reproductive tract,
    213–214
  endometrium disorders, 217–218
  selection of syndromes with abnormal female
    reproductive tract anatomy, 214t
Fetal chromosome analysis, 223
*FGFR*3 gene, 46
Fibrillin gene (*FBN*), 180–182
  *FBN*1 gene, 40–41
  mutation testing, 182
Fibrillinopathies, 180–183
Firmicutes, 110–111
First-line test, 39
FISH. *See* Fluorescence in situ hybridization
  (FISH)
FLAP. *See* 5-Lipoxygenase-activating protein
  (*FLAP*)
Fluorescence in situ hybridization (FISH),
  147–148, 213
Fluorophores, 299–300
5-Fluorouracil (5-FU), 129
FM. *See* Full mutation (FM)
*FMR*1 gene, 215
FMR1. *See* Fragile X mental retardation 1
  (FMR1)
FMRP. *See* Fragile X mental retardation protein
  (FMRP)
Follicle stimulating hormone (FSH), 214–215
Follow-up policies, 192b–193b
ForenSeq DNA Signature Prep Kit, 304
Forensic genetics
  and genomics, 306b
  NGS solutions in, 300–301
    first commercial NGS kits for, 303–305
    forensic markers, 301
    new frontiers in forensic genetics, 305
    NGS platform/assay, 301–302
    STR sequencing, 302–303
Forensic genetics, first commercial NGS kits
  for, 303–305
Forensic medicine, genomic applications in
  DNA sequencing, 295
  NGS, 297–299
    solutions in forensic genetics,
      300–305

pyrosequencing, 296
sequencing methods, 296f
single molecule sequencing, 299–300
work flow for high-throughput
sequencing, 297f
Fort Lauderdale Agreement, 30
Fragile X mental retardation 1 (FMR1), 56
Fragile X mental retardation protein
(FMRP), 56
Fragile X syndrome (FXS), 54, 56
disease manifestations, 56
DNA methylation, 56
FMR1, 56
FMRP, 56
full-mutation CGG repeat length, 56
human embryonic stem cells, 57
human-induced pluripotent stem cells, 57
hydroxymethylation, 57
PM, 57
Fragile X syndrome type A (FRAXA), 148,
215–217
Fragile X–associated tremor/ataxia syndrome
(FXTAS), 57
Frataxin gene (FXN gene), 53f, 54
gene repression in FRDA, 54–55
FRAXA. See Fragile X syndrome type A
(FRAXA)
FRDA. See Friedreich ataxia (FRDA)
Friedreich ataxia (FRDA), 54
FXN gene repression in, 54–55
FSH. See Follicle stimulating hormone (FSH)
FSHD. See Facioscapulohumeral dystrophy
(FSHD)
5-FU. See 5-Fluorouracil (5-FU)
5-FU–induced toxicity, 129
Full mutation (FM), 56
Functional
DNA elements, 168
evidence, 46
genomics, 5–6, 9–10
FXN gene. See Frataxin gene (FXN gene)
FXS. See Fragile X syndrome (FXS)
FXTAS. See Fragile X–associated tremor/ataxia
syndrome (FXTAS)

G

G-protein coupled receptors (GPCRs), 252
G6PD. See Glucose-6-phosphate dehydro-
genase (G6PD)
GA4GH. See Global Alliance for Genomics and
Health (GA4GH)
Gamete selection, 221
Gas chromatography–MS based metabonomic
method (GC–MS based metabonomic
method), 95
Gastric cancer, 94
Gastric phylotypes, 111–112
Gastroesophageal reflux disease (GERD),
250–251
Gastrointestinal health, 18
Gastrointestinal tract (GI tract), 18
GATK. See Genome Analysis Toolkit (GATK)
GC. See Guanine-cytosine (GC)
GC–MS based metabonomic method. See
Gas chromatography–MS based

metabonomic method (GC–MS based
metabonomic method)
GDFs. See Growth differentiation factors
(GDFs)
GDP. See Guanosine diphosphate (GDP)
Gefitinib (Iressa), 24
Gencode, 165
Gene(s), 1–2
dosage effect, 215
expression, 218
regulation, 51–53
flow, 8
gene-driven approaches, 9
gene-panel sequencing test, 20
ontology, 84
panels, 22–23
testing, 38–39
Genetic code. See Genome
Genetic counseling, 156, 201
case study, 205–206
counseling issues, 203
confidentiality and privacy of genomic
data, 203–204
family communication, 205
healthy genome, 205
secondary findings, 204–205
uncertainty, 203
elements, 202–203
ethos and principles, 201–202
practicalities of counseling for genomic
tests, 206
patient autonomy, 210
patient emotions, 210
relationship, 210
for testing for genetic conditions,
207b–209b
similarities and differences in, 206t
Genetic diseases, 140–141
MFS
diagnosis, 182–183
to fibrillinopathies, 180–183
management, 183
treatment, 183
Molecular approach to genetic disease
nosology, 179–180
molecular pathology–based disease classifications, 180t–181t
RASopathies
clinical features, 186t
diagnosis of genetic disorder, 186–187
emergence, 184–186
malignancy risk in, 185–186
Genetic disorders
targeted therapeutics development for, 23–24
treatment, 239
Genetic Information Nondiscrimination Act, 25
Genetic testing, 37, 264, 311–312
in cancer predisposition syndromes, 264
cancer surveillance, 265
pharmacological management, 266
prophylactic surgery, 265–266
risk information as therapy, 264–265
cost-effectiveness, 133
Genetic(s), 1–2
alternations, 20
diagnostic testing, 241

distance, 7
drift, 8
factors, 121, 218–219
in IVF, 221–222
Genetics File, 244
information, 209
laboratory techniques, 228
measurement, 6–7
studies, 313
testing and mainstreaming, 281
and NGS, 281–282
nondetection of causative genetic changes,
282–283
variants, 312
variation, 255–256
effect on compound screening, 252–253
Genetically obese mice (ob/ob mice), 113
Geneticist health providers, 240–241
Genome, 1–4, 191. See also Human genome
assemblers, 16–17
data deluge, 78–79
databases, 165
sequencing, 16–17, 172
in newborn healthcare, 21
in other research areas, 22
variation, 7–8
as laboratory tool, 176
Genome Analysis Toolkit (GATK), 43
Genome architecture disorders, 150. See also
Epigenetic diseases
chromosomal rearrangements, 150
contiguous gene syndrome, 154–155
contiguous gene syndromes as genomic
disorders, 151t
disorders with trinucleotide repeats
expansion, 153t
hereditary motor and sensory neuropathy, 153
human genome, 155
location of four classes of triplet repeats in
human diseases, 155f
lower legs and feet in Charcot-Marie-Tooth
disease, 154f
Mendelian genomic disorders, 152t
polyglutamine tract, 156f
Genome of the Netherlands Project (GoNL), 31
Genome Reference Consortium (GRC), 39
Genome Sequencer 20, 296–297
1000 genomes project, 18–19
Genome-wide association studies (GWASs),
125, 127, 131, 239, 263–264, 313
Genomic analysis, 221
complexity of genome outputs, 192b–193b
confidentiality, 192b–193b
consent, 192b–193b
for genome testing in clinical practice,
191–195
data interpretation problem, 197–198
diagnostic vs. population screening, 198
difficulties in clinical interpretation, 198b
genomic medicine, 197–198
data management, 192b–193b
of DNA, 107–108
familial aspects, 192b–193b
IFs, 192b–193b, 195–197
management, 195–197
incidental genomic findings, 195t–196t

Genomic analysis (*continued*)
    public perception of genetics/genomics,
        192b–193b
    recontacting/follow-up policies, 192b–193b
    research/clinical boundary, 192b–193b
    "right to know or not know", 194b
Genomic anatomy, 161–170. *See also* Special
        genomic structures
    functional DNA elements, 168
    human chromosomes, 163f
    noncoding, RNA-only genes, 166, 168
        miRNAs, 168
        rRNA genes, 166–167
        snoRNAs, 168
        snRNAs, 168
        tRNA genes, 167–168
    number of nucleotides per chromosome, 162t
    observed genomic segments, 165f
    Pie chart of fractions of genome sequenced,
        162f
    protein-coding genes, 165–166, 167f
        gene density per chromosome, 165f
        size, 166
    repetitive elements, 168–169
    segmental duplications, 169–170, 170f
    transcription regulation regions, 168
    UCSC genome browser, 164f
Genomic data, 102
    confidentiality and privacy, 203
        data sharing and undiagnosed patient, 204
        reanalyzing stored genomic data, 204
        sharing genomic data for advancement of
            knowledge, 204
    controlled-access databases and pertinent
            data access committees
        dbGaP, 31–32
        EGA, 30–31
    ethical and legal challenges
        consent, 32–33
        data producers and data users' interests and
            benefits, 33–34
        downstream uses and enforceability, 33
    mining and databases, 29
        DACs, 29–30
        underlying principles, policies, and
            guidelines, 30
Genomic imprinting disorders. *See* Epigenetic
        diseases
Genomic medicine, 10, 197–198, 237. *See
        also* Precision medicine; Reproductive
        medicine; Stratified medicine
    competencies in, 241–242
    practice, 238
        diagnosis, 239
        prevention, 238–239
        therapy, 239–240
Genomic technologies, 179
    applications, 17
    cancer
        diagnosis, 22–23
        genome atlas, 22
        genomics, 22
    computational and information technologies,
        16–17

cost per human genome, 16f
future model for genomic and precision
        medicine, 25f
genome sequencing
    in newborn healthcare, 21
    in other research areas, 22
genomic education, 25
in medicine and health, 15
microbiome and human health, 18
NIPT, 19
PM, 23–25
policy and regulatory issues, 25
postnatal diagnosis, 20
prenatal diagnosis by whole genome
        sequencing of "jumping libraries",
        19–20
sequencing technologies, 16
1000 genomes project and structural
        variations in human genomes, 18–19
Genomic tests, 239
    practicalities of counseling for, 206
        patient autonomy, 210
        patient emotions, 210
        relationship, 210
        for testing for genetic conditions,
            207b–209b
Genomic(s), 1–2, 238, 242, 247–248
    approach, 285
    characterization, 103–104
    disorders, 145
        classification, 146t
        complex genomic diseases, 156–157
        genome architecture disorders, 150–155
        genomic imprinting disorders, 145–149
        molecular mechanisms, 150f
        with trinucleotide repeats, 155–156
    to drug discovery
        choosing best drug targets, 250–251
        complex trait genetics, 251
        drug-specific targets approach, 252
        drugs with targets by human genetics, 251t
        genetic variation effect on compound
            screening, 252–253
        single gene disorders/traits, 251–252
    education, 25
    genomic-based approaches, 24
    information, 102–103, 312
    laboratory techniques, 228
    medical school, 242
    postgraduate education, 243
    premedical education, 242
    profiling, 24
        tests, 22–23
    residency education, 242
    revolution, 15
    science, 16, 25, 311
    sequencing, 202–203, 209
    variability, 172
        CNVs, 174–175
        insertion/deletion polymorphisms, 174
        inversions, 175–176
        SNPs, 172–174, 173f
        SSRs, 174, 175f
    variation measurement, 6–7

Genotype–phenotype correlations, 84–85,
        280–281
GERD. *See* Gastroesophageal reflux disease
        (GERD)
Germline DNA mutations, 128–130. *See also*
        Somatic DNA mutations
    estrogen receptor antagonists, 129
    poly(ADP-ribose) polymerase inhibitor, 129
    pyrimidine analogues, 129
    thiopurines, 128–129
    topoisomerase I inhibitors, 129
    urate oxidase enzymes, 129–130
GI tract. *See* Gastrointestinal tract (GI tract)
Gleevec. *See* Imatinib
Global Alliance for Genomics and Health
        (GA4GH), 44, 85, 204, 290
α-Globin subunits, 62
Glucose-6-phosphate dehydrogenase (G6PD),
        129–130
Golm Metabolome database, 90–93
GoNL. *See* Genome of the Netherlands Project
        (GoNL)
Gorlin syndrome, 266b
Governance of dbGaP, 31–32
GPCRs. *See* G-protein coupled receptors
        (GPCRs)
GRC. *See* Genome Reference Consortium
        (GRC)
Green fluorescent protein-tagged *HP*1
        (HP1-GFP), 52
*GRK*4 3SNP haplotype, 128
Group discrimination, 93
Growth differentiation factors (GDFs), 217
GTP. *See* Guanosine triphosphate (GTP)
Guanine-cytosine (GC), 162–163
Guanosine diphosphate (GDP), 184
Guanosine triphosphate (GTP), 184
Gut microbiome, 96, 111, 115
    diseases associated with variation in,
        112–115
        correlation in human-associated
            microbiota, 116t
        CVD, 114–115
        IBD, 112
        major depressive disorder, 114
        obesity, 112–113
        other diseases, 115
        T2D, 113–114
Gut microbiota, 96
GWASs. *See* Genome-wide association studies
        (GWASs)

H

*H1bK*26, 62
*H3K27*methylation. *See* Histone H3 lysine 27
        methylation (*H3K27*methylation)
H3K4 methylation. *See* Histone H3 lysine 4
        methylation (*H3K4* methylation)
H3K4. *See* Methylation of lysine 4 of histone
        3 (H3K4)
H3K9 methylation. *See* Histone H3 lysine 9
        methylation (*H3K9* methylation)
Haemophilus, 110–111

Haploid, 4
Haploinsufficiency mutations, 183
Haplotype, 173–174, 174f
Hardy-Weinberg equilibrium, 172–173
HBM. *See* Human baby microbiota (HBM)
HCM. *See* Hypertrophic cardiomyopathy
    (HCM)
HD. *See* Huntington disease (HD)
HDACs. See Histone deacetylases (*HDACs*)
Health and disease, 107, 112
Healthy genome, 205
Healthy human microbiome, alterations in, 110
    diversity and abundance of microbes, 110
    gut microbiome, 111–115
    oral microbiome, 110–111
    skin microbiome, 111
Healthy oral bacterial consortium, 18
HeLa Genome Data Access Working Group,
    32–33
Helicobacter pylori (*H pylori*), 111–112
    exposure, 95–96
Helicos Biosciences, 299–300
HeliScope platform, 299–300
Hemoglobinopathies, 287
*HER*2 overexpression, 130
Herceptin. *See* Trastuzumab
Hereditary factors, 1–2
Hereditary leiomyomatosis and renal cell
    carcinoma (HLRCC), 265
Hereditary motor and sensory neuropathy, 153
Hereditary neuropathy with liability to pressure
    palsy (HNPP), 150–153
Heredity, 1–2
HERVs. *See* Human endogenous retrovirus
    sequences (HERVs)
Heterochromatin, 10, 51–52
Heterochromatin protein 1 (*HP*1), 51–52
Heteroplasmy, 144–145
Heterozygotes, 172–173
HGMD. *See* Human Gene Mutation Database
    (HGMD)
HGP. *See* Human Genome Project (HGP)
HHMI. *See* Howard Hughes Medical Institute
    (HHMI)
High-fat diet, 113–114
High-resolution magic-angle spinning ¹H NMR
    spectroscopy, 94
High-throughput screens, 248
High-throughput sequencing methods, 296–297
    work flow for, 297f
HiSeq 2500 system, 299
Hispanic Americans, 313
HIST1HIA. See Histone H1a (*HIST1HIA*)
Histone code, 147
Histone deacetylases (*HDACs*), 52
Histone H1a (*HIST1HIA*), 166
Histone H3 lysine 27 methylation (*H3K*27meth-
    ylation), 52
Histone H3 lysine 4 methylation (*H3K*4
    methylation), 52
Histone H3 lysine 9 methylation (*H3K*9
    methylation), 52, 147
Histone methyltransferase enhancer of zeste
    homologue 2 (*EZH*2), 62

Histone modifications, 147
HLA system. *See* Human leukocyte antigen
    system (HLA system)
HLRCC. *See* Hereditary leiomyomatosis and
    renal cell carcinoma (HLRCC)
5-hmC. *See* 5-Hydroxymethylcytosine (5-hmC)
HMG-CoA. *See* 3-Hydroxy-3-methylglutaryl–
    coenzyme A (HMG-CoA)
HMP. *See* Human Microbiome Project (HMP)
HNPP. *See* Hereditary neuropathy with liability
    to pressure palsy (HNPP)
Homozygotes, 172–173
Homozygous, 304–305
Howard Hughes Medical Institute (HHMI), 242
HP1-GFP. *See* Green fluorescent protein-tagged
    *HP*1 (HP1-GFP)
*HP*1. *See* Heterochromatin protein 1 (*HP*1)
HPO. *See* Human Phenotype Ontology (HPO)
Human baby microbiota (HBM), 97
Human centromeres, 170–171, 171f
Human diseases, 7–8, 140
    genetic and genomic pathology, 159t
    taxonomy, 158t
Human embryonic stem cells, 57
Human endogenous retrovirus sequences
    (HERVs), 169
Human epidermal growth factor receptor 2, 94
Human Gene Mutation Database (HGMD), 39
Human genome, 1, 155, 237. *See also* Genome
    functional genomics, transcriptomics, and
        proteomics, 9–10
    genomic anatomy, 161–170
        functional DNA elements, 168
        noncoding, RNA-only genes, 166–168
        protein-coding genes, 165–166
        regions of transcription regulation, 168
        repetitive elements, 168–169
        segmental duplications, 169–170, 170f
    hereditary factors, genes, genetics, and
        genomics, 1–2
    human genomics for socioeconomic
        development, 11–12
    mitochondrial genome, 8–9
    sequence, 228
    special genomic structures
        genome variation as laboratory tool, 176
        genomic variability, 172–176
        human centromeres, 170–171, 171f
        human telomeres, 171, 171f
        mitochondrial genome, 172
        short arms of human acrocentric
            chromosomes, 171–172
    structure and organization of nucleic acids, 2–4
    1000 genomes project and structural
        variations in, 18–19
    translational human genomics, 10–11
    variation and human disease, 5
        DNA and gene content of human reference
            genome, 6t
        genome variation and human disease, 7–8
        human nuclear and mitochondrial genomes
            comparison, 7t
        measuring genetic and genomic variation,
            6–7

nuclear genes, 5–6
Human Genome Project (HGP), 5, 150, 161,
    250, 285
Human genomics for socioeconomic
    development, 11–12
Human leukocyte antigen system (HLA
    system), 133
    alleles, 131
    Class II alleles, 131
Human metabolic characterization, 90
Human microbiome, 107–108, 110
    initiative, 16–17
Human Microbiome Project (HMP), 103–104
Human nutrition, 96
Human Phenotype Ontology
    (HPO), 41–42, 84
Human studies, 249b
Human telomeres, 171, 171f
Human Variome Project, 47
Human-associated microflora, 109
    gut microbial community, 109–110
    interplay of human gut microbiome and
        human genome, 110f
    microbiome role, 109
Human-induced pluripotent stem cells, 57
Huntington disease (HD), 155–156
3-Hydroxy-3-methylglutaryl–coenzyme A
    (HMG-CoA), 250–251
Hydroxymethylation, 57
5-Hydroxymethylcytosine (5-hmC), 57
Hypercholesterolemia, 238
Hypertrophic cardiomyopathy (HCM), 38, 40

I

IBD. *See* Inflammatory bowel disease (IBD)
IBS. *See* Irritable bowel syndrome (IBS)
ICCs. *See* Inherited cardiovascular conditions
    (ICCs)
ICF syndrome. *See* Immunodeficiency,
    centromeric region instability, and facial
    anomalies syndrome (ICF syndrome)
ICGC. *See* International Cancer Genome
    Consortium (ICGC)
ICSI. *See* Intracytoplasmic sperm inoculation
    (ICSI)
ID. *See* Intellectual disability (ID)
IDAC. *See* International Data Access
    Committee (IDAC)
IEL. *See* Isolated ectopia lentis (IEL)
IFs. *See* Incidental findings (IFs)
IISNPs. *See* Individual identification SNPs
    (IISNPs)
IL. *See* Interleukin (IL)
Illumina, 80, 299, 304
    sequencing technology, 107–108
Imatinib, 24, 253, 256–257
    mesylate, 130
IMDs. *See* Inborn metabolic disorders (IMDs)
Immune system, 111
Immune-mediated adverse drug reactions, 131
    abacavir, 131
    allopurinol, 133
    HLA alleles association, 132t

Immunodeficiency, centromeric region instability, and facial anomalies syndrome (ICF syndrome), 58
  ATRX, 59–60
  DNA hypomethylation in, 60
  DNA replication, 60
  DNMT3B3 isoform, 59
  loss of DNA methylation, 60
  mutation responsible for, 60
  reduced DNA methylation levels, 59
  total knockout, 59
Immunogenic capsular region, 103
Immunohistochemistry technique, 130
Imprinting, 145–147, 219–221
in *trans* variant, 45
In vitro fertilization (IVF), 85, 221
  gene-expression profiles, 222
  preimplantation genetic testing and screening, 222
Inborn metabolic disorders (IMDs), 93–94
Incidental findings (IFs), 47–48, 191, 192b–193b, 195–197, 204
  management, 195–197
    clinical utility of results, 197b
    uncertainties, 196–197
Incidental genomic findings, 195t–196t
Indels. *See* Insertions and deletions (Indels)
Individual identification SNPs (IISNPs), 303–304
Individualized therapy, 256
Induced pluripotent stem cell (iPSC), 51
Infection, 130–133
  IBD, 131
  immune-mediated adverse drug reactions, 131–133
    abacavir, 131
    allopurinol, 133
    HLA alleles association, 132t
    rheumatology, 130–131
Inflammation, 130–133
  IBD, 131
  immune-mediated adverse drug reactions, 131–133
    abacavir, 131
    allopurinol, 133
    HLA alleles association, 132t
    rheumatology, 130–131
Inflammatory bowel disease (IBD), 20, 96, 109, 112, 131
Informatics, 10, 11f
Information technology (IT), 16–17
Informative pharmacogenetic label, 122
Inheritance, 45
Inherited cancers, 261
  cancer predisposition gene identification, 261–263
  genes, 266
    genotype–phenotype correlation, 280–281
    phenotypic effects of mutations, 266–280
  high, medium, and smaller risks, 263–264
  oncogenes, 261–262
  TSGs, 261–262
Inherited cardiovascular conditions (ICCs), 40
"Inhibiting chromosomes", 261
INR. *See* International Normalized Ratio (INR)

Insertions and deletions (Indels), 38
  insertion/deletion polymorphisms, 174
Insulin receptor 1 substrate (IRS1), 113–114
Intellectual disability (ID), 37, 41
Interleukin (IL), 96–97
International Cancer Genome Consortium (ICGC), 29
International Data Access Committee (IDAC), 30–31
International Mito-Patients, 82
International Normalized Ratio (INR), 126
International Society of Forensic Genetics (ISFG), 302
Intersociety Coordinating Committee for Practitioner Education in Genomics (ISCC), 241
Intestinal metabolome, 96–97
Intestinal microflora study, 96
Intracytoplasmic sperm inoculation (ICSI), 219
Introns, 2–4
Invasive nontyphoidal *Salmonella* (iNTS), 103
Inversions, 175–176
"Inverted Champagne bottle" appearance, 153
Ion AmpliSeq technology, 303–304
Ion PGM System, 303–304
iPSC. *See* Induced pluripotent stem cell (iPSC)
Iressa. *See* Gefitinib (Iressa)
Irinotecan, 129
Irritable bowel syndrome (IBS), 97, 112
IRS1. *See* Insulin receptor 1 substrate (IRS1)
ISCC. *See* Intersociety Coordinating Committee for Practitioner Education in Genomics (ISCC)
ISFG. *See* International Society of Forensic Genetics (ISFG)
Isolated ectopia lentis (IEL), 182
IT. *See* Information technology (IT)
Ivacaftor, 24
IVF. *See* In vitro fertilization (IVF)

J

Janus kinase (JAK), 252
JMML. *See* Juvenile myelomonocytic leukemia (JMML)
Jumping libraries, prenatal diagnosis by whole genome sequencing of, 19–20
Juvenile myelomonocytic leukemia (JMML), 185

K

KEGG COMPOUND database, 90–93
KEGG PATHWAY database, 90–93
KRAS mutations, 23

L

Lachnospiraceae, 115
Lactobacilli, 115
Lactobacillus, 113, 115–116
  L. rhamnosus, 114
Large homologue 5 (*Drosophila*) (*Dlg5*), 251
LCME. *See* Liaison Committee on Medical Education (LCME)
LCR. *See* Low-copy repeat (LCR)

LD. *See* Linkage disequilibrium (LD)
LDL. *See* Low-density lipoprotein (LDL)
LDLC. *See* Low-density lipoprotein cholesterol (LDLC)
LDS. *See* Loeys-Dietz syndrome (LDS)
Lean phenotype. *See* Obese phenotype
Learn phase, 250
Leflunomide, 131
Legius syndrome, 184
Leukemia inhibitory factor (LIF), 218
LH. *See* Luteinizing hormone (LH)
Li-Fraumeni syndrome, 280
Liaison Committee on Medical Education (LCME), 242
Library pool, 298
LIF. *See* Leukemia inhibitory factor (LIF)
lincRNAs. *See* Long intergenic noncoding RNAs (lincRNAs)
LINEs. *See* Long interspersed nuclear elements (LINEs)
Linkage disequilibrium (LD), 174, 174f
Linkage studies, 251
Lipid-lowering drugs, 127
  statins, 127
Lipopolysaccharide (LPS), 113
5-Lipoxygenase-activating protein (*FLAP*), 251
Liquid biopsy, 23
Literature databases, 47
Living organisms, 2
Locus Reference Genomic sequences (LRG sequences), 44
Loeys-Dietz syndrome (LDS), 40
Long intergenic noncoding RNAs (lincRNAs), 168
Long interspersed nuclear elements (LINEs), 169, 174
Long QT interval gene (*LQT* gene), 198b
Long terminal repeat (LTR), 169
"Looked for" findings, 47
Low-copy repeat (LCR), 150
Low-density lipoprotein (LDL), 95
Low-density lipoprotein cholesterol (LDLC), 127
  reduction, 127
LPS. *See* Lipopolysaccharide (LPS)
LQT gene. *See* Long QT interval gene (*LQT* gene)
*LQT*1, 198b
LRG sequences. *See* Locus Reference Genomic sequences (LRG sequences)
LTR. *See* Long terminal repeat (LTR)
Luteinizing hormone (LH), 214–215
Lynch syndrome, 266, 280b, 282–283

M

Machine learning algorithms, 93
Madison Metabolic Consortium database, 90–93
MAF. *See* Minor allele frequency (MAF)
Maintenance of certification (MOC), 243
Major depressive disorder (MDD), 95
Major histocompatibility complex (MHC), 255
Malaria Genomic Epidemiology Network (MalariaGEN), 29

Malassezia, 111
  M restricta, 111
Male factor infertility, 218. *See also* Female
     infertility
  congenital anomalies of male reproductive
     tract, 218–219
  genes on Y chromosome, 220t
  mitochondrial genes in, 221
  molecular genetics, 219
     epigenetics and epigenomics in, 219–221
  new genomic applications in, 221
  prevalence and phenotypes of chromosomal
     abnormalities, 219t
  Y Chromosome abnormality/deleted, 220f
Management committee (MC), 31
March of Dimes (MOD), 288
Marfan syndrome (MFS), 40, 180–183. *See also*
     RASopathies
  diagnosis, 182
     differential, 182–183
  management, 183
  treatment, 183
Mass spectrometry (MS), 89
Massively parallel sequencing, 296–298
Matchmaker Exchange, 48, 85
Matrix metalloproteinases (MMPs), 218
MBD proteins. *See* Methyl binding domain
     proteins (MBD proteins)
MBNL. *See* Muscle blind protein (MBNL)
MC. *See* Management committee (MC)
MCAT. *See* Medical College Admission Test
     (MCAT)
MDD. *See* Major depressive disorder (MDD)
MDG. *See* Millennium Development Goal
     (MDG)
*MECP*2 gene, 60
Medical College Admission Test
     (MCAT), 242
Medical education landscape, 242
  medical school, 242
  postgraduate education, 243
  premedical education, 242
  residency education, 242
Medical exome, 39
Medical genetics, 240, 285
Medical genomics education, 243
  continuing medical education, 244
  medical school, 243–244
  residency education, 244
Medical health aspects of genetics and
     genomics
  chromosomal disorders, 141–142
  classification of genetic disorders, 140t
  disease spectrum, biological pathways, and
     genotypes, 157–159
  genetic factors in human disease, 140f
  genetic information, 139–140
  genomic disorders, 145
     classification, 146t
     complex genomic diseases, 156–157
     disorders with trinucleotide repeats,
        155–156
     genome architecture disorders, 150–155
     genomic imprinting disorders, 145–149
     molecular mechanisms, 150f

human disease
  genetic and genomic pathology in, 159t
  taxonomy, 158t
human diseases, 140
human mitochondrial DNA molecule, 145f
Karyotype of female with Down syndrome,
     141f
Mendelian disorders, 142
mitochondrial genetic disorders, 144–145
philosophy of medicine, 139
polygenic disorders, 142–144
Medical school, 242–244
Medical services incorporating genetics and
     genomics, 290
  advances in genetic testing, 290–293
  Global Alliance for Genomics and Health,
     290
  personalized medicine, 292
  psychosocial and ethical concerns, 291
  therapeutics, 293
Meiosis, 4, 5f
MEN2. *See* Multiple endocrine neoplasia type
     2 (MEN2)
Mendelian disorders, 142
  pedigree appearances in, 143f
Mendelian genomic disorders, 152t
6-Mercaptopurine (6-MP), 128–129
MERRF. *See* Mitochondrial encephalopathy
     with ragged-red muscle fibers (MERRF)
messenger RNA (mRNA), 10, 55–56, 222, 301
Metabolic processes, 107
Metabolites, 89, 95
Metabolome, 89
Metabolomics, 89–90, 115, 221
  biomarker detection, 93
  biomarkers for psychiatric diseases, 94–95
  cancer metabolomics, 94
  in environmental and public health, 95–96
  group discrimination, 93
  in human diseases, 90–93
  human nutrition, 96
  IMDs, 93–94
  intestinal metabolome, 96–97
  metabolomic resources, 91t–92t
  methods applied in metabolomic research, 90t
  oversimplified model of database
     relationship, 93f
Metabonate, 95
Metabonomics, 89
Metagenomic insight into human microbiome
  alterations in healthy human microbiome,
     110
     diversity and abundance of microbes, 110
     gut microbiome, 111–115
     oral microbiome, 110–111
     skin microbiome, 111
  approach used in metagenomic analysis of
     microbial communities, 108f
  human-associated microflora, 109–110
     gut microbial community, 109–110
     interplay of human gut microbiome and
        human genome, 110f
     microbiome role, 109
  Illumina sequencing technology, 107–108
  NGS, 107–108

novel therapeutic strategies, 115–116
     World Health Organization, 115–116
  study of human microbiome, 108
Metagenomic(s), 107–108, 115
  analysis, 108
  methods, 96
Metformin, 114
Methicillin-resistant *Staphylococcus aureus*
     (MRSA), 102
Methotrexate (MTX), 130
Methyl binding domain proteins (MBD
     proteins), 52
Methylation, 282–283
Methylation of lysine 4 of histone 3 (H3K4),
     147
5,10-Methylenetetrahydrofolate reductase
     (*MTHFR*), 129
MFS. *See* Marfan syndrome (MFS)
MHC. *See* Major histocompatibility complex
     (MHC)
Mice-model experiments, 221
Microarray(s), 221
  gene-expression profiling, 217
  technology, 221
Microbes, 18, 107, 110
Microbial genomics
  AMR, 101
  control and prevention, 101–103
  treatment and diagnostics, 103–105
     genomic characterization, 103–104
     vaccine research, 104
     WGS phylotyping assay, 104
  WGS, 101–102
Microbiome, 18, 112
  and gastrointestinal health, 18
  microbiome-gut-brain axis, 114
  and oral health, 18
  pediatric microbiome, 18
Microbiomics, 103–104
MicroRNAs (miRNAs), 10, 168
Microsatellites, 174
Microscopic analysis, 261
Millennium Development Goal (MDG), 103
Mini-sequencing, 295
MinION
  platform, 300
  sequencer, 299–300
Minisatellites, 174
Minor allele frequency (MAF), 172
miRNAs. *See* MicroRNAs (miRNAs)
Miscarriages, 222
Missense mutation, 4
MitoAction, 82
MitoCanada, 82
Mitochondrial diseases, 78–79, 81–82, 85, 145
  databases/analysis and translational
     platforms, advocacy and data exchange
     groups, 83t
  diagnosis, 75–80
  nongenetic diagnostic tests for, 77t–78t
mitochondrial DNA (mtDNA), 8, 75, 144, 172,
     172f, 221, 295, 306b
Mitochondrial donation, 85
Mitochondrial encephalopathy with ragged-red
     muscle fibers (MERRF), 147f

Mitochondrial genes in male factor infertility, 221
Mitochondrial genetic disorders, 144–145
  genetic classification, 146t
Mitochondrial genome, 6, 8–9, 172
Mitochondrial genomics
  clinical evaluation methods for diagnosis, 85f
  DNA testing, 79–80
  genome data deluge, 78–79
  mitochondrial disease diagnosis, 79–80
  mitochondrion, 76f
  mtDNA, 75
  NGS, 79–80
  NGS–based diagnostic assay design for mitochondrial disorders, 80–81
  nongenetic diagnostics for mitochondrial dysfunction, 75–78
  road ahead, 85
  semantic data standards for community collaboration, 84–85
  translational advancements, 81–84
Mitochondrial Medicine Society (MitoSOC), 82
Mitochondrial Research Society, 82
mitochondrial RNA (mtRNA), 2–4
Mitosis, 4, 5f
MitoSOC. See Mitochondrial Medicine Society (MitoSOC)
Mixed polymorphisms, 175–176
MLPA. See Multiplex ligation-dependent probe amplification (MLPA)
6-MP. See 6-Mercaptopurine (6-MP)
MMPs. See Matrix metalloproteinases (MMPs)
MOC. See Maintenance of certification (MOC)
MOD. See March of Dimes (MOD)
Molecular epidemiology, 104
Molecular genetic(s)
  of male factor infertility, 219
    epigenetics and epigenomics in, 219–221
    testing, 37
Molecular medicine, 227–228, 230–231
Moore's law, 16
Mosaicism, 282
Mouse genome, 168
MPD. See Myeloproliferative disorder (MPD)
mRNA. See messenger RNA (mRNA)
MRSA. See Methicillin-resistant Staphylococcus aureus (MRSA)
MS. See Mass spectrometry (MS)
mtDNA. See mitochondrial DNA (mtDNA)
16.5-kb mtDNA molecule, 144
MTHFR. See 5,10-Methylenetetrahydrofolate reductase (MTHFR)
mtRNA. See mitochondrial RNA (mtRNA)
MTX. See Methotrexate (MTX)
Multicistronic RNAs, 9
Multifactorial disorders. See Polygenic disorders
Multifactorial/polygenic diseases, 157
Multiple endocrine neoplasia type 2 (MEN2), 262
Multiplex ligation-dependent probe amplification (MLPA), 38
Muscle blind protein (MBNL), 55–56
Mutation, 8, 45

Mutational mechanism, 46
Mutations in mtDNA, 9
Mycobacterium leprae (M. leprae), 102
Mycoplasma pneumonia (M. pneumoniae), 104
Myeloproliferative disorder (MPD), 185
Myo-inositol, 94
Myosins, 40
Myotonic dystrophy (DM), 155–156
  DM1 and DM2, 55
Myotonic dystrophy, 55–56. See also Facioscapulohumeral dystrophy (FSHD)

N

n-of-1 trial method, 84
NAHR. See Nonallelic homologous recombination (NAHR)
NAMDC. See North American Mitochondrial Disease Consortium (NAMDC)
National Cancer Institute (NCI), 22, 32
National Center for Advancing Translational Sciences (NCATS), 82
National Center of Medical Genetics, 286
National Human Genome Research Institute (NHGRI), 21, 241
  GWAS, 314f
National Human Genome Research Institute
  GWAS, 313–315, 314f
Natural selection, 8
NCATS. See National Center for Advancing Translational Sciences (NCATS)
NCI. See National Cancer Institute (NCI)
ncRNAs. See Noncoding RNAs (ncRNAs)
nDNA. See Nuclear DNA (nDNA)
Neonatal intensive care units (NICUs), 20
Neurofibromatosis type 1 (NF1), 184
Neurofibromatosis type 2 (NF2), 282
Neurophysiological studies, 153
Neutral mutations, 4
Newborn healthcare, genome sequencing in, 21
Newborn screening, 238
Next generation sequencing (NGS), 15–16, 37, 75, 79–80, 107–108, 192b–193b, 281b, 296–297
  DNA sequence, 298–299
  experimental design, 299
  and genetic testing, 281–282
  library building and sequencing strategies, 298f
  library pool, 298
  NGS–based diagnostic assay design for mitochondrial disorders, 80–81
  PCR reactions, 297–298
  read length, 299
  solutions in forensic genetics, 300–301
    first commercial NGS kits for, 303–305
    forensic markers, 301
    new frontiers in forensic genetics, 305
    NGS platform/assay, 301–302
    STR sequencing, 302–303
  technologies, 262–263
Next–next generation sequencers, 296–297
NF1. See Neurofibromatosis type 1 (NF1)
NGS. See Next generation sequencing (NGS)

NHGRI. See National Human Genome Research Institute (NHGRI)
NICUs. See Neonatal intensive care units (NICUs)
NIH. See US National Institutes of Health (NIH)
NIPND. See Noninvasive PND (NIPND)
NIPT. See Non–invasive prenatal testing (NIPT)
NMR spectroscopy. See Nuclear magnetic resonance spectroscopy (NMR spectroscopy)
NOD2. See Nucleotide-binding oligomerization domaincontaining 2 (NOD2)
Nonallelic homologous recombination (NAHR), 150
Noncoding, RNA-only genes, 166, 168
  miRNAs, 168
  rRNA genes, 166–167
  snoRNAs, 168
  snRNAs, 168
  tRNA genes, 167–168
Noncoding RNAs (ncRNAs), 10
Nongenetic diagnostics for mitochondrial dysfunction, 75–78, 77t–78t
Nongeneticist health providers, 240–241
Noninvasive PND (NIPND), 223
Non–invasive prenatal testing (NIPT), 19
Nonprocessed pseudogenes, 166
Nonsense mutation, 4
Non–small cell lung cancer (NSCLC), 231, 257
Noonan syndrome (NS), 184
  genotype–phenotype correlations, 185t
  with loose anagen hair, 185
Nor, 6–7
Normal diploid cells, 6
Normal transmitting male (NTM), 156
North American Mitochondrial Disease Consortium (NAMDC), 82
Novel therapeutic strategies, 115
  World Health Organization, 115–116
NS. See Noonan syndrome (NS)
NSCLC. See Non–small cell lung cancer (NSCLC)
NTM. See Normal transmitting male (NTM)
Nuclear DNA (nDNA), 75, 144
Nuclear genes, 5–6
Nuclear genome, 6
Nuclear magnetic resonance spectroscopy (NMR spectroscopy), 89
Nucleic acids, 2
  cell division, 4
  codons, 2–4
  comparative genomics, 4
  DNA, 2
  human chromosomes, 3f
  peptide chain synthesis, 4f
  proteins, 4
  steps in mitosis and meiosis, 5f
  Watson-Crick model, 3f
Nucleotide conservation, 45
Nucleotide-binding oligomerization domain-containing 2 (NOD2). See Caspase recruitment domain-containing protein 15 (CARD15)

Nucleotides, 296
Nutrigenetics, 96
Nutrigenomics, 96
Nutrimetabolomics, 96
ob/ob mice. *See* Genetically obese mice
    (ob/ob mice)

O

Obese phenotype, 113
Objective structured clinical examinations
    (OSCEs), 244
Office of Rare Disease Research (ORDR), 82
-*Omics* paradigm, 9f, 10
OMIM. *See* Online Mendelian Inheritance in
    Man (OMIM)
Oncogenes, 156–157, 261–262
Oncology, 128, 248. *See also* Cardiology;
    Rheumatology
    germline DNA mutations, 128–130
    pharmacogenomic markers, testing for, 128
    somatic DNA mutations, 130
"One-size-fits-all-approach" treatments, 227
Online learning systems, 244
Online Mendelian Inheritance in Man (OMIM),
    39, 47, 142
Ontology, 84
Oogenesis, 214–215
Open reading frame (ORF), 57, 108f
"Opening up exome", 39
Operational taxonomic unit (OUT), 108f
Oral anticoagulants, 126–127
Oral health, 18
Oral microbiome, 110–111
ORDR. *See* Office of Rare Disease Research
    (ORDR)
ORF. *See* Open reading frame (ORF)
Organ systems, 131
Organ-system diseases, 140
Organic acid disorders, 94
OSCEs. *See* Objective structured clinical
    examinations (OSCEs)
Osmoprotectants taurine cyanate and nitrate
    (*OTCN*), 251
OUT. *See* Operational taxonomic unit (OUT)
Overarching principles of genomic data sharing,
    30
Ovulation, 214–215
Oxagen, 252
Oxalobacter formigenes (*O. formigenes*), 115
Oxford Nanopore, 299–300

P

P-glycoprotein efflux transporter (ABCB1
    efflux transporter), 127
PacBio platform, 299–300
Pacific Biosciences, 299–300
PAGE-OM. *See* Phenotype and Genotype
    Object Model (PAGE-OM)
PAPPA. See Pregnancy-associated plasma
    protein A (*PAPPA*)
Parent–offspring analysis, 39–40
PARP inhibitor. *See* Poly(ADP-ribose)
    polymerase inhibitor (PARP inhibitor)

Patient autonomy, 210
Patient emotions, 210
Patient population studies, 250
PBMCs. *See* Peripheral blood mononuclear
    cells (PBMCs)
PC. *See* Phosphatidylcholine (PC)
PCA. *See* Principal component analysis (PCA)
PCI. *See* Percutaneous coronary intervention
    (PCI)
PCOS. *See* Polycystic ovarian syndrome
    (PCOS)
PCR. *See* Polymerase chain reaction (PCR)
PCR-based capture method, 297
PD variability. *See* Pharmacodynamic
    variability (PD variability)
PDE4D. See Phosphodiesterase 4D (*PDE4D*)
Pearce v United Bristol Healthcare NHS Trust,
    193–194
Pediatric microbiome, 18
Percutaneous coronary intervention (PCI), 127
Peripheral blood mononuclear cells (PBMCs),
    54–55
Personalized medicine, 228, 231f
Pertinent data access committees, 30–32
PEV. *See* Position effect variegation (PEV)
PFGE. *See* Pulsed-field gel electrophoresis
    (PFGE)
PGD. *See* Preimplantation genetic diagnosis
    (PGD)
PGS. *See* Preimplantation genetic screening
    (PGS)
Pharmaceutical companies, 247, 250
Pharmacodynamic variability (PD variability),
    253–254
Pharmacogenetics, 121
    anticoagulants, 127
    to drug development, 253
        genotype–phenotype correlations, 254t
        PD variability, 253–254
        pharmacokinetic variability, 254–255
    genetic marker implementation into clinical
        practice, 134f
    improving drug response in patients, terms
        for, 122f
    pharmacogenetic guidelines, 124
        clinical pharmacogenetics implementation
            consortium, 125t
    pharmacogenetic study design, 124–133
        clinical areas, 126–133
        GWAS, 125–126
Pharmacogenomic(s), 23–24, 121, 122f, 248,
    292, 312. *See also* Pharmacogenetics
    information in drug labeling, 122–124
        drugs list, 122, 123t–124t
        pharmacogenetic guidelines, 124
    markers, 128–129
        thiopurines, 128–129
    studies, 253
Pharmacogenomics Knowledge Implementation
    Group (PharmGKB), 122, 124
Pharmacokinetic parameters (PK parameters),
    249
Pharmacokinetic variability, 254–255
Pharmacological management, 266

PharmGKB. *See* Pharmacogenomics
    Knowledge Implementation Group
    (PharmGKB)
Phase 1 translation research, 11
Phase 2 translation research, 11
Phase 3 translation research, 11
Phase 4 translation research, 11
Phase I trials, 249
Phase Ib studies, 249
Phase II trials, 249
Phase III trials, 250
Phase IV trials, 250
Phenocopies, 264
Phenotype and Genotype Object Model
    (PAGE-OM), 84
Phenotypes, 113
    phenotype-driven approaches, 9
Phenprocoumon, 126–127
Philadelphia chromosome, 130, 261
Phosphatidylcholine (PC), 115
Phosphodiesterase 4D (*PDE4D*), 251
Piwi-associated RNAs (piRNAs), 10
PK parameters. *See* Pharmacokinetic
    parameters (PK parameters)
Platelet glycoprotein GPIIb/IIIa, 127
PM. *See* Precision medicine (PM); Premutation
    (PM)
PML nuclear bodies. *See* Promyelocytic nuclear
    bodies (PML nuclear bodies)
PML-RARα. *See* Promyelocytic leukemia–
    retinoic acid receptor-α (PML-RARα)
*PMP*22, 153–154
PND. *See* Prenatal diagnosis (PND); Prenatal
    genetic diagnosis (PND)
POF. *See* Premature ovarian failure (POF)
Polar amino acids, 94
Policy issues, 25
Poly(ADP-ribose) polymerase inhibitor (PARP
    inhibitor), 129
Polycomb group repressor complexes (PRCs),
    62
Polycystic ovarian syndrome (PCOS), 217
Polygenic disorders, 142–144
Polymerase chain reaction (PCR), 52, 79–80,
    222
    PCR–based mtDNA sequencing method, 295
Polymorphisms, 7, 130–131
Populations, 7
    databases, 46–47
    genomics, 311
        cross-cutting issues, 315–316
    science, 311, 315
    screening, 198
    variation, 313
Porphyromonas gingivalis (*P. gingivalis*),
    110–111
Position effect variegation (PEV), 52–53
Postgraduate education, 243
Postnatal diagnosis, 20
*POT*1. *See* Predisposition gene (*POT*1)
Poverty, 285
PRCs. *See* Polycomb group repressor
    complexes (PRCs)
Prebiotics, 97

Precision medicine (PM), 23, 121, 227–229, 258. *See also* Stratified medicine
  genomic profiling and counseling, 24
  hypothetical scenario, 229–230, 229f
  molecular, genetic, and genomic revolutions in medicine, 227–228
  personal genetic information, 230
  personalized medicine, 231f
  Precision Medicine Initiative, 24–25
  targeted therapeutics development
    for cancers, 24
    for genetic disorders, 23–24
Preclinical testing, 248–249
Predicting safety
  adverse drug reactions, 257
  drug response modification with genetic polymorphisms, 255t
  improving disease classification, 256–257
  individualized therapy, 256
  type A adverse events, 255–256
  type B adverse events, 255
Predisposition gene (*POT*1), 263b
Pregnancy in women, 183
Pregnancy-associated plasma protein A (*PAPPA*), 223
Preimplantation genetic diagnosis (PGD), 222, 264–265
Preimplantation genetic screening (PGS), 222
  and screening, 222
Premature ovarian failure (POF), 215
  autosomal abnormalities in, 217
  familial, 215
  FRAXA, 215–217
  genetic aspects, 216t
  genetics, 215
  PCOS, 217
Premedical education, 242
Premutation (PM), 57
Prenatal diagnosis (PND), 141–142, 264–265, 286, 288
  recent advances in, 223
    aneuploidy pregnancy, 223
    cell-free fetal DNA, 223–224
    by whole genome sequencing of "jumping libraries", 19–20
Prenatal genetic diagnosis (PND), 223
Prenatal genetic services, 286–287
Prevotella, 110–111
  P. copri, 111
Primary nerve tissue, 55
Principal component analysis (PCA), 93
Principal components, 93
Probiotic therapy, 97
Probiotics, 108, 113–116
Problem-based learning, 243
Processed pseudogenes, 166
Prokaryotes, 2
Promyelocytic leukemia–retinoic acid receptor-α (PML-RARα), 157
Promyelocytic nuclear bodies (PML nuclear bodies), 60
Prophylactic surgery, 265–266
Propionibacterium, 111
  P. acnes, 111
Propranolol, 183
Proteins, 4

protein-altering variants, 45
protein-coding genes, 165–166, 167f
  gene density per chromosome, 165f
  size, 166
Proteomics, 9–10
Protooncogenes, 261–262
Provision of medical and health genetics and genomics
  epidemiological factors, 285
  genetic services and public health genetic programs, 286
  health interventions in genetics, 287
  medical genetics, 285
  medical services incorporating genetics and genomics, 290–293
  public health programs in genetics and genomics, 287–290
Pseudogenes, 166
Pseudomonas enterocolitis (*P. enterocolitis*), 103
4 P's of Medicine, 230
Psychiatric diseases, biomarkers for, 94–95
Psychiatry, 133
Psychosocial tool, 202
Public and population health genomics
  Alzheimer dementia, 313–315
  breast cancer, 311
  colorectal cancer, 312–313
  Crohn disease, 313
  cross-cutting issues of population genomics, 315–316
  cystic fibrosis, 315
  population genomics, 311
Public education, 287, 289
Public health programs, 287
  community-based health educational programs, 289
  consanguineous marriages, 288
  ethical principles and cultural diversity, 289
  genetic factors, 290
  in genetics and genomics, 287
  prenatal diagnosis, 288
Public perception of genetics/genomics, 192b–193b
"Pull" factors, 233–234
Pulsed-field gel electrophoresis (PFGE), 153–154
"Push" factors, 233–234
Pyrimidine analogues, 129
Pyrosequencers, 302
Pyrosequencing, 80, 296, 298–299

Q

Quality control (QC), 43
quantitative fluorescence PCR (QF-PCR), 223

R

Radical therapeutic strategies, 55
Randomized clinical trial, 126–127
"Rapid sequencing" technique, 21
RAS genes, 184
RAS/MAPK pathway, 184–185
RASopathies. *See also* Marfan syndrome (MFS)
  clinical features, 186t
  diagnosis of genetic disorder, 186–187

  emergence, 184–186
  malignancy risk in, 185–186
*RB*1 gene, 262b
RCC. *See* Renal cell carcinoma (RCC)
Read length, 299
Reciprocal engagement model (REM), 202
Recontacting policies, 192b–193b
Reduced folate carrier 1 (RFC1), 130–131
Reference genome, 43–44
Reflex test. *See* Second-line test
Regulatory issues, 25
REM. *See* Reciprocal engagement model (REM)
Renal cell carcinoma (RCC), 281b
Repetitive elements, 168–169
Reponema denticola (*R. denticola*), 110–111
Reproducibility, 300–301
Reproductive medicine, 213. *See also* Genomic medicine; Precision medicine; Stratified medicine
  female infertility, 213
    abnormal ovulation disorders, 214–217
    congenital anomalies of reproductive tract, 213–214
    endometrium disorders, 217–218
    selection of syndromes with abnormal female reproductive tract anatomy, 214t
  genetic factors in IVF, 221–222
  male factor infertility, 218
    congenital anomalies of male reproductive tract, 218–219
    mitochondrial genes in, 221
    molecular genetics, 219–221
    new genomic applications in, 221
  recent advances in prenatal diagnosis, 223–224
Research ethics, 30
Research/clinical boundary, 192b–193b
Residency education, 242, 244
Residency review committee (RRC), 242
Restriction fragment length polymorphism analysis (RFLP analysis), 295
Retinoblastoma, 262b
Retrovirus-like LTR transposons, 169
Rett syndrome, 60–61
Reversible Terminator Sequencing, 80
RFC1. *See* Reduced folate carrier 1 (RFC1)
RFLP analysis. *See* Restriction fragment length polymorphism analysis (RFLP analysis)
Rhabdomyolysis, 127
Rhesus D factor (RhD), 223–224
Rheumatology, 130–131. *See also* Cardiology; Oncology
  disease-modifying antirheumatic drugs, 130–131
  response to TNF antagonists, 130
Ribonucleic acid (RNA), 2
  gain of function mechanism, 55–56
  sequencing, 297
ribosomal RNA (rRNA), 10
  genes, 166–167
  sequencing and analysis, 107–108
"Right to know or not know", 194b
Risk information as therapy, 264–265
Rituximab, 130
RNA. *See* Ribonucleic acid (RNA)

Roseburia, 114–115
  R. intestinalis, 113
RRC. *See* Residency review committee (RRC)
rRNA. *See* ribosomal RNA (rRNA)
rs10757274 genotyping, 315

S

S-enantiomer, 126
Salmonella typhi genome, 104
Sanger DNA sequencing, 38
Sanger sequencing method, 295
SBE. *See* Single base extension (SBE)
SCA. *See* Spinocerebellar ataxia (SCA)
SCFAs. *See* Short-chain fatty acids (SCFAs)
SCID. *See* Severe combined immunodeficiency (SCID)
Scientific Foundations for Future Physicians, 242
Sclerosteosis, 252
Sclerostin (*SOST*), 251–252
SDHB. See Succinate dehydrogenase subunits B (*SDHB*)
SDHD. See Succinate dehydrogenase subunits D (*SDHD*)
Sebum, 111
Second generation sequencers, 296–297
Second-line test, 39
Secondary findings, 204–205
Segmental duplications, 169–170, 170f
Segregation analysis, 45–46
Semantic data standards for community collaboration, 84–85
Semiconductor sequencing, 298–299
Sepiapterin reductase gene (*SPR* gene), 20
Sequencing, 295
  depth, 297
  technologies, 16
Severe combined immunodeficiency (SCID), 252
Sex chromosomes, 2
Sex-chromosome pair, 141
Sherman paradox, 156
Shigella flexneri (*S. flexneri*), 103
Shigella sonnei (*S. sonnei*), 103
Short Arms of human acrocentric chromosomes, 171–172
Short interspersed nuclear elements (SINEs), 169, 174
Short sequence repeats (SSRs), 174, 175f
Short tandem repeats (STRs), 306b
  assays, 295
  sequencing, 302–303
    analysis using ForenSeq Universal Analysis Software, 304f
    analysis using Torrent Suite Server, 303f
    D12S391 alleles, 302t
Short-chain fatty acids (SCFAs), 112–113
Shotgun sequencing, 224, 297, 300–301
Sidaway case, 193–194
Silico analysis of DNA, 104
SINEs. *See* Short interspersed nuclear elements (SINEs)
Single base extension (SBE), 295
Single gene disorders/traits, 251–252
Single gene testing, 38

Single molecule sequencing, 299–300
Single nucleotide polymorphisms (SNPs), 7, 17, 38, 79, 101, 125, 157, 172–174, 215, 295, 306b
  analysis using ForenSeq Universal Analysis Software, 304f
  analysis using Torrent Suite Server, 303f
Single nucleotide proteins (SNPs), 251
Single nucleotide variants (SNVs), 20, 38, 79–80
Single-gene disorders. *See* Mendelian disorders
SJS. *See* Stevens-Johnson syndrome (SJS)
Skin, 111
Skin microbiome, 111
SLCO1B1 gene, 127
Small nuclear RNAs (snRNAs), 10, 168
Small nucleolar RNAs (snoRNAs), 10, 168
Smith-Magenis syndrome (SMS), 150–153
SMRM. *See* Society for Mitochondria Research and Medicine, India (SMRM)
SMS. *See* Smith-Magenis syndrome (SMS)
snoRNAs. *See* Small nucleolar RNAs (snoRNAs)
SNPs. *See* Single nucleotide polymorphisms (SNPs); Single nucleotide proteins (SNPs)
snRNAs. *See* Small nuclear RNAs (snRNAs)
SNVs. *See* Single nucleotide variants (SNVs)
Social science, 311
Society for Mitochondria Research and Medicine, India (SMRM), 82
"Solenoid" helical fibers, 51
SOLiD, 80
Somatic DNA mutations, 130. *See also* Germline DNA mutations
  B-raf/MEK/ERK pathway inhibitors, 130
  monoclonal antibodies, 130
  receptor tyrosine kinase inhibitors, 130
Somatic mosaicism, 282
SOST. See Sclerostin (*SOST*)
Special genomic structures. *See also* Genomic anatomy
  genome variation as laboratory tool, 176
  genomic variability, 172
    CNVs, 174–175
    insertion/deletion polymorphisms, 174
    inversions, 175–176
    SNPs, 172–174, 173f
    SSRs, 174, 175f
  human centromeres, 170–171, 171f
  human telomeres, 171, 171f
  mitochondrial genome, 172
  short arms of human acrocentric chromosomes, 171–172
Spectral databases, 90–93
Spermatogenesis, 218–219, 221–222
Spinocerebellar ataxia (SCA), 155–156
SPR gene. *See* Sepiapterin reductase gene (*SPR* gene)
SSRs. *See* Short sequence repeats (SSRs)
Staphylococcus, 111
  S epidermidis, 111
Statin-induced myopathy, 127

Stevens-Johnson syndrome (SJS), 131, 312
"Sticky" DNA, 54
"Stimulatory chromosomes", 261
Stratified medicine, 227, 230–231, 247, 256–257. *See also* Genomic medicine; Precision medicine; Reproductive medicine
  challenges for, 232–233
  criteria for, 231t
  in current medical practice, 232t
  future, 233–234
  molecular, genetic, and genomic revolutions in medicine, 227–228
  NSCLC, 231
  in planning and executing treatment for cancers, 233f
  to planning and selection for targeted therapy, 234f
Streptococcus, 110–111
STRs. *See* Short tandem repeats (STRs)
Structural Genomics Consortium, 252
Structural variations (SVs), 17–19
Succinate dehydrogenase subunits B (*SDHB*), 149
Succinate dehydrogenase subunits D (*SDHD*), 149
Sulfasalazine, 131
Support vector machines (SVMs), 93
Suv39h histone methyltransferase, 51–52
SVMs. *See* Support vector machines (SVMs)
SVs. *See* Structural variations (SVs)
Symendogenous compounds, 95
Symxenobiotic metabolites, 95
Syndromes, 248
Systems biology, 10

T

T2D. *See* Type 2 Diabetes Mellitus (T2D)
Tamoxifen, 129
Tandem repeats of sequence (TTAGGG)n, 171
Tannerella forsythia (*T. forsythia*), 110–111
Targeted (re)sequencing, 297
Targeted therapeutics development for genetic disorders, 23–24
Tay-Sachs disease screening, 238
TCA cycle. *See* Tricarboxylic acid cycle (TCA cycle)
TCGA. *See* The Cancer Genome Atlas (TCGA)
Teaching and training medicine in genomic era
  competencies in genomic medicine, 241–242
  geneticist and nongeneticist health providers, 240–241
  genomic medicine, 237
  genomics and medical education landscape, 242–243
  genomics integration into medical practice, 237–242
  medical genomics education, 243–244
  practice of genomic medicine, 238
    diagnosis, 239
    prevention, 238–239
    therapy, 239–240
  training paths in medical genetics, 240t
Termination of pregnancy (TOP), 288
Tetrahydrobiopterin, 20

TGF-β. *See* Transforming growth factor β (TGF-β)
*TGFBR*1 genes, 40–41
*TGFBR*2 genes, 40–41
α-Thalassemia X-linked mental retardation syndrome (ATR-X syndrome), 59–62
Thalassemias, heterozygotes, 8
The Cancer Genome Atlas (TCGA), 22
Therapeutic(s), 293
    approaches, 241
    regimens, 227
Therapeutics of Rare and Neglected Diseases (TRND), 82
Therapy for genetic disorders, 239–240
Thermo Fisher Scientific, 303–304
Thiopurine-S-methyltransferase (*TPMT*), 128–129, 131, 254
Thiopurines, 128–129
    pharmacogenomic markers, 128–129
Third generation
    sequencers, 296–297
    sequencing, 299
Three-dimensional organization of genome (3D organization of genome), 52–53
Threshold effect, 142–144
Thymidylate synthase (*TYMS*), 129
Time-of-flight MS (TOFMS), 94
Titin (*TTN*), 40, 166
TLRs. *See* Toll-like receptors (TLRs)
TMAO. *See* Trimethylamine N-oxide (TMAO)
TNF. *See* Tumor necrosis factor (TNF)
TOFMS. *See* Time-of-flight MS (TOFMS)
Toll-like receptors (TLRs), 113–114
TOP. *See* Termination of pregnancy (TOP)
Topoisomerase I inhibitors, 129
Toxic epidermal necrolysis, 312
Toxicogenomics, 248–249
Toxicology studies, 248–249
TPMT. *See* Thiopurine-S-methyltransferase (*TPMT*)
Tracking transmission of diseases, 103
Transcription regulation regions, 168
Transcriptional repressor domain (TRD), 61
Transcriptomics, 9–10
transfer RNA (tRNA), 10, 75
    genes, 167–168
Transforming growth factor β (TGF-β), 180
Translational advancements, 81–84
Translational genome research, 10–11
Translational human genomics, 10–11
Transmission of hereditary factors, 1–2
Transxenobiotic compounds, 95
Trastuzumab, 24, 128, 130, 253, 256–257
TRD. *See* Transcriptional repressor domain (TRD)
Tricarboxylic acid cycle (TCA cycle), 94
Trimethylamine N-oxide (TMAO), 115
Trinucleotide repeats, 54
    disorders with, 155–156
    trinucleotide repeats expansion, disorders with, 153t
Trio analysis. *See* Parent–offspring analysis
Triplet repeats. *See* Trinucleotide repeats

Trisomic rescue mechanism, 147–148
Trisomy 21. *See* Aneuploidy pregnancy
tRNA. *See* transfer RNA (tRNA)
TRND. *See* Therapeutics of Rare and Neglected Diseases (TRND)
Tryptophan, 114
TSGs. *See* Tumor Suppressor Genes (TSGs)
TTN. *See* Titin (*TTN*)
Tumor cells, 94
Tumor necrosis factor (TNF), 130–131
Tumor Suppressor Genes (TSGs), 261–262
Tumor-associated cfDNA, 23
"Two-hit" hypothesis, 157
TYMS. *See* Thymidylate synthase (*TYMS*)
Type 1 CMTD (CMT1), 153–154
Type 2 Diabetes Mellitus (T2D), 113
Type A adverse events, 255–256
Type B adverse events, 255

U

UC. *See* Ulcerative colitis (UC)
UCSC. *See* University of California Santa Cruz (UCSC)
UGT1A1. *See* Uridine diphosphate glucuronosyltransferase isoform 1A1 (*UGT1A1*)
UK100K Genomes Project, 42
    bioinformatics pipelines, 43
    clinical bioinformatics, 42
    clinical phenotyping, 42–43
    data sharing, 48
    databases and literature, 46–47
    De Novo mutations, 46
    DNA sequence variants interpretation, 44–45
    functional evidence, 46
    incidental findings, 47–48
    inheritance and mutation type, 45
    literature databases, 47
    mutational mechanism and variant spectrum, 46
    nucleotide conservation, 45
    OMIM, 47
    protein-altering variants, 45
    reference genome, 43–44
    segregation analysis, 45–46
    variants causing aberrant messenger RNA splicing, 45
Ulcerative colitis (UC), 96
Ultraperformance liquid chromatography–electrospray ionization, 94
UM. *See* Uveal melanoma (UM)
UMDF. *See* United Mitochondrial Disease Foundation (UMDF)
Unified Medical Language System, 84
Uniparental disomy (UPD), 38, 147–148
United Mitochondrial Disease Foundation (UMDF), 81–82
University of California Santa Cruz (UCSC), 164f
3′ Untranslated region (UTR), 55
uPA. *See* urokinase-type plasminogen activator (*uPA*)

Urate oxidase enzymes, 129–130
UPD. *See* Uniparental disomy (UPD)
Urease gene representation, 109–110
Uridine diphosphate glucuronosyltransferase isoform 1A1 (*UGT1A1*), 129
Urinary metabolites, 95
urokinase-type plasminogen activator (*uPA*), 218
US Food and Drug Administration (FDA), 24, 122, 239, 254, 312–313
US Medical Licensing Examination (USMLE), 242
US National Institutes of Health (NIH), 21, 30
US NHGRI Clinical Sequencing Exploratory Research program, 22
USMLE. *See* US Medical Licensing Examination (USMLE)
UTR. *See* 3′ Untranslated region (UTR)
Uveal melanoma (UM), 281b

V

Vaccine research, 104
Vagus nerve, 114
Variable expressivity, 179
Variable number tandem repeats (VNTRs), 62
Variant calling, 43
Variant databases, 46
Variant spectrum, 46
Variants of unknown significance (VUSs), 39, 290
Variation in gut flora, 112
VariO, 84
Vemurafenib, 24, 130, 253–254
Very–low-density lipoprotein (VLDL), 95
VHL. *See* Von Hippel-Lindau disease (VHL)
Vibrio cholera (*V. cholera*), 102–103
Viral resistance, 257
Virtual gene panels, 38–39
"Virtual panel" technique, 282
Vismodegib, 266b
Vitamin K receptor, 255–256
VKORC1 gene, 126
VLDL. *See* Very–low-density lipoprotein (VLDL)
VNTRs. *See* Variable number tandem repeats (VNTRs)
Von Hippel-Lindau disease (VHL), 265
VUSs. *See* Variants of unknown significance (VUSs)

W

Warburg effect, 94
Warfarin, 126
Watson-Crick model, 3f
WDR5-containing *H3K4* methyltransferase complexes, 52
Web-based platforms, 84
Wellcome Trust Case Control Consortium, 251
WES. *See* Whole exome sequencing (WES)

WGS. *See* Whole genome sequencing (WGS)
WHO. *See* World Health Organization (WHO)
Whole exome, 38–39
Whole exome sequencing (WES), 20, 228, 281b, 282
data, 42
Whole genome, 38–39
analysis, 44, 191
Whole genome sequencing (WGS), 15, 20, 39, 101, 191, 192b–193b, 228, 263b, 266. *See also* Next generation sequencing (NGS)
data, 104

Workforce, 237
World Health Organization (WHO), 115–116, 287

X
X-linked dominant disorder, pedigree with, 143f
X-linked inhibitor of apoptosis deficiency, 20
X-linked recessive spinal bulbar atrophy, 155–156
Xalkori. *See* Crizotinib
Xenobiotic compound, 95
*XPNPEP*2 gene, 127–128

Y
Y chromosome, 223
Yersinia pestis (*Y. pestis*), 102

Z
Zelboraf. *See* Vemurafenib
Zero-mode waveguide (ZMW), 299–300
Ziagen. *See* Abacavir
Zinc finger 9 gene (*ZNF*9 gene), 55